IFIP Advances in Information and Communication Technology 655

IFIP – The International Federation for Information Processing

IFIP was founded in 1960 under the auspices of UNESCO, following the first World Computer Congress held in Paris the previous year. A federation for societies working in information processing, IFIP's aim is two-fold: to support information processing in the countries of its members and to encourage technology transfer to developing nations. As its mission statement clearly states:

IFIP is the global non-profit federation of societies of ICT professionals that aims at achieving a worldwide professional and socially responsible development and application of information and communication technologies.

IFIP is a non-profit-making organization, run almost solely by 2500 volunteers. It operates through a number of technical committees and working groups, which organize events and publications. IFIP's events range from large international open conferences to working conferences and local seminars.

The flagship event is the IFIP World Computer Congress, at which both invited and contributed papers are presented. Contributed papers are rigorously refereed and the rejection rate is high.

As with the Congress, participation in the open conferences is open to all and papers may be invited or submitted. Again, submitted papers are stringently refereed.

The working conferences are structured differently. They are usually run by a working group and attendance is generally smaller and occasionally by invitation only. Their purpose is to create an atmosphere conducive to innovation and development. Refereeing is also rigorous and papers are subjected to extensive group discussion.

Publications arising from IFIP events vary. The papers presented at the IFIP World Computer Congress and at open conferences are published as conference proceedings, while the results of the working conferences are often published as collections of selected and edited papers.

IFIP distinguishes three types of institutional membership: Country Representative Members, Members at Large, and Associate Members. The type of organization that can apply for membership is a wide variety and includes national or international societies of individual computer scientists/ICT professionals, associations or federations of such societies, government institutions/government related organizations, national or international research institutes or consortia, universities, academies of sciences, companies, national or international associations or federations of companies.

More information about this series at https://link.springer.com/bookseries/6102

Robert Nowak · Jerzy Chrząszcz ·
Stelian Brad (Eds.)

Systematic Innovation Partnerships with Artificial Intelligence and Information Technology

22nd International TRIZ Future Conference, TFC 2022
Warsaw, Poland, September 27–29, 2022
Proceedings

Editors
Robert Nowak
Warsaw University of Technology
Warsaw, Poland

Jerzy Chrząszcz
Warsaw University of Technology
Warsaw, Poland

Stelian Brad
Technical University of Cluj-Napoca
Cluj-Napoca, Romania

ISSN 1868-4238 ISSN 1868-422X (electronic)
IFIP Advances in Information and Communication Technology
ISBN 978-3-031-17290-8 ISBN 978-3-031-17288-5 (eBook)
https://doi.org/10.1007/978-3-031-17288-5

This Springer imprint is published by the registered company Springer Nature Switzerland AG
The registered company address is: Gewerbestrasse 11, 6330 Cham, Switzerland

Preface

The Theory of Inventive Problem Solving (TRIZ) is popular creative thinking and systematic organizing knowledge methodology. It is a scientifically grounded and structured approach for forecasting the evolution of technological systems, and includes numerous tools and methods for product and process innovation, enabling an increase in creative and inventive productivity.

The European TRIZ Association (ETRIA) World Conference TRIZ Future (also known as the International TRIZ Future Conference, TFC) is traditionally a fundamental venue for disseminating information about TRIZ. TFC has reached its 22nd edition (TFC22) which took place in Warsaw, Poland, during September 27–29, 2022. The conference was organized at the Faculty of Electronics and Information Technology, Warsaw University of Technology, in Warsaw, Poland by the Artificial Intelligence Division. The conference motto is ‘Systematic Innovation partnerships with AI and IT’, which higlights the main idea of this event: to join TRIZ with Information Technology (IT) and Artificial Intelligence (AI).

Information Technology (IT) and Artificial Intelligence (AI) achieve many goals. It is worth noting that thoughtlessly comparing artificial intelligence to human intelligence may be a source of misunderstandings and confusion. Computers are only machines that, to some extent, should be viewed as fast-acting calculators that can only perform simple operations on binary numbers. Despite the huge successes, scope and scale of their applications, it is not without reason that computers are sometimes called ‘stupid automatons’. It is impossible for any program executed by a computer to become a unit autonomous enough to execute something other than its creator, a computer scientist, originally designed. The success of IT and AI is the success of the creativity of computer scientists, which now needs the support of TRIZ.

After disrupting the traditional format of the TFC 2020 and TFC 2021 conferences, TFC 2022 was provided in a hybrid form, which included seeing each other face-to-face as well as participating online.

Despite the focus on joining TRIZ and IT and AI, other topics typically linked with TRIZ and protagonists of previous TFC editions were still accepted. All 39 papers published in this book were at least triple peer-reviewed by several members of the Scientific Committee, and the authors had the chance to improve their papers based on these reviews before the final decision. The papers were presented during the TRIZ Future Conference 2022, a lively discussion followed the presentation.

August 2022

Robert Nowak
Jerzy Chrząszcz
Stelian Brad

Organization

General Chair

Jerzy Chrząszcz	Warsaw University of Technology, Poland

Scientific Chair

Robert Nowak	Warsaw University of Technology, Poland

Program Committee

Stelian Brad	Technical University of Cluj-Napoca, Romania
Denis Cavallucci	INSA Strasbourg, France
Jerzy Chrząszcz	Warsaw University of Technology, Poland
Sebastian Koziołek	Wroclaw University of Technology, Poland
Pavel Livotov	Offenburg University, Germany
Robert Nowak	Warsaw University of Technology, Poland
Tomasz Starecki	Warsaw University of Technology, Poland
Paweł Wawrzyński	Warsaw University of Technology, Poland
Cezary Zieliński	Warsaw University of Technology, Poland

Scientific Committee

Rafał Biedrzycki	Warsaw University of Technology, Poland
Robert Bembenik	Warsaw University of Technology, Poland
Rachid Benmoussa	Cadi Ayyad University, Marrakesh, Morocco
Yuri Borgianni	Free University of Bozen-Bolzano, Italy
Hicham Chibane	INSA Graduate School of Science and Technology of Strasbourg, France
Michał Chwesiuk	Warsaw University of Technology, Poland
Marco De Carvalho	Universidade Tecnológica Federal do Paraná, Brazil
Roland De Guio	INSA Graduate School of Science and Technology of Strasbourg, France
Sébastien Dubois	INSA Graduate School of Science and Technology of Strasbourg, France
Oleg Feygenson	International TRIZ Association, Korea
Lorenzo Fiorineschi	University of Florence, Italy
Grzegorz Protaziuk	Warsaw University of Technology, Poland
Tomasz Gambin	Warsaw University of Technology, Poland
Piotr Gawrysiak	Warsaw University of Technology, Poland
Hans-Gert Gräbe	University of Leipzig, Germany
Maciej Grzenda	Warsaw University of Technology, Poland

Claudia Hentschel	Hochschule für Technik und Wirtschaft Berlin, Germany
Mariusz Kaleta	Warsaw University of Technology, Poland
Włodzimierz Kasprzak	Warsaw University of Technology, Poland
Jacek Komorowski	Warsaw University of Technology, Poland
Maciej Ławryńczuk	Warsaw University of Technology, Poland
Lorenzo Maccioni	Free University of Bozen-Bolzano, Italy
Nicolas Maranzana	Ecole Nationale Supérieure d'Arts et Métiers Paris, France
Oliver Mayer	Bayern Innovativ GmbH, Germany
Jacek Misiurewicz	Warsaw University of Technology, Poland
Mieczysław Muraszkiewicz	Warsaw University of Technology, Poland
Toru Nakagawa	Osaka Gakuin University, Japan
Stephane Negny	Toulouse Graduate School of Chemical Materials and Industrial Engineering, France
Piotr Pałka	Warsaw University of Technology, Poland
Valeriy Prushinskiy	Samsung Electronics, Korea
Federico Rotini	University of Florence, Italy
Davide Russo	University of Bergamo, Italy
Dominik Ryżko	Warsaw University of Technology, Poland
Łukasz Skonieczny	Warsaw University of Technology, Poland
Fernando Solano-Donado	Warsaw University of Technology, Poland
Christian Spreafico	University of Bergamo, Italy
Christian Thurnes	Hochschule Kaiserslautern, Germany
Jarosław Turkiewicz	Warsaw University of Technology, Poland
Jacek Wytrębowicz	Warsaw University of Technology, Poland

Organizing Committee

Urszula Adamiec	Warsaw University of Technology, Poland
Krystian Chachuła	Warsaw University of Technology, Poland
Łukasz Chorchos	Warsaw University of Technology, Poland
Owais Khanday	Warsaw University of Technology, Poland
Gabriela Miączyńska	Warsaw University of Technology, Poland
Farhan Muhammad	Warsaw University of Technology, Poland
Katarzyna Nałęcz-Charkiewicz	Warsaw University of Technology, Poland
Łukasz Neumann	Warsaw University of Technology, Poland
Marek Sieczkowski	Warsaw University of Technology, Poland
Justyna Zgórzak	Warsaw University of Technology, Poland

Partners and Sponsors

Conference Partners

Silver Sponsor

Contents

New Perspectives of TRIZ

Mathematical Modelling and Formalization of TRIZ: Trimming for Product Design

Chris Edward[1](✉), Jane Labadin[1], and Narayanan Kulathuramaiyer[2]

[1] Faculty of Computer Science and Information Technology, Universiti Malaysia Sarawak, 94300 Kota Samarahan, Sarawak, Malaysia
chrised@sains.com.my, ljane@unimas.my

[2] Institute of Social Informatics and Technological Innovation, Universiti Malaysia Sarawak, 94300 Kota Samarahan, Sarawak, Malaysia
nara@unimas.my

Abstract. This work aims to formalize TRIZ modelling framework and trimming techniques through mathematical notations to lay the foundations for rigorous analysis of TRIZ as a Science of Innovation. Mathematical modelling has been employed to formalize the heuristic models of trimming. A case study was presented to demonstrate the use of the proposed modelling scheme. The paper has demonstrated the correlation of TRIZ modelling framework and trimming techniques with well-established mathematical fields such as Formal Logic, Set Theory and Graph Theory. It presents initial efforts in formalizing the functional analysis and trimming techniques as a rigorous formal approach. The acceptance of systematic innovation as a scientific discipline that can be supported by knowledge systems and can be connected to mathematical models remains a dream. This work provides directions for inquiry into this non-trivial endeavour. The value of this work will see future computational models for supporting systematic innovation. The real-life use case demonstrates the powers and gaps with regards to Genrich Altshuler's modelling of product innovation using heuristics.

Keywords: Data dictionary · Graph modelling · Trimming principles · Trimming iteration · Semantic invariant · Formal Logic · Set Theory · Graph Theory

1 Introduction

The *Theory of Inventive Problem Solving* (Russian acronym: TRIZ), an international exact science of creativity developed by G.S. Altshuller and his colleagues during the erstwhile U.S.S.R. reign circa 1946–1985, is a problem solving method based on logic and data – not intuition – which could potentially accelerate a project team's ability to solve problems creatively. Due to its systemic structural framework and algorithmic approach, TRIZ offers repeatability, predictability, and reliability. In contrast to brainstorming techniques which tend to rely on spontaneous and intuitive creativity of

Published by Springer Nature Switzerland AG 2022
R. Nowak et al. (Eds.): TFC 2022, IFIP AICT 655, pp. 3–16, 2022.
https://doi.org/10.1007/978-3-031-17288-5_1

individuals or groups, the TRIZ methodologies rely on the study of the patterns[1] of problems and solutions – more than three million patents have been examined through empirical analysis in order to discover patterns behind the breakthrough solutions for various tough challenges. Altshuller's heuristic mental model proposed that *there exists a set of fundamental universal principles of creativity that form the basis for all creative innovations*. In addressing this imperfect knowledge, there is a need to translate the knowledge into a formalized model as we hypothesize that if *all* of these fundamental principles could be identified *completely* and codified *consistently*, they could be taught to anyone to make the creative process more systematic and predictable. A similar aim was discussed by Lau [1] on his proposed framework for developing a TRIZ implementation model in China. His empirical approach towards adopting TRIZ has given insights to properly formalizing existing TRIZ principles and techniques. In contrast, our study employs an analytical approach in formalizing TRIZ modelling specifically on trimming principles via the use of graph theory.

There are other research in similar pursuit such as the work of Nordlund [2] who provided case studies concerning the applications of Axiomatic Design theory towards a manufacturing process, micro and nano product design, software design, and socioeconomic systems design. The theory prescribes nominative rules to adhere in a design process. Efimov-Soini & Chechurin [3] introduces a new ranking method for component-function-component duplet chain structures in a system. Overall, their proposed method is better than the classical method of ranking and Miao Li's linear convolution method of ranking. Just like previous functional model ranking methods, the proposed method is intended to aid in trimming using existing TRIZ trimming rules by proposing candidate duplet chains to be trimmed based on their low rankings. The same authors then in [4] presented an innovative approach to objectively rank functions (between components) in a dynamic system model.

Motivated by [1–4] and as highlighted in [5], there is a need to formalize TRIZ in some ways and to eliminate ambiguity in interpretations where it is deemed necessary. This paper, therefore, presents a formalization of TRIZ trimming principles, of which is chosen amongst others due to it being canon in classical TRIZ literature, so that it be (mathematically) defined formally, and the trimming process, based on the rigorous trimming principles definition, reflects iterative and algorithmic nature. This paper focuses on Device Trimming which is an important instrument particularly in the area such as product design and patent circumvention [13]. Trimming is an analytical tool that can eliminate one or more components in a system while still maintaining the main useful functionality.

2 Model Formulation

Contrasting from TRIZ diagram-based modelling approach, we introduce a novel algebraic notation and graph modelling, based on Graph Theory, that could be considered

[1] Dr. Keith Devlin, emeritus mathematician at Stanford University and director of the Stanford Mathematics Outreach Project [6], referred Mathematics as the science of patterns [7]. Interestingly, there seems to be massive overlaps and parallels between Mathematics and TRIZ; just as TRIZ explores patterns of problems and solutions.

a shorthand notation to the existing TRIZ modelling techniques without compromising the conceptual and logical integrity. Additionally, once the algebraic notation graph modelling approach are defined, we introduce six new trimming principles as well as the formal mathematical formulation of the new principles using the acquainted algebraic notation.

Before introducing the algebraic notation, we first introduce the corresponding data dictionary (cf. Table 1) which provides a brief definition of m components and n functions found within a systems model [8], where $m, n \in \mathbb{N}$.

Table 1. Generic example of data dictionary.

Components	Functions
c_1: component_1	f_1: function_1
c_2: component_2	f_2: function_2
...	...
c_m: component_m	f_n: function_n

For example, the data dictionary for a "playing ball" system constitute two components namely "hand" and "ball" with two functions "hold" and "push", and they can be defined as shown in Table 2. We notated the sub-components within the indices of the component.

Table 2. Playing ball data dictionary.

Components	Functions
c_1: hand	f_1: hold
$c_{1.1}$: thumb	f_2: push
$c_{1.2}$: index finger	
$c_{1.3}$: middle finger	
$c_{1.4}$: ring finger	
$c_{1.5}$: little finger	
c_2: ball	

Based on the "Playing Ball" data dictionary on Table 2, the function model for "hand hold ball" can be notated as the following mathematical construct:

$$c_1 \xrightarrow{f_1(c_2)} c_2 \tag{1}$$

$$c_1 \left\{ \begin{matrix} c_{1.1} \\ c_{1.2} \\ c_{1.3} \\ c_{1.4} \\ c_{1.5} \end{matrix} \right\} \xrightarrow{f_1(c_2)} c_2 \tag{2}$$

As defined in Table 2, the "hand" component consists of sub-components thumb, index finger, middle finger, ring finger and little finger. These sub-components are encapsulated in braces as shown in Eq. (2). Here, the function model still reads "hand hold ball" as all the sub-components of the hand are contributing to the function called 'hold.' If one of the sub-component performs a function to another component, then the sub-component must be stated before the function and separated by a colon (:). For instance,

$$c_1 \left\{ \begin{matrix} c_{1.1} \\ c_{1.2} \\ c_{1.3} \\ c_{1.4} \\ c_{1.5} \end{matrix} \right\} \xrightarrow{c_{1.2}:f_2(c_2)} c_2 \tag{3}$$

which reads "index finger push ball". Notice in the above, a function corresponds to an edge or an arrow, indicating that an edge must have a function and vice versa.

2.1 Trimming Principles

Let V be the set of components and sub-components in the data dictionary, F be the set of functions in the data dictionary, and g_1, g_2, ..., and g_i be the set of ordered pairs $((a, b), f_1)$, $((a, b), f_2)$, ..., $((a, b), f_i)$ such that $f_1 = g_1((a, b)), f_2 = g_2((a, b))$, ..., and $f_i = g_i((a, b))$ respectively, where $a, b \in V$ and $i \in \mathbb{N}$. Here,

$$g_1 = \{((a, b), f_1) \text{ if and only if } a \text{ function}_1 b\},$$

$$g_2 = \{((a, b), f_2) \text{ if and only if } a \text{ function}_2 b\},$$

$$\ldots$$

$$g_i = \{((a, b), f_i) \text{ if and only if } a \text{ function}_i b\}$$

Define the relation, E as the union of the set of ordered pairs $g_1, g_2, \ldots, g_i$, that is $E = \{g_1 \cup g_2 \cup \ldots \cup g_i\}$. Now let each element in E be called $e_1, e_2, \ldots, e_n$, therefore we have $V = \{c_1, c_2, \ldots, c_m\}$, $E = \{e_1, e_2, \ldots, e_n\}$ and $F = \{f_1, f_2, \ldots, f_i\}$ where $i, m, n \in \mathbb{N}$. Finally, let $G = (V, E)$ be a graph, and define the trim function and append function as trim(x, y): $x \notin y$ and append(x, y): $x \in y$ respectively. Based on these algebraic notations and the definitions of trim and append functions, together with graph theory, the following principles are defined.

Principle 1 (Non-functionality). *If* $\exists c \in V$ *such that* $\deg(c) = 0 \Rightarrow \text{trim}(c, V)$

Meaning: *"If a node in the graph does not execute or receive any function to or from another node nor itself, then trim the node."*

Principle 2 (Redundancy). *If* $c_1, c_2, c_3 \in V, f \in F$ *and* $((c_1, c_3), f), ((c_2, c_3), f) \in E \Rightarrow \text{trim}(((c_1, c_3), f), E)$

Meaning: *"If two nodes in the graph also execute the same function to the same node, then trim one of the edges."*

Principle 3 (Transitivity). *If* $c_1, c_2, c_3 \in V, f \in F$ *and* $((c_1, c_2), f), ((c_2, c_3), f) \in E \Rightarrow \text{trim}(((c_1, c_2), f), E), \text{trim}(((c_2, c_3), f), E), \text{append}(((c_1, c_3), f), E)$

Meaning: *"If two nodes in the graph also execute the same function to the next node in the 3-node chain, then trim each edges in the chain and append the edge from the first node to the last node."*

Principle 4 (Obsolescence). *If* $c_1, c_2 \in V, f \notin F$ *and* $((c_1, c_2), f) \in E \Rightarrow \text{trim}(((c_1, c_2), f), E)$

Meaning: *"If a function in the graph becomes obsolete, then trim all the edges having the obsolete function in the graph."*

Principle 5 (Transfer). $\text{Transfer}(e_1, e_2) : \text{trim}(e_1, E), \text{append}(e_2, E)$ where $e_1, e_2 \in E$

Meaning: *"Transferring one function between a pair of nodes to another pair of nodes in the graph is permissible; trim the edge between a pair of nodes then append the edge to another pair of nodes."*

Principle 6 (Generalisation). *Let* $f_1, f_2, \ldots, f_i \in F$ *where* $i \in \mathbb{N}$. *If* $f_1 = g$ *in context* y_1, $f_2 = g$ *in context* y_2, ..., *and* $f_i = g$ *in context* y_i *regardless if* $g \in F$, *define a piecewise function called* g. *Thus,*

$$g = \begin{cases} f_1 \text{ if } y_1 \\ \ldots \\ f_i \text{ if } y_i \end{cases}$$

Then, substitute f_1 *in context* y_1, f_2 *in context* y_2, ..., *and* f_i *in context* y_i *with* g.

Meaning: *"If at least two different functions in some context are equivalent to some function, then each of the functions can be generalised by that same function."*

To illustrate the principles 1–6, we consider a system with a data dictionary defined as in Table 3.

Table 3. Illustration data dictionary.

Component	Function
c_1: left hand	f_1: grip
c_2: right hand	f_2: deform
c_3: plier	f_3: hang
c_4: string	f_4: levitate
c_5: blower	Φ: other functions performed by the node
c_6: ping pong ball	

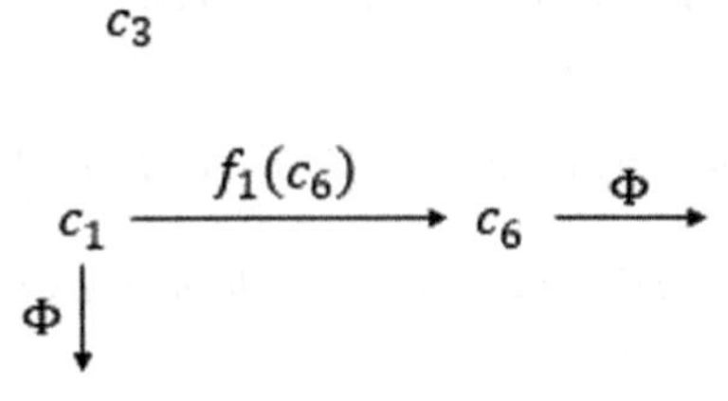

Fig. 1. Example of a graph model construct of three components in a system where principle 1 can be applied.

Figure 1 depicts a graph model of the components "plier" and function model of "left hand grip ping pong ball". Here, $\deg(c_3) = 0$ while the degree of the other nodes (i.e. components c_1 and c_6) are non-zero. We may apply principle 1 and thus trim c_3.

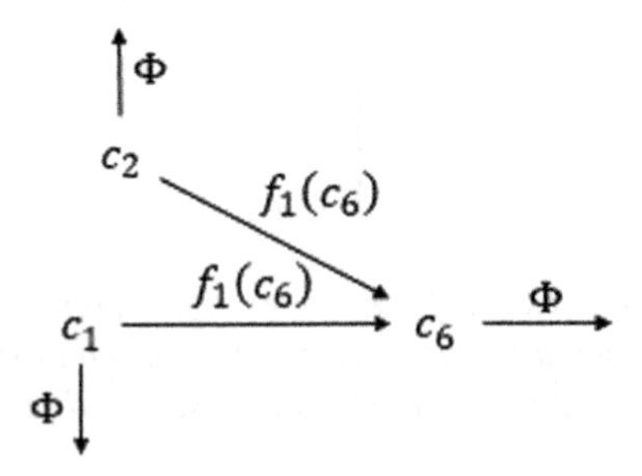

Fig. 2. Example of a graph model construct of three components in a system where principle 2 can be applied.

A function model for "left hand and right hand grip ping-pong ball" is formulated into a graph shown in Fig. 2 where the edges $((c_1, c_6), f_1)$ and $((c_2, c_6), f_1)$ are executing the same function, f_1, to the same node, c_6. We can then apply principle 2, and thus trim either one of the edges.

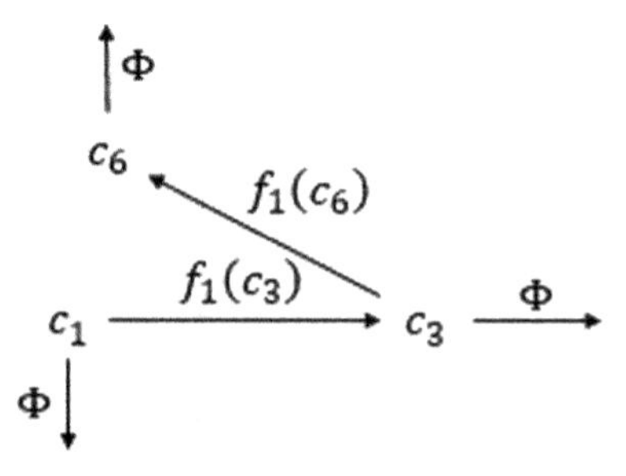

Fig. 3. Example of a graph model construct of three components in a system where principle 3 can be applied.

"left hand grip plier and plier grip ping-pong ball" formed a graph as in Fig. 3 which illustrates the nodes c_1, c_3 and c_6 forming a 3-node chain with edges $((c_1, c_3), f_1)$ and $((c_3, c_6), f_1)$ and are executing the same function to the next node in the chain. Therefore, based on Principle 3, we can trim both edges and append edge $((c_1, c_6), f_1)$.

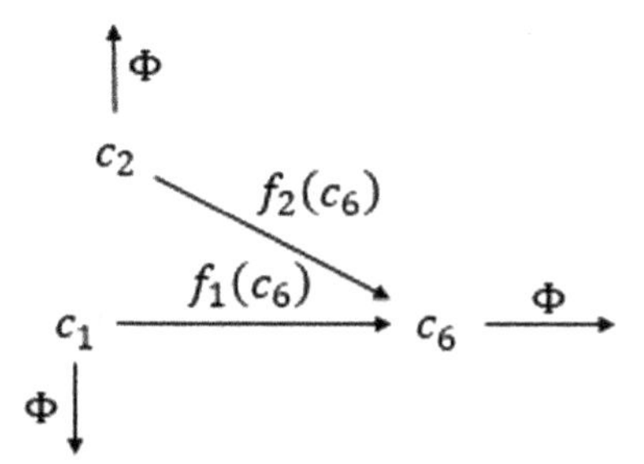

Fig. 4. Example of a graph model construct of three components in a system where principle 4 can be applied.

Consider in Fig. 4 where the function f_2 becomes obsolete and thus $f_2 \notin F$. We may apply principle 4, and thus trim edge $((c_2, c_6), f_2)$ and all other edges with f_2.

Figure 5 depicts a graph showing three functions acting on the same component. Here, we want to trim edge $((c_4, c_6), f_3)$ but node c_6 must receive function f_3. Hence, we apply principle 5 and trim edge $((c_4, c_6), f_3)$ between the pair of nodes c_4 and c_6, then append another edge $((c_3, c_6), f_3)$ between the pair of nodes c_3 and c_6.

Figure 6 depicts functions f_1, f_3 and f_4 which are similar. Therefore, we apply principle 6 and generalised the functions by the same function, g: hold. Thus substituting the functions f_1, f_3 and f_4 with g then we get a ubiquitious case of redundancy. We then may apply principle 1 and trim the relevant edges in the graph.

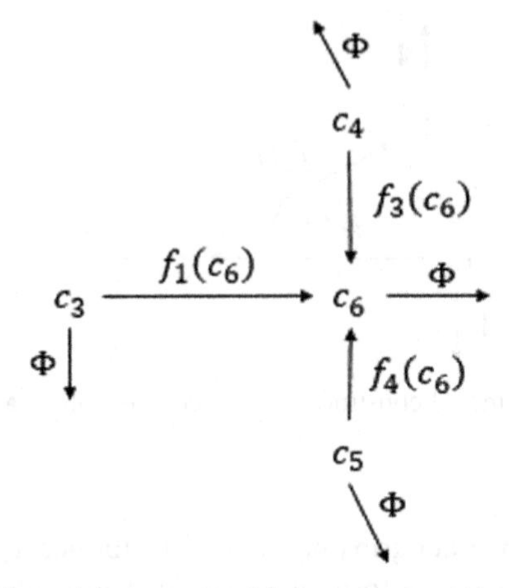

Fig. 5. Example of a graph model construct of four components in a system where principle 5 can be applied.

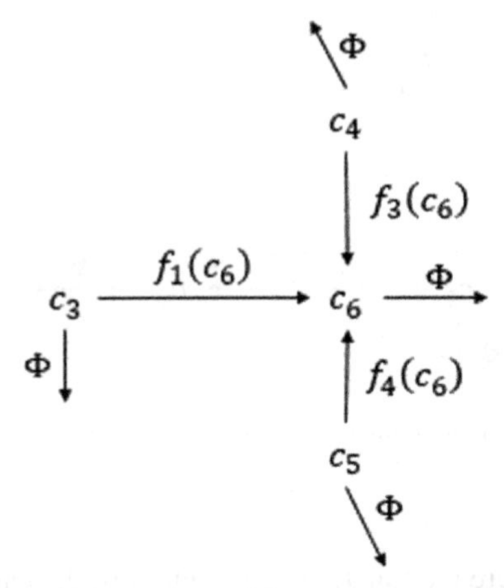

Fig. 6. Example of a graph model construct of four components in a system where principle 6 can be applied.

3 Case Study

To further illustrate the algebraic notation, graph modelling and the trimming principles in practice, we consider a case study on face mask.

Figure 7 shows the face mask system modelled using TRIZ function model. Following our defined algebraic notations and graph modelling, the function model of the face mask system shown in Fig. 7 can be represented into a graph model shown in Fig. 8. As formulated earlier, the data dictionary for the face mask system can be defined as in Table 4.

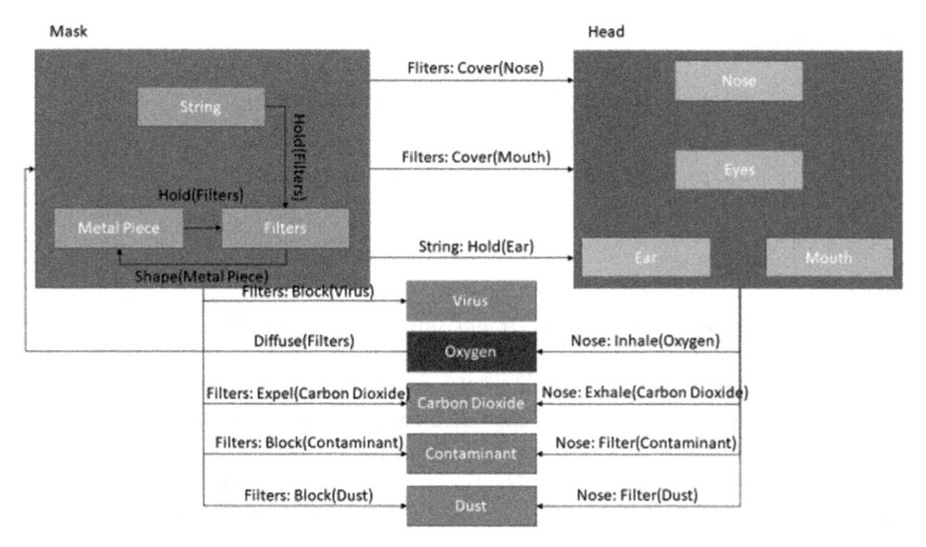

Fig. 7. Face mask system using TRIZ function model.

Table 4. Face mask system data dictionary.

Components	Functions
c_1: mask	f_1: hold
$c_{1.1}$: string	f_2: grip
$c_{1.2}$: filters	f_3: block
$c_{1.3}$: metal piece	f_4: diffuse
c_2: head	f_5: shape
$c_{2.1}$: nose	f_6: cover
$c_{2.2}$: mouth	f_7: inhale
$c_{2.3}$: ear	f_8: exhale
$c_{2.4}$: eyes	f_9: expel
v_0: virus	f_{10}: filter
a_1: Oxygen (O_2)	
a_2: Carbon Dioxide (CO_2)	
a_3: contaminant	
a_4: dust	

Based on the data dictionary in Table 4, let V_1 be the set of components and sub-components found in the data dictionary, F_1 be the set of functions found in the data dictionary, $p_1, p_2, \ldots, p_{10}$ be the set of ordered pairs $((a, b), f_1)$, $((a, b), f_2)$,…,$((a, b), f_{10})$, such that $f_1 = p_1((a, b)), f_2 = p_2((a, b)), \ldots,$ and $f_{10} = p_{10}((a, b))$ respectively, where

$a, b \in V_1$. Here,

$$p_1 = \{((a, b), f_1) \text{ if and only if } a \text{ function}_1 b\},$$

$$p_2 = \{((a, b), f_2) \text{ if and only if } a \text{ function}_2 b\},$$

$$\ldots$$

$$p_{10} = \{((a, b), f_{10}) \text{ if and only if } a \text{ function}_{10} b\}$$

Define the relation E_1 as the union of the set of ordered pairs p_1, p_2,…,p_{10}, thus $E_1 = \{p_1 \cup p_2 \cup \ldots \cup p_{10}\}$. Now let each element in E_1 be called $e_{1.1}$, $e_{1.2}$,…, $e_{1.15}$, therefore we have $V_1 = \{c_1, c_{1.1}, c_{1.2}, c_{1.3}, c_2, c_{2.1}, c_{2.2}, c_{2.3}, c_{2.4}, v_0, a_1, a_2, a_3, a_4\}$, $E_1 = \{e_{1.1}, e_2, \ldots, e_{1.15}\}$ and $F_1 = \{f_1, f_2, \ldots, f_{10}\}$. Lastly, let $G_1 = (V_1, E_1)$ be a graph of the face mask systems, and $\xi_1 = \{V_1 \cup F_1\}$ representing the type of nodes and edges found within the face mask systems.

Before any trimming iterations are done, i, the initial state of the face mask system graph is set to $i = 0$ and is depicted in Fig. 8.

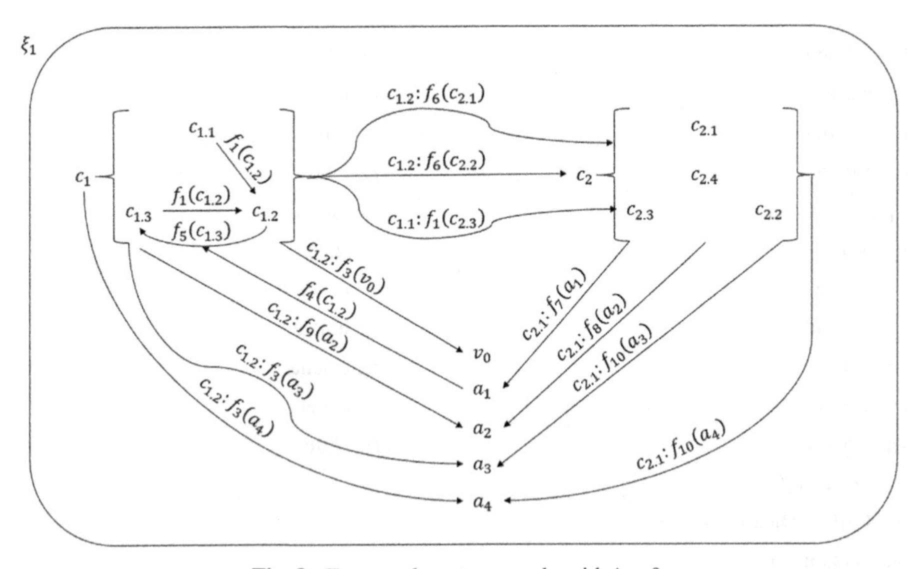

Fig. 8. Face mask system graph, with $i = 0$.

Once the initial state of the face mask graph is set, we now apply the trimming principles. We apply principle 1 to node $c_{2.4}$ because $\deg(c_{2.4}) = 0$. Nodes $c_{1.1}$, $c_{1.2}$ and $c_{1.3}$, and edges $((c_{1.1}, c_{1.2}), f_1)$ and $((c_{1.3}, c_{1.2}), f_1)$ form the redundancy case as in principle 2. We decide to trim edge $((c_{1.3}, c_{1.2}), f_1)$. The graph now has been trimmed neatly as shown in Fig. 9.

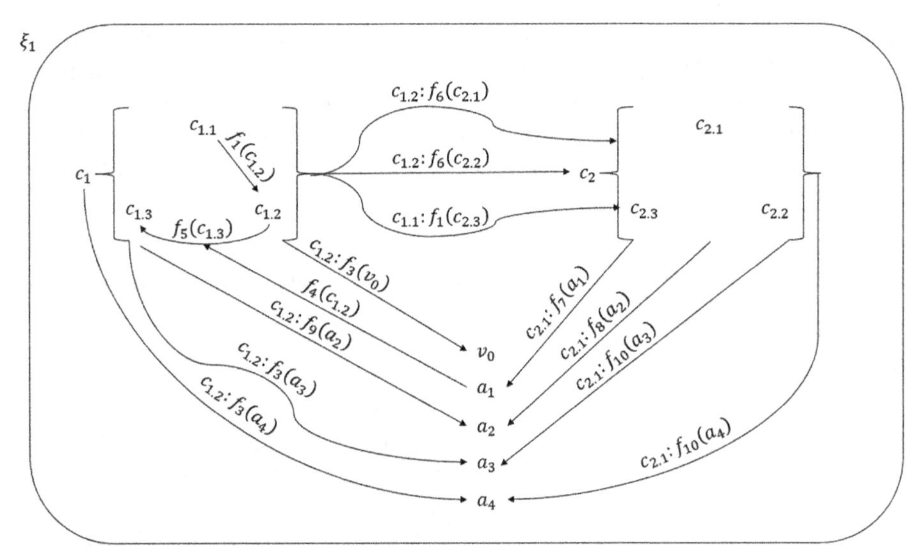

Fig. 9. Face mask graph after trimming, $i = 1$.

4 Analysis and Discussion

Inspired by Formal Logic, Set Theory and Graph Theory, advantage of the TRIZ Mathematical formalization, algebraic notation and novel TRIZ graph modelling includes: invariant of a semantic interpretations, open to one-to-many interpretation, simpler and quicker modelling, and such proper mathematical formalization enables TRIZ concepts to be programmable.

Unlike standard TRIZ model which declares an instance of an object that comprise of components (including subcomponents) and interactions between components (functions) which are tagged based on nouns and verbs, the proposed graph-based modelling does not declare an instance of such object and is invariant of what that component or function actually is. It is more concerned with which component it is and how the components interacts with each other (function).

The use of logical formalisms is not new in the field of Artificial Intelligence. The expressive capacity of logic, augmented by the structuring capability of semantic networks (a form of graph-representation) has been adopted widely by AI researchers. However, the proposed formal models have the advantage of providing a formalism for the formulation of ideal models based on a principled approach to trimming. The proposed framework also mainly suggests structural features as means of selecting trimming conditions.

5 Results

For the sake of completeness, we map existing TRIZ trimming rules to our trimming principles. We chose the term 'principle' as nomenclature to describe our graph-based trimming techniques, despite the fact it has no difference with the word 'rule' in Mathematics and Physics [11], purely for distinction purposes only. Unlike TRIZ trimming rules which is derived from empirical research of 200,000 over patents and real-world physical systems, our trimming principles are based on the mathematical study of networks and discrete structures [9], thus it is entirely structure-based (Table 5 and Table 6).

Table 5. Comparing the fundamental TRIZ trimming rules, A, B and C, to the trimming principles.

No.	Trimming Rules	Trimming Principle
1	**Rule A**: The functions (thus its carrier) can be trimmed if the object of the function is trimmed [10, 12].	**Non-functionality**
		Redundancy
		Transitivity
		Obsolescence
		Transfer
		Generalisation
2	**Rule B**: The functions carrier can be trimmed if the object of the function can perform the useful function by itself [10, 12].	**Non-functionality**
		Redundancy
		Transitivity
		Obsolescence
		Transfer
		Generalisation
3	**Rule C**: The functions carrier can be trimmed if another existing component in the system or super system can perform the useful function by the current function carrier [10, 12].	**Non-functionality**
		Redundancy
		Transitivity
		Obsolescence
		Transfer
		Generalisation

Highly Comparable | Slightly Comparable | Least Comparable

Based on the above table, we can observe that, in general, there are more than one trimming principle of ours that correspond to each existing TRIZ trimming rules because the rule exhibits at least some elements of the principles. This suggest perhaps the existing fundamental and extended trimming rules may not be sufficiently fundamental to be considered the basis trimming techniques in TRIZ as some – e.g., Rule D – is based on an external factor – e.g., 'new/niche market' – that is arguably subjective in nature. For TRIZ to become a hard science of innovation, we want to minimize as much as possible assumptions and fuzzy human factors to allow for rigorous, objective-based analysis.

Table 6. Comparing the extended TRIZ Trimming Rules, X, D and E, to the trimming principles.

No.	Trimming Rules	Trimming Principle
1	**Rule X**: The functions carrier can be trimmed if its useful function is trimmed or not needed [10].	**Non-functionality**
		Redundancy
		Transitivity
		Obsolescence
		Transfer
		Generalisation
2	**Rule D**: Function carrier can be trimmed if a new or niche market can be identified for the trimmed product [10].	**Non-functionality**
		Redundancy
		Transitivity
		Obsolescence
		Transfer
		Generalisation
3	**Rule E**: Function carrier can be trimmed if the function can be performed better by a new/improved part providing enhanced performance or other benefits [10].	**Non-functionality**
		Redundancy
		Transitivity
		Obsolescence
		Transfer
		Generalisation

Highly Comparable   Slightly Comparable   Least Comparable

6 Conclusion

The efforts to formalize TRIZ to turn it into the hard science of innovation calls for frameworks based on the adoption of rigorous mathematical models. In this paper, we have introduced the data dictionary, algebraic notation, graph modelling and trimming principles which are attempts to mathematically formalize TRIZ models in general and the trimming model. Trimming is seen as a valuable tool not only for its ability to retain functionality with a much more compact (or minimal) system design, but also trimming provides a way of thinking whereby product innovation triggers can be identified by suggesting a component(s) to trim.

The case study presented via mathematical modelling, applying Formal Logic, Set Theory and Graph Theory, is just a starting point in this direction. Our proposed framework in mapping TRIZ trimming rules to the formulated trimming principles, is seen to be an important contribution. In future works we will explore the formalization of other tools which includes process trimming.

Acknowledgement. The authors thank Universiti Malaysia Sarawak for the support rendered for this research. We truly appreciate it.

References

1. Lau, D.K.: An emperical mathematical approach for TRIZ implementation model - "LEADS" in China for innovative product development. Int. J. Pure Appl. Math. 121–133 (2005)
2. Nordlund, M., Lee, T., Sang, G.K.: Axiomatic Design: 30 Years After. https://dspace.mit.edu/bitstream/handle/1721.1/107378/Kim_Axiomatic%20design.pdf (2015)
3. Efimov-Soini, N.K., Chechurin, L.S.: Method of ranking in the function model. Procedia CIRP **39**, 22–26 (2016)
4. Efimov-Soini, N., Chechurin, L., Renev, I., Elfvengren, K.: Method of Time-dependent TRIZ Function Ranking, pp. 1–5. Elsevier, Warsaw (2016)
5. Borgianni, Y., Matt, D.T.: Applications of TRIZ and axiomatic design: a comparison to deduce best practices in industry. Procedia CIRP **39**, 91–96 (2015)
6. Devlin, K.: Keith Devlin at Stanford University. https://web.stanford.edu/~kdevlin/ (2021). 30 Oct 2021
7. Devlin, K.: Mathematics: The Science of Patterns : The Search for Order in Life, Mind, and the Universe, 1st edn. Henry Holt and Company, New York (1996)
8. Coronel, C., Morris, S., Rob, P.: Database Systems: Design, Implementation, and Management, 9th edn. Cengage Learning, Boston, Massachusetts (2011)
9. Carlson, S.C.: Graph Theory. https://www.britannica.com/topic/graph-theory (2020). Accessed 2 May 2022
10. Daniel Sheu, D., Hou, C.T.: TRIZ-based systematic device trimming: theory and application. Procedia Eng. **131**, 237–258 (2015)
11. Physics Forum: What's the difference between principle, law, rule, theorem and equation? https://www.physicsforums.com/threads/whats-the-difference-between-principle-law-rule-theorem-and-equation.779245/ (2014). Accessed 1 May 2022
12. GEN-TRIZ: MATRIZ – Level 1 Training Manual, 2019th edn. GEN-TRIZ, Newton (2019)
13. Ikovenko, S.: Design for Patentability®. DFP Institute, Boston (2019)

Improving the TRIZ Creative Engineering Methodology to Take into Account the Notion of the Value of the Idea

Serge Tremblay(✉) and Mickaël Gardoni

École de Technologie Supérieure, Montréal, Québec H3C 1K3, Canada
serge.tremblay.3@ens.etsmtl.ca

Abstract. Most of the time, the notion of the value of an idea, if not ignored by the company, seems to be established toward the end of the technical development of the idea during the resolution of a problem or in the final stages of the innovation management process for the development of products or services. Generally speaking, the birth of ideas with or without the use of creativity methods consists in the resolution of a technical problem, i.e., the technical aspect of the idea. This paper proposes an experiment using a creative process integrating a TRIZ approach and seeking to demonstrate that taking into account the notion of value, expressed by value proposition statements, in the evaluation of ideas for selection purposes influences the selection of the best idea that satisfies the client's needs, i.e., the value aspect, but neglects the technical aspect that the client expects. Currently, the TRIZ creative engineering methodology is used to solve mainly the technical aspect of a problem, among others the technical and physical contradictions of technological processes and technical systems. When using a TRIZ approach to find and discover solutions to a problem, it is proposed to associate the value aspect with the technical aspect of the ideas or solutions generated. Thus, in parallel to the search for solutions, be able to associate a set of value proposition statements also generated during the process.

Keywords: TRIZ · Problem solving · Notion of the value · Idea selection · Decision-making

1 Introduction

1.1 Formulating the Problem

Context. The concept of value appears in many fields, such as the arts, philosophy, and linguistics. It can also be found in various scientific fields, such as engineering science. In this last field, the focus is placed on the technical systems and technological processes for things like artificial or natural objects and activities required or desired by individuals [1].

Most often, if a company does not entirely overlook an idea, its value is perceived toward the end of the technical development phase. Accordingly, to develop the best

Published by Springer Nature Switzerland AG 2022
R. Nowak et al. (Eds.): TFC 2022, IFIP AICT 655, pp. 17–28, 2022.
https://doi.org/10.1007/978-3-031-17288-5_2

idea, the notion of value should be determined prior to the development stages, i.e., in the area if the idea (idea generation, idea development, idea selection process [2]).

Gap Statement. When developing new products (or services), a company is faced with solving technical problems regarding the product's design. Currently, the TRIZ creative engineering methodology is mostly used to resolve the technical aspects [1]. The method does not address the notion of value which could determine the choice of the best idea from among a number of generated ideas. The technical aspects of a problem are expressed, in part, by the technical and physical contradictions of technological processes and technical systems [1].

The following observation is essential: the difficulty with this research stems from the value or value aspect of the idea, the definition of which is non-existent or inadequate.

Objectives. The primary purpose of this article is to illustrate the possibility of influencing the decision support process when selecting an idea by considering the notion of value, i.e., by associating the value aspect with the technical aspect of an idea or solution. When an adapted TRIZ method is used to search for and identify solutions to a problem, it is then that a set of value proposition statements [3, 4] are linked to the technical aspect. These proposition statements are identified in conjunction with the problem-solving process.

1.2 Creating Value in Innovation

Creating Value and the Concept of Innovation

Authors Carlson, Polizzotto, and Gaudette [5] contend that the survival and growth of a company tend to be influenced by its ability to create value. They point out that a company that fails to create value is experiencing difficulties. For authors, value creation is described as a process enabling a company to meet its customers' needs and thereby encourage innovation.

They complete by describing two critical elements of the value creation process: "Important customer and market opportunities, to create a significant market impact"; and "Value propositions, as the starting point for all value creation activities." [5].

The second key element seems to be the means of expressing value creation.

Lindic and da Silva [4] define the value proposition as follows: "A value proposition describes how a company's offer differs from its competitors and explains why customers buy from the company.“ [4]

In this way, value creation is shown to be closely linked to the concept of innovation. The Oslo Manual version 2018 [6] defines two critical components in the concept of innovation, namely "the role of knowledge as a basis for innovation, novelty, and utility, and value creation or preservation as the presumed goal of innovation.“ [6]

The Customer's Needs

To successfully meet the customer's needs, it appears necessary to gain an understanding of how they perceive value.

The overall value is an aggregate approach that combines the perceived benefits and sacrifices to obtain the client's perceived value. In other words, it represents the contrast between various elements used to evaluate the benefits and those used to evaluate the sacrifices. Together, these evaluations assess the overall value of a product when it is purchased. [7–13, 16].

Rivière and Mencarelli [16, 18] point out that most authors recognize the existence of a number of different characteristics of value. They suggest the following criteria:

- Value stems from making a comparative judgment: it is the result of a relative judgment made by a consumer about an object [14, 16, 17];
- Value is personal: it is commonly recognized in literature on the topic that value is subjective and personal as opposed to being objectively determined by vendors [11, 16–20];
- Value is both contextual and dynamic: many authors share the opinion that perceived value would vary based on the type of product being purchased and the type of usage the individual has in mind for the product [11, 14–16, 21, 22].

Therefore, to meet the customer's needs and address these criteria, it is essential to focus on the value aspect when developing a product or service, i.e., the customer's perception of value. These criteria also reveal the extent of the options available when developing attractive offers.

Problem and Problem Solving

Now, it is worth defining the notion of a problem allowing to meet the customer's needs.

A problem is a gap between existing and desired situations [1]. This gap provides many opportunities to solve the customer's problem. Of course, the gap can also be filled by adding, changing, or removing certain functions carried out by technological processes and technical systems [1], and by adding, changing, or removing "potential applications" that make up the value proposition [23]. In terms of value and from a technical point of view, the customer's problem is solved as the actions have turned the initial situation into a desired situation or closer to the one desired through desired potential applications. As a result, the customer perceives the offering's value and learns of the solutions provided by the functions [1, 23].

Problem-solving by identifying both the desired functions and the desired potential applications [1, 23] is closely tied to a company's ability to understand the customer's point of view [4].

The Notion of the Idea

The notion of the idea embodies more than its definition. It involves both the creativity behind the idea and how it is expressed.

There are two definitions of the word idea. The first is from the dictionary: "Someone's opinion or belief about something." [24] and the second by authors Koen, Bertels, and Kleinschmidt [25]: "An idea is the most embryonic form of the new product or service." [25]

As such, the expression of an idea will be regulated by the area of the idea. The latter then organizes the exitance of the idea, from its birth to when it is selected [2]. If selected as the most promising solution, the idea will be developed into a concept and eventually materialize and lead to a unique result.

Accurately determining the value of an idea without further study of its potential and impact remains highly challenging. The perceived value of an idea can only be perfected during the development of the concept, thereby eliminating or reducing uncertainties, ambiguities, and risks.

As mentioned, ideas tend to be fragile. Besides this, another feature of a new idea is amazement. This suggests a certain strength and boldness to an idea capable of driving change. The opposing notions of fragility and amazement suggest a connection between ideas and change. Without the notion of amazement, the relationship between these two things is not functional [26].

Decision Support and Selection of the Idea

Ultimately, there remains only to select a single idea from among the set that best solves the problem. The best idea is selected based on the technical aspect of the problem (product or service). However, if the value aspect of the selected idea is overlooked the client may be dissatisfied. The customer expects a solution to the technical problem and, more importantly, expects the idea to add value.

The selection process occurs toward the end of the creative process, where there is not enough time between the conception and selection of the idea [27]. The idea selection process aims to identify one or more ideas that have the most potential according to a set of specifications (identification, description, business values, features, feasibility, etc.) and eventually be developed to perform specific tasks [2].

Developing the Technique

Savransky [1] states that "Any application of TRIZ starts with understanding of the technical system or technological process and the situation in which the problem appears." [1]

In order to accomplish the project and its primary function, both the process of developing the technique and the description of the parts of the technique involve identifying and describing the technological process(es), technical systems, functions, subsystems and super-systems, attributes, etc. [1]

The application of the TRIZ method makes it possible to find solutions to problems involving technological processes and technical systems, i.e., the technical aspect of the problem. Therefore, each part of the technique constitutes the technical aspect.

The Notion of Value

Next, let us examine another component supporting the goals of the research problem, i.e., the notion of value. To begin, we must first define the meaning of "value": "The regard that something is held to deserve; the importance, worth, or usefulness of something." [30]

This definition introduces an important point supporting the notion of value: "The regard that something is held to deserve" [30] establishes a connection between someone and something. While the "something" may appear relatively straightforward and unchanging, the complex nature of this relationship becomes apparent when we examine the "someone," namely due to the breadth of their perception of value and their judgement of its significance.

Having defined the word value in itself, we recall the value proposition definition. Lindic and da Silva [4] define the value proposition as follows: "A value proposition describes how a company's offer differs from its competitors and explains why customers buy from the company." [4]

It is important to remember that the value proposition is an expression of value creation. For Osterwalder and Pigneur [3], "The value proposition is the reason why customers turn to one company over another. It solves a customer problem or satisfies a customer need." [3] The authors add: "In this sense, the value proposition is an aggregation, or bundle, of benefits that a company offers customers." [3]

The customer purchases a product or service from a company because it meets their needs. They recognize that the company has provided them with value to influence their purchasing decisions. As such, value creation is expressed by developing text-based statements that form the value proposition. Together, these statements constitute the value aspect.

1.3 Considering the Notion of Value

The Adapted TRIZ Method

Here, Cavallucci's [28] TRIZ method, is adapted to include the notion of value. There are only two possible spheres in the initial problem-solving approach [1, 29], namely that of initial or specific realities and abstraction or generic realities. The author's approach includes a middle sphere, that of technological realities [28]. Figure 1 illustrates a flow beginning at the "Initial situation" box and then flowing to the following boxes: "Well-defined problem," "Problem model," "Solution model," "Solution concept" and "Detailed solution." For the flow to move vertically upwards and downwards, certain steps such as formulation, modeling, interpretation, and construction are necessary to process the information from the previous box and enable the flow from one sphere to the next. The creative space is the mind's ability to generate ideas or solutions with the help of creative tools.

In order to demonstrate how the notion of value is considered, the vertical downward boxes are duplicated to form a new section of downward vertical boxes parallel to the original boxes respecting the three spheres. Each box is renamed to reflect a solution-value pairing, i.e., "Solution-value model," "Solution-value concept," and "Solution-values detailed."

Progression of the Method by the Participants in the Creative Process

The participants in the group will follow the method to solve a problem. The left side of the method (Fig. 1), the upward vertical boxes, is used to identify and state the problem, while the right side is used to capture and process ideas or solutions, as well as value proposition statements.

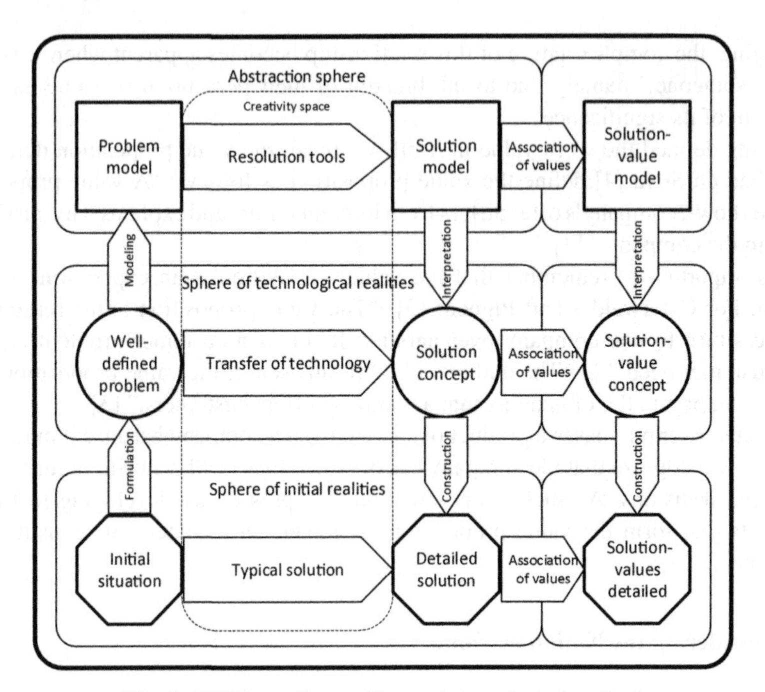

Fig. 1. TRIZ creative problem-solving adapted method

The process begins at the "Initial Solution" box. Participants take note of the problem and its scope, developing an initial understanding. They can generate ideas using trial-and-error methods [1] from the creative space of the initial realities sphere.

Continuing the process, a problem formulation identifies the problem by describing the initial situation, the main useful function, the work of the system, the parts of the technique, the levels of the systems and the ideality. In addition, they can identify known solutions and other relevant information. Being in the sphere of technological realities, they can generate ideas with the help of TRIZ tools such as the 9 boxes, the Size-Time-Cost operators, the operative Zone and Period, the resources, etc. [1, 29]

Continuing the process towards the sphere of abstraction, a model of the defined problem is created. By developing a problems graph and partial solutions, participants construct technical and physical contradictions and substance-field analysis. From these constructs, TRIZ tools such as the 9 boxes, the Contradiction Matrix, the 11 Separation Principles, the 76 Standard Solutions of Invention Problems, the 9 Evolution Laws, etc. [1, 29], are used to generate ideas.

In this process, the creativity space is about generating ideas, but also value proposition statements. Figure 1 has two parallel descending verticals. This depiction involves associating the value aspect identified by value proposition statements with the technical aspect identified by the parts of the technique of the ideas or solutions. These associative processes are depicted by the arrows "Association of values."

Table 1 depicts an example of solutions identified and which value proposition statements are suggested by association.

Table 1. Creative results table

No	Solution record	Value proposition
1	Solution 1	Value proposition 3 Value proposition 4
2	Solution 2	Value proposition 1 Value proposition 2 Value proposition 5
3	Solution 3	Value proposition 6
n	Solution n	Value proposition 3 Value proposition 4 Value proposition 6 Value proposition n

This table captures all the ideas and solutions, as well as the value propositions generated in the creativity space and in the three spheres. It remains for the participants to process and select the best idea. Thus, the solution models or solution-value models need to be interpreted by giving them a description and meaning and, if necessary, assigning them viable technologies. Thus, these solution models or solution-value models become solution concepts or solution-value concepts in the sphere of technological realities. A construction of these concepts allows to complete the association of the value aspect with the technical aspect and to select the best concept (idea, solution). Then, to develop, demonstrate, verify and validate that the solution solves the problem. Thus, a detailed solution or detailed value-solution is available in the sphere of initial realities. All that remains is to implement it [1, 29].

Influencing the Selection of the Best Idea

As a general rule, selecting the best solution (Table 1, column "Solution record") to solve the problem is done by evaluating the technique of each solution. The technical resources mostly carry out this assessment during the problem-solving process. This raises the question of whether the technique-based selection of the best solution truly addresses the customer's needs or those of the end-user.

The assessment of the proposed solutions must be based as much on the value aspect as on the technical aspect in order to demonstrate the importance of considering the notion of value. In other words, during the assessment, the value propositions associated with the solutions (Table 1, column "Solution record" and "Value proposition") ought to influence the assessment with a view to selection.

2 Experimental Methods

2.1 Defining the Problem and the Objective

The experiment's goal is to determine whether the inclusion of value (value proposition statements) impacts the selection of the best idea or solution from among those generated using a creative process that includes an adapted TRIZ method (Fig. 1). The primary focus of the TRIZ method was originally to produce ideas that would solve the technical aspect of the problem since the value aspect of the problem did not exist. Value is not (or rarely) considered in problem-solving. Given that experimentation falls within the area of the idea, the term idea will be used to define the methodology.

2.2 Methodology Design

Applying a creative process integrating an adapted TRIZ method (Fig. 1) involves using a research methodology to explore and confirm its applicability in real-life situations. The selected research class is qualitative descriptive research combined with the qualitative descriptive study [31].

The case study was conducted for a manufacturer of commercial equipment in the food industry. The company wanted to remain ahead of the competition by improving its innovative processes. The company is interested in applying a creative process integrating an adapted TRIZ method (Fig. 1) to solve problems. They are also looking to create value for their customers.

The company is considered a small business and has an organizational structure consistent with that of a small business.

The study population is composed of at least one manager and employees involved in product and service innovation, development, and design. The company nominated a group of participants to be assigned to the study and considers that they will be able to follow up after the experimental period.

Three operational concepts were selected to measure and collect data [31]. The first concept to measure is the creative process. Participant observation was deemed the most suitable way to measure certain characteristics of the creative process. The second concept to measure is the participants in the group. Participant observation was deemed the most suitable way to measure the participants' interest and involvement. The third concept to measure involves generating ideas and value proposition statements [31]. The ideas generated by the participants are recorded in a Creative Results Table. Non-participant type observations seem to be the correct technique to measure certain aspects of the generated ideas and value proposition statements [31].

The suggested measurement techniques provide a more general and chronological account of the researcher's observations. Field Notes and logbooks are selected as the data collection method [31].

A set of data collection instruments were developed for recording observations.

Observations are listed in a Qualitative Data Collection Table and include boxes for noting observations (observations, impressions, conversations, experiences [31]) and boxes for adding observation points. This table has three groups of observation points concerning the three measurement concepts.

Two interconnected tables were developed to collect data on ideas and value proposition statements: the Creative Results Table and the Results Prioritization Table. The first table is used to collect data during the creative process, and the second is to prioritize the data from the first table after the fact.

During the creative process, the participants generate ideas and value proposition statements. These are listed in the Creative Results Table provided for this purpose.

Following this, an assessment of relevance is carried out in the Results Prioritization Table. This enables each idea to be evaluated according to a predefined set of criteria (selected strategically to align with the company's strategy). In order to test whether or not the inclusion of value can truly influence participants' evaluation of ideas, this experiment features a two-phase assessment process: the first phase does not consider the notion of value, and the second phase does.

Based on the set of observations recorded during the creative process, those concerning the participants affect the validity of the creative process and the data collected about the ideas and value proposition statements. The participants' interest and time involvement are taken into account to determine the level of value of the experiment [31].

As stated, the experiment offers a two-part process for evaluating the ideas. A comparative analysis ensures the thoroughness of the two assessment processes. Comparing the evaluation results and prioritization between the two assessment processes completes the analysis.

The interpretation of the results helps to uncover the meanings of the data analysis. As such, the results must present a rationale, a direction, a value, while also providing an explanation [31]. From these interpretations, a conclusion is developed.

2.3 Expected Results

The data collected through observation is used to discuss, confirm and support the analysis results on the impact of the notion of value on the final selection of an idea.

The type of results expected should support the goal of the experiment. The results of the two-part process of evaluating ideas, i.e., the first phase that does not consider the notion of value, and the second phase that does, clearly defines the process of selection. In other words, out of a set number of exercises in the creative problem-solving process, a number of solutions resulted in different choices between the two assessment phases. Using the value proposition to deliver value to the customer may be more interesting when suggesting a less attractive technical solution to the problem.

2.4 Limits of the Methodology

There are limitations to the experiment that can influence its credibility [31]. For example, the company's area of activity may lead to different results for different areas and therefore affect the ability to replicate the experiment [31]. As a result, selecting a field that requires a certain level of technology for product or service development and that requires advanced creative problem-solving methods seems essential.

The second limitation is the completion of a small number of creative processes. The creative process should ideally be carried out several times to solve different problems and ensure that a large amount of data is collected. Collecting a tiny amount of data could create doubt on the influence of the value aspect in selecting the best solution.

Another limitation is the participants' limited or non-existent ability and experience in using a creative process that integrates an adapted TRIZ Method (Fig. 1). The underlying premise for this experiment was that participants had no experience in using the TRIZ method.

Lastly, the creative chaos involved in carrying out a creative process constitutes an important managerial limitation. An overly rigid structure will constrain creativity by confining the participants to the problem space and hampering the discovery of an ideal solution. Creativity requires freedom of action to break through psychological inertia [1].

3 Discussion and Conclusion

By associating the value proposition statements with the ideas, the influence of the former guides the participants involved in the creative process to select the best idea. The significance of the results will reassure the participants that the customer is receiving the best service from the solution that most benefits them. Results should reflect that the customer's point of view and needs occupied a prominent place in the participants' minds during the creative process.

Ultimately, this type of experiment calls for a situation where a creative problem warranting the use of a creative process is solved. In other words, the problem must present the possibility of identifying technical or physical contradictions and have a high difficulty ratio [1]. As a result, considering the notion of value in certain problem-solving situations with lower difficulty ratios appears to be more difficult to apply, even unnecessary. Following this experiment, several possible areas for further research can be explored. A study of the relationship between the combined value aspect and the technical aspect, and the problem's difficulty ratio in order to discover if there is a limit to the application of the notion of value. Lastly, applying the notion of value to other area of activity and to different creative processes to assess the notion of value in the decision support process.

References

1. Savransky, S.D.: Engineering of Creativity: Introduction to TRIZ Methodology of Inventive Problem Solving. CRC Press LLC, Boca Raton, Fl (2000)
2. Alexe, C.G., Alexe, C.M., Militaru, G.: Idea management in the innovation process. Netw. Intell. Stud. **2**, 143–152 (2014)
3. Osterwalder, A., Pigneur, Y.: Business Model Generation: A Handbook for Visionaries, Game Changers, and Challengers. John Wiley & Sons Inc., Hoboken, NJ (2010)
4. Lindic, J., da Silva, C.M.: Value proposition as a catalyst for a customer focused innovation. Management Decision **49**, 1694–1708 (2011). https://doi.org/10.1108/00251741111183834
5. Carlson, C., Polizzotto, L., Gaudette, G.R.: The "NABC's" of value propositions. IEEE Eng. Manag. Rev. **47**, 15–20 (2019). https://doi.org/10.1109/EMR.2019.2932321
6. OCDE/Eurostat: Oslo Manual 2018: Guidelines for Collecting, Reporting and Using Data on Innovation. OCDE Publishing, Measurement of Scientific, Technological and Innovation Activities, Paris, France (2019). https://doi.org/10.1787/9789264304604-en
7. Lai, A.W.: Consumer values, product benefits and customer value: a consumption behavior approach. Adv. Consum. Res. **22**, 381–388 (1995)
8. Aurier, P., Evrard, Y., N'Goala, G.: Comprendre et mesurer la valeur du point de vue du consommateur. Recherche et Applications en Marketing (French Edition) **19**, 1–20 (2004). https://doi.org/10.1177/076737010401900301

9. Rivière, A.: Vers un modèle de formation de la valeur perçue d'une innovation : le rôle majeur des bénéfices perçus en amont du processus d'adoption. Recherche et Applications en Marketing **30**, 5–27 (2015). https://doi.org/10.1177/0767370114549908
10. Grewal, D., Monroe, K.B., Krishnan, R.: The effects of price-comparison advertising on buyers' perceptions of acquisition value, transaction value, and behavioral intentions. J. Mark. **62**, 46–59 (1998). https://doi.org/10.2307/1252160
11. Zeithaml, V.A.: Consumer perceptions of price, quality, and value: a means-end model and synthesis of evidence. J. Mark. **52**, 2–22 (1988). https://doi.org/10.1177/002224298805200302
12. Filser, M., Plichon, V.: La valeur du comportement de magasinage. Statut théorique et apports au positionnement de l'enseigne. Revue Française de Gestion **30**, 29–44 (2004)
13. Merle, A., Chandon, J.-L., Roux, E.: Comprendre la valeur perçue de la customisation de masse. Une distinction entre la valeur du produit et la valeur de l'expérience de co-design. Recherche et Applications en Marketing (French Edition) **23**(3), 27–50 (2008). https://doi.org/10.1177/076737010802300301
14. Holbrook, M.B.: The nature of customer value: an axiology of services in the consumption experience. In: Rust, R., Oliver, R. (eds.) Service Quality: New Directions in Theory and Practice, pp. 21–71. SAGE Publications, Inc., 2455 Teller Road, Thousand Oaks California 91320 United States (1994). https://doi.org/10.4135/9781452229102.n2
15. Holbrook, M.B.: Introduction to consumer value. In: Holbrook, M.B. (ed.) Consumer Value: A Framework for Analysis and Research, pp. 1–28. Routledge, London, UK (1999)
16. Rivière, A., Mencarelli, R.: Vers une clarification théorique de la notion de valeur perçue en marketing. Recherche et Applications en Marketing **27**, 97–123 (2012). https://doi.org/10.1177/076737011202700305
17. Sinha, I., DeSarbo, W.S.: An integrated approach toward the spatial modeling of perceived customer value. J. Mark. Res. **35**, 236–249 (1998). https://doi.org/10.1177/002224379803500209
18. Woodruff, R.B.: Customer value: the next source for competitive advantage. J. Acad. Mark. Sci. **25**, 139–153 (1997). https://doi.org/10.1007/BF02894350
19. Day, E., Crask, M.R.: Value assessment: the antecedent of customer satisfaction. J. Consum. Satisfaction, Dissatisfaction Complaining Behav. **13**, 52–60 (2000)
20. Sánchez-Fernández, R., Iniesta-Bonillo, M.-A.: Consumer perception of value: literature review and a new conceptual framework. J. Consum. Satisfaction, Dissatisfaction Complaining Behav. **19**, 40–58 (2006)
21. Sheth, J.N., Newman, B.I., Gross, B.L.: Why we buy what we buy: a theory of consumption values. J. Bus. Res. **22**, 159–170 (1991). https://doi.org/10.1016/0148-2963(91)90050-8
22. Gardial, S.F., Clemons, D.S., Woodruff, R.B., Schumann, D.W., Burns, M.J.: Comparing consumers' recall of prepurchase and postpurchase product evaluation experiences. J. Consum. Res. **20**, 548–560 (1994). https://doi.org/10.1086/209369
23. Le Loarne, S., Blanco, S.: Management de l'innovation. Pearson France, Montreuil, France (2012)
24. Pearson_Education_Limited: Longman Dictionary of Contemporary English. Pearson Education Limited, Essex, Angleterre (2014)
25. Koen, P.A., Bertels, H.M.J., Kleinschmidt, E.: Effective practices in the front end of innovation. In: Kahn, K.B., Kay, S.E., Slotegraaf, R.J., Uban, S. (eds.) The PDMA Handbook of New Product Development, pp. 117–134. John Wiley & Sons Inc., Hoboken, NJ (2013)
26. Kourilsky, F.: Du désir au plaisir de changer: Comprendre et provoquer le changement. Dunod, Paris, France (2004)
27. Naggar, R.: Créativité et R&D dans une entreprise exploitante. Séminaire sur les parcours d'idées et le co-design, HEC Montréal, Montréal, QC (2012)

28. Cavallucci, D.: Contribution à la conception de nouveaux systèmes mécaniques par intégration méthodologique, p. 232. Université Louis Pasteur, Strasbourg, France (1999)
29. Haines-Gadd, L.: TRIZ For Dummies. John Wiley & Sons Ltd, West Sussex, UK (2016)
30. Stevenson, A.: Oxford Dictionary of English. Oxford University Press, Oxford, UK (2010)
31. Fortin, M.-F., Gagnon, J.: Fondements et étapes du processus de recherche: méthodes quantitatives et qualitatives. Chenelière Éducation inc., Montréal, Qc (2016)

VA++ - The Next Generation of Value Analysis in TRIZ

Christian Iniotakis[✉]

Ulm University of Applied Sciences, Prittwitzstr. 10, 89075 Ulm, Germany
christian.iniotakis@thu.de

Abstract. Around twenty years ago, two methods relevant for systematic innovation and improvement - TRIZ and Value Analysis - have been merged in a specific way, which was then incorporated in software packages about innovation and became part of certified TRIZ education as well. An intended key purpose of this nowadays established method of Value Analysis in TRIZ is to identify system components of low ideality - or value, respectively -, i.e. parts of the system, that do not give a satisfactory functional contribution in relation to their cost. Hence, Value Analysis should point out the sweet spots for subsequent improvement, innovation or even patent circumvention activities. Unfortunately, the commonly used standard approach for Value Analyis in TRIZ, which is based on a function ranking algorithm, leads to results, that are inconsistent and not trustworthy in general. This work illustrates these shortcomings and explains their origins. Derived from key requirements necessary for a meaningful concept, VA++, a new advanced approach for Value Analysis in TRIZ, is presented and validated.

Keywords: TRIZ · Value analysis · Systematic innovation · Function analysis · Function ranking · Ideality · Algorithm · CAI · Trimming

1 Introduction

Ideality is a key concept in TRIZ, which is supposed to indicate, if a system - such as a product, a process, or even a business model - will prove successful under competitive circumstances on the long run. Ideality is typically described by the ratio of functionality divided by costs (cf. e.g. [1, 2]). Here, functionality is a measure for all the useful functions, i.e. the main function(s) and also further additional ones, the system delivers to the supersystem or its environment, respectively. Roughly speaking, functionality expresses, how good the system is considered to be in total. The system costs describe, how expensive the system is. Depending on the perspective and the actual problem setting, the costs might cover different aspects. For example, the costs could be represented by material and assembly costs, by the market price, or also by a much more general concept. In the latter case, usage and long-term maintenance costs might also be taken into account, as well as negative side effects and harms done by the system. Since ideality expresses, how good a system is per general costs to bear, it obviously is very similar to the well-known value concept in the framework of Value Analysis according to Miles

Published by Springer Nature Switzerland AG 2022
R. Nowak et al. (Eds.): TFC 2022, IFIP AICT 655, pp. 29–38, 2022.
https://doi.org/10.1007/978-3-031-17288-5_3

[3]. As a consequence, ideality and value are often used interchangeably. However, the notion 'value' has several more general meanings and might be interpreted in different ways by different individuals, whereas the term 'ideality' is quite unique. For this reason, the latter one will be predominantly used for the rest of this work. It is important to realize, that most TRIZ methods have the effect or are designed to increase ideality, e.g. by solving problems or improving the system. Thus, there is a significant implicit relation between TRIZ and Value Analysis, and a deeper methodical combination is very promising in general [4]. The focus of this work, however, is set on the approach of an explicit integration of Value Analysis into TRIZ based on Function Analysis.

2 Function Analysis

Function Analysis is one of the most important TRIZ methods. It allows to analyze systems in detail and is found in many newer typical TRIZ textbooks, e.g. in [1, 2]. Together with the proposal, to extent Function Analysis also to spatio-temporal parameters, Litvin et al. [5] give more details about its historic evolution. The first step of Function Analysis is to divide the system into basic parts called components. Likewise the environment is represented by its relevant parts, the so-called supersystem components. Then, all of these components are checked for the relevant functions, they perform on each other, in a systematic way. Identified functions are expressed in a very compact manner - typically a verb only -, which just represents the underlying and sometimes very complex functional details in form of a shorthand description. Applying Function Analysis to a given system eventually results in a complete picture containing all the system and involved supersystem components and their relevant functional interplay. Based on that, it becomes clear from the functional perspective, how the system works, in particular how it manages to provide its key functionality in form of the main functions, and how each component of the system contributes to it internally. For illustrative purposes, a simplified example showing the result of Function Analysis for a drone for autonomous package delivery is depicted in Fig. 1.

3 VA in TRIZ

Within TRIZ, a key reason for explicitly applying value analysis methods to a given system is to determine the individual values or idealities, respectively, of its components. The intention is to find suitable sweet spots for system improvement and innovation. For example, a common general approach is to identify components of low ideality as trimming candidates. Trimming denotes a TRIZ method, which has the goal to get rid of one or more system components, while the remaining system is still providing its functionality unaffectedly (cf. e.g. [1, 2, 6]). Obviously, this only works out, if the remaining components are thoroughly enabled to take on the functions of the trimmed ones, or if these functions are rendered unnecessary. In any case, this involves good ideas, and often requires very innovative system changes.

A component of low ideality is a natural candidate for trimming, if the system ideality should be increased in an efficient way, because it doesn't contribute much for its costs. Experts with a good intuition and deep knowledge of the system are sometimes able to

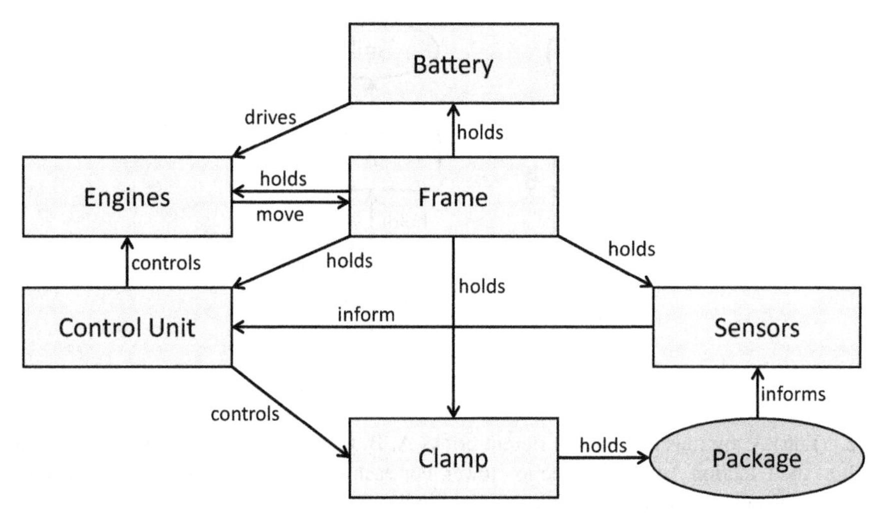

Fig. 1. Sample result of Function Analysis applied to a drone for autonomous package delivery. The drone is represented by six top level system components (frame, control unit, engines, battery, sensor, and clamp), whereas the relevant supersystem component is the package. Note, that the main function of the drone to "transport package" is not explicitly stated here, but implicitly results from the full functional interplay: After the control unit knows about the delivery location, e.g. from an address written on the package, it controls the engines to move the frame to the desired position. The frame itself holds all of the other system components, in particular the clamp holding the package. Thus, the package eventually arrives where it should.

directly point out components, which are likely to have a low individual ideality. Since in practice, unfortunately, those experts are not always available or able to do so, or there are several experts of different opinions, there is a strong need to put the ideality estimation on a more profound and systematic basis. Currently, there is one major approach known to a significant part of the TRIZ community, which was developed for that very purpose. It takes the results of Function Analysis as a key input, and is supposed to return the individual idealities of the system components as a main result. Actually, this specific algorithmic approach is also denoted as 'Value Analysis' in TRIZ, but will regularly be abbreviated as VA in the following, to avoid confusion with the more general methodical framework of Value Analysis in a wider sense. Standard VA as an established procedure can be broken down into three steps:

- Step 1: An empirical algorithm performs a so-called function ranking. Based on this, a functional value for each function is determined, which serves as a measure for its functionality.

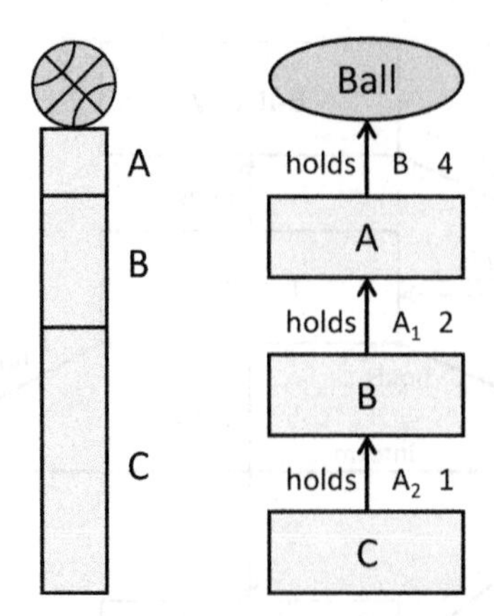

Fig. 2. (Left) A toy tower of three different bricks A, B, C, constructed to hold a ball. (Right) Result of the Function Analysis of the toy tower. For each function, both the function rank and the functional value according to standard VA are indicated.

- Step 2: Likewise, a functional value for each component is determined by summing up the functional values of all the functions performed by it.
- Step 3: The functional value of each component is taken as the basis for the final analysis. For example, the ideality of each component is directly evaluated by dividing its functional value by its costs, or the components are compared by simply placing them in a two-dimensional diagram of functional value over cost.

Involved in automated or semi-automated innovation software [7–9], this standard VA is supposed to provide "sophisticated Value Analysis algorithms to diagnose the weak elements" [9: 2]. As a consequence, it is also implemented in CAD software to extent it towards computer-aided invention (CAI) capabilities [10] and incorporated in processes for patent circumvention [11]. Furthermore, standard VA is part of renowned TRIZ education and certification according to the International TRIZ Association MATRIZ (cf. e.g. [6]) and applied in typical TRIZ innovation projects, e.g. [12, 13]. There are proposals for extending it, for example to include the impact of harmful functions [11] or to take care of different time settings [13]. Regarding Step 1, the most established original algorithm for function ranking will be used as a reference for standard VA in the following. It is described in the Appendix in more detail. There are some works with specific changes in ranking the functions or the actual way to calculate their functional value out of the function ranking (e.g. [11, 14]), but nevertheless a common feature still is, that the functions themselves and most of their underlying properties are completely ignored. In other words, all of the algorithms and procedures only take into account, that there is a functional relation of a given type, but in the end they do not really care about what this function actually is and does.

If the first two steps of standard VA are applied to the autonomous drone based on its Function Analysis as depicted in Fig. 1 for illustrative purposes only, some of the results are quite unexpected and remarkable. For example, the control unit controls both the engines and the clamp, and it is reasonable to assume the first function to be much more complex, demanding, and important than the latter one. The VA function ranking indicates just the opposite: The functional value for controlling the clamp (4) is two times higher than the one for controlling the engines (2). Regarding the components, the functionality of the frame (12) is calculated to be even higher than the functionalities of control unit (6), engines (3), and battery (2) summed together. Moreover, also the clamp (6) itself even yields the same functional value as the control unit, and excels that of engines and battery summed together.

4 Testing VA

In the following, the established VA approach is tested on a very simple toy model system: A child builds a tower of bricks in order to hold a ball at its top. The higher the tower, the better. The model is chosen that simple on purpose, to allow an immediate and objective verification of results, which does not depend on deep expertise or long personal experience with the system. In contrast, highly complicated systems with huge numbers of components and functions are often subject to individual interpretation or bear an intrinsic complexity, such that a serious and objective verification of results is much harder or even impossible. Certainly, this is one important reason, why the shortcomings of standard VA in TRIZ have not been clearly recognized in previous applications.

As a first test scenario, the child is considered to build a tower of exactly three different bricks A, B, and C, such that the ball is held at a height of 70 cm above the floor. The bricks are different in length and cost only, all other brick properties are assumed to be fully identical. Both the tower and the result of its Function Analysis are depicted in Fig. 2. In addition, the function rank and the functional values according to Step 1 of standard VA are already attached to each function. The key characteristics of the three bricks, namely their individual lengths and their individual costs, are listed in Table 1, and according to Step 2, each brick gets the functional value of the function it performs. Finally, the ideality of each brick can be derived according to Step 3 by simply dividing its functional value by its costs. Note, that the idealities are given in dimensionless units, normalized to the maximum value. According to standard VA in TRIZ the result is clear: Brick C is supposed to have the lowest ideality by far, and would be recommended for trimming. In contrast, the idealities of brick B and A are assumed to excel that of brick C by a factor of two, or even three, respectively. Since ideality should indicate, what you get for the money, the recommendation of standard VA does not even work qualitatively, but fails completely. Obviously, the result contradicts common sense and basic logical thinking, which tells us the exactly opposite ideality ranking as seen in the right column of Table 1: Brick C has the highest ideality, it contributes 40 cm to the tower height for costs of 3 €. If this is set to 100%, brick B has an ideality of 50%, contributing only half of that for the same costs. Finally, brick A contributes 10 cm for 4 €. In relation to brick C, this yields an ideality of $10/4 \cdot 3/40 = 3/16$, which is approximately 19%.

In a hypothetical subsequent trimming scenario, standard VA would pose the challenge, to enable brick A and/or brick B to make up for the height of brick C. Even if an innovative solution can be found, its implementation must not involve an additional cost impact higher than 3 € - the costs of the trimmed brick C - to still result in a net

Table 1. Characteristics of the three bricks A, B, and C. The two right-hand columns show the brick ideality in dimensionless, normalized units, according to standard VA in contrast to the ideality by pure common sense and logical thinking.

Brick	Length	Cost	Functional value	Ideality (VA)	Ideality
A	10 cm	4 €	4	100%	~19%
B	20 cm	3 €	2	~67%	50%
C	40 cm	3 €	1	~33%	100%

increase of the overall tower ideality at the end. In contrast, guided by and based on the real idealities, the bricks A and B could simply be replaced by a second brick of type C, yielding a new two component tower which is even higher (+10 cm), drastically cheaper (−4 €), and thus clearly more ideal than before. This option - to increase the overall tower ideality by a factor of ~190% due to rather simple straightforward measures - relies on brick C only. Therefore it would certainly be overlooked following standard VA, which proposes to get rid of exactly that particular brick.

As a second test scenario, the child is assumed to build the tower with a set of N identical bricks. Since the costs are also the same for each brick, the ideality of a given brick is fully proportional to its functionality. With n denoting the position of the brick in the tower counted in a bottom-up direction, it is straightforward to see that the functionality according to Steps 1 and 2 of standard VA is given by a functional value of n for the brick at position $1 \leq n < N$, and a functional value of $N + 1$ for the brick at the top position $n = N$. As a consequence, each of the identical bricks exhibits a different ideality. In normalized units, the ideality ranges from $1/(N + 1)$ for the bottom brick at position $n = 1$ to 100% for the brick at the top position $n = N$. Therefore, the factor between the calculated idealities of the top and the bottom brick is $N + 1$, which obviously scales with the total number of bricks N. If the system of bricks should be improved by tackling the bricks of lowest ideality according to standard VA in TRIZ, there would be a distinct and strong recommendation to primarily focus on the bottom brick, while the top brick should not be touched at all. Since we speak about totally identical bricks, which are all involved in a comparable way and could even be interchanged in their positions without any system change, this result is not only weird, but again simply wrong. In contrast, all bricks need to have equal idealities, such that it's not possible to differentiate between them in this simple test case. For illustrative purposes, Table 2 shows the resulting numbers for a sample tower consisting of exactly 9 identical bricks.

With standard VA in TRIZ failing already on these simple test scenarios, there is no reasonable or obvious argument, why it should perform better in general, when the system and its functional interplay are even more complicated. The above example of the autonomous drone might serve as an illustrative indication for this as well.

Table 2. A set of $N = 9$ fully identical bricks of the same individual costs are piled up to a tower. According to standard VA the normalized ideality of the bricks covers a full range depending on the brick position, as depicted in the second line. The (maximum) ideality of the top brick ($n = 9$) is higher than the (minimum) ideality of the bottom brick ($n = 1$) by a factor of 10 ($N + 1$). This result is again in contrast to the ideality by pure common sense and logical thinking, demanding the same ideality for all the identical bricks here.

Brick pos. $n \rightarrow$	1	2	3	4	5	6	7	8	9
Ideality (VA)	10%	20%	30%	40%	50%	60%	70%	80%	100%
Ideality	100%	100%	100%	100%	100%	100%	100%	100%	100%

5 What's Wrong with VA?

It might seem obvious, that the VA function ranking algorithm as used according to Step 1 and described in more detail in the Appendix causes the problems. As discussed above, it determines functional values of each function out of the functional structure only, while not taking into account the real functional content and situation. This clearly has to affect the reliability and quality of the final results of standard VA in a negative way. It should be pointed out, that this problem also holds for all algorithmic variants, which do not rely on concrete properties and circumstances of the identified functions. As an actual example, simply adapting values or involving weighting factors in the functional ranking algorithm is not sufficiently appropriate to resolve this issue.

Moreover, a key message of this work is, that in Step 2 of VA, which is undisputed sofar, there is a second problematic issue, that lies even deeper from a conceptional perspective. Not only in Mathematics and Theoretical Physics it is an important concept that the modelling of a given system could typically be done in various valid ways, while a meaningful result about the system must not depend on the individual choice of the model eventually. In our case, the system might be described by choosing different sets of components, and these components might themselves be decomposed into sub components or recombined to groups again. No matter, if these modelling steps are necessary for a proper system description, seem impractical or inappropriate, or are even performed purely virtually: A profound concept has to ensure that the functional value of a component is not at all affected by those modelling operations around it, be it directly involved or not. Expressed in a more theoretical way, a meaningful functional value of a component is required to be invariant under any type of valid component operations, such as e.g. combination and decomposition. This requirement also directly ensures the functional value of a system being invariant under scaling of the model, i.e. the actual level of system description. As a key result of this consideration, the simplest concept for the functional value δ_C of a component C, which meets this type of requirement, is found to be

$$\delta_C = \phi_{C,out} - \phi_{C,in}.$$

Here, $\phi_{C,out}$ is the total functional value of all functions performed *by* C, and $\phi_{C,in}$ represents the total functional value of all functions performed *on* C, analogously. In contrast, the functional value of a component according to Step 2 of standard VA is determined by

$$\delta_{C,VA} = \phi_{C,out},$$

and does not fulfill the requirement in general, leading to arbitrary or wrong results in most cases. It should be mentioned, however, that on some occasions the results coincide by chance. In the special case of a component, which does not require a (relevant) functional input, for example, its functional value is only determined by its functional performance. Also the difference in the functional values of two competing systems happens to be correct, as long as both of them rely on similar functional input from the surrounding supersystem components. So it turns out, that due to conceptual requirements, the functional value of a component should generally be determined by its net functional contribution and not by its absolute functional performance.

6 Validating VA++

Now, how do results for the simple brick tower test cases change, if Value Analysis is applied correctly? First of all, the functional values of all involved functions need to be determined properly, taking into account their actual content, and not by an algorithm, which neglects that. For the concrete system of the brick tower, we only deal with functions that hold either the next brick or the ball in a similar way. What differs, however, is the absolute height level (e.g. above the floor). In the first test scenario of three different bricks, brick C holds brick B at 40 cm, brick B holds brick A at 60 cm, and brick A holds the ball at 70 cm. In the second test scenario of N identical bricks, the brick at position n simply provides the "hold"-function at the level of $n \cdot h$, where h is the height of a single brick. These individual height levels provided by the functions are suitable for measuring their corresponding functional values. As a next step, functional values of each brick have to be determined based on its net functional contribution, instead of its functional performance. It is easy to see, that the functional contribution of each brick is to increase the level of the holding function exactly by its specific height. Evaluated in units of our free choice, the results may be written as $\delta_A = 10$ cm, $\delta_B = 20$ cm, and $\delta_C = 40$ cm in the first test case, and $\delta_n = h$ for each of the identical bricks in the second test case. Setting these functional values in relation to the brick costs, the correct idealities are perfectly reproduced for both test scenarios. So the explicit use of Value Analysis in TRIZ is expected to work fine, as long as the afore-mentioned two issues about the evaluation of functions and contribution of components are properly taken care of. Correcting these two shortcomings of VA results in VA++, the improved next generation of Value Analysis in TRIZ.

7 Resume and Outlook

This work shows, that standard VA in TRIZ does not give trustworthy results, for two different reasons. The first reason is, that functions are evaluated by an algorithm, which lacks system insight. The second reason is to use an inappropriate concept for determining the functional value of a component. The first issue has to be addressed by evaluating the functional value of functions on the basis of system knowledge, at least at an qualitative level. The second issue can simply be solved by changing to a different, meaningful concept, which takes into account the net functional contribution of a component and not its absolute functional performance. Following this corrected approach results in an improved next generation of Value Analysis in TRIZ termed VA++. Due to its solid conceptional footing, VA++ is not only supposed to enable access to correct results, but it is also expected to be of a much higher robustness with respect to typical questions in system modelling, that appear rather often in practical application. To give some concrete examples, the results of VA++ are unaffected by the pure modelling decision to attach two or more functions to one arrow, and also by the explicit incorporation of self loops in the function model. Furthermore, by derivation, VA++ is fully consistent with valid TRIZ operations, such as trimming of all types, and grouping and decomposition of components as well.

Appendix: Function Ranking and Function Values

According to Step 1 of standard VA in TRIZ, the functional value of each useful function is derived from an empirical function ranking algorithm, which is summarized here in short ([8, 9], also cf. [6]): Each main function gets the highest function rank denoted as basic, B. Any other function acting on the same object as a main function is considered to be basic, too. Each function acting on a component, which itself performs a basic function, is of rank A_1, the second highest function rank. Each remaining function acting on a supersystem component is also of rank A_1. For all the remaining functions, which are system internal only, the function rank is A_{n+1} if they act on a component, which itself performs a function of rank A_n. Thus, this procedure works stepwise, and eventually attaches the function ranks B, A_1, ..., A_M in descending order to all of the involved functions. Here, M is a natural number which generally depends on the concrete functional pattern between the components of the system. Finally, function values are assigned to the functions, starting with the lowest function rank: $A_M \rightarrow 1$, $A_{M-1} \rightarrow 2$, ..., until $A_1 \rightarrow M$. Since basic functions are supposed to be especially valuable, they get the function value $B \rightarrow M + 2$.

References

1. Mann, D.: Hands-on Systematic Innovation. 2nd edn. IFR Press (2010)
2. Gadd, K.: TRIZ for Engineers: Enabling Inventive Problem Solving. Wiley (2011)
3. Miles, L.D.: Techniques of Value Analysis and Engineering. 3rd edn. Lawrence D. Miles Value Foundation (2015)
4. Cooke, J.: Improving the value of products and processes by combining value analysis techniques and lean methods with TRIZ. J. Eur. TRIZ Assoc. – Innovator **1**(1), 27–36 (2014)
5. Litvin, S., Feygenson, N., Feygenson, O.: Advanced function approach. Procedia Eng. **9**, 92–102 (2011). https://doi.org/10.1016/j.proeng.2011.03.103
6. Litvin, S., Ikovenko, S., Lyubomirskiy, A., Stevenson, K.: MATRIZ - Level 1 Training Manual. GEN-TRIZ (2019)

7. Devoino, I.G., Koshevoy, O.E., Litvin, S.S., Tsourikov, V.: Computer based system for imaging and analyzing a process system and indicating values of specific design changes. Patent No. US 6,202,043 B1 (2001)
8. Arel, E.T., Verbitsky, M., Devoino, I., Ikovenko, S.: TechOptimizer Fundamentals. Invention Machine Corporation (2002)
9. Arel, E.T.: TechOptimizer 4.0 User Guide. Invention Machine Corporation (2003)
10. Chechurin, L.S., Wits, W.W., Bakker, H.M., Vaneker, T.H.J.: Introducing trimming and function ranking to SolidWorks based on function analysis. Procedia Eng. **131**, 184–193 (2015). https://doi.org/10.1016/j.proeng.2015.12.370
11. Li, M., Ming, X., He, L., Zheng, M., Xu, Z.: A TRIZ-based trimming method for patent design around. Comput. Aided Des. **62**, 20–30 (2015). https://doi.org/10.1016/j.cad.2014.10.005
12. Adunka, R.: Function analysis for electronic products. In: Proceedings of the TRIZ Future Conference 2010, pp. 165–171. Bergamo University Press (2010)
13. Wessner, J.: Value analysis as practiced in TRIZ-based function analysis with time steps. TRIZ Rev.: J. Int. TRIZ Assoc. - MATRIZ **2**(1), 33–43 (2020)
14. Efimov-Soini, N.K., Chechurin, L.S.: Method of ranking in the function model. Procedia CIRP **39**, 22–26 (2016). https://doi.org/10.1016/j.procir.2016.01.160

Using MBSE for Conflict Managing TRIZ on a Systems Engineering Level

Tim Julitz[1](✉), Manuel Löwer[1], and Tim Katzwinkel[2]

[1] Product Safety and Quality Engineering, University of Wuppertal, Gaußstr. 20, 42119 Wuppertal, Germany
julitz@uni-wuppertal.de

[2] Institute for Product Innovations, Bahnhofstr. 15, 42651 Solingen, Germany

Abstract. The objective of the presented research is to find a method to identify and resolve high level conflicts with TRIZ applications in systems engineering on a parameter level. These conflicts are caused by several competing TRIZ problem solving patterns (innovative principles) on lower system levels. Here, interference parameters result from the physical adoption of the general innovative basic principles in a specific application context (e.g. vibration, heat, additional energy consumption). To manage these conflicts on a parameter level, the system engineering point of view is matched with the physical parameter level in the engineering design of a product. Within an experimental research approach, an exemplary MBSE system model of a technical device is used to resolve parameter conflicts on a higher system level. The conflicts only arise on a higher system level consideration, which makes computer-aided conflict identification essential to design engineers. The presented method does help system engineers to foresee conflict potentials of innovative principles on a system level and to control them accordingly using a MBSE system model approach.

Keywords: MBSE · Systems engineering · Systematic innovation · Methodology · Engineering design

1 Introduction

1.1 Motivation

A holistic and creative problem analysis in combination with innovative solution finding are crucial for the development of marketable and innovative products in a strongly changed business environment [1]. For this purpose, companies use different methods and tools from TRIZ [2, 3]. TRIZ contains a comprehensive collection of methods and tools for innovative problem solving. However, the problems are only considered partially and not in the overall system context. A holistic view of the system is provided by Systems Engineering (SE). Existing approaches dealing with the combination of SE and TRIZ only consider the interaction between functions and components [4]. But if contradictions are investigated on the system level, individual parameters on the component level must be taken into account. This leads to the necessity of combining TRIZ with a parameter-based approach of model-based systems engineering (MBSE).

Published by Springer Nature Switzerland AG 2022
R. Nowak et al. (Eds.): TFC 2022, IFIP AICT 655, pp. 39–49, 2022.
https://doi.org/10.1007/978-3-031-17288-5_4

1.2 Problem Statement

Technical contradictions result from uneven development of subsystems [5]. This phenomenon can be observed when attempting to identify the technical contradictions of a system. Contradictions can be identified at the system level. Usually, their cause can only be determined by considering the subsystem level. This becomes relevant in the case of interdisciplinary product development according to SE. Different disciplines work independently on subsystems while also using TRIZ tools to solve their individual problems. This leads to the chance that interactions between the solutions and their effects on the system level are not taken into account (see Fig. 1).

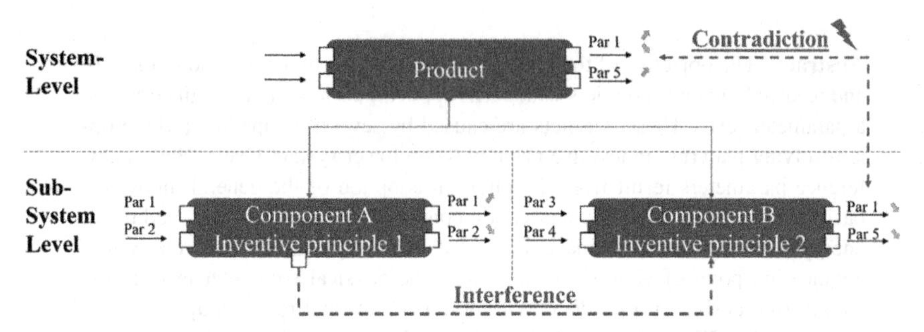

Fig. 1. Interreferences between TRIZ IPs and resulting contradiction on the system level

E.g., components A and B are developed independently and different inventive principles are applied. The overall product has two target values which are characterized by the parameters (Par) 1 and 5, which are influenced by the output of component A and B. A is responsible for an increase of Par 1 while B causes Par 1 to decrease as well as Par 5. The result is a contradiction at the system level. The inventive principle (IP) 1 has a negative interference on the IP 2 which causes the decrease of the parameters. In this context, the following challenges (CH) can be derived.

CH 1. A comprehensive understanding of the system at the top level does not exist across disciplines. E.g., electrical engineers have no deep knowledge in the field of mechanical engineering which could lead to negative effects among different discipline dependent subsystems.
CH 2. Contradictions are only visible at the system level. But they arise at the parameter level. Developers of the subsystems are not aware of the impact of their solutions on the system level which could lead to contradictions.
CH 3. The ancillary variables of IPs negatively affect other IPs and the target variables of the system. E.g., dynamization according to IP 15 may be accompanied by heating or vibration, which can have a negative influence on other parameters.

These challenges lead to the research questions: Can technical contradictions in product systems be regulated down to the parameter level? Which IGPs contradict each other and cannot be used in combination? How can TRIZ and MBSE be combined at the parameter level?

2 State of the Art

Models are simplified representations of reality. In mechanical engineering, different models of a product can be generated, each providing a different view on the same product (e. g. product requirements, product structure, physical description, etc.). Model-based systems engineering (MBSE) aims at a formalized product description. The aim of this modeling is to support the product data management throughout the entire product life cycle [6]. MBSE does focus on a parametric system model of the product including all different stakeholder perspectives in one data model [7].

Bielfeld et al. showed the principle connection of the TRIZ method with an MBSE model [8]. With the help of the Demand Compliant Design (DeCoDe) MBSE approach, which combines the system views requirements, functions, processes and components, a product model was developed respecting the requirements of generic systems engineering. Here, technical contradiction problems according to TRIZ could be considered on a functional level, including the structure of the entire product system. The interferences of components and their effects on a system level were then investigated and quantified using real product application examples [9]. The quantification was performed in detail by weighting critical components, without proceeding from the functional system level to the detailed parameter level. Consequently, the combination of the DeCoDe approach with TRIZ is only suitable for complexity management with respect to high-level systems structures and functional hierarchies within the system [10].

However, a quantitative evaluation of the effects of interferences of individual IGPs to the total system requires a consideration of the specific parameter level in the systems engineering model (see 1.2).

3 Methodology and Approach

To identify inconsistencies between IPs, a system is modeled at the parameter level. The potential IP contradictions are analyzed in advance with a matrix in pairwise comparison of IPs. The assessment of the impact of each parameter on a system-level contradiction is then performed using an MBSE + DOE approach.

3.1 Contradiction Matrix of the Inventive Principles

IPs implemented at the component level have the potential to conflict at a higher system level, as explained in Sect. 1.2. In a pre-study, the commonly known 40 IPs of TRIZ have been compared pairwise according to their main working principles. For example, a combination of IP 1 “fragmentation, segmentation” and IP 5 “combining, merging” within a product system can lead to conflicts due to their opposite characteristics. The found potential contradictions with other IPs are documented in Fig. 2.

In the presented matrix, the three attributes “no contradiction”, “contradiction conceivable/unclear” and “strong contradiction” have been used to qualitatively assess the contradiction potential. If IPs are used in a system that contradict each other according to this table, it can be assumed that there will be target conflicts at the parameter level.

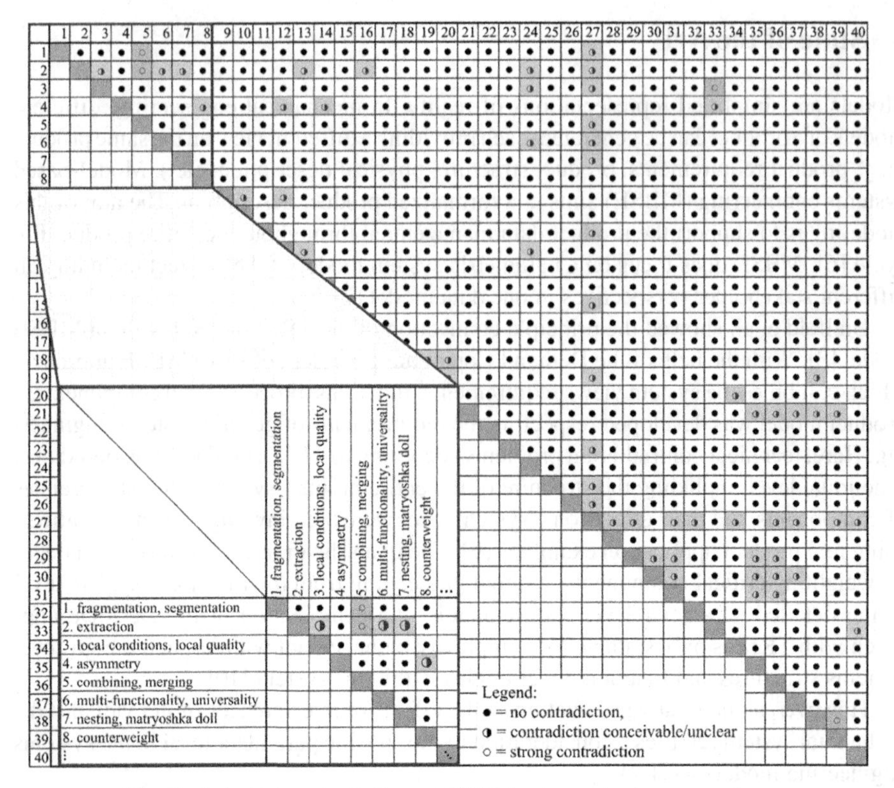

Fig. 2. Potential contradictions of TRIZ Ips in a pairwise comparison

The pre-study shows that potential contradictions (126) are not very numerous throughout the total number of possible combinations of IP pairs (1600). However, their occurrence is spread throughout the total field of the combination matrix. Therefore, the potential risk of contradictions on a superior system level could not be located to a specific cluster of IPs. Consequently, a general system engineering approach has to be found to deal with the problem of IP contradiction throughout the whole product system.

3.2 Implementation of a Model-Based Systems Engineering Approach

In addition to the fundamental conflict potential of the operating principles of individual IPs, disturbance variables or secondary variables also occur in technical implementations. This is due to the fact that the specific boundary conditions of the product and its environment must also be taken into account for the physical implementation of an IP. For example, the application of IP 29 "pneumo/hydro designs" always requires the safeguarding of a certain sealing capability or the potential disturbance variable leakage.

Since the specific disturbance variables result from the specific application in the product context, the classic TRIZ application requires the expertise of the product designer. Unfortunately, this expertise is no longer part of the knowledge of a system engineer for more complex, domain-distributed products. In order to enable a conflict resolution on system level for early phases of the development process, a SysML based MBSE database of innovative principles was created. In this database, known disturbance variables from documented use cases are stored for the IPs. With the help of this MBSE database, disturbance variables and conflicting goals can be identified in the early stages of product development on a system level.

Besides the knowledge database about potential disturbance variables, the application of the MBSE principle in the context of TRIZ offers the further advantage that the design of the system can be systematically accompanied down to the physical parameter level and the intensity of the disturbance influences can be numerically evaluated.

Based on the system engineering with the SysML-based MBSE approach, the product structure is now refined with the involvement of all domain-specific experts and stakeholders.

3.3 Trade Study on Parameter Level

Since all input and output variables as well as disturbance variables are always recorded in a defined context, the methodical MBSE approach can then be used in later phases of the product development process to evaluate the parameter influences. Here, the first step is to formulate the physical relationships in the specific product context (expert knowledge) and then to perform a computer-aided system-wide evaluation by means of a sensitivity analysis of the parameter relationships. With the help of established computer-aided tools, this last step can be performed directly from the MBSE model according to the Design of Experiments approach.

4 Explorative Study

The method presented in Sect. 3 is now demonstrated in the context of an experimental study on a real product model. For this purpose, the technical system is represented as an MBSE model using the SysML language within the software tool Cameo Systems Modeler (CSM) from Dassault Systems Inc. Based on a real-life parameter conflict, the application of the IGP database during the concept phase of the product development process is first demonstrated. Then, the final parametric system model is subjected to a sensitivity analysis using the ModelCenter software from the company Phoenix Integration.

4.1 Technical System Description

The example product used in this paper is a fanless laptop for compact and lightweight mobile use (ultrabook). The main requirements of this product can be stated as low weight, high battery runtime and quiet operation. In real-life operation, however, the so-called coil whine occurs in such devices from time to time, which can be attributed to the high-frequency oscillation of inductive coils in circuits on the mainboard or graphics card [11]. Depending on the user's hearing, this coil whining is annoying and contradicts the requirement of low-noise operation. At the same time, the cause of coil whining is located at the level of a specific subcomponent of an electronic circuit, which consequently represents a complex system engineering problem.

While modeling the product example in a SysML MBSE model, some simplifications have been made. The laptop is not described in its complete product structure, but reduced to the relevant components. This reduces the modeling effort and makes it easier to understand the application example. The following figure shows the top-level product structure (Fig. 3).

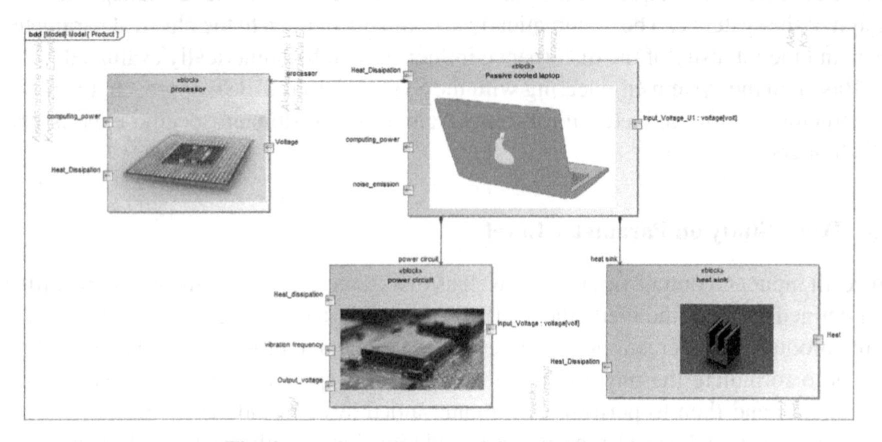

Fig. 3. Product structure of the MBSE model

Furthermore, the parameter relationships are reduced to simple mathematical formulas. The real acoustic correlations can in principle also be mapped mathematically correctly in the system, but they increase the degree of complexity many times over without contributing significantly to the understanding of the method. Finally, only the actually relevant IGPs are mapped in the database in the following. For this purpose, the relevant parameter conflict was defined according to the TRIZ methodology as follows [12]:

Parameter to be Improved: Adaptability/Versatility. This is based on the consideration that the individual components (e.g. processor) have a versatile voltage requirement over the service life, while the power supply is provided via a constant voltage.

Deteriorating Parameter: Time Loss. A component-dependent adjustment of the voltage at the external power supply of the laptop requires a time-discrete adjustment of the respective states, which means that other components cannot be supplied simultaneously with their individual demand.

Using the updated TRIZ contradiction matrix of Mann, the following innovative principles have been identified as potential solutions, see Table 1.

In the next section, the methodical use of the MBSE model in product development at an early stage (preliminary design) and then at a later stage (design refinement) is discussed.

Table 1. Suitable innovative basic principles from TRIZ matrix 2010 [12].

IP Nr.	Name
15	Dynamisation
28	Replacement of mechanical principles
29	Pneumo/hydro designs
35	Parameter changes

4.2 MBSE Concept Modeling

According to the methodological approach from Sect. 3.2, a database in the form of a SysML package was created in the MBSE system model, which contains named IPs. Each IP was created with a solution-neutral input or output as well as specific potential disturbance variables. Here, the disturbance variables were identified based on a literature search of validated IP solutions. For example, implementation examples for the IP "Dynamization" can be found in the literature, where the disturbance variables "Increased component variance", "Vibrations", "Volume change" and "Waste heat" were identified [13].

This form of database is solution-neutral and can be imported or exported across projects. The problem to be solved is created within the system model as a black box in the form of a block element in the internal block diagram (ibd) of the product. Here, the actual inputs and outputs, as far as known, are concretized. In a way, this step represents the preliminary work of the engineer.

In order to now use individual suitable IPs as solutions of the black box, the IP blocks can be used (corresponds to type change). Here, a comparison of the two-element information takes place, whereby the developer is confronted with the potential disturbance variables of the specific IP documented in the database (Fig. 4).

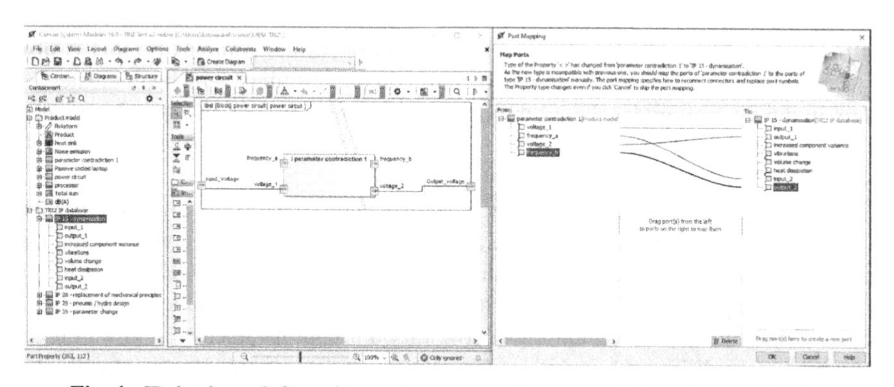

Fig. 4. IP database (left) and type change revealing potential interferences (right)

In this step, the engineer is provided with specific database knowledge, on the basis of which previously unknown potential disturbance variables are visualized and now need to be evaluated in the system context. This process is redundant for different IPs, so an iterative evaluation of different solution approaches for the parameter conflict is also possible.

4.3 MBSE Parameter Study

In the further elaboration (design refinement) of the product concept, solution-specific physical relationships between the respective parameter flows within the overall product architecture must now be worked out. With the help of expert knowledge, concrete formula constraints are determined and entered in the system model (parametric diagram). Figure 5 shows an example of the modeling of heat dissipation and noise emission for the given product example.

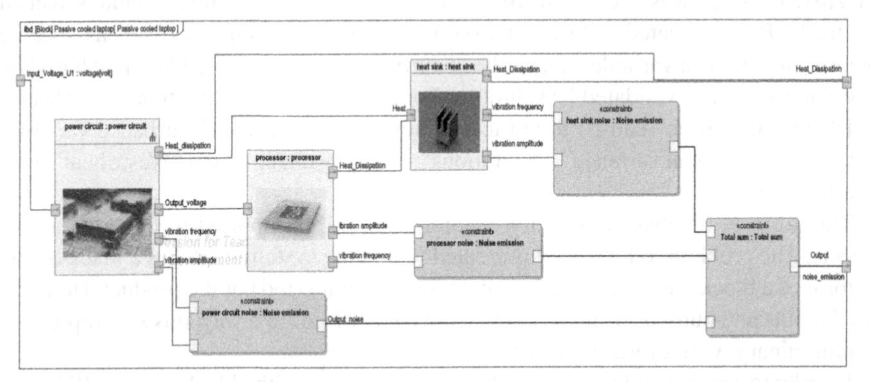

Fig. 5. Example of heat dissipation and noise emission parameter constraints within the MBSE model

Due to the complexity of the parameter interrelationships, it is not yet possible for the system engineer to objectively assess at this point which parameters influence the final target variables and in what quality. In addition to the complexity, this is also due to the domain-specific expertise of the individual constraints, which are created by experts in the system model shared throughout the company (e.g., electrical engineering, mechanics, IT). A sensitivity analysis (Fig. 6) is now carried out to counter this circumstance.

For this purpose the program extension ModelCenter of the company Phoenix Integration is used. By coupling the SysML parameter relationships and constraints, a numerical sensitivity analysis can now be performed according to the DOE approach. For this purpose, the problem of noise generation is modeled in a mathematically simplified way. The DOE tool of ModelCenter can now vary all input parameters in given ranges and interval steps. With each variation, the calculation is performed automatically. Finally, a graphical comparison shows the relevant influencing variables for the specific loudness values (Fig. 6), showing that the power circuit vibrations variables (frequency and amplitude) have the biggest influence for the maximum noise emission in the mathematical model.

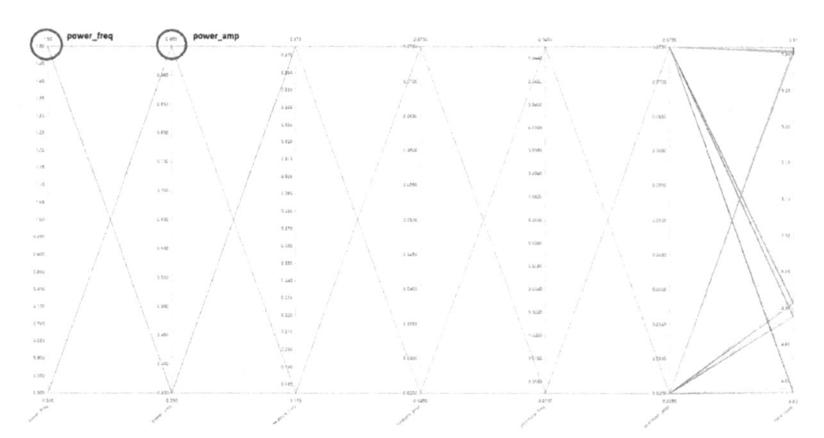

Fig. 6. Sensitivity analysis carried out with ModelCenter using the DOE approach

It becomes apparent that the disturbance variable "vibrations" of IP 15 "dynamization" has a significant influence on the target variable "noise emission" by influence of the component "power circuit". Consequently, this influence must be minimized in the subsequent design (e.g. by damping, changing the material or changing the IP).

5 Discussion

The presented method enables a computer-aided integration of expert knowledge about the IP into the product development process. The disturbance variables stored in the database are important indicators for design decisions of the system engineer. The IP database makes the system engineer more independent of his TRIZ expertise and also provides him with domain-specific disturbance information.

In contrast, the initial effort for MBSE system modeling in general and the specific effort for creating the IP database can be classified as high to very high. This problem is inherent to the SysML-based MBSE approach and is amplified by the required parameter-based level of detail of the model.

Moreover, both the SysML-based system design and the IP database maintenance require the collaboration of domain-specific experts.

Therefore, the presented approach is particularly suitable for products of medium to high complexity in those companies that implement an IT infrastructure for collaborative system model maintenance across domains.

Finally, the actual disturbance variables of each IP depend on the concrete implementation of the generic solution approaches, so that in early concept phases no generally valid statement about the relevant disturbance variables can be made at first. Rather, the database contains a collection of all previously known possible disturbance variables in the context of the respective IP. Therefore, the product developer must make a context-sensitive adaptation to the specific use case and company context here. On the other hand, the IP database can be easily transferred to further development projects or other product types due to its generic content, enabling knowledge-based engineering.

While for the IP database is particularly suitable for early concept phases in product development, the presented method for sensitivity analysis of the disturbance variables first requires a complete capture of the parameter interrelationships in the form of calculation rules. While this approach seems particularly attractive for adaptation and variant designs, the initial modeling effort could be a significant hurdle for innovative new developments.

6 Conclusion and Outlook

In this paper we have discussed the research questions: Can technical contradictions in product systems be regulated down to the parameter level? A method for a systematic and data-based identification of potential contradictions on a parameter level has been presented using a SysML-based MBSE approach in combination with a DOE sensitivity analysis strategy. The method has been demonstrated using a technical product example with a commonly known contradiction problem.

Within the experimental approach, the presented method seems feasible for products of mid-range complexity. However, the approach requires a solid IT-infrastructure and a MBSE oriented development process for industrial application feasibility. In Addition, the database approach needs a continuous enhancement and refinement in the future.

The research of TRIZ contradiction problems in systems engineering and the refinement of an application oriented MBSE methodology is topic of future projects at the chair for product safety and quality engineering at the university of Wuppertal.

References

1. Brown, S.L., Eisenhardt, K.M.: Competing on the Edge: Strategy as Structured Chaos. Harvard Business School Press, Boston. ISBN 0-875847544
2. Ilevbare, I.M., Probert, D., Phaal, R.: A review of TRIZ, and its benefits and challenges in practice. Technovation **33**(2–3), 30–37 (2013). https://doi.org/10.1016/j.technovation.2012.11.003
3. Moehrle, M.G.: How combinations of TRIZ tools are used in companies - results of a cluster analysis. R and D Management **35**(3), 285–296 (2005). https://doi.org/10.1111/j.1467-9310.2005.00390.x
4. Bielefeld, O.: Development of a Methodology for a Model-Based and Holistic Failure Analysis. Dissertation, University of Wuppertal, Germany (2021). https://doi.org/10.25926/s3ne-nc15
5. Orloff, M.A.: Inventive Thinking Through TRIZ. A Practical Guide. Springer, Berlin Heidelberg (2006). https://doi.org/10.1007/978-3-540-33223-7
6. Weilkiens, T.: Systems Engineering with SySML/UML, Elsevier (2011)
7. INCOSE: Systems Engineering Vision 2020: V2.03. http://oldsite.incose.org/productspubs/pdf/sevision2020_20071003_v2_03.pdf. Last Accessed 12 May 2017
8. Bielefeld, O., Sizikov, V., Schlüter, N.: Research of the possibilities for using and linking TRIZ methods with systems engineering. In: Benmoussa, R., De Guio, R., Dubois, S., Koziołek, S. (eds.) TFC 2019. IAICT, vol. 572, pp. 174–186. Springer, Cham (2019). https://doi.org/10.1007/978-3-030-32497-1_15

9. Bielefeld, O., Sizikov, V., Schlüter, N., Löwer, M., Katzwinkel, T., Schleicht, A.: Quantification of influences between components, functions and process usage stages by linking TRIZ methods and systems engineering. In: Cavallucci, D., Brad, S., Livotov, P. (eds.) TFC 2020. IAICT, vol. 597, pp. 292–303. Springer, Cham (2020). https://doi.org/10.1007/978-3-030-61295-5_24
10. Stühler, B., Bielefeld, O.: Untersuchung der Kopplungsmöglichkeiten Zwischen Generic Systems Engineering und dem TRIZ-Ansatz. Lehrstuhl Produktsicherheit und Qualität, Wissenschaftstag (2020)
11. Harper, C.: What Is Coil Whine – What does it sound like – how to fix it. In: CGdirector. https://www.cgdirector.com/fix-coil-whine/. Last Accessed 29 May 2022
12. Mann, D.: Matrix 2010: Re-updating the TRIZ Contradiction Matrix. IFR Press, Clevedon (2017)
13. Chang, H.-T., Chen, J.L.: Eco-Innovative examples for 40 TRIZ inventive principles. In: The TRIZ Journal, 1–16 (2003)

TRIZ-Based Approach in Co-creating Virtual Story-Maps

Narayanan Kulathuramaiyer(✉) and Timothy George Mintu(✉)

Institute of Social Informatics and Technological Innovations, University of Malaysia Sarawak (UNIMAS), 94300 Kuching, Sarawak, Malaysia
nara@unimas.my, timothygeorgemintu@gmail.com

Abstract. TRIZ based models are particularly instrumental in formulating knowledge-based solutions in a variety of areas. The knowledge engineering capacity gained by systematic approaching problems according to the TRIZ structuring and modelling of problems, enables a powerful mechanism for drilling into the core conflicting or operating zone of the problem. This research then explores the knowledge engineering capacity of TRIZ to enable inventive solutions to solve even complex socio-technical problems. This paper presents a TRIZ-based methodology in the participatory design of shaping community-based virtual tourism programmes for indigenous communities living in the highlands of Borneo. In this paper, we demonstrate digital story-maps as a platform for unlocking tacit knowledge and giving indigenous communities a capacity to promote the uniqueness of their culture and heritage. Based on the initial TRIZ based framing of the problem, the use of digital story-maps has given rise to a systems-approach that has managed to bring out untold stories. These models have also supported the characterization of parameters of the virtual story-map solutions.

Keywords: TRIZ · Virtual tourism · Story-maps · Indigenous knowledge-base · Inventive principles

1 Introduction

1.1 Introduction to Story-Maps

A story-map is a system that is developed to fulfill a purpose of sharing information with a pinpoint accuracy in the geospatial field [3]. The term *story-map* has been popular since the release of StoryMapJS in 2013 [7]. The tool highlights a place of interest as a marker on a map while being able to parse multimedia contents such as text and video in a slideshow format. In essence, it can give the viewer a better understanding of the geographic environment in the story.

The methods of collecting stories with a visual representation is not a new thing. Based on Scherf [17], its aim in identifying communities' assets and using the knowledge to boost tourism sectors has been successfully adopted. System development is led by the communities with the stakeholders giving their opinions that in the end can result

Published by Springer Nature Switzerland AG 2022
R. Nowak et al. (Eds.): TFC 2022, IFIP AICT 655, pp. 50–60, 2022.
https://doi.org/10.1007/978-3-031-17288-5_5

in improved development. The mapping of an area is done by evaluating a frequently visited place and it is considered a hotspot therefore putting the marker on the map.

A study for the indigenous people of Sarawak also has been done in the past to incorporate cultural values in emoji messaging system. The TRIZ-based socio-technical model has been proposed to aid them in describing the actual workflow for designing and incorporating emojis in the messaging system [4].

We can see that there have been efforts to serve as a visual catalyst to enhance an operation. This paper will also explore the use of co-creation and participatory approach methods as a guide for us to shape the workflow of this research.

As a society that is surrounded by technology, people are getting more familiar in using GPS *(global positioning system)* for navigations in our daily lives, for a variety of purposes. On the same note, the use of GPS in storytelling has offered a potential market opportunity in the field of digital tourism during the state of the coronavirus epidemic across the world.

1.2 Indigenous People of Sarawak

This project is aimed at the digital enabling of tourism leadership in the indigenous areas in Sarawak, primarily, in the Heart of Borneo. Our target has been to support community-based tourism initiatives by empowering these remote communities in co-creating story-maps through a partnership with researchers from the Institute of Social Informatics and Technological Innovations. The project has been explored as a community-university partnership project working closely with communities living in Bario, Bakelalan and Long Lamai. The community-based knowledge extraction and organization capacity target the development of a connected story-mapping system to support virtual tourism for remote Bornean communities.

The people in these areas are of diverse backgrounds comprising of Lun Kelabit and Lun Bawang an Penan communities. The economy was mainly sourced from their agriculture with unique tourism products such as Bario rice and other crops such as pineapples. Other than that, they are also known for their handicrafts which consists of traditional attires and beadworks [6].

In Bakelalan, their economic strength comes from the industrialization of the mountain salt. It was hundreds of years ago the salt came from the spring located in the mountain where villagers noticed that there are animals drinking water from the source [9]. But, even before the discovery of the salt springs, their economy was focused mainly on agriculture.

Meanwhile, in Long Lamai, the Penan people in majority. They are the people who in past are practicing nomadic lifestyle. Nowadays, they have their settlements while some are still practicing semi-nomadic lifestyle [18]. In the case of the Penans, their livelihood is very dependent on the jungle in terms of sources of food and shelter. Nowadays, they are still practising handicraft and slowly adopting agriculture in their livelihood which will be their new economic strength.

1.3 Indigenous Stories into Story-Maps

The stories from the indigenous communities are often documented by visiting researchers and curious visitors in the form of digital prints or other multimedia formats.

Collecting stories in story-maps format is fairly a new concept in the hearts of Borneo, though a tremendous potential has been observed in the cultural heritage and rich biodiversities. In terms of the contents, the story-map is often depicted in the perspective of the story-map authors instead of the related communities. Thus, the emphasis on local capacities for value-creation in the tourism industry by harnessing on local knowledge and traditions has been explored.

2 Story-Map as a Tool for Virtual Tourism

2.1 Virtual Tourism

In recent days, COVID-19 has impacted the tourism sector the most, but there are still ways to recover from this with the help of current technologies [1]. To move forward in this tourism industry, we are seeing the tourism sectors are adapting to technology [16].

According to Kayumovich [8], virtual tourism is a method of transforming the tourism industry by implementing information technologies using mobile technologies, internet, or even 3-dimensional methods.

Since the impact of a pandemic on the tourism industry is known to be susceptible to the economy, virtual tourism becomes the next step to be adopted by the tourism industry and travel agencies to stay sustainable in the coming years.

2.2 Related Works in Employing Story-Map in the Area of Tourism

A storymap is a web-platform tool used to represent the stand-alone resources, created with deep thoughts with the purpose of showing information in the form of text, image, video and it also provides the functionality of map markers to show the reader the geographical information of a story [3].

The use of geospatial technology has helped residents and visitors obtain the characteristics of their surrounding area [12]. Furthermore, the use of storymaps will increase the promotional growth of the tourism industry while providing visualization of the potential tourism sites [11].

Other than that, the combination of visual and narrative methods will be used to showcase the everyday life of the local community [10]. It is an effective method to show outsiders aspects of their sustainability practices which leaves an impact to the outsider's opinion on the community. The process is also supported by web-based GIS technology which can be very useful to the rural community in engaging with stakeholders [3, 14].

Therefore, the use of storymaps in tourism will shape the future of the industry and will also help struggling tourism sites to attract more potential tourist.

2.3 StorymapJS by Knightlab

The first alpha release of the StorymapJS was around 2013 and it was marked as a new tool for storytelling [7] (see Fig. 1).

One of the main reasons why we chose StorymapJS instead of ArcGIS Storymap is because it is open-sourced, highly manageable and has third-party extensions for the multimedia contents. Other than that, the map is highly customizable with the use of third-party services.

Fig. 1. Snapshot of the community culture preservation site in Bario

3 Co-creating Storymaps Powered by a TRIZ Approach

3.1 Integrating TRIZ with Storymaps

The integration of TRIZ with the problem faced in the co-designing of storymap requires an elaborate planning with the structural story-telling while helping to preserve the community values and the way of life. TRIZ tools were initially adopted in problem modelling to identify the key problems for abstract storytelling tasks.

This chapter will cover the use of *Cause-and-effect chain analysis* as a way to identify the key problems. Then, we performed *Component Analysis* to formulate a *Function Model* highlighting the relationships and interactions between the components and the use modelling of contradictions helped to identify the possible inventive principles to be used.

3.2 Co-creation and Participatory Practices in a Community

In a study done with the full involvement and participation of a community, careful steps were required as a way to put the community members and organizers in charge of the content creation and in the purpose driven site management.

By adapting the co-creation practices in [5], we start by building connections for connectivity and collaborative customs. A collaborative project design was adopted together with locals, and it goes down to the community consultations, oral history interviews and storytelling.

Through the participatory practices, we can adopt several participatory methods which included informal description practices, preservation practices, post-custodial practices and using social media as a forum for community participation. Furthermore, we conducted targeted interviews with community leads to collect multiple narratives.

3.3 Cause and Effect Chain Analysis

In adopting TRIZ at a problem modeling stage the cause-and-effect chain analysis [2] was applied to acquire the characteristics and requirements in the design of story-maps as a potential virtual tourism and as a business enabler tool for these communities (see Fig. 2).

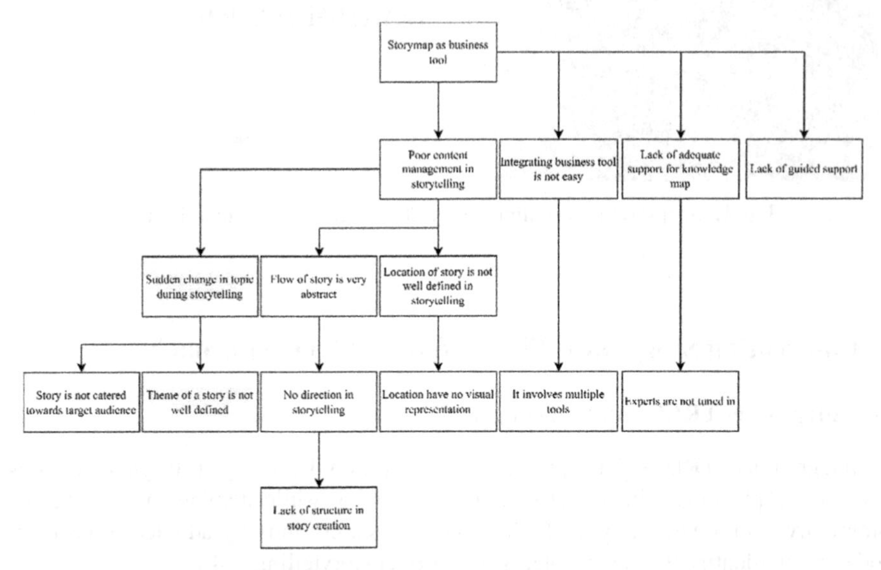

Fig. 2. Cause and effect chain analysis

3.4 Component Analysis and Function Analysis

Component Analysis was performed to identify all the possible entities within the system design phase. The relationship between the entities were identified during interaction modelling. After performing the function model, function analysis was then performed to determine the strength of each connection between the entities (see Fig. 4). The outcomes of the function analysis provided linkage to formulate the design model [13] (see Fig. 3).

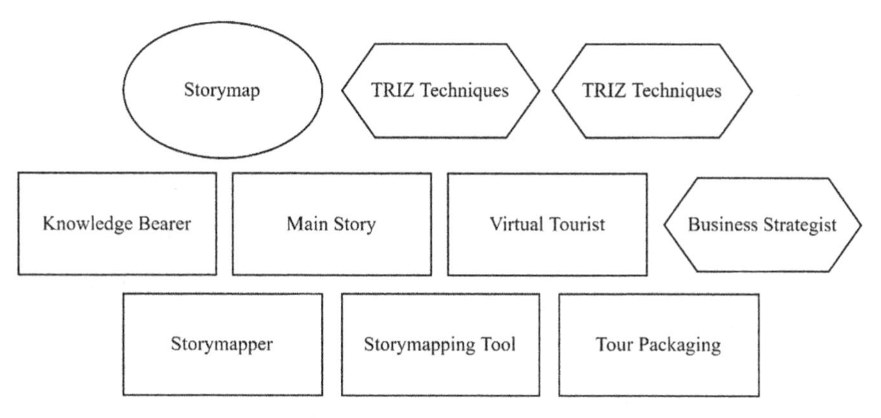

Fig. 3. Component analysis

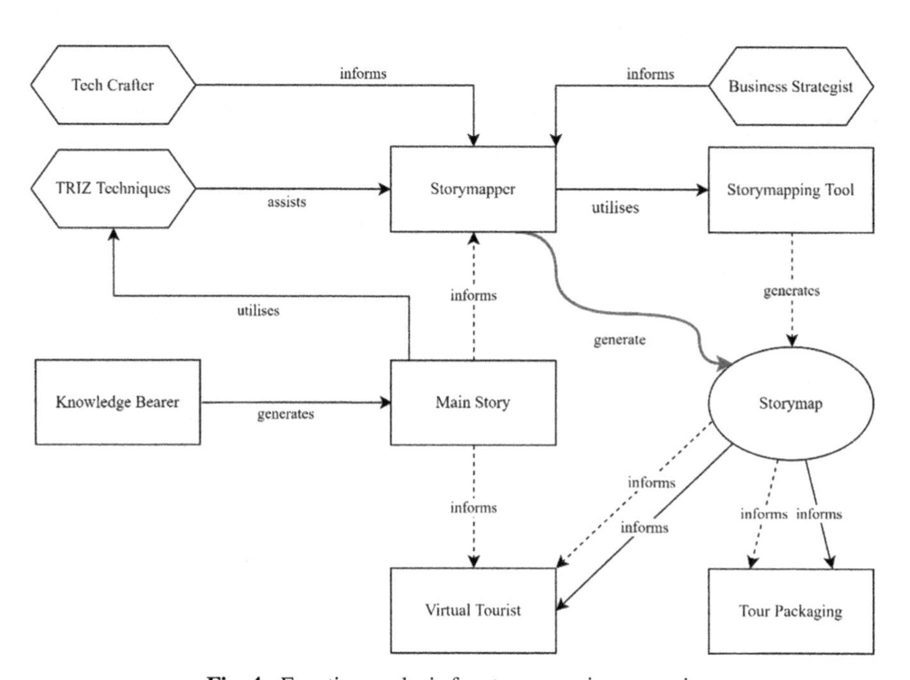

Fig. 4. Function analysis for story mapping scenario

3.5 Engineering Contradictions

This model defines the improving and worsening parameters as components that forms the engineering contradiction. The use of engineering contradiction tools enhances the understanding and logical model involving the key parameters [15].

The next step to determined the parameters of the given condition. Lastly, we can see the suggested 40 principles as our pivot point to resolve the particular contradiction. The table below are the collection of generated If-then-but statement based on the previous analysis (see Table 1).

Table 1. Generation of following engineering contradictions

Engineering Contradiction #1		
IF	Storymap is used as a way of capturing and preserving communities' traditional knowledge	**Parameter**
THEN	The interaction between knowledge capture will be instantaneous	**39. Productivity**
BUT	Story is limited to audience's attention span	**35. Adaptibility or Versatiliy**
Suggested Principles	26. Instead of reading the text, audio aid is added 17. Adding another dimension of the business process as additional layer 19. Include short dialogues periodically between the long ones 1. Divide contents into sections of the main story	
Engineering Contradiction #2		
IF	Story is catered towards target audience,	**Parameter**
THEN	the content will appear to be useful towards the audience	**18. Illumination Intensity**
BUT	it leads to slower story generation	**9. Speed**
Suggested Principles	10. Storymaker generate rough story outline for target audience 13. Determine what target audience wants instead of creating stories on the fly 19. Create scenarios where during periodic changes in activities can be captured by using templates	
Engineering Contradiction #3		
IF	Theme of the story is well defined	**Parameter**
THEN	the content of the story will be concise	**13. Stability of the object**
BUT	storytelling will show less expression	**18. Illumination intensity**
Suggested Principles	32. Apply chromatic storytelling during sessions with kindergarten children 3. Engage with community inputs when highlighting indigenous flavor in stories 27. Use of cheap objects as props and backdrops to maximize immersion 16. Use a platform where exaggeration in storytelling is encouraged	
Engineering Contradiction #4		
IF	A story is created with a solid structure	**Parameter**
THEN	the flow of a story will be organized	**39. Productivity**
BUT	knowledge bearer will have to follow the structure	**35. Adaptability or Versatility**
Suggested Principles	1. Divide the structure into different sections 35. Make the story to be more flexible for knowledge bearer by allowing free flow	

(*continued*)

Table 1. (*continued*)

Engineering Contradiction #5			
IF	Tourist is fully involved in designing the story-map		**Parameter**
THEN	We can see what other tourists want to see		**29. Manufacturing Precision**
BUT	The suggestions from tourists could disrupt the flow of story		**31. Object Generated Harmful Effect**
Suggested Principles	17. Rather than asking what tourist wants to see, put tourist in the perspective of the storyteller 34. Use only acceptable opinions from tourist while preserving all ideas		
Engineering Contradiction #6			
IF	Business strategist is involved in storymap development		**Parameter**
THEN	Storymapper can create a richer story		**29. Manufacturing Precision**
BUT	Creating content together is not always easy		**39. Productivity**
Suggested Principles	10. Expose the information to the business strategist 32. Story-mapper and business must always be transparent in knowledge exchange		

The use of functional models and TRIZ modelling tools have provided insights into the core modelling of systemic elements relating to the software design steps. As we embarked on the process of building pilot models of story-maps in leading to various tourism products and services, the guidance has been instrumental in enabling an insightful modelling.

4 Discussions and Future Works

4.1 Implementation of TRIZ in Indigenous Community

In the effort of simplifying the methods, we have subsequently produced a model that is easier for the community to visually understand our design and concepts. This enabled us to translate the TRIZ modeling outcomes to serve as requirement specification and design descriptors for a high-level software design process (see Fig. 5).

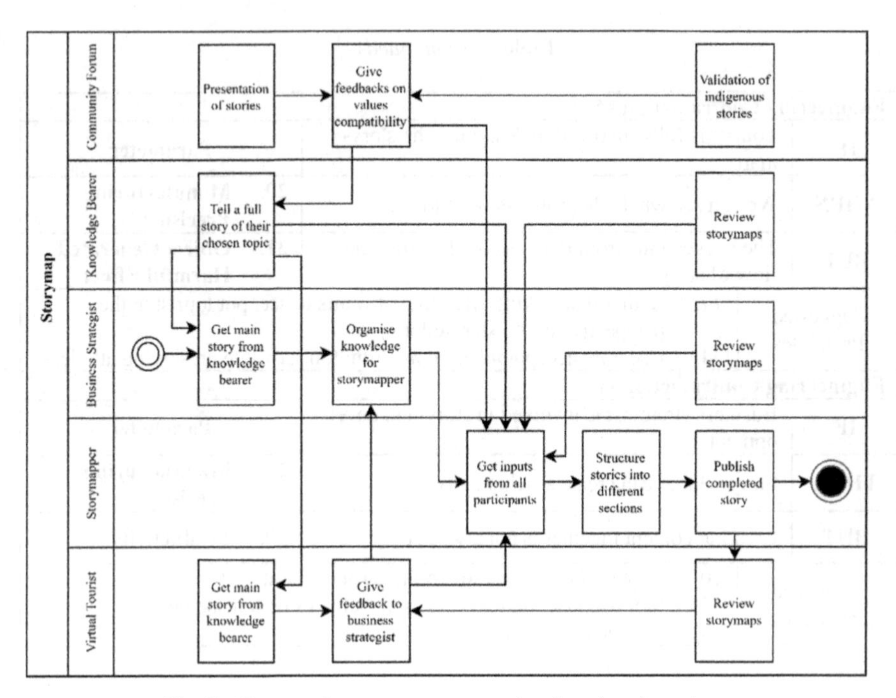

Fig. 5. Community story-maps overview for virtual tourism.

The results of our initial co-creation, co-design and interactive co-development has demonstrated the value of TRIZ modelling tools in the area of software and content development. The need to learn from experiences of TRIZ knowledgebases in guiding the interactions and component formulation has insightful in the initial models.

4.2 TRIZ Components in Generating Storymaps

In generating storymaps, several TRIZ tools such were used from the beginning until the end of the procedure. In the table below, the importance of each TRIZ tools usage are justified for it is very crucial for each step of the process (see Table 2).

Table 2. Importance of TRIZ tools in storymapping

TRIZ Tools	Importance in Storymapping
Cause-and-effect Chain Analysis	In Chapter 3.3, we constructed the storymap as a business tool for the paper. By performing CECA on this point, we observe underlying problems that relates to the flow and structure of the storymap.
Component Analysis	Next, in Chapter 3.4, component analysis was done to lay out all possible components (component, supersystem) that affects the buildup of the storymap.
Function Analysis	Subsequently, in Chapter 3.4, all possible functions of the components were defined with an assessment of the nature of the relationship (excessive, useful, harmful or insufficient).
Engineering Contradiction	Engineering contradictions were represented as a technical contradictions. Based on function analysis in Chapter 3.4, we identified the improving and worsening parameter IF – represents the problem model element (story-teller, story structure, business value) based on function analysis THEN – Improving parameter for the targeted systems design BUT – Worsening parameter for the targeted systems design
39 Parameters	The improving and worsening parameters were mapped to a selected engineering 39 Parameters.
Original Engineering Contradiction Matrix	After the Parameters were identified, the contradiction matrix was referenced to identify a set of recommended partial solutions with selected Inventive Principles as triggers.
40 Principles	40 Inventive Principles were then used as triggers to guide problems solving.

Acknowledgements. This study was funded through the project; Ministry of Higher Education (*Kementerian Pengajian Tinggi*) and *Formulation of TRIZ-based Sustainability-Oriented Innovation Model for Indigenous Knowledge Management* (I03/FRGS/2009/2020). The authors are thankful for the indigenous people from Bario, Bakelalan and Long Lamai who as our partners and local champions have provided insights about the real story that needed to be told in the story-map. Authors are also very grateful for the dedication and support from the AHRC project partners; *Local Heritage and Sustainability:Promote Reflection and Sharing Within and Across Communities* (GL/I03/AHRC/2021).

References

1. Abbas, J., Mubeen, R., Iorember, P.T., Raza, S., Mamirkulova, G.: Exploring the impact of COVID-19 on tourism: transformational potential and implications for a sustainable recovery of the travel and leisure industry. Current Research in Behavioral Sciences **2**, 100033 (2021)

2. Abramov, O.Y.: TRIZ-based cause and effect chains analysis vs root cause analysis. In: Proceedings of the TRIZfest-2015 International Conference, Seoul, South Korea, pp. 288–295 (Sep 2015)
3. Esri: What Can You Do with a Story Map?. [online] Available at: https://www.esri.com/about/newsroom/arcuser/what-can-you-do-with-a-story-map 2022. Accessed 5 Mar 2022
4. Goh, C.H., Kulathuramaiyer, N.: TRIZ based conceptual design framework for future emoji system. In: MyTRIZ Conference, p. 51 (Dec 2020)
5. Grant, K.A.: Affective collections: exploring care practices in digital community heritage projects (2020)
6. Janowski, M.: Beads, prestige and life force among the Kelabit of Sarawak (1998)
7. Joe, G.: Announcing StoryMapJS developer release — a new tool for storytellers. [online] Northwestern University Knight Lab. Available at: https://knightlab.northwestern.edu/2013/10/17/announcing-storymapjs-developer-release-a-new-tool-for-storytellers 2022. Accessed 18 Mar2022
8. Kayumovich, K.O.: Prospects of digital tourism development. Economics **1**(44) (2020)
9. Mail, R.: The Lun Bawangs are certainly worth their salt. The Borneo Post. Retrieved 2 April 2022 (2 Nov 2014)
10. Marshall, D.J., Smaira, D., Staeheli, L.A.: Intergenerational place-based digital storytelling: A more-than-visual research method. Children's Geographies **20**(1), 109–121 (2022)
11. Matondang, F.: Application of story maps techniques in visualizing the tourism potential of lake toba in north sumatera province. Sustainability: Theory, Practice and Policy **1**(2), 188–199 (2021)
12. Mínguez, C.: Teaching tourism: urban routes design using GIS story map (2020)
13. Muenzberg, C., Michl, K., Heigl, H., Jeck, T., Lindemann, U.: Further development of TRIZ function analysis based on applications in projects. In: International Design Conference-DESIGN 2014 (2014)
14. Rawat, P., Anuar, K.A., Yusuf, J.E.W., Loftis, J.D., Blake, R.N.: Communicating and co-producing information with stakeholders: examples of participatory mapping approaches related to sea-level rise risks and impacts. In: Communicating Climate Change, pp. 79–96). Routledge (2021)
15. Royzen, Z.: Solving contradictions in development of new generation products using TRIZ. The TRIZ Journal (1997)
16. Saura, J.R., Reyes-Menendez, A., Palos-Sanchez, P.R.: The digital tourism business: A systematic review of essential digital marketing strategies and trends. Digital Marketing Strategies for Tourism, Hospitality, and Airline Industries, pp. 1–22 (2020)
17. Scherf, K.: Deep mapping as a cultural mapping process and a creative tourism driver: Two examples. Creative Tourism: Activating Cultural Resources and Engaging Creative Travellers **111** (2021)
18. Solhee, H., Langub, J.: Challenges in extending development to the Penan community of Sarawak. In: Lieth, H., Lohmann, M. (eds.) Restoration of Tropical Forest Ecosystems. Tasks for vegetation science, vol 30. Springer, Dordrecht. https://doi.org/10.1007/978-94-017-2896-6_23

Bridging Two Different Domains to Pair Their Inherent Problem-Solution Text Contents: Applications to Quantum Sensing and Biology

Nicolas Douard[1,2](✉), Ahmed Samet[2], George Giakos[1], and Denis Cavallucci[2]

[1] Department of Electrical and Computer Engineering, Manhattan College, 3825 Corlear Avenue, New York, NY 10463, USA
nicolasdouard@gmail.com

[2] ICUBE/CSIP Team, INSA Strasbourg, 24 Bd de la Victoire, 67000 Strasbourg, France

Abstract. The multifaceted purpose of this study is to explore the potential of fusing quantum sensing with bioinspired-based principles toward efficient solutions aiming to amplify innovation. The morphological, functional, and biochemical parameters of the biological retina, integrated with parallel and decentralized vision-sensing architectures coupled with neuromorphic computing and polarization principles, would yield unparalleled new-generation domains of knowledge. This paper exposes a roadmap for such a research and investigates how various techniques from Artificial Intelligence could pave the way of a future where TRIZ, assisted with AI, could accelerate innovation.

Keywords: Biomimetism · TRIZ · Neuromorphic · Contradictions · Artificial Intelligence

1 Introduction

Classical methodologies, such as the Theory of Inventive Problem Solving (TRIZ) [1] are aimed at providing designers and engineers with resolutions to contradictions found in inventive problems, so that to not only accelerate the design process but assist them in achieving world-class performance improvements. However, classical methodologies, including TRIZ, may be inadequate in addressing the emerging problems of the quantum industry due to the new frontiers of knowledge and scale. As a matter of fact, we are seeing the emergence of a wide array of novel sensors relying on properties of quantum physics.

Quantum sensors use the properties of quantum physics, describing phenomena at the atomic scale. Quantum states, which can be manipulated by scientists today, are extremely sensitive to the slightest environmental disturbance. It is on this very principle that quantum sensors are based, explaining their exceptional

Published by Springer Nature Switzerland AG 2022
R. Nowak et al. (Eds.): TFC 2022, IFIP AICT 655, pp. 61–69, 2022.
https://doi.org/10.1007/978-3-031-17288-5_6

sensitivity to minuscule signals of different natures, whether it is the gravitational attraction of an object located underground, or magnetic fields emitted by our brain [2].

For instance, let's consider atom interferometers, in which a cloud of atoms is cooled by laser at very low temperatures, having allowed the development of ultra-stable atomic gravimeters for fine detection in the subsoil, allowing the prospection and management of natural resources; or the gradiometers allowing to perform measurements despite the ground vibrations in the case of exploration of the subsoil prior to construction, and to save long and expensive preliminary studies [3].

Another example is Thales, the French technology giant that is developing several quantum technologies for different fields of use, seeking to miniaturize the device as much as possible using an innovative technique: atom chips. Here, it is still a question of laser-cooled atoms, but instead of dropping them, they are magnetically trapped on a chip, where they are manipulated with the help of radio waves produced by electrical micro-wires. The objective being to design an inertial sensor with a volume of only one liter by 2030 [4].

Other extremely promising quantum sensors are Nitrogen Vacancy (NV) centers, microscopic defects lodged inside synthetic diamonds, capable of detecting very weak magnetic fields. This allows the measurement of the magnetic field of a material with a resolution of a few tens of nanometers. Introduced by Vincent Jacques from the Charles Coulomb Laboratory in 2012, his team made the first magnetic images by NV center microscopy. Since this demonstration, several start-ups have embarked on the development of these new microscopes and the first commercial prototypes are now available for sale [5].

The fields of application related to quantum sensors are plethoric and concern all domains, including health. In parallel, applications of quantum computing are growing at a very fast pace, for example, from quantum computers to quantum algorithms and even the development of the quantum Internet [6].

To increase the capacity of researchers to approach these novel topics and help innovation, the approaches of resolution by analogy between domains are proving insufficient, especially for the most difficult challenges. Thus, the potential contribution of the TRIZ method, as currently practiced in industrial R&D, and which favors the exchange between disciplines, will only be very limited. If the classic TRIZ method manages level 2 or even level 3 innovations, in the sense of Altshuller, it is less effective in level 4 or even 5 innovations. The desired breakthrough innovations require extensive knowledge management in order to find solutions without long periods of testing. It is therefore necessary to invent new, more efficient resolution processes. If we recognize TRIZ's ability to orchestrate the inventive process, such orchestration is only possible if the informational elements are targeted in such a way as to be matched to the problems presented in the form of tradeoffs, or contradictions.

The research topic emerging from this analysis will consist in working on the implementation of a methodology capable of amplifying and accelerating the contribution of TRIZ in order to identify and propose innovative methods for a given need related to quantum detection, to pair problems relating with quantum detection with solutions in a vast field of existing knowledge belonging to other fields.

This work builds in part on the high-level holistic view developed by Val Tsourikov and presented at TRIZCON 2019 [7], while opening on other perspectives. This introductory work paves the way for a more detailed analysis of this problem, especially, how to tackle the development of technologies that rely on quantum effects to higher levels of maturity using TRIZ-related methodologies.

2 TRIZ and Domains of Knowledge

At the intersection of sensing and computing, bioinspired computer vision systems are a promising avenue for developing more adaptive, faster, low-power, robust, and agile robotic vision solutions [10–13]. The term "bioinspired vision" encompasses vision structures ranging from human cognitive vision architectures to arthropod and marine animal species. Arthropods, certain marine invertebrates, and human cognitive vision exhibit unique vision capabilities that complement each other. For instance, neuromorphic computing is a bioinspired computing technology emulating the human brain that provides parallel, distributed computer architectures fast computing, and communication systems at high processing speed while operating at low bandwidth, low memory, and low storage [8,9]. Similarly, polarimetric vision, which certain animal species exhibit, provides unparalleled detection and classification capabilities for a variety of scenarios. Due to potentially high background scatter, polarized imaging can provide high-contrast images in low-light and cluttered environments while providing information about the geometric, material, molecular, and chemical composition of the object. Bioinspired image processing architectures that combine the capabilities of human cognition, such as computations and memories that emulate neurons and synapses, integrated through neuromorphic computing, together with the polarization vision capabilities of certain animal species, have the potential to revolutionize and lead to the next generation of highly efficient artificial intelligence image processing systems [8,9,13,19].

Bioinspired-based domains could be articulated as "solution-driven method" or "problem-driven method" [25]. The "solution-driven method" considers a bioinspired-based system that performs a function that will be served as a starting point. The process is concentrated on abstracting bioinspired-based systems so that the designer can then use the functional model to inspire quantum sensing concepts. In contrast, the problem-driven method considers that there is a specific function, or attribute that the engineer wishes to perform. The process is concentrated on pinpointing which biological systems are needed to be considered for inspiration and further implementation. The relationship between TRIZ and domains of knowledge proposed in this paper is shown in Fig. 1.

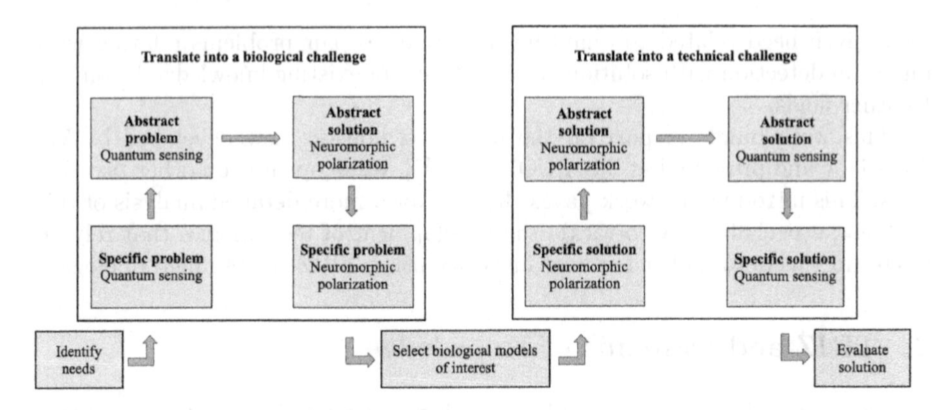

Fig. 1. Relationship between TRIZ and domains of knowledge.

3 Case Example: Patent Reference US20130308132A1

In order to illustrate the potential of TRIZ as a means to pave the way toward bioinspired-based applications, the patent "System and method for polarimetric wavelet fractal detection and imaging" (US patent 20130308132A1) developed by Giakos and coworkers is considered. The application of specialized deep learning information retrieval algorithms [20], namely, SummaTRIZ [26], predicts with a 20% probability that this patent solves a contradiction between the parameters "noisy images" and "localization in time and frequency domains", as shown on Fig. 2. To solve this contradiction, the TRIZ matrix proposes inventive principle 13 "other way around", among several other, as shown on Fig. 3. An extractor of the semantically closest inventive principles, namely, Finder, predicts this patent relies on principle 13, as shown in Fig. 4. This path developed intuitively by the inventors of the patent could have been marked out by the TRIZ methodology. This constitutes a retroactive proof that the hypothesis of building artificial TRIZian paths is possible in the promising view of what the algorithms developed by Cavallucci and coworkers presently enable.

4 Problem Definition, Challenges, and Solutions

4.1 Bioinspired Computing

Von Neumann computer architectures have traditionally been highly efficient at performing computationally intensive tasks, but they are not capable of recognizing, analyzing, and classifying large amounts of data.

In contrast, neuromorphic computing potentially offers bio-inspired capabilities such as parallel processing and human cognition. Unlike devices based on semiconductors, the human brain has the potential to learn, understand, and recognize images, all while consuming very little energy. A key property of human cognitive vision systems, inherited from neuromorphic computer vision, is their

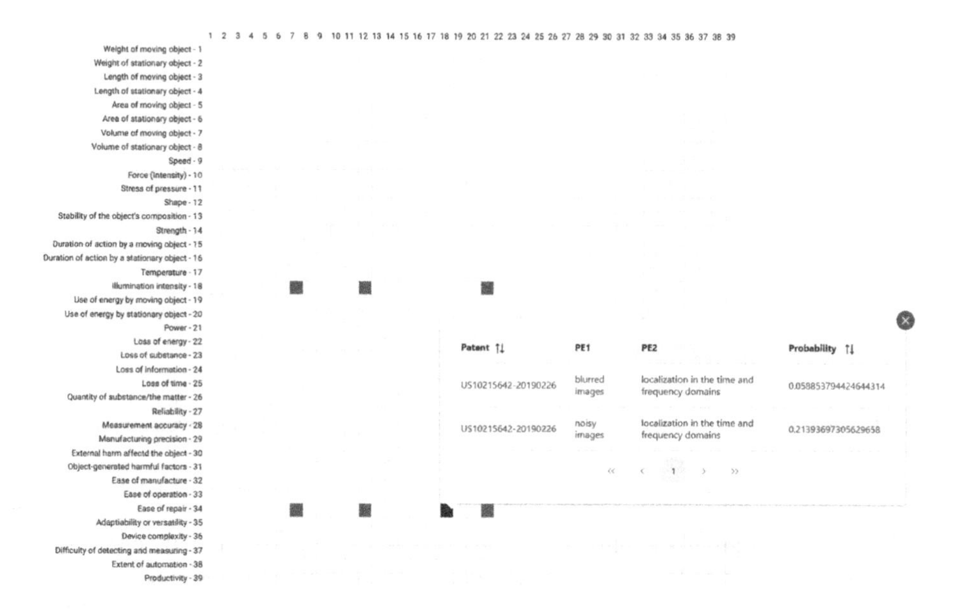

Fig. 2. Screenshot of SummaTRIZ prototype for heatmap extraction of contradictions [27].

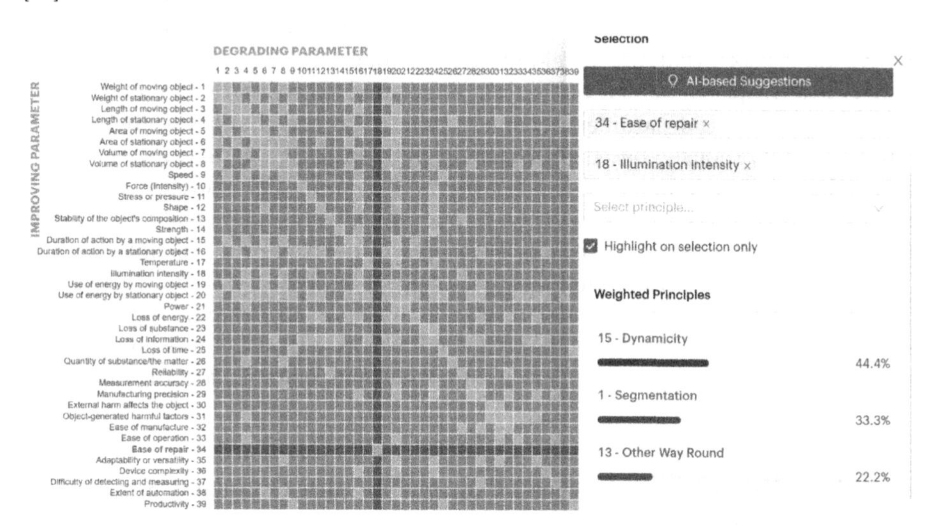

Fig. 3. Screenshot of matrix software tool for statistical manipulation of Altshuller's matrix.

inherent redundancy and ability to focus on the most relevant part of the scene and retrieve salient features [8,9,13–19]. These salient features contribute to fast visual processing and lead to visual functions such as detection, localization, and tracking while reducing the amount of information processed by the system, as well as the bandwidth and storage used.

1 Scrape the data
Url: https:// Scrape
Title: Title
Content: A method for imaging a target object illuminated by a light source at a plurality of aspect angles, the method comprising: generating an angular Mueller Matrix for each one of the aspect angles measured using the light source
Image: Url
Pub. Date: 2022-05-09 22:57:07
Inv. Principles: 13 - Other Way Round × 17 - Another Dimension × 14 - Spheroidality ×

Fig. 4. Screenshot of Finder: an extractor of the semantically closest inventive principles from any sentence using Doc2Vec.

In parallel, significant efforts are being made to mirror human cognition by exploring the development of deep learning architectures such as convolutional neural networks that mimic the connectivity and adaptation of synapses, along with machine learning and deep learning algorithms. Neuromorphic or retinal vision sensors mimic and implement models of biological vision systems [8,9,13–18]. Event-based dynamic vision sensors (DVS) are a class of neuromorphic vision sensors that use pixel-autonomous detection of temporal contrast. The DVS is inspired by the biological retina and allows the consideration of an additional physical parameter, namely time. DVS-based systems are able to operate asynchronously based on different light intensity variations. In other words, this type of image sensor is sensitive in dynamically evolving scenarios and responds directly to changes, i.e., temporal contrast, individual pixels, and near real-time. Pairing DVS sensors with optical polarimetric principles yields the so-called Polarimetric Dynamic Vision Sensor p(DVS). p(DVS) are aiming to further enhance the local spatial-temporal contrast and dynamic range of the DVS imaging sensors and can be used in conjunction with efficient deep learning algorithms, enhanced target detection, tracking, and motion pattern classification.

4.2 Links Between Bioinspired Computing and Quantum Sensing

The example provided by bioinspired computing appears to be a promising approach to addressing contradictions in the field of quantum sensing. As a means to amplify innovation and assist with the resolution of problems relating to quantum sensing, we investigate the use of Natural Language Processing Techniques [20–24] to first highlight contradictions in the quantum sensing domain so as to facilitate the pairing with a potential scientific text of biology as a solution.

The high-level vision involves two stages: the first stage would be to isolate the target sites where information is found in the appropriate fields. The idea is to proceed to real-time mapping of the information feed from the latest discoveries in these fields (publications, journals, and target magazines) so that they constitute a reservoir of potential solutions to contradictions of the studied field that the second stage will build. The second stage would thus consist in mapping the problems of the target domain, which is quantum sensing, by analyzing the existing literature through a TRIZian reading (using the contradiction formalism) and annotating the texts with the help of domain experts in order to create a training set. In order to achieve this, the proposed methodology first involves the extraction of sentences relating to the problem to be solved, followed by the translation of these elements into engineering parameters, and finally, the transformation of these elements into a set of tradeoffs, or contradictions. Finally, as the two stages converge, the problems formulated as contradictions will have to be matched with information extracted from biology, the domain that can constitute potential solutions to these contradictions.

Toward this high-level vision, as a means to achieve the first stage, prior work on Semi-Open Relation Extraction shows its high potential in supporting a new kind of Information Extraction (IE) system, with the intent to guide readers to the central information in scientific documents. Semi-Open Relation Extraction distinguishes itself from traditional approaches, namely, narrow IE and open IE, in that it combines the output of both. In doing so, it overcomes the shortcomings of open IE systems, which do not perform well on the long and complex sentences encountered in scientific texts, as well as the shortcoming of narrow IE systems, which only extract a fraction of the information captured. Applications to biology show Semi-Open Relation Extraction enables extracting arguments that are central to biology texts outperforming traditional approaches, where the retained extractions are significantly more often informative to a reader [23].

Toward the second stage of the high-level vision, and as a means to analyze current literature relating to quantum sensing, the target domain, prior work toward automatic information retrieval constitutes a potential exploratory axis. Specifically, the application Doc2vec and Cosine Similarity, as a means to simplify the Inverse Problem Graph (IPG) process, a lean-based method meant to formulate problems in the initial analysis phase of the inventive design process [24]. In conjunction with this approach, Semi-Open Relation Extraction can also be used as a means of extraction of trade-off relations, or contradictions [23]. Prior work in connection with information extraction from patents using machine learning algorithms in the context of TRIZ [20–22] shows methodological axes that can help overcome the limited amount of annotated data available. Beyond approaches based on unsupervised learning algorithms such as Latent Dirichlet Analysis (LDA) that might be inadequate for complex extraction tasks, and beyond very small supervised learning algorithms such as Support Vector Machine (SVM) or Multi-Layer Perceptron (MLP), deemed inefficient in that context [20], deep learning supervised algorithms show potential be specialized for very low volumes of annotated data. As a parallel path to solve the labeled data scarcity problem, semi-supervised learning with Generative Adversarial Networks (GAN), which combines a document classifier and a sentence-level classifier inside a GAN for patent documents understanding also constitutes a promising axis to extract motivating problems, or contradictions.

The applied methodology and future directions of our efforts are shown on Table 1.

Table 1. Applied methodology and future directions.

Identification	Definition		Alternative generation	Choice of a solution	Implementation and testing			
Define the needs/challenges, tradeoffs	Abstract the quantum sensing problem	Translate into a bioinspired-based challenge	Identify potential bioinspired/AI models	Select the bioinspired/AI models of interest	Abstract bioinspired-based strategies	Translate into quantum sensing solutions	Implement the initial solution, testing	Evaluate

5 Conclusion

The purpose of this study is to explore the potential of fusing quantum sensing with bioinspired-based principles toward efficient solutions aiming to amplify innovation. As a result, it would yield unparalleled new-generation domains of knowledge, therefore amplifying the contribution of TRIZ toward accelerating innovation.

References

1. Altshuller, G.S., Shapiro, R.V.: About a technology of creativity. Questions Psychol. **6**, 37–49 (1956)
2. Fresillon, C.: Sensors, the other quantum revolution. CNRS News (2021)
3. Alzar, C.L.G.: Atom interferometers warm up. Physics **10**, 41 (2017)
4. Travagnin, M.: Cold atom interferometry for inertial navigation sensors. JRC Technical Reports (2020)
5. Rondin, L., Tetienne, J.-P., Hingant, T., Roch, J.-F., Maletinsky, P., Jacques, V.: Magnetometry with nitrogen-vacancy defects in diamond. Rep. Progress Phys. (2014)
6. Gefen, T., Rotem, A., Retzker, A.: Overcoming resolution limits with quantum sensing. Nat. Commun. **10**, Article no. 4992 (2019)
7. Tsourikov, V.: AI Software that Creates new Inventions. TRIZCON2019 (2019)
8. Douard, N., Surovich, M., Bauman, G., Giakos, Z., Giakos, G.: A novel cognitive neuromorphic polarimetric dynamic vision system (pDVS) with enhanced discrimination and temporal contrast (2018). https://doi.org/10.1109/CPEM.2018.8500952
9. Giakos, G., et al.: Integration of bioinspired vision principles towards the design of autonomous guidance, navigation, and control systems. In: IISA 2018, International Conference Information/Intelligence/Systems/Applications, Zakynthos, Greece (2018)

10. Kartheek Medathatia, N.V., Neumann, H., Masson, G.S., Kornprobst, P.: Bioinspired computer vision: towards a synergistic approach of artificial and biological vision. Comput. Vis. Image Underst. **130**, 1–30 (2016)
11. Cho, D., Lee, T.: A review of bioinspired vision sensors and their applications. Sens. Mater. **27**(6), 447–463 (2015)
12. Herculano-Houzel, S.: The human brain in numbers: a linearly scaled-up primate brain. Front. Hum. Neurosci. **3**(31) (2009)
13. Giakos, G.C., Quang, T., Farrahi, T., et al.: Bioinspired polarization navigation sensor for autonomous munitions systems. In: Proceedings of the SPIE 8723, Sensing Technologies for Global Health, Military Medicine, and Environmental Monitoring, pp. 87231H–87231H-11 (2013)
14. Indiveri, G., Horiuchi, T.K.: Frontiers in neuromorphic engineering. Front. Neurosci. **5**(118) (2011)
15. Mahowald, M., Mead, C.: Silicon Retina. In: Analog VLSI and Neural Systems, pp. 257–278. Addison-Wesley VLSI Systems Series, San Francisco (1989)
16. Lichtsteiner, P., Posch, C., Delbruck, T.: An 128 × 128 120 dB 15 us latency temporal contrast vision sensor. IEEE J. Solid-State Circuits **43**(2), 566–576 (2008)
17. Benosman, R., Ieng, S.H., Clercq, C., Bartolozzi, C., Srinivasan, M.: Asynchronous frameless event-based optical flow. Neural Netw. **27**, 32–37 (2012)
18. Sarpeshkar, R.: Neuromorphic and biomorphic engineering systems. In: Research Review, McGraw-Hill Yearbook of Science & Technology. McGraw-Hill, New York (2009)
19. Nowak, M., Beninati, A., Douard, N., Giakos, G.C.: Polarimetric dynamic vision sensor p(DVS) principles. IEEE Instrum. Measur. Mag. **23**, 18–23 (2020)
20. Guarino, G., Samet, A., Cavallucci, D.: Patent specialization for deep learning information retrieval algorithms. In: Borgianni, Y., Brad, S., Cavallucci, D., Livotov, P. (eds.) TFC 2021. IAICT, vol. 635, pp. 162–169. Springer, Cham (2021). https://doi.org/10.1007/978-3-030-86614-3_13
21. Guarino, G., Samet, A., Nafi, A., Cavallucci, D.: PaGAN: generative adversarial network for patent understanding. In: IEEE International Conference on Data Mining (ICDM) 2021, pp. 1084–1089 (2021). https://doi.org/10.1109/ICDM51629.2021.00126
22. Berdyugina, D., Cavallucci, D.: Automatic extraction of potentially contradictory parameters from specific field patent texts. In: Borgianni, Y., Brad, S., Cavallucci, D., Livotov, P. (eds.) TFC 2021. IAICT, vol. 635, pp. 150–161. Springer, Cham (2021). https://doi.org/10.1007/978-3-030-86614-3_12
23. Kruiper, R., et al.: In Layman's Terms: Semi-Open Relation Extraction from Scientific Texts. ACL (2020)
24. Hanifi, M., Chibane, H., Houssin, R., Cavallucci, D.: Problem formulation in inventive design using Doc2vec and Cosine Similarity as Artificial Intelligence methods and Scientific Papers. Eng. Appl. Artif. Intell. (2022). https://doi.org/10.1016/j.engappai.2022.104661
25. Fayemi, P.-E., Maranzana, N., Aoussat, A., Bersano, G.: Bioinspired design characterization and its links with problem solving tools. In: International Design Conference-Design 2014, Dubrovnik - Croatia, 19–22 May 2014 (2014)
26. Guarino, G., Samet, A., Nafi, A., Cavallucci, D.: SummaTRIZ: summarization networks for mining patent contradiction. In: 19th IEEE International Conference on Machine Learning and Applications, Miami, USA (2020)

Inventive Design Solutions for the Complex Socio-technical Problems in Preserving Indigenous Symbolic Visual Communication

Chu Hiang Goh[1(✉)] and Narayanan Kulathuramaiyer[2]

[1] Department of Graphic Communication, School of the Arts, Universiti Sains Malaysia, Penang, Malaysia
goh@usm.com

[2] Institute of Social Informatics and Technological Innovations, Universiti Malaysia Sarawak, Kota Samarahan, Malaysia

Abstract. Visual symbolic communication systems such as emojis are increasingly important to facilitate casual communications and spontaneous information exchange in our daily lives. However, the use of such systems poses dangers to the preservation of local visual symbolic languages as practised by many indigenous and culturally rich local communities. This research aims at developing a local cultural value-based visual communication system for indigenous people in the Malaysian Borneo states of Sabah and Sarawak. The design of such systems requires systematic analysis and identifying the core issue in solving complex socio-technical system problems. Meaningful engagement with different community levels, the sustainability of local knowledge, and cultural values were the primary considerations in designing a culture-preserving model. By utilising the Law of System Completeness of TRIZ, and the engagement of the interaction of supersystems, a conceptual model that can map and analyse indigenous symbolic visual communication systems was developed. This modelling approach has provided numerous insightful ideas for transforming global communication approaches to be sensitive to the cultural needs of indigenous communities.

Keywords: Indigenous Knowledge Communication System (IKCS) · TRIZ · Law of System Completeness · Information Communication Technology for Development (ICT4D)

1 Introduction

The instant messaging system (IMS) in mobile phones is prevalent in our daily lives. Symbolic visual communication systems such as emojis have become a popular way of communication across mobile platforms to deliver emotional expression, gesture and action in a nonverbal and non-text manner. Emojis fill the need for adding non-verbal cues in digital communication about the intent and emotion behind a message. However, there is a further depth to emoji usage as language, suggesting that we are returning

Published by Springer Nature Switzerland AG 2022
R. Nowak et al. (Eds.): TFC 2022, IFIP AICT 655, pp. 70–85, 2022.
https://doi.org/10.1007/978-3-031-17288-5_7

language to an earlier stage of human communication [1]. Emoji resembles the form of communication typical of a natural language. One can trace it back to find its similarity with prehistoric pictographic found in cave drawings. It is an intuitive communication and expression of human emotions [2]. Similarly, the indigenous communities have used symbolic language, relying on visual symbolism to communicate for many years, far before the invention of emoji.

The use of mobile phones and instant messaging services (IMS) in the indigenous community have increased recently. The universality of emojis poses a significant problem to the sustainability of the indigenous cultural values as it imposes western cultural and social behaviour hegemony on the indigenous society. The younger generation is inclined to use digital devices and learn to communicate as modern people do on social media. This development poses a generational divide between the younger and older generation of the indigenous community. Youths are dislocating the routine and their roles in social interaction in the community. The standardised emoji is also incompatible with expressing the specific gesture, human action, and emotional expression in the specific cultural intent of the indigenous people. Interactions with remote rural communities in Borneo over the years have revealed the devastating effect of such an outside-in communication medium on the cultural resilience of the indigenous communities. The traditional way of cultural communication among the indigenous people in the natural environment and the inheritance of indigenous knowledge from the older generation to the younger generation is under threat.

Sarawak and Sabah are the eastern states of Malaysia, located on Borneo Island. It has big groups of multi ethnic indigenous communities such as the Kenyah, Kayan, Kedayan, Murut, Punan, Bisayah, Kelabit, Berawan and Penan. Many indigenous groups live in the interior of Sabah and Sarawak. They use symbolic visual language to communicate and share their unique indigenous knowledge. Visual symbols constructed using twigs, leaves, and other natural resources are an effective way of communication in the rainforest. Such a communication method is used to convey the message of indigenous people and pass their unique knowledge about the natural environment and climate change. The loss of their indigenous knowledge communication signifies the loss of their precious cultural heritage and their knowledge about the rainforest and climate change.

The efforts of Information and Communication Technology for Development (ICT4D) would bring positive aspects of development to the community while sustaining indigenous knowledge. However, the traditional way of ICT4D must be practised cautiously by directly implementing and developing ICT applications for indigenous people with modern technology. The linear mechanistic notion of intervention in the indigenous community cause problems when a technology is introduced from outside the local context.

Based on the above factors, the authors initialled a research project aiming to develop a local cultural value-based visual communication system for indigenous people through a co-creation initiative with the indigenous community. In order to solve the complex socio-cultural and technical problems, TRIZ methods were adapted in the systematic modelling of the indigenous knowledge communication systems. This article presents the initial part of this ongoing project which uses the TRIZ Law of Completeness to map and analyse the indigenous symbolic visual languages in identifying the essential components in the systems.

2 Indigenous Knowledge Communication Systems

Indigenous knowledge is a way of knowing, and the knowledge is generated and accumulated over generations of living in an environment. It allows communities to make sense of their living world [3]. Indigenous knowledge differs from the modern world's international or "exogenous" knowledge in a number of ways. As opposed to the universal "exogenous" knowledge [4], such as science and history in modern educational institutions like we learned in schools and universities, indigenous knowledge emerges naturally and intuitively in the natural environment. The indigenous knowledge systems involve observing natural processes and constructing knowledge carefully by adapting natural resources such as plants, animals and other natural materials in their living environment. Therefore, indigenous knowledge is location-specific, value-laden, and closely linked to the local culture of a specific indigenous group. Indigenous knowledge consists of a vast repository of indigenous "know-how" and experience regarding their way of living, understanding of nature and environment, social interaction, and spiritual beliefs. It is transmitted among the community members to inform, educate, and entertain them through overt and covert communicative practices known as indigenous knowledge communication systems (IKCS) [3, 5]. Each Indigenous communication system is distinctive to a specific indigenous group. It functions to generate, transfer and share knowledge and cultural values within the indigenous social system. The communication system is self-constructed, locally owned and controlled. It is a crucial factor in the sustainability of indigenous cultural heritage, values and identity.

IKCS are unique and operate within and outside modernity [3]. Direct application of modern technologies and research models into the study of indigenous communication systems is inappropriate, and it could jeopardies the idea of cultural sustainability of the indigenous values and heritage. Through our observation and the review in the related literature [3, 5–8], the IKCS have the following common characteristic attributes, which some modern researchers have ignored.

2.1 The Indigenous Knowledge Communication Systems are Dynamic

The traditional ways of looking at indigenous knowledge as static and frozen in time and space, which can be captured, digitalised and preserved as artefacts, must be reconsidered. This approach overlooked the dynamism of culture itself, as culture is malleable, shifting, contextual, and a situational set of meanings and ideas that can change according to perspectives [9]. Therefore, the IKCS supporting the process of producing and exchanging indigenous knowledge are dynamic and not static. A modelling tool that can capture the systems' dynamism is essential.

2.2 The Indigenous Knowledge Communication Systems are Self-sustained

Indigenous communities make decisions based on their existing knowledge and experience. For outsiders, indigenous knowledge may look inconsistent and seen to be based on superstition. However, the indigenous communities see it as logical in their world-view perspective, valuable and practical to their living needs. Rhoades and Bebbington [8] observed how local farmers in Peru generate new knowledge every day through trial-and-error to develop new techniques and adaptations to suit a changing economic and biophysical environment. Similar things happen to the indigenous people, whereby their knowledge constantly evolves and adapts to their living environment changes. Their knowledge of climate change is a shred of living evidence of this. The design of indigenous communication systems is always self-construct, self-regulated and self-controlled in supporting the sustainability of their knowledge.

2.3 The Indigenous Knowledge Communication Systems are Complex

Indigenous knowledge involves knowledge–practice–belief complexity [10], including knowledge about local land, animals, and plants. It also includes the foundations of rules and norms about interacting with the natural environment. Nevertheless, it includes a worldview which denotes how the person or group interacts with the world. How the indigenous sees shapes how people make observations, make sense of them, and learn. It is a multi-layered and multi-dimensional holistic worldview [6]. Communication systems that perform specific social functions within the IKCS encompass language, naming and classification systems, resource use practices, ritual, spirituality and worldview [11]. It carries the complex attributes of the indigenous knowledge system. Indigenous study scholars [10, 12] refer to holism as the key characteristic of indigenous knowledge. The examination of indigenous knowledge needs to view it as one continuum of knowledge whose distinct characteristics are only observable under different contexts. Contextualisation is essential in the study of IKCS [3]. It is inadequate to study indigenous knowledge communication by looking at a single perspective as practised by many modern researchers. It should be treated holistically; the whole is more than the sum of its parts, or a completeness approach, as bestowed in the TRIZ Law of Completeness.

3 Indigenous Symbolic Visual Communication Systems in Sabah and Sarawak

Languages are a significant element in the IKCS. Besides oral and written languages, some indigenous groups use visual symbols to communicate. Aboriginals in Australia use message sticks and rock painting as a means of their communication systems. Similarly, in the Sabah and Sarawak states of Malaysia in Borneo, indigenous groups use symbolic visual languages to conduct asynchronous communication in the rainforest. Two significant indigenous groups' symbolic visual languages were identified as the subject of study in this research project. It is the visual language called *Tatanda* of the Murut people in Sabah and the *Oroo'* of Penan people in Sarawak. Both symbolic visual languages of *Tatanda* and *Oroo'* use message stick to convey the message. They share some similarities in constructing the message stick using natural materials available in the rainforest. A message stick is a vertical stick made of a tree branch with a height of one to three meters which acts as a placeholder for various symbols made from carved twigs, barks and leaves representing information and messages the indigenous groups communicate in the forest. The message stick is usually placed beside the jungle track for easy visibility by the indigenous travelling in the forest.

Clefts are cut in the stick to hold various symbols and arranged in different sequences and combinations to represent different narratives. However, the *Tatanda* and *Oroo'* are different in the contexts of usage. The Murut *Tatanda* is used for hunting successes, land claims and boundary rights, warning of traps and other hazards and notice of feasts and weddings and also represent communications to the spirits on graves and altars [13]. *Oroo'* is used in forest travelling or *Toro*. *Toro* refers to the journey to the forest, and most of the time, it is carried out by two indigenous groups. Travelling in the forest could be a hunting trip, food gathering or migration. The first group always consists of the young and stronger man who will lead and explore the forest, followed by women and children in the second group. *Oroo*'s sign language is used as a way to communicate among these two groups.

4 Information Communication Technology for Development (ICT4D)

IKCS can operate parallelly with the exogenous communication system. They can form the information environment for both exogenous and indigenous communities [5]. It is always the objective of the works in ICT4D to preserve and promote indigenous knowledge while narrowing the digital gaps between the indigenous and exogenous communities. This objective can only be achieved by introducing information communication technology to the indigenous communities. However, implementing and developing ICT applications must be vigilant, attentive, and respectful of indigenous wisdom. The linear mechanistic notion of top-down intervention in the indigenous community should be avoided. It could cause a social change in a deterministic way [14]. Anna Bon [7, 15] suggested that a new version of ICT4D 3.0 would be implemented to sustain the indigenous cultural values. ICT4D 3.0 should move beyond the traditional technology-centric thinking toward more participation of the stakeholders. ICT4D 3.0 is an open-ended process and not centrally controlled. The best practice could be a co-creation approach involving the local indigenous community. The designer would act as a facilitator to guide the local community in constructing and contextualising the action and its actual meaning, cooperating to solve a real-world problem in their specific context.

5 TRIZ for Complex Socio-technical Problems

TRIZ was initially developed as pure engineering science based on the statistical research of patents and other sources of technical information. However, there are also many research and publications in the non-technical area in recent years that indicate the competence of TRIZ in providing innovative solutions to the area outside of the engineering and technical domain. Analysis conducted by Zlotin [16] on research and projects in the non-technical area identified the following two approaches where TRIZ can be applied in the non-technical studies:

- Transfer TRIZ patterns, problem-solving tools and algorithms into non-technical areas, identifying their applicability and adapting them to the new area.
- Transfer patterns from other areas into TRIZ, identifying their applicability and adapting them to TRIZ.

Souchkov [17] proposed that the similar thinking patterns used in the technical system can be applied in business, arts or social systems. Both Mann [18] and Souchkov [19] demonstrated that the TRIZ technical system is viable for non-technical systems. TRIZ even was used in the efforts for cultural heritage preservations as carried out by Fiorineschi [20] to provide a short and non-representative overview of the possible support that the systematic design methodologies can provide in cultural heritage tasks. Kulathuramaiyer [14] highlights the possibility of contextualising and adapting patent or even new patentable ideas by applying the TRIZ inventive principles systematically across domains of study. TRIZ has therefore demonstrated the potential for solving complex problems in a structured manner through its systematic modelling.

6 The Law of System Completeness

Altshuller introduced the Law of System Completeness in 1979. In the classical TRIZ model, a complete technical system must consist of four key components: (1) Engine, (2) Transmission, (3) Control Unit and (4) Working Unit. Each of these components must provide a relevant function and must be combined into a system. As defined in the Glossary of TRIZ by MATRIZ 2018 [21], the engine unit converts energy to a specific type required to operate a working unit. The working unit refers to the function acting on a product for which the technical system has been developed. The control unit controls the energy supply to the other parts of the technical system and coordinates their operation. The transmission relates to a flow of energy required to operate a working unit which is the output area of the systems. The completeness in TRIZ law requires that all components are present in the system and integrate and function as one continuum. "If any component is missing, the technical system does not exist; if any component fails, the system does not survive", as quoted by Darrell Mann [22]. This ideality perspective of TRIZ is parallel with the holistic perspective of IKCS. It provides the foundation for developing the conceptual model in this research project.

6.1 The Whole is More Than the Sum of Its Parts

Theoretically, Gräbe [23] pointed out, "Man is the only creative productive force; it must be and remain the subject of development. Therefore, the concept of full automation, according to which the human is to be eliminated from the process gradually, misses the point!" as he argues about the law of displacement of humans from technical systems. He further highlighted that the problem of the concept of full automation would trigger an ecological crisis in the planetary dimension. His view of engineering system design should be embedded into the world of technical systems as he quoted the idiom of "The whole is more than the sum of its parts", which was well accepted in this project. Hence, a technical system is a way of a socio-technical system. A similar argument and proposition of solving a complex problem in TRIZ should move to a broader perspective of "a whole" rather than a single dimension of the technical system from Czinki [24]. These go in line with our view in this project. We believe there is no distinction between the earlier models developed in our prior time. A socio-technical system was also involved, but the social components have been subdued to a large extent and not brought forward in the point of attention. In the technical systems of TRIZ, the human was never a separated component. It means that the engagement of the local community and the interaction with supersystems needs to be explicitly formulated.

6.2 The Law of Supersystem Completeness

Valeri Souchkov's proposition of expanding the new TRIZ law called The Laws of Supersystem Completeness [25] provides a promising direction for our project: develop a conceptual model with a direct engagement with socio-technical systems and direct interaction with human components with the indigenous communities. The new law is based on the Law of System Completeness which is the subset in the categories of Law of Statics. Souchkov [25] argued that the complete technical system described in the classical TRIZ does not exist alone. A higher system must be involved that utilises the technical system's function. The engine receives the energy source from an outer supersystem, and the product of the working unit provides a proper function to a specific target. Supersystems are involved in utilising the function of a complete technical system.

Souchkov's [25] proposal defines a supersystem as "a system which interacts (through physical or informational links) with a technical system at each phase of the technical system's lifecycle". It indicates that multiple supersystems exist in a technical system and interact with each component to facilitate its function. Such supersystems should not be confused with the term environment. Gräbe [23] further elaborated that it is a specific system with its own language and logic, and the relationship of the supersystem-system is similar to the relationship system component. Any supersystem also acts functionally in the system's perspective, and "a supersystem is nothing more than a special kind of component, a neighbouring component."

Souchkov [25] postulated that the engagement of the supersystem in the technical system would make the technical system more "complete" and avoid problems like a mismatch between a product with its supersystem and the immaturity of the supersystem concerning the system's target and use. Gräbe [23] further echoed that the supersystem would be a complement of the "totality of the world" as his proposition of a technical system is, in a way, a socio-technical system, as discussed in the early section.

The above proposition sheds light on our proposed conceptual model, which is able to map and analyse the indigenous symbolic visual languages. Therefore, the engagement of supersystems with various resources interacting with the proposed technical system is crucial. It provides a unique supersystem-system relationship in a "complete" socio-technical system.

7 Conceptual Modelling for the Indigenous Symbolic Visual Communication Systems

Mundy [5] postulated some parallels between indigenous and exogenous communication based on the source-message-channel-receiver model. Manyozo [3] drew a theoretical framework of IKCS and its capacity to integrate with exogenous knowledge systems. In this project, we intend to map and analyse the different components and attributes of the complex IKCS via the powerful tools of TRIZ. Furthermore, we can understand how the indigenous symbolic visual communication systems function in a systematic order as provided by the TRIZ tools. This path is the way to help us develop future indigenous symbolic visual communication systems that demonstrate the parallel co-existent indigenous and exogenous communication systems.

The function model of Shannon & Weaver's [26] basic communication model, which is a well-received basic communication model for most of the exogenous communication studies, was first adopted and used as a comparison to the IKCS function model (see Fig. 1). The TRIZ function model was then used as a means to elaborate on the actual interactions. The IKCS function model (see Fig. 2) shows the importance of indigenous knowledge and contextualisation parameters in determining the definition and interpretation of indigenous communication messages.

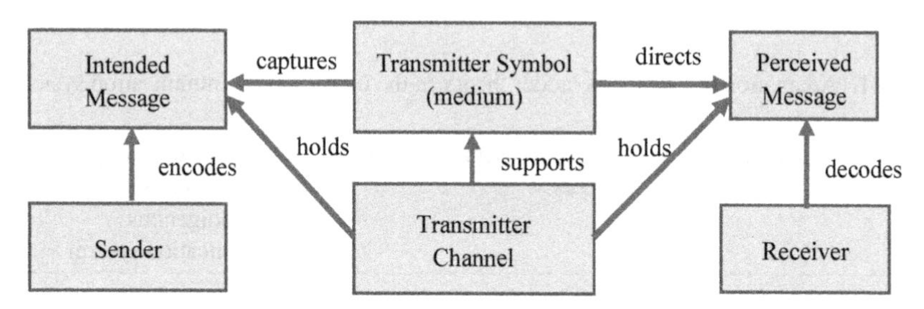

Fig. 1. Shannon & Weaver's basic communication function model

Based on the Laws of System Completeness, a conceptual model for the indigenous symbolic visual communication system was developed (See Fig. 3). The technical system demonstrated the interaction of the technical system components with the outer supersystems. There are internal sub-systems within the components, which are the ingredients [25] in the components. Components of the model were adapted from Ronald Stamper's [27] semiotic ladder theory, which is an extension of the traditional semiotic components of syntactic, semantics and pragmatics. The additional components of the social, physical, and empirics are essential to represent the human and material per-spectate of sign and symbol construction. This approach helps develop our model as cultural aspects, and material use is crucial in the indigenous symbolic visual language construct. Cultural components replaced the social world component in this context. Table 1 shows the adaptation of Stamper's Ladder Theory as components of the indigenous communication system we are developing.

The developed conceptual model (see Fig. 3) of the Indigenous Symbolic Visual Communication Systems adapted from the TRIZ technical systems has four components

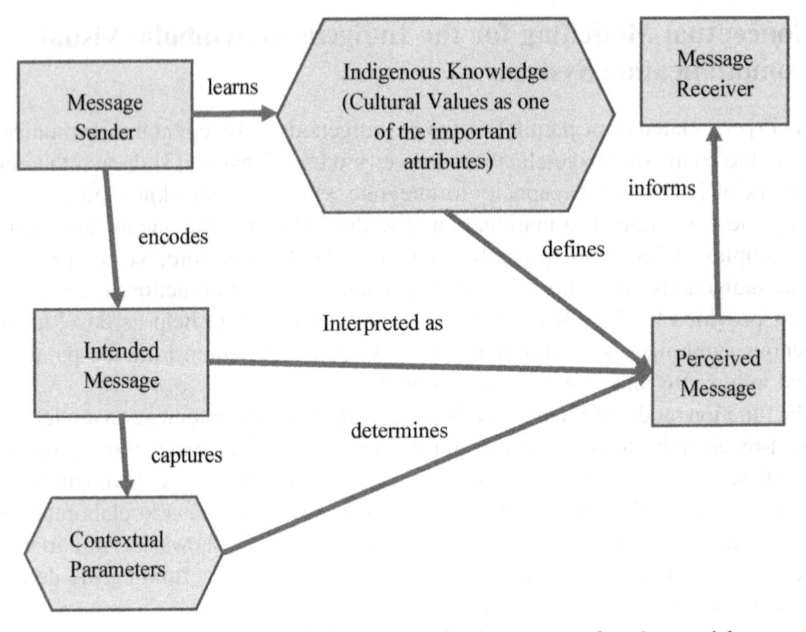

Fig. 2. Indigenous knowledge communication systems function model

Table 1. Adaptation of Stamper's ladder theory in the indigenous communication system

Traditional semiotic divisions	Stamper's semiotic ladder theory	Adaptation of Stamper's ladder theory as components of the indigenous communication system
	Physical	Physical aspect
	Empirical	Statistical properties
Syntactic		Message structure
Semantic		Message meaning
Pragmatic		Message usage
	Social	Cultural values

of the conceptual model, which are: (1) System Control, (2) Intended Message, (3) Transmitter, and (4) Perceived Message. Four supersystems respectively support the system, and they are (1) Indigenous Knowledge, (2) System Resource, (3) Living Space, and (4) Target Audience. The inventory of the developed model is shown in Table 2.

7.1 System Control

This system control determines the system's parameters and governs the whole indigenous communication process. Cultural values and contextual parameters control how the

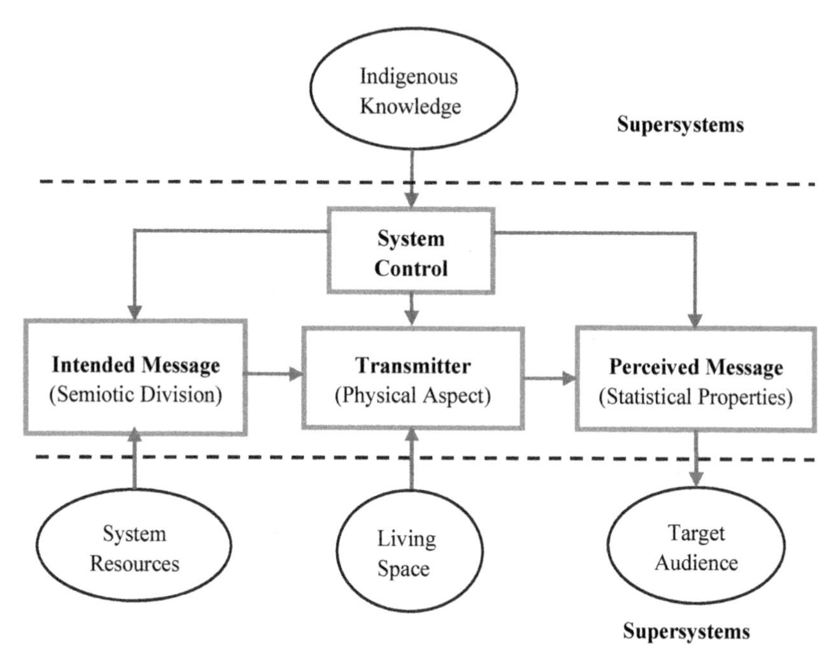

Fig. 3. Conceptual model for the indigenous symbolic visual communication systems

whole process of indigenous symbolic communication occurs effectively. In our earlier finding [28], we discovered that ambiguity is one of the problems in symbol and emoji interpretation due to different cultural coding and values. The same sign forms may be used in different places, but they have different meanings depending on the context and cultural background. The component has two sub-systems: (1) Indigenous Cultural Values and (2) Contextual Parameters. It will involve the integration with the indigenous knowledge as the supersystem.

7.2 Intended Message

The key components in the system will be the intended message adapted from the engine components of the TRIZ technical systems. It represents the original message the indigenous message sender intended to communicate. It encompasses the encoded information of events, news, and instructions they want to convey to others in the community. It gets the energy source from the supersystems, the system resource consisting of the social and cultural activities of the community.

7.3 Transmitter

The transmitter component of the indigenous communication system is unique as it represents the transmission process of the intended message to the perceived message through the special transmitter channels as used by the indigenous message sender, which is the symbolic visual language construction using natural materials. They gather natural

Table 2. The inventory of the developed conceptual model

System	System components in TRIZ technical system	Indigenous communication system components (adapted from TRIZ)	Sub-systems	Supersystems
Indigenous symbolic visual communication design systems	Control unit	System control	1. Cultural aspect	Indigenous knowledge
			1.1 Indigenous cultural value	
			1.2 Contextual parameters	
	Engine	Intended message	1. Semiotic division of indigenous communication	System resource
			1.1 Syntactic (structure)	
			1.2 Semantic (meaning)	
			1.3 Pragmatic (usage)	
	Transmission	Transmitter	1. Physical aspect	Living space
			1. 1 Natural material used	
			1. 2 Message stick construct	
	Working unit	Perceived message	1. Statistical properties	Target audience

materials found in the forest to construct messaging symbols. It could be in the form of a folded leaf, tree branches, twigs and animal parts as practised by the indigenous group of Penan and Murut people in Sabah and Sarawak. The extended semiotic component of the physical aspect with the sub-systems of natural material used and the sequence of message stick construct is identified as parallel with the transmission component of the original TRIZ technical systems. The Living space as the supersystem refers to the natural living environment of the indigenous people where they gather the materials to construct the communication symbols in the forest.

7.4 Perceived Message

The perceived message in the conceptual model parallels the working unit in TRIZ technical systems. It refers to the message perceived and interpreted by the other end of the indigenous communication, the receiving side, and how the symbolic message is decoded. The message decoding process's effectiveness depends on the shared cultural values and contextual parameters that govern the indigenous communication. The end message receiver is the supersystem of the indigenous or target audience in the conceptual model.

8 Snapshots of *Tatanda* and *Oroo'* Symbolic Visual Language

We studied both the indigenous symbolic visual languages in our project. It consists of *Tatanda* symbolic language of the Murut and *Oroo'* symbolic visual language of the Penan. Figure 4 shows the wedding message stick used by the Murut, and Fig. 5 shows the forest navigation symbolic visual message used by the Penan.

The Murut wedding message stick is a practice carried out by the Murut indigenous in Sabah to provide information regarding weddings in the community. It is at the forest track entrance leading to the wedding celebration. The message stick has multiple clefts which hold multiple symbols to give a narrative of the wedding event. It informs the targeted audience about the type of food served, the amount of food served, and the wedding present. It also has empty clefts at the bottom of the stick to allow wedding-attended families to provide their family symbols as the 'signature' of their attendance at the wedding.

The *Oroo'* symbol message in the figure represents a single message of forest navigation activities which tells the narrative of "I am going to the old hut which can be found as you walk this direction". Both symbolic languages may not represent a parallel communication genre and meaning. However, it provides a snapshot demonstrating

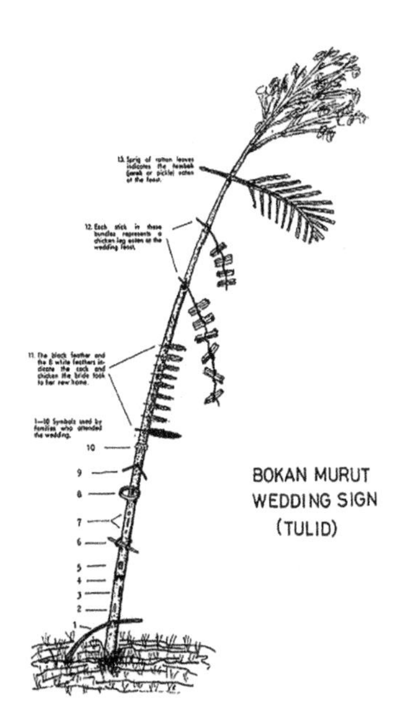

Fig. 4. Murut *Tatanda* wedding message stick. (Image source from Burrough [13])

that the developed conceptual model adapted via the TRIZ technical system is viable in providing a systematic analysis and mapping of the IKCS. The details of mapping both languages' snapshots based on the developed conceptual model are shown in Table 3.

Fig. 5. Penan *Oroo'* forest navigation symbol. (Image courtesy of Franklin George)

Table 3. Snapshots of *Tatanda* and *Oroo'* symbolic visual language mapping

Adaptation of Stamper's ladder theory in the indigenous Comm. system	Musrut Tatanda Sabah		Penan Oroo' Sarawak
Physical aspect	Material used	Twigs, rattan leaves, sticks, chicken feathers, family identity symbols	Leaves, stick
	Message stick construct	Non-linear Sequence	Non-linear sequence
Statistical properties	Baukan Murut, central region of Sabah		Nomadic Penan communities, Sarawak
Message structure and relationship	Intuitiveness, representation, familiarity		Intuitiveness, representation, familiarity
Message meaning	1. Tradition pickle (tamboh) eaten in the feast		I am going to the old hut, which can be found as you walk in this direction
	2. Amount of chicken eaten		

(continued)

Table 3. (*continued*)

Adaptation of Stamper's ladder theory in the indigenous Comm. system	Musrut Tatanda Sabah	Penan Oroo' Sarawak
	3. Amount of chicken brought by the bride to the new home	
	4. Families "sign" to show which family attended the wedding	
Message usage	Wedding announcement	Jungle navigation
Cultural values	Benefit to all, sharing	Immediacy

9 Insightful Ideas

This article provided insightful ideas on how the inventive design solution of TRIZ is viable in transforming global communication approaches to be sensitive to the cultural needs of indigenous communities. These could set the path for future endeavours to use TRIZ in non-technical and cultural heritage preservation research projects. Following are the finding ideas from our practice:

- The technical system thinking pattern of TRIZ can be applied in non-technical systems with some adaptation of the systems model.
- A higher system, the supersystem, must be involved to govern the technical system's function.
- The idea of 'completeness' in TRIZ Law of System Completeness parallels indigenous knowledge's holistic perspective. It is theoretical viable and practically applicable.

10 Conclusion

The conceptual model developed has provided substantial results in mapping and analysing the essential components of an indigenous symbolic visual language. We identified the components of Murut's *Tatanda* and Pena's *Oroo'* symbolic visual language. It further set the path for our primary research process of developing a local cultural value-based visual communication system for indigenous people in the Malaysian Borneo state of Sarawak. It also demonstrated that the proposition of Souchkov's Law of Supersystem Completeness is viable in providing an inventive solution for complex socio-technical problems. Although the technical aspect of the socio-technical system has been highlighted as similar to the technical system, in reality, one cannot deny the involvement of the human component, which most of the time subsists in the supersystems. The human component is an essential element to consider in developing the TRIZ technical system in the context of solving problems for complex socio-technical systems. This new approach contributes to the resilience and revitalisation of indigenous symbolic visual language. The research is significant in the sustainability of indigenous knowledge and cultural value.

References

1. Alshenqeeti, H.: Are emojis creating a new or old visual language for new generations? A socio-semiotic study. Adv. Lang. Lit. Stud. **7**(6) (2016)
2. Marcel, D.: The Semiotics of Emoji: The Rise of Visual Language in the Age of the Internet. Bloomsbury Publishing (2016)
3. Manyozo, L.: The context is the message: theory of indigenous knowledge communication systems. Javnost The Public **25**(4), 393–409 (2018). https://doi.org/10.1080/13183222.2018.1463351
4. Mundy, P., Lin Compton, J.: Indigenous communication and indigenous knowledge. Dev. Commun. Rep. 74 (1991)
5. Mundy, P.: Indigenous knowledge and communication: current approaches. J. Soc. Int. Dev. (1993)
6. Bala, P., Kulathuramaiyer, N., Eng, T.C.: Digital socio-technical innovation and indigenous knowledge. In: Recent Advances in Knowledge Management, IntechOpen (2022)
7. Bon, A.: Intervention or Collaboration? Rethinking Information and Communication Technologies for Development. Pangea, Amsterdam (2019)
8. Rhoades, R., Bebbington, A.: Farmers Who Experiment: An Untapped Resource for Agricultural Research and Development. Intermediate Technology Publications Ltd (ITP) (1995)
9. Wade, P.: Cultural Identity: Solution or Problem? Institute for Cultural Research London, UK (1999)
10. Barker, P.G., Yazdani, M.: Iconic Communication. Intellect Books (2000)
11. Haverkort, B., van't Hooft, K., Hiemstra, W.: Ancient roots, new shoots: endogenous development in practice. Appropriate. Technol. **30**(2), 62 (2003)
12. Pottier, J.: Negotiating local knowledge: an introduction. In: Pottier, J., Bicker, A., Sillitoe, P. (eds.) Negotiating Local Knowledge, pp. 1–29. Pluto Press, London (2003)
13. Burrough, P.A.: Message sticks used by Murut and Dusun people in Sabah. J. Malays. Branch R. Asiat. Soc. **48**(2) (228), 119–123 (1975)
14. Kulathuramaiyer, N.: Human Versus Machine Intelligence. Unimas Publisher, Kuching (2019)
15. Bon, A., Akkermans, H.: Rethinking Technology, ICTs and Development: Why It is Time to Consider ICT4D 3.0. The Network Institute VU University, Amsterdam, (2014)
16. Zlotin, B., Zusman, A., Kaplan, L., Visnepolschi, S., Proseanic, V., Malkin, S.: TRIZ beyond technology: the theory and practice of applying TRIZ to non-technical areas. TRIZ J. **6**(1) (2001)
17. Souchkov, V.: Innovative problem solving for social applications: a structured approach. In: Proceedings of the IADIS International Conference Web-Based Communities and Social Media, Prague (2013)
18. Mann, D.: Laws of system completeness. TRIZ J. May (2001)
19. Souchkov, V., Hoeboer, R., Van, M., Zutphen, M.: TRIZ for business: application of RCA+ to analyse and solve business and management problems. In: ETRIA TFC 2006 (2006)
20. Fiorineschi, L., Barsanti, R., Cascini, G., Rotini, F.: Application of systematic design methods to cultural heritage preservation. In: IOP Conference Series: Materials Science and Engineering. IOP Publishing (2020)
21. Xtriz.com Homepage: http://www.xtriz.com/publications/glossary.htm. Last accessed 10 Apr 2022

22. Care, I., Mann, D.: Using Mindmap with TRIZ. TRIZ J. Jan (2001)
23. Gräbe, H.-G.: Men and their technical systems. In: Internationa TRIZ Conference, pp. 399–410. Springer International Publishing (2020)
24. Czinki, A.P.D.I., Hentschel, C.: Solving complex problems and TRIZ. Procedia CIRP **39**, 27–32 (2016)
25. Souchkov, V.: the law of supersystem ccmpleteness. In: Proceedings of the 13th MATRIZ TRIZfest-2017 International Conference, pp. 399–406, Kraków, Poland (2013)
26. Shannon, C.E., Weaver, W.: The mathematical theory of communication. University of Illinois Press (1949)
27. Stamper, R.: New directions for systems analysis and design. enterprise information systems, pp. 14–39. Springer (2000)
28. Goh, C.H., Kulathuramaiyer, N.: Developing an indigenous cultural values based emoji messaging system: a socio-technical systems innovation approach. In: Proceeding of 12th ACM Conference on Web Science Companion, Southampton, United Kingdom (2020)

Exploitation of Causal Relation for Automatic Extraction of Contradiction from a Domain-Restricted Patent Corpus

Daria Berdyugina and Denis Cavallucci(✉)

ICUBE/CSIP, INSA of Strasbourg, 24 Boulevard de la Victoire, 67084 Strasbourg, France
dberdyugina@etu.unistra.fr, denis.cavallucci@insa-strasbourg.fr

Abstract. Altshuller contradiction matrix is one of the most popular tools among TRIZ practitioners, especially beginners, due to its simplicity and intuitive design. However, scientific and technological progress induces the constant appearance of new scientific vocabulary, which lower accuracy when using this static tool from the end of the sixties. Some attempts to rebuild the matrix or update it has been made within the past four decades but without any successful legitimation due to the lack of scientific proof regarding its relevance. Our recent findings in the use of Natural Language Processing (NLP) techniques allow the creation of a methodology for automatic extraction of the necessary information for establishing a domain-restricted contradiction matrix. In this paper, we relate a technique that exploits the internal language semantic structure to mine the causal relation between terms in patent texts. Moreover, the subject or domain restriction for a patent collection allows observing the links between extracted information at the over-text level. Such an approach relies on inter-and extra-textual features and permits a real-time extraction of contradictory relations between elements. These extracted elements could be presented in matrix form, inspired by The Altshuller contradiction matrix. We postulate that such a representation allows the construction of a state of the art in each domain, which will facilitate the use of TRIZ to solve contradictions within it.

Keywords: TRIZ · NLP · Contradiction matrix · Text-mining · Automatic extraction

1 Introduction

The Contradiction Matrix (CM) remains a popular tool among a lot of TRIZ practitioners because of its simplicity. Despite the existence of other tools developed by the experts (for example, ARIZ85C [1] or Vepole [2]), the CM does not lose popularity. Created in 1969, this tool is largely used, notably, by less highly experienced TRIZ users.

Nonetheless, due to today's rapid scientific advancements, the tool's efficiency is called into question. Since the establishment of Altshuller's matrix, many new technologies, methods, and even whole fields of study have evolved. The CM is out of date

Published by Springer Nature Switzerland AG 2022
R. Nowak et al. (Eds.): TFC 2022, IFIP AICT 655, pp. 86–95, 2022.
https://doi.org/10.1007/978-3-031-17288-5_8

because the terminology is becoming outdated. Another issue is its general nature. For newcomers to TRIZ, forming a relevant contradiction and connecting terminology from their field of expertise to technical data utilized in the CM might be challenging, resulting in a misinterpretation of its use.

The IDM (Inventive Design Method) was developed to overcome some of TRIZ's drawbacks, one of which was the lack of a codified ontology. Another disadvantage of the IDM-resolved grounding theory is the difficulty of performing any computation on abstract ideas [3]. Problems, partial solutions, and parameters are the three fundamental core principles of IDM. However, focusing on the parameters is required for CM reconstruction on the user's patent corpus because they provide summarized information about the technologies specified in the patents. Furthermore, the parameters define CM's "borders."

As previously stated, the CM's author examined around 40,000 of the most innovative patent texts [4]. The ability of automatic text-mining tools has become crucial in recent years. Patents are a plentiful and numerically accessible source of inventive data. As a result, we intend to make use of a corpus of patent texts from the most recent field of research to develop a CM based on the user's selection of patents. Patent analyzing apps are becoming increasingly popular among industry and engineers, thus it's critical to identify appropriate methodologies and processing tools. In any patent-related operation, the better the processing instrument employed, the better the results.

Because of their extensive and complex language structures and unusual style, patent documents are frequently difficult to comprehend. This is owing to the patent text's dual nature, which is both a legal and a technical document aimed at protecting the inventor and identifying the invention's limitations. As a result, given the large number of patents issued each day, it is more practical for businesses and scientists to have a more compact TRIZ-based representation form for patents.

The standards for doing an automatic patent analysis are exceedingly high. It necessitates some knowledge of domain-specific technologies and information retrieval methods. This type of analysis is both expensive and difficult to perform. It's difficult to do such a common analysis scenario by hand.

However, the modern approaches and techniques of Computer Science and NLP permits performing such analysis and extraction faster. Notably, the transformers, a Deep Learning model that employs the self-attention process, differentially ranks the significance of each component of the input data (for example, BERT[1] [5]), facilitates the task of adaptation of the model for a specific task.

Our research goal is to determine the best method for automatically extracting parameters from domain-specific corpus to populate the CM's rows and columns.

By analyzing the text of patents, we concluded that the contradictions, expressed in texts of patents are usually not fully disclosed, i.e., only one parameter of a contradiction is clearly expressed, but the second parameter is either hidden or located elsewhere in the text. However, this second parameter could be found thanks to the internal logic of the text structure.

In this article, we present an overview of IDM approaches, as well as a literature analysis on transformers' learning methodologies (2). Following that, we describe a

[1] Bidirectional Encoder Representations from Transformers (BERT).

method that is developed for the automatic extraction of IDM-related information from patents, particularly the parameters (3). Then, employing the approaches, mentioned above, we outline the methodology for improving and refining the process of extracting parameters (4). The findings of our experimental efforts are then presented in (5).

2 State of Art

This chapter discusses the TRIZ, Inventive Design Method ideas, and NLP methods used to attain our goal to better understand the methodology of the current research.

2.1 TRIZ

TRIZ [6, 7] arose from a careful examination of a hundred thousand patent documents. According to Genrich Altshuller, the birth of innovation is conditional on the presence of objective rules and the observance of universal principles.

His investigation led him to the conclusion that some patterns of difficulties repeat irrespective of industrial domains, regardless of the industry to which a specific artifact belongs.

Genrich Altshuller also distinguishes between a tiny amplitude invention, which may be characterized as a modest innovation requiring little work, and discoveries that need significantly more time and effort to obtain. His descriptions of the trial-and-error approach are tied to these efforts.

This progression from low-amplitude innovation to scientific discovery has resulted in the establishment of five levels of inventiveness [8].

TRIZ incorporates a variety of essential principles that may be used to conceive and solve any problems, as well as obtain a solution without compromising.

The most important of these principles are the rules of Engineering Systems Evolution, contradiction, resources, and ideality [9].

In the context of TRIZ, contradiction is the basis of any problem formulation.

This term reflects the idea that increasing the value of one parameter leads to a decrease in the value of another parameter in a system [10]. It is important to emphasize that, according to TRIZ, any inventive problem must be expressed as a contradiction.

Further to that, the problem is described generally, and the search for an uncompromising solution to this contradiction is done through meditation by analogy utilizing the innovative concepts of TRIZ [11].

2.2 Inventive Design Method

The current study aims to extract IDM-related information from a domain-specific corpus. Parameters are of particular significance to us because they reflect the elements of contradiction. We have discussed the important topics above for explanation purposes:

issues, partial solutions, parameters, and contradictions.

A major part of our laboratory's work is based on TRIZ, which was invented by Genrich Altshuller.

The TRIZ-based Inventive Design Method (IDM) expands on the limitations of the grounding theory, notably the absence of a defined ontology, which makes it impossible to perform any computation on its abstract concepts. Thanks to IDM, this ontology has been created.

The problem-solving process, according to IDM, consists of four steps [12]:

1. Information extraction, comprising "problems" and "partial solutions";
2. Contradiction formulation;
3. Solving each key contradiction;
4. Selection of the most suitable solution concept.

For making TRIZ effective for industrial innovation, the IDM provides a functional formulation of the contradiction notion. The contradiction, according to this definition, is "[...] characterized by a set of three parameters and where one of the parameters can take two possible opposite values Va and $\overline{Va}$" [10]. Hence, it is necessary to provide a definition for two types of parameters.

There are two types of parameters: action parameters (AP) and evaluation parameters (EP).

The first, AP, "[...] is characterized by the fact that it has a positive effect on another parameter when its value tends towards and that it harms another parameter when its value tends towards [...]. (i.e., in the opposite direction)" [10].

The EP "[...] can evolve under the influence of one or more action parameters" and which allows to "evaluate the positive aspect of a choice made by the designer" [10].

The graphical representation of the contradiction notion is provided below.

$$AP \overset{Va}{\underset{\overline{Va}}{=}} \begin{pmatrix} EP_1 & EP_2 \\ -1 & 1 \\ 1 & -1 \end{pmatrix} \tag{1}$$

During the application of the solution to the technical contradiction, the physical contradiction may arise.

Physical contradiction reveals a system characteristic when the application of action in a physical condition results in contradictory requirements for its opposite characteristic [13].

2.3 Transformer Model

Transformers [14] have taken NLP by storm since its introduction in 2017, delivering greater parallelization and better modeling of long-term dependencies.

Transformers, like recurrent neural networks (RNNs), are built to handle sequential input data, such as natural language, for tasks like translation. This type of model does not always treat the input data in order. On the other hand, the attention mechanism permits obtaining the context for any place of the input phrase. For example, for the text input, it is not needed for such a model to process the beginning of the phrase before the end of its phrase. I.e., it determines the context that provides the meaning for each word in a phrase. This feature allows shortening the training duration [14].

Introduced in 2017, this model starts to replace recurrent neural networks models (for example, long short-term memory), notably for NLP tasks [15]. The main feature of transformers, additional parallelisation, permits training on larger datasets than previous models. Hence, the pre-trained systems (for example, BERT or GPT[2] that are trained on large datasets [16]) have been created. The architecture of such systems allows a fine-tuning for a specific NLP task.

BERT [5] is the most well-known Transformer-based model; it achieved cutting-edge performance in a variety of benchmarks and remains a must-have baseline [17].

The standard BERT workflow consists of two stages: pre-training and fine-tuning. Pre-training tasks include masked language modeling (MLM, which predicts randomly masked input tokens) and next sentence prediction (NSP, which predicts if two input phrases are close to each other). One or more fully-connected layers are generally placed on top of the final encoder layer to fine-tune for downstream applications [17]. It is important to note that there are a lot of versions of BERT, provided by Google[3] and HuggingFace [15]. These versions include 'large' and 'base' variants, which are differentiated by numbers, layers, and hidden state size.

3 Methodology

In this chapter, we describe the methodology applied to achieve the goal of automatic extraction of inventive information (parameters and their probability to be in contradiction) from the domain-specific patent corpus.

3.1 General Description of Methodology

By analyzing the text of patent documents, we concluded that two parameters in contradiction are linked by the causal relation. For example:

> *The mold has an excellent thermal conductivity since it is formed from a metal material. When a mold of a metal material is used and when the mold temperature is set at a level considerably lower than the deflection temperature under load of a crystalline thermoplastic resin as described above, the molten crystalline thermoplastic resin filled in the cavity begins to be cooled as soon as it is brought into contact with the cavity wall of the mold. As a result, an amorphous layer or a low-crystallinity fine crystal layer is formed as a surface of the molded article. Such a layer is generally called a skin layer. The molded article having the skin layer has a problem in that it is greatly degraded in surface properties* [18].

In this example, we could find an expressed contradiction between *excellent thermal conductivity* and *degraded surface properties*, which are two EP. The AP here is *the mold.* These two EPs are located far from each other. However, there is an intertextual link between them, which is asserted by causal relations. I.e., when the AP is expressed as a cause that influences two EPs (effects) and when one of the EP has a negative connotation

[2] Generative Pre-trained Transformer.

[3] https://github.com/google-research/bert.

and the second EP has a positive connotation, in this situation we could suppose that two effects are EPs and they are in a contradiction situation. Figure 1 illustrated the described idea.

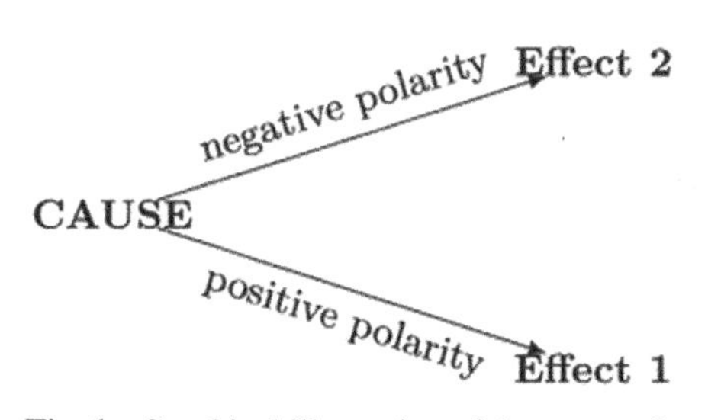

Fig. 1. Graphical illustration of the approach.

For the task of polarity calculation, we employ the Flair model [19], which was trained to detect the polarity of the phrases.

3.2 Extraction Model

In order to perform the above-mentioned extraction, it is needed to fine-tune a model, which could be capable to recognize the causal relation. As it is described in the previous section, we employ the transformer model (BERT and ALBERT) in order to extract the domain-specific parameters and their probability to be in contradiction. In order to perform this task, it is needed to apply a fine-tuning step.

For performing this step, we took an annotated corpus for the classification of semantic relation between pairs of nominals [20]. This corpus[4] contains not only annotation for causal relations, but also a lot of semantic relations, such as Component-Whole or Member-Collection. The example of annotation is presented in Table 1.

Table 1. Example of the original annotated corpus.

sentence (string)	relation (class label)
The <e1>burst</e1> has been caused by water hammer <e2>pressure</e2>	1 (Cause-Effect(e2, e1))

However, for our research, it is important to extract not only the nominals but also the adjective, which could define the polarity of a parameter. Thus, we were obliged to re-annotate this corpus, considering only the Cause-Effect part (Table 2).

Classification of the contradictions.

By performing the first experiments with the extraction model, we infer that the quantity of extracted parameters exceeds the limitation of the understanding, i.e., there is a lot of noise, expressed by the extracted terms, which are not the parameters.

[4] Accessible in https://huggingface.co/datasets/sem_eval_2010_task_8.

Table 2. Example of the re-annotated corpus.

Sentence: 1, The, DT, N
Sentence: 1, burst, NN, E
Sentence: 1, has, VBZ, N
Sentence: 1, been, VBN, N
Sentence: 1, caused, VBN, N
Sentence: 1, by, IN, N
Sentence: 1, water, NN, C
Sentence: 1, hammer, NN, C
Sentence: 1, pressure, NN, C

In order to overcome this difficulty, we apply another model for contradictions classification.

Using the output of the causal extraction, we annotate 7,268 pairs of candidates. In Table 3, we present an example of the annotated corpus, where label 0 represents the absence of contradiction and label 1 represents the presence of contradiction.

Table 3. Example of annotation for sentence-pair classification task

Idx	Label	Sentence1	Sentence2
Num 18	0	'Occurrence of sink mark'	Knock'
Num 19	1	'Occurrence of sink mark'	'Excellent mass productivity'

4 Case Study

In this section, we present the application of our approach in a concrete case study.

To validate our methodology, we have been using a set of 10 patents, having the subject of the creation of hydrophobic paper. The corpus statistic is presented in Table 4.

Table 4. Corpus statistics

Number of patents	Words	Symbols
10	127,612	775,121

Thanks to our extraction model, we extract 1,051 pairs of contradictions candidates. However, by applying the second model of contradictions classifications, we achieve to restrict this number by 28 candidates. We restricted the score of the probability by >0.45. The score is relatively low because the classification model is trained based on another domain corpus. Based on a restricted set of candidates, the experts approve 17 pairs of contradictions.

However, in the list of extracted pairs not classified as a potential contradiction, there is a set of pairs that have been approved by the experts to be in contradiction.

Our methodology has been built in an API that allows users to examine a domain-specific corpus of patent documents to construct the matrix representation of opposing parameters. The work is still in progress.

5 Discussion and Results Evaluation

In the present section, we discuss the results of the implementation of our methodology.

In the context of classification of extracted contradictions, our approach has an important drawback which is manifested in the presence of the noise. However, with more cases of the use of our methodology, we aim to collect more examples of valid contradiction pairs.

The extraction of pairs of candidates has equally its drawbacks. For the present day, we did not test other models for causal relation mining, for example, RoBERTa [21]. Moreover, we still lack annotated data from patent texts.

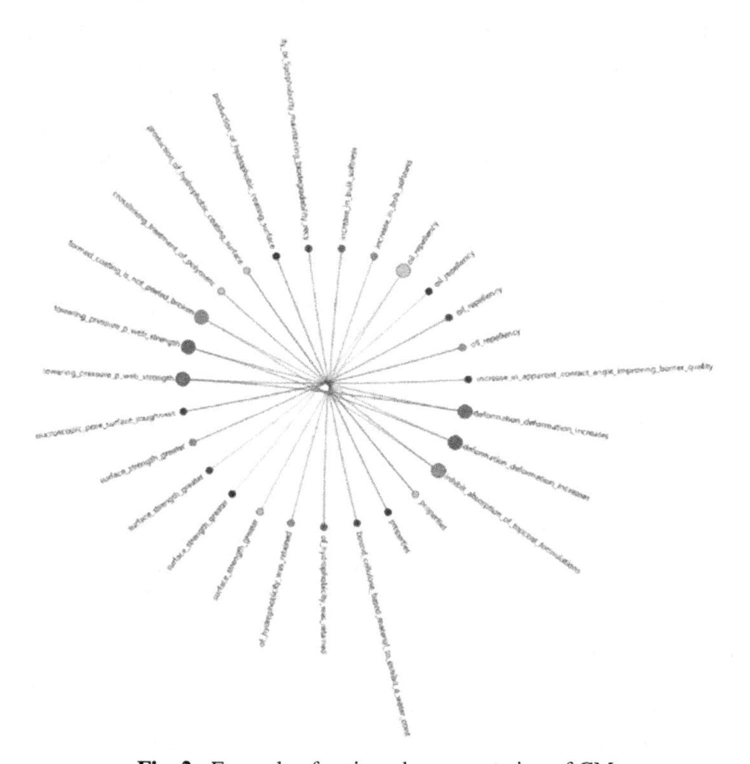

Fig. 2. Example of reviewed representation of CM

However, the present results remain promising in terms of parameters and contradictions identification. As our experiment shows, we achieve correctly identify 17 contradictions from 28.

It is important to note that the parameters, identified thanks to our method, could not be in contradiction with others parameters. That is the reason, why we decide to replace the classical matrix representation for our tool with a chord diagram. The example of reviewed CM is shown in Fig. 2. The base for this diagram is the list of validated pairs of parameters from the case study.

6 Conclusion

In the present article, we introduce the method for the automatic extraction of contradictions from the domain-specific patent corpus. Our methodology is based on IDM, for which a contradiction is one of the core elements in the inventive problem-solving process.

On a higher scale, our approach attempts to simplify the problem-solving process by automating the extraction of each required element in order to populate the ontology. The automated method not only allows each TRIZ-user to save time, and to gain a greater representative content from all corpus data.

Presently, we are working on dressing the "borders" of the CM. However, for future work, we consider linking the extracted contradictions to any source permitting solving them, notably the inventive principles. Moreover, the source of solution could be found in the claims of patent texts or any other sources of inventive information, such as scientific articles.

In a longer perspective, we intend to use additional sources of innovative information, such as scientific articles, to fill in the gaps left by other sources of extraction.

References

1. Marconi, J.: ARIZ : the algorithm for inventive problem solving. Triz J. [Online]. Available: https://triz-journal.com/ariz-algorithm-inventive-problem-solving/ (Apr 1998). Accessed 08 Apr 2021
2. Dubois, S., Lutz, P., Rousselot, F., Caillaud, E.: A formal model for the representation of problems based on TRIZ. In: International Conference on Engineering Design, ICED 05, Melbourne, Australia, pp. NA. [Online]. Available: https://hal.archives-ouvertes.fr/hal-00340989 (Aug 2005)
3. Souili, A., Cavallucci, D.: Automated extraction of knowledge useful to populate inventive design ontology from patents. In: TRIZ – The Theory of Inventive Problem Solving, pp. 43–62. Springer (2017). https://doi.org/10.1007/978-3-319-56593-4_2
4. Altshuller, G.: 40 Principles: TRIZ Keys to Innovation. Technical Innovation Center, Inc. (2002)
5. Devlin, J., Chang, M.-W., Lee, K., Toutanova, K.: BERT: pre-training of deep bidirectional transformers for language understanding. ArXiv181004805 Cs. [Online]. Available: http://arxiv.org/abs/1810.04805 (May 2019). Accessed 29 Apr 2022
6. Altshuller, G.: Creativity As an Exact Science. Taylor & Francis (1984)

7. Altshuller, G., Altov, G.: And Suddenly the Inventor Appeared: TRIZ, the Theory of Inventive Problem Solving. Technical Innovation Center, Inc. (1996)
8. Souchkov, V.: Differentiating among the five levels of solutions. Online TRIZ J. (2007)
9. Cavallucci, D.: Contribution à la conception de nouveaux systemes mécaniques par integration methodologique. These de doctorat, Université Louis Pasteur (Strasbourg) (1971–2008). [Online]. Available: http://www.theses.fr/1999STR13238 (1999). Accessed 07 Apr 2021
10. Rousselot, F., Zanni-Merk, C., Cavallucci, D.: Towards a formal definition of contradiction in inventive design. Comput. Ind. **63**(3), 231–242 (2012). https://doi.org/10.1016/j.compind.2012.01.001
11. Zlotin, B., Zusman, A., Kaplan, L., Visnepolschi, S., Proseanic, V., Malkin, S.: TRIZ beyond technology: the theory and practice of applying TRIZ to nontechnical areas. TRIZ J. **6**(1) (2001)
12. Cavallucci, D.: From TRIZ to Inventive Design Method (IDM): towards a formalization of Inventive Practices in R&D Departments. In: Japan TRIZ Symposium 2012, pp. 2 (2012)
13. Mann, D.: Re-thinking physical contradictions #1: technical problems. Triz J. [Online]. Available: https://triz-journal.com/re-thinking-physical-contradictions-1-technical-problems/ (Mar 2018). Accessed 08 Apr 2021
14. Vaswani, A., et al.: Attention is all you need. Advances in Neural Information Processing Systems, **30**. [Online]. Available: https://proceedings.neurips.cc/paper/2017/file/3f5ee243547dee91fbd053c1c4a845aa-Paper.pdf (2017)
15. Wolf, T., et al.: Transformers: state-of-the-art natural language processing. In: Proceedings of the 2020 Conference on Empirical Methods in Natural Language Processing: System Demonstrations, Online, pp. 38–45 (Oct 2020). https://doi.org/10.18653/v1/2020.emnlp-demos.6
16. Open Sourcing BERT: State-of-the-Art Pre-training for Natural Language Processing: Google AI Blog. http://ai.googleblog.com/2018/11/open-sourcing-bert-state-of-art-pre.html. Accessed 30 Apr 2022
17. Rogers, A., Kovaleva, O., Rumshisky, A.: A primer in BERTology: what we know about how BERT works. ArXiv200212327 Cs. [Online]. Available: http://arxiv.org/abs/2002.12327 (Nov 2020). Accessed 29 Apr 2022
18. Yako, T., 八箇毅: Insert molding of plate glass fitted product. JPH06246782A. [Online]. Available: https://patents.google.com/patent/JPH06246782A/en (06 Sep 1994). Accessed 29 Apr 2022
19. Akbik, A., Bergmann, T., Blythe, D., Rasul, K., Schweter, S., Vollgraf, R.: FLAIR: an easy-to-use framework for state-of-the-art NLP. In: Proceedings of the 2019 Conference of the North American Chapter of the Association for Computational Linguistics (Demonstrations), Minneapolis, Minnesota, pp. 54–59 (Jun 2019). https://doi.org/10.18653/v1/N19-4010
20. Hendrickx, I., et al.: SemEval-2010 task 8: multi-way classification of semantic relations between pairs of nominal. In: Proceedings of the Workshop on Semantic Evaluations: Recent Achievements and Future Directions - DEW '09, Boulder, Colorado, pp. 94 (2009). https://doi.org/10.3115/1621969.1621986
21. Liu, Y., et al.: RoBERTa: a robustly optimized BERT pretraining approach. ArXiv190711692 Cs. [Online]. Available: http://arxiv.org/abs/1907.11692 (Jul 2019). Accessed 30 Apr 2022

Systems, Resources, and Systemic Development in TRIZ

Hans-Gert Gräbe(✉)

InfAI, Leipzig University, Leipzig, Germany
graebe@infai.org

Abstract. In TRIZ theory, resources play an important role when it comes to operate a systemic solution. It is only in this phase that the resource identified in the detailed solution plan as "any type of tangible or intangible matter that can be used to solve an inventive problem" must prove itself in practice. Conceptual distinctions such as "role definition" and "role occupation", which are central for the management of human resources, play only a subordinate role in the TRIZ resource conception. In this paper, the close connection of the terms resource and component with systemic operating conditions is analysed in more detail and it is shown which influence, for example, the management and reproduction of scarce resources has on systemic development processes in a supersystem. The resulting questions are compared with corresponding theoretical approaches from component software in order to work out the significance of higher-level abstraction concepts such as component models, component architectures or middleware.

It is proposed to bundle these overarching questions of the interplay of independent third parties providing resources in the huge real "world of technical systems" and thus constituting resource management structures in an new area *Resource Management Analysis* in the TRIZ theory corpus.

Keywords: Systemic approach · Resource · Operating conditions · Place and content · Interfaces · Component models

1 The Aim of This Paper

The aim of this paper is to analyse the concept of a resource in the conceptual framework of TRIZ in more detail and in particular to analyse its relationship to systemic operating conditions. Further, the TRIZ resource concept is compared to resource concepts in the conceptual foundations of component systems in technical domains with a focus on Component Software.

Developed engineering disciplines are characterised by the extensive use of components that are developed, produced and offered by independent third parties, thus bringing the terms *resource* and *component* close together. It was

Published by Springer Nature Switzerland AG 2022
R. Nowak et al. (Eds.): TFC 2022, IFIP AICT 655, pp. 96–108, 2022.
https://doi.org/10.1007/978-3-031-17288-5_9

only with the transition to software components that computer science developed from an "Art of Programming" into an engineering discipline and thus embarked on this path of a systemic development towards a mode of production based on the deep division of labour, which other engineering disciplines had already taken before. The theoretical concepts that accompanied this development towards Component Software are therefore of particular interest, as experiences from other engineering domains were taken up and further elaborated during this development.

The connection between a viable resource concept and the interrelation of systemic development processes of technical functionalities and their bundling in technical systems over longer periods of time is presented as a specific form of "organisation of material" in the sense of [14, p. 98], which institutionalises itself in patterns, norms, standards, component models and finally as "state of the art". In particular, it is shown the significance of *component models* and *component frameworks* in developed component architectures for a qualified development of systems of resource management. The explanations deepen the view on the conceptual foundations of TRIZ developed in [2].

In this paper, such approaches are discussed only in the scope of reuse of technical functionalities in cooperative action *within* larger companies. In another paper [3], this question is discussed for cross-company cooperation and the importance of open architectures is elaborated.

The results may remain unsatisfactory insofar as only *questions* of an appropriate conceptualisation are raised. It is suggested that TRIZ theory be extended to include the instrument of a *Resource Management Analysis*, which links classical and Business TRIZ and addresses the interrelation between problems of short-term operational resource provision and long-term development of resource availability.

2 The Concept of a Resource in TRIZ

In [21] Wessner collected some common definitions of the resource concept from various TRIZ schools. All these definitions focus more on the availability than on the structure and material composition of resources.

In general, in TRIZ the resource concept is mostly used in an intuitive way, without making any effort to establish a precise conceptual foundation. In the *TRIZ Body of Knowledge* [6], for example, the word "resource" only appears in item 1.6 as "substance-field resource" and in the title of four publications [10], [11,13,22] listed there. In common TRIZ glossaries [5,9,16,18,19], only [16] and [9] explain the term *resource*. In [16] it is defined as "any type of tangible or intangible matter that can be used to solve an inventive problem: time, space, substances, fields, their properties and parameters, etc.". The object of *Resource Analysis* is derived from it as "examination of resources available in the technical system and its supersystem in order to compile a list of resources that can be used for solving a particular inventive problem" (ibid.).

In Sect. 4.2 of the textbook [4], Resource Analysis is discussed on 13 pages and with material, field-like, spatial, temporal, informational and functional resources six types of resources are distinguished. Specific *qualitative* determinations of such "substances and fields" as resources play almost no role in the classification proposed in [4, p. 51–52] according to *value* (free, not expensive, expensive), *quality* (harmful, neutral, useful), *quantity* (unrestricted, sufficient, insufficient) and *readiness for use* (ready, to be modified, to be developed). Such qualitative determinations in the sense of the fulfilment of *specifications* are, however, essential in more complex technical contexts in order to ensure the *operation* of a specific functional property, which is to be provided by the systemic context.

Matvienko estimates in [9] the current state of conceptual penetration of the notion *resource* as follows:

> A resource is in general a set of systemic properties of an object not previously used to solve some inventive problem. It is not defined in TRIZ in any way, although there are numerous methodologies for finding resources, resource tables, resource lists, etc.
> Despite the obvious abundance of methodical literature on the subject, the search and use of resources in specific conditions of practice always remains subjective, because no problem solver can ever reliably know whether or not a given system property has been used before to solve this inventive problem.
> As a rule, in one and the same technical system, acting under the same problem conditions using the same methods, different solvers find completely different resources, often even of different epistemological level.

The aim of a systemic modelling of a problematic situation in TRIZ is not only, and not so much, to develop a functionality that solves the problem *potentially*, but to develop the solution up to its practical operational use. For such a practical operation, however, *operating conditions* must exist or be established, which include the use of prestructured resources, which "exist or can be easily produced" only to the extent as this is provided by a developed market structure.

Such a *use* of resources both in artefact form (as objects) and functional resources (components) is an essential point of the *implementation* of a solution plan and the subsequent operation of the solution, i.e. of the "system to be", in the interplay with the operation of both the components and the neighbouring systems. This qualitatively and quantitatively determined availability of substance, energy and information is closely connected to the concept of resources, but requires much more structure than just the (better) exploitation of "any available type of tangible or intangible matter".

3 On the Systemic Approach

The systemic approach is one of the central methodological elements of TRIZ theory. As a problem-solving methodology it unfolds its advantages if it is possible to work out a *contradiction* within the requirements, to delimit this contradiction spatiotemporally in an *operative zone*, to demarcate it from an *environment* in a systemic way and to analyse the problem more precisely within such a well delimited system.

Through such a threefold delimitation the horizon of consideration is focussed – by demarcation from the outside against an *environment*, by internal delimitation against *components* and by limiting the *relations* between these components to be considered to essential ones, see [2] for details.

Typically, TRIZ methodology is limited to describing a solution *plan* for transforming a "system as is" into a "system to be" and does little focus on the implementation of that plan. The four phases Define, Select, Generate and Evaluate, into which the TRIZ solution process is divided in [8], end with such a taylored solution plan only.

In this section, the relationship between such a solution plan and its implementation is examined in more detail. It will become apparent that the TRIZ concept of *ideality* corresponds to the concept of *pure function*, whose "viability" (a term from [14]) only emerges and can emerge in the course of practical implementation through connection with a "viable environment".

3.1 Systems and Emergent Functions

Systems are characterised by the fact that they realise *emergent functions* which cannot be reduced to individual parts of the system, but result from the interaction of these parts [12, p. 17]. For a system considered from the outside as a Black Box, such a *main useful function* as *main parameter of value* is in the foreground. Usefulness, expediency and purposefulness embed the (technical) system into larger socio-cultural contexts and justify the existence of the system itself.

On the respective system level, therefore, the appropriate arrangement and interplay of these *relations* play a leading role, whereby a distinction is to be made between the dimensions of structural and processual organisation. In the interrelationship of both dimensions the fundamental contradiction of every systemic approach does manifest itself – the contradiction between decomposability and unity in the categorical part-whole relationship.

Petrov [12] emphasises that for analytical purposes, the system *must* be disassembled, but it can be operated only in assembled state. This inherent contradiction between decomposability and wholeness does not end at the boundary of the system: the operation of the assembled system in turn requires a qualitatively and quantitatively determined *throughput* of substance, energy and information. Even if the decomposition of a system into its parts provides important insight into its functioning, only in assembled state the system can be operated and thus unfold its specific functionality. In this sense every systemic approach reduces in a certain way to a *conditional mind game.*

In the TRIZ notion of a *minimal technical system*, a *tool* acts on an *object* (workpiece) to be processed in order to transform it into a *useful product.* The concept of the *ideal system* [4, p. 40] considers the tool as a purely functional property, the effect of which to intentionally change the state of the workpiece to a useful product is achieved without any additional efforts and any wear of the tool. In other words, it is not the *real* tool but the *imagination of the tool* that creates the required action in such an *ideal machine.*

3.2 Systems and Their Operating Conditions

This is, of course, only an ideal picture, since in addition to the structural design the *operation* of the system and thus a *throughput* of substance, energy and information through the system is required in a qualitatively and quantitatively determined form. This aspect is somewhat underexposed in TRIZ, as the usefulness of a system is primarily defined in terms of its *main useful function* [4, p. 40], i.e. in its *potential* usefulness.

For the *real* usefulness, the mentioned three types of throughput must be organised, i.e. the system must have *resources* at its disposal for its operation. In the classical understanding of a *complete technical system* [7, 4.2], [19, p. 9] the energy throughput is centered on the tool, the throughput of substance transports the workpieces and the throughput of information is directed to the control of the action. Thus, in any case, the concept of a resource is understood in [4, p. 51] and also [19, p. 7] as "means that can be used to solve a problem."

The understanding of the relationship of action conveyed here is asymmetrical. An active tool has a state-changing effect on a passive workpiece, while retaining its own functionality and – ideally – without undergoing a state change itself. In substance-field models this understanding is replaced by a more symmetrical model of a field-mediated action between two substances. At the same time, in the systemic abstraction, the materiality of the tool is pushed back further from the tool to the action of a field, and a component concept is prepared as proposed by Szyperski [17] for Component Software. There, *components* are basically conceptualised as *stateless* with all resulting consequences. In contrast to this, *objects* are conceptualised as state-bearing units of instantiation to maintain a certain standardisation of workpieces required for a repeated application of a function within a production process.

Such an approach also corresponds well with the widespread organisation of production processes, where a distinction is made between operating and maintenance mode. In the operating mode, the focus is on the functional properties of the tool, while in the maintenance mode its material properties are focused. As an independent technical system in a narrower sense, only the operating mode is modelled as the target of a "problem solution". The maintenance mode is part of the supersystem, which is concerned with the *reproduction* of the tools as *resources* used in the operating mode. In the (classical) operating mode the focus is on the use of tools and the material throughput of workpieces, which are thereby transformed into useful products, in many cases *technical artefacts*, which are either further processed as semi-finished products in a following technical system or enter into such contexts as tools themselves. In both cases the useful product is a *resource* for further systemic processes.

This roughly outlines what must be conveyed by the concept of a resource in a systemic context. As already explained above there exists a whole variety of resource concepts proposed by different TRIZ schools. Let us take a closer look at Souchkov's definition in [4, p. 51], where a resource is understood as "a means, a tool to carry out an action or to make a process take place" and equipment, money funds, raw material, energy or even people (human resources) are mentioned as examples of resources. Souchkov also sees *Resource Analysis* as an essential component of TRIZ with two goals:

- Analysis of the resources that are to be *treated or consumed* in the course of a process,
- and analysis of the resources that can be *used* to carry out the process or to solve the problem,

i.e. he distinguishes resources of the first kind, which undergo state-changing transformations as *workpieces* and resources of the second kind, which are used as tools to *mediate* these state changes.

3.3 Systemic Development and Problem Solving

While the focus of our considerations so far has been on the operating conditions of a *given* technical system, TRIZ is about problem solving and thus it is concerned with the design of viable technical systems in a *systemic development process*. For this purpose the role of Resource Analysis is defined more precisely in [4, p. 51]:

> A technical system has different resources at its disposal for the completion of its function. A function can only be completed using suitable resources. Resources are therefore elementary building blocks of a problem solution. The skilful use of resources distinguishes an efficient from an inefficient system.

The question of systemic operating conditions is thus reversed – it is not about what conditions are *required* for the operation of a particular system, but what kind of system under *given* operating conditions promises an efficient problem solution. The focus thus shifts from the operating conditions of an existing system to the question of a systemic development under given conditions. This systemic development can cover a complete genesis of a system from the scratch, when vague technical solution concepts have to be detailed and developed into a full size practical solution. In most cases, however a working technical system already exists, in which deficiencies have to be overcome, often resulting from changes in operating conditions. Such a conception of the development of a "system as is" to a "system to be" is the core of the TRIZ ontology project [18], which aims to further sharpen TRIZ conceptualisations.

In both approaches, a *sustainable* problem solution requires the *sustainable* availability of the necessary resources in the "environment" to operate the system. Hence in the next section the structure of this "environment" is detailed in which these resources are to be found.

4 The World of Technical Systems

The operational demand of a technical system is fixed in the form of *specifications* as requirements to the "environment", which must be fulfilled for the *operation* of the (assembled) system. The "reduction to the essentials" that characterises the systemic approach is, as already stated above, only a *conditional mind game* that presupposes a sufficiently powerful *environment* as given, in which the necessary resources can be allocated to fulfil the operating conditions.

However, this environment consists of similarly structured systems. Hence the *coupling* of these specifications comes into focus. Technically these specifications are transformed into *interface definitions*, and the specifications are divided into input and output specifications in order to differentiate which resources a system requires for operation and which it produces and makes available to other systems. Those interface definitions are a moment of decomposition of the unity, because it affects *two* systems that are evolving separately. In the simplest case the *agreement* on the interface definition takes place in a *supersystem* which covers both systems. Altshuller's development laws of "'energetic conductivity' of a system", of "coordination of the rhythms of the parts of a system", of "transition to a supersystem" and to a certain extent also of "transition from the macro-level to the micro-level" [1, p. 72–74] address different aspects of this problem of coordination of interfaces.

4.1 Components, Interfaces, Component Models

Sommerville [15, ch. 6.4] emphasises the importance of such interface specifications for the development of software systems that "need to interoperate with other systems that have already been developed and installed in the environment" (ibid). The same perspective is significant when large systems are to be created in a cooperative development process and for this a decomposition into subsystems is required that are to be developed independently of each other [15, ch. 10.2].

Such component-based development scenarios are of growing importance over the last 20 years and developed to an established approach in Software Engineering, even if no reusable components from third parties are used [15, p. 477]. Systemic development manifests itself as a concurrent process of parallel in time developments and unfolding of subsystems, which is controlled by a socio-technical supersystem of project coordination.

In the V-Modell XT [20], for example, a process model of software development widely used in Germany, the requirements elicitation and system specification are carried out in this supersystem in cooperation between the client and contractor. It concludes with the requirements specification as a detailed (legally binding) agreement between both sides. This part of the process is similar to part 1 of ARIZ-85C. It is followed by the definition and development of the *architecture* and the *design* of the system including the *component specifications* as a prerequisite and reference for the parallel development of the individual components. At the end of the development process, these pieces are separately tested in *component tests* and based on an appropriate *integration strategy* assembled into the overall system. The behaviour of the whole system with regard to the functional and non-functional requirements is validated in various *system tests.*

Sommerville [15, p. 477] emphasises that this development process in turn requires a more extensive socio-technical infrastructure with

1. *independent components* that can be fully configured via their interfaces,
2. *standards for components* that simplify their integration,
3. a *middleware*, which supports the component integration with software
4. and a *development process* that is designed for component-based software engineering.

Components are thus conceptually integrated into an overarching *component model*, which essentially ensures the technical interoperability of different components beyond concrete interface specifications and thus forms a moment of unity in the diversity of the components. However, this unity extends not only to the model, but also to the operating conditions of the components (as "viable" functional property provided by the middleware) as well as to their socio-technical development conditions (as a partial formalisation of the development process). This frame constitutes as *component framework* [17, ch. 9] a socio-technical supersystem as an "environment" of components that were created according to the specifications of that component model. At that supersystem level a subdivision of functional properties to be used or to be developed into *core concerns* and *cross cutting concerns* allows for further synergetic effects of a division of labour also on higher levels of abstraction, such as the *CORBA services*, which themselves have component character, but are provided by the CORBA platform as *services* (i.e. as "living components") [17, ch. 13.2].

4.2 Functional and Attributive Properties

The explanations show that systemic development processes even within a single company working on component-based foundations are interweaved in many ways and cannot be described solely on the level of lines of development of individual technical systems. Szyperski [17] shows clearly that the component approach is an approach of reuse that is not limited to the (possibly modified) abstract reuse of the technical functionality of a problem solution, but always reuses components together with their operating conditions as *services* and thus not detached from their environment.

For this, Shchedrovitsky's distinction between functional and attributive properties in the categorial relation of part and whole, as well as the distinction between the notions of *part* and *element* are essential. This cannot be elaborated here in more detail due to lack of space and is reduced to the quotation of essential points in the words of Shchedrovitsky himself.

> *Elements* are what a unity is made up of, so an element is a part inside the whole, which functions inside the unity, without as it were being torn

> out of it. A simple body, a part, is what we have when everything has been disassembled and is laid out separately. But elements only exist within the structure of *connections*. So an element implies two principally different types of properties: its properties as material, and its functional property derived from connections.
> In other words, an element is not a part. A part exists when we mechanically divide something up, so that each part exists on its own as a simple body. An element is what exists in connections within the structure of the whole and functions there. [...]
> *Functional properties* belong to an element to the extent that it belongs to the structure with connections, while other properties belong to the element itself. If I take out this piece of material, it preserves its *attributive properties*. They do not depend on whether I take it out of the system or put it into the system. But functional properties depend on whether or not there are connections. They belong to the element, but they are created by a connection; they are brought to the element by connections. [14, p. 93–94]

4.3 Functional Properties and Ideality

In the TRIZ methodology of the genesis of a system, these functional properties as "usefulness for others" are in the foreground. An engine as itself is not interesting, but only as an engine that drives a vehicle and is therefore "useful". The terms *usefulness* and *harmfulness* play an important role in TRIZ alongside the objectives of profitability and efficiency as socio-cultural guiding principles. With the concepts of *Ideality* and *Ideal Machine* [4, ch. 4.1] a mental construct of anticipation of the functional properties of a system stands at the beginning of its genesis. "The ideal machine is a solution in which the maximum utility is achieved but the machine itself does not exist." [4, p. 40]. The ideal machine is therefore pure functionality, pure "connection" in the sense of Shchedrovitsky, without any resource-related underpinning. Nonetheless, that fictitious idea, reminiscent of the fairy tale of Cockaigne, is central to TRIZ, for it develops a strong orientation towards the intended usefulness and thus has a socio-cultural guiding effect.

4.4 Place and Content

In the further system genesis, this conceptual frame of functional properties has to be filled with suitable resources [4, ch 4.2]. The systemic concept turns out to be a kind of magnifying glass, under which the combination of the functional properties, filling the "connections" with resources can be followed. To describe this composition process Shchedrovitsky distinguishes the concepts *place* and *content*.

> An *element* is a unity of a place and its content - the unity of a functional place, or a place in the structure, and what fills this place.

> A *place* is something that possesses functional properties. If we take away the content, take it out of the structure, the place will remain in the structure (assuming that the structure has a conservative and rigid nature), held there by connections. The place bears the totality of functional properties. The *content* by contrast is something that has attributive functions. Attributive functions are those that are retained by the content of a place, when this content is taken out of the given structure. We never know whether these are its properties from another system or not. Now we might take something out as content, but it is in fact tied to another system, which, as it were, extends through this place. [14, p. 94]

The search for resources as "content" is constitutive for the process of confinement in the course of the implementation of the system that is to be developed from the pure functionality of the ideal machine. This corresponds to Altshuller's first law of development of "completeness of the parts of a system": "The necessary condition for the viability of a technical system is the existence of the main parts of the system *and* their minimal functionality (i.e. viability – HGG)." [1, p. 72]

However, the thing viewed with the magnifying glass as a connection of place and content remains a "dead body", because "a living being has no parts" [14, p. 91]. Beyond the connection of place and content an operational process dimension is essential for a living system. It is not enough to insert the plug ("place") of an electrical appliance into the socket ("content") to bring the appliance "to life". The fit of plug and socket guarantee a certain minimum compliance, but to operate the device, the socket itself must be "alive" and make electrical energy available in precisely specified quality and quantity. The resource plugged in as "content" at this "place" requires an at least rudimentary system of resource "lifecycle" management.

5 Systemic Development Processes in a Modern Society

This is a typical phenomenon of a modern society, in which the electricity comes from the socket and the milk from the shop. The division of labour in such a modern mode of production leads to the emergent phenomenon of social unity and stratification of the reproduction of infrastructural conditions.

In a developed country, one can rely on electricity coming out of the socket and can use it at any time for devices that run on electrical power, provided that the technical standards such as operating voltage and power consumption are adhered to and a suitable plug-socket combination is used. The existence, reliability and robustness (resilience) of such an infrastructure has a significant influence on the way people organise their daily lives. Even in a less developed country where a continuous supply of electrical power is not guaranteed, it is still possible to use electrical devices. However, a coordination effort is required to match the availability of electrical power and the working processes in which the electrical equipment is used. Altshuller's "Law of coordinating the rhythm of the parts of a system" [1, p. 73] is thereby seemingly reversed into its opposite

– the more perfect the infrastructure, the less there is a need for coordination with that black box of power supply. Nevertheless the law is not invalidated, because the stable availability of electricity as a resource requires a sophisticated management *inside* the power supply system.

These requirements of coordination grow even more markedly in the transition from classical electricity supply systems with clearly defined base loads and unidirectional power distribution to modern systems of decentralised power generation based on "renewable energies". The cascade of trends from coordination, controllability and dynamisation [7, p. 6] is becoming increasingly effective and, with smart meter concepts, also reaches the end consumer, who is thus raised to a more comfortable level of rhythmic coordination.

These developments in the electricity supply system, however, are in turn dependent on a digital infrastructure, in which machine-readable descriptions of control information circulate. Evolutionary technological development in the web as one area of technology leads to disruptive changes in this power supply system as another area of technology. The future will show whether those reserves of control potential beyond the (present) limits of the power supply system will be used or whether the *systemic decoupling* associated with an unconditional stable power supply as *anti-trend* to increasing coordination has a socially higher value.

6 Summary

In modern component architectures, the concepts of resource and component move closer together. In a "world of technical systems", artificial artefacts combine functional and material properties that link their usefulness for a certain purpose in a structural system design of a more complex unit with the guarantee of operation, if the necessary operating conditions are provided in the "living" operating environment. However, this fundamental capability of a socially provided resource infrastructure, which is also legally fixed in the concept of the "state of the art", requires an actice reproduction. The management of scarce resources and the preemptive development of resource pools are essential forms of collective action that extend beyond the narrow horizon of individual companies.

In the term *Resource Management System*, the concepts of resource and component are equally present. However, socio-technical abstractions on a higher process level such as component model, middleware, component architecture, etc. are required to describe corresponding operating conditions of such a "system of systems".

In concrete technical domains, such conceptual worlds have long been developed and are waiting to be included and generalised in the methodological toolkit of TRIZ. "Components are for composition" [17, ch. 1.1] is a short definition by Szyperski and those *rules of composition* in turn constitute a diversity of socio-technical development processes corresponding to the diversity of component models, which provide different environments of systemic development processes of concrete components. Szyperski, for his part, analyses in [17] this diversity of compatibilities and incompatibilities of different component models and identifies different levels of abstraction for the reuse of concepts that go beyond the use of prefabricated components. In his 20-year-old book he already emphasises a diversity of conceptual notations as

> the growing importance of component deployment, and the relationship between components and services, the distinction of deployable components (or just components) from deployed components (and, where important, the latter again from installed components). Component instances are always the result of instantiating an installed component – even if installed on the fly. Services are different from components in that they require a service provider. [17, p. xvii]

In our modern "world of technical systems" the question of resources to be used in a systemic problem-solving context has to cover the condition that resources are both offered and required in a highly pre-structured form. These pre-structures rely on *standards*, are the basis for *component models*, and are supported in that "world of technical systems" by "living" *technical infrastructures.*

Trends of increasing coordination, controllability and dynamisation [7] refer not only to system-internal development lines, but also to the coordination *between* systems which are developed, offered and operated by independent third parties. The systemic development of such infrastructural frameworks, for example, of the power supply system, as supersystem has to take into account the relations of *mutual interdependency* of such independent third parties in a modern industrial mode of production and thus forces of socio-cultural self-organisation on the inter-company level of such a supersystem as target of a forthcoming TRIZ concept of a *Resource Management System.*

References

1. Altshuller, G.S.: Creativity as an Exact Science. Quoted from the Russian edition. Moscow, Sov. Radio (1979)
2. Gräbe, H.-G.: Technical Systems and Their Purposes. In: Mayer, O. (ed.) TRIZ-Anwendertag 2020, pp. 1–13. Springer, Heidelberg (2021). https://doi.org/10.1007/978-3-662-63073-0_1
3. Gräbe, H.-G.: Components as resources and cooperative action. Accepted for Publication in the Proceedings TRIZ-Anwendertag 2022
4. Koltze, K., Souchkov, V.: Systematic Innovation Methods (in German). Hanser (2017)
5. Lippert, K., Cloutier, R.: TRIZ for digital systems engineering: new characteristics and principles redefined. Systems **7**, 39 (2019)
6. Litvin, S., Petrov, V., Rubin, M., Fey, V.: TRIZ body of knowledge. MATRIZ Website (2012)
7. Lyubomirskiy, A., Litvin, S., Ikovenko, S., Thurnes, C. M., Adunka, R.: Trends of engineering system evolution (TESE). TRIZ Consulting Group (2018)
8. Mann, D.: Hands-On Systematic Innovation for Business and Management. IFR Press (2007)

9. Matvienko, N.N.: TRIZ Encyclopedia (in Russian). https://triz.org.ua/works/ws72.html
10. Petrov, V.M.: Principles of the Theory of Resource Utilization. Leningrad (1985)
11. Petrov, V.M.: A technology of resource utilization. - Theory and Practice of Teaching Engineering Creativity. Abstracts of Scientific Papers, pp. 55–56. UDNTP, Chelyabinsk (1988)
12. Petrov, V.: Laws and Patterns of Systems Development (in Russian). Independent Publishing (2020)
13. Royzen, Z.: Specific Features of Resources Utilization for Problem Solving and Improving Obtained Solutions. Kishinev (1986)
14. Shchedrovitsky, G. P.: Selected works. A guide to the methodology of organisation, leadership and management. In: Khristenko, V.B., et al. (eds.) Methodological School of Management. Bloomsbury Publishing (2014)
15. Sommerville, I.: Software Engineering. Citations Based on the 8th German edition. Pearson Studium (2007)
16. Souchkov, V.: Glossary of TRIZ and TRIZ-related terms. The International TRIZ Association, MATRIZ 2018 (2014). 1st edn. http://www.xtriz.com/publications/glossary.htm
17. Szyperski, C.: Component Software. 2nd edn. ACM Press (2002)
18. TRIZ Glossary of the TRIZ Ontology Project. https://triz-summit.ru/onto_triz/100/
19. VDI: Norm 4521, Blatt 1. Inventive Problem Solving with TRIZ. Fundamentals, Terms and Definitions, September 2021
20. Weit e.V.: V-Modell XT. Release 2.3 (in German) (2020). http://weit-verein.de/
21. Wessner. J.: Resource-oriented search. In: Mayer, O. (ed.) Proceedings TRIZ-Anwendertag 2020, pp. 93–105 and 106–113. Springer, Heidelberg (2021). https://doi.org/10.1007/978-3-662-63073-0_9
22. Zlotin, B.L., Vishnepolskaya, S.V.: Use of resources in search for new engineering solutions. Kishinev (1985)

The Use of Publicly Available Image Search Engines to Find Solution Ideas Efficient Use of TRIZ Information Resources

Jochen Wessner(✉)

TRIZ Campus, 73728 Esslingen, Germany
jochen.wessner2@de.bosch.com

Abstract. The present paper proposes a set of steps to search for data and generate relevant information to come up with solution ideas for a given problem or product improvement. Among others, the terms data and information are defined and an overview of the current state-of-the-art, which contains the TRIZ tool Function-Oriented Search and the description of possible search engines, is given. For getting relevant search results and generating information a workflow is suggested. This workflow is based on pictorial analogy to a sketched problem situation and the use of standard internet browsers with image search functionality. The task of improving the design of a tea press illustrates the workflow and its steps. The obtained solution ideas show the feasibility of the proposed workflow and lead to the conclusion that even the single use of pictorial analogy yields useful results.

Keywords: TRIZ · Pictorial analogy · Image search engines · Information resources

1 Introduction

The World Wide Web provides a lot of publicly accessible data. This data can be easily retrieved using internet search engines. Nevertheless, creating relevant information from this data to come up with solutions to a specific problem or develop specific innovation ideas could be a difficult and tedious task.

Data, information and relevancy are defined in the English Dictionary [1] in the following way:

- Data: (singular datum) a quantity, condition, fact, or other premise, given or admitted, from which other things or results may be found
- Information: notice, knowledge acquired
- Relevancy: pertinent, applicable, bearing on the matter in hand, apposite

The acquisition of knowledge after the retrieval of facts could be made more efficient if the number of retrieved facts already match a promising search or solution direction. The concept of analogy might be a suitable to path to follow and perform a directed search with internet search engines. In [2] different concepts of analogy are stated:

Published by Springer Nature Switzerland AG 2022
R. Nowak et al. (Eds.): TFC 2022, IFIP AICT 655, pp. 109–120, 2022.
https://doi.org/10.1007/978-3-031-17288-5_10

- Pictographic (same appearance within pictures, images, sketches, drawings)
- Nomologic (same rules, laws, principles)
- Functional (same functions)

Ideally, the applied analogy is not only pictographic, i.e., not only based on the same appearance within pictures, images, sketches or drawing, but show functional and nomologic analogy as well. However, it could be assumed that even pictographic analogy alone might yield better result than a completely undirected search. A TRIZ tool applying functional analogy to find solutions is Function-Oriented Search [3]. In [4] Cavalucci and Feygenson propose an algorithm to use the obtained data to create information as a resource for improving a system. Function-Oriented Search and the algorithm of Cavalucci and Feygenson focus on the generalisation of the system's main function. The term of the search is very general and the search itself makes no use of an abstract sketch which could be drawn.

The following chapter of the paper proposes a methodology to carry out a search for solutions with the help of present internet search engine using conceptual sketches or sketches representing the principle of a concept.

2 The Usage of Picture/Image Search

Sketches of solutions could be found in a variety of sources. Figure 1 shows some of these sketches with different levels of abstraction. On the left-hand side details from an engineering drawing displaying standard components, in the middle the drawing of piston together with the piston kinematic [5] and on the right-hand side a pulley [6].

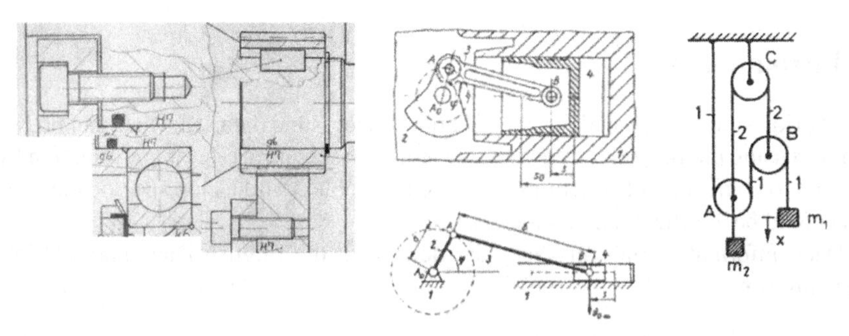

Fig. 1. Representation of design elements with different levels of abstraction – left: screws, ball bearing and gear with feather key; middle: piston [2]; right: pulley [3]

In [7] Herrig offers some more sketches which focus on the principal layout of a design (see Fig. 2). On the left-hand side, a gear is displayed and in the middle and on the right-hand side an umbrella in different states of operation. The system is reduced to rigid areas and joints. These joints define the possible dynamics within the system but do not cast any light on user interaction (if present) and on loads and their value.

In [8] Herrig also presents an image of a workflow containing eight steps to develop specific solutions for a specific problem. The steps are illustrated in Fig. 3. In the first

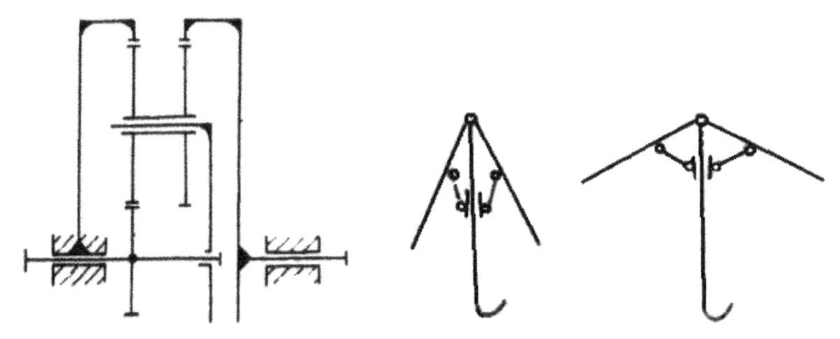

Fig. 2. Sketches of different principles of design – left: gear [7]; middle: umbrella (closed); right: umbrella (open) [7]

four steps the level of abstraction of the problem is increased until two contradicting parameters are found. Having reached this level, the next four steps deal with the solving of the problem. First, the contradiction is solved and from that a principal layout is created. This is made more specific by defining components of a structure fulfilling the needed functionality.

The general idea of the workflow is to reduce the amount of information to the necessary, crucial parts, which cause the problem. The focus is to answer the following question: What exactly is the problem? However, the TRIZ concepts of operational space and operational time, i.e., where and when the problem exists is not explicitly required or referenced. On the other hand, moving from the result state Z2 to Z3 and from result state Z6 to Z7 (German: Zustand) the application of graphical representation is clearly visible. In Fig. 4 a direct search for solutions with a feasible sketch of the problem is proposed. This problem sketch is the result of a problem analysis which could be carried out using the TRIZ tools Cause-Effect-Chains Analysis and Function Analysis to be sure that the right problem is solved.

Moving from state Z2 to Z7 by conducting an image search tries to solve the problem by pictographic analogy. The TRIZ tool Function-Oriented Search uses functional analogy to identify a leading industry in which the solution of the general problem is vital for the system. This is a great difference, and the question is, if pictographic analogy alone would be sufficient to find useful solutions.

Figure 5 shows a proposed workflow, which defines the steps to perform a picture search. After a deep analysis of the problem, the specific sketch is used to generate an abstract sketch. This is done by a human expert but could possibly be supported by feature extraction software in the future. The abstract sketch is a model of the problem containing the relevant information with reduced complexity to perform a useful search of analogue images using an existing search engine. This limitation to the relevant features increases the chance to get good results. The search results are evaluated either by the same expert or by a small team to reduce subjectivity. This is done by comparing it to the problem situation. In case there is no fit, the abstract sketch could be adapted, and the search is rerun until a satisfying fit is achieved. The manual iteration is stopped when a suitable solution is found with the help of the pictographic analogy. Then, the problem solving could move on to generating the specific solution. The step of creating

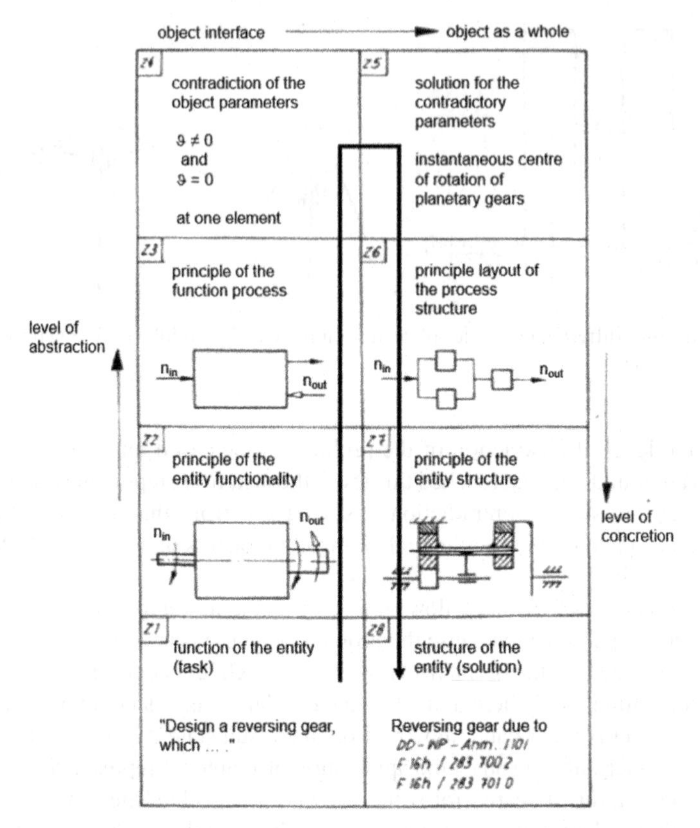

Fig. 3. Workflow from a task to a solution [8, translated by the author]

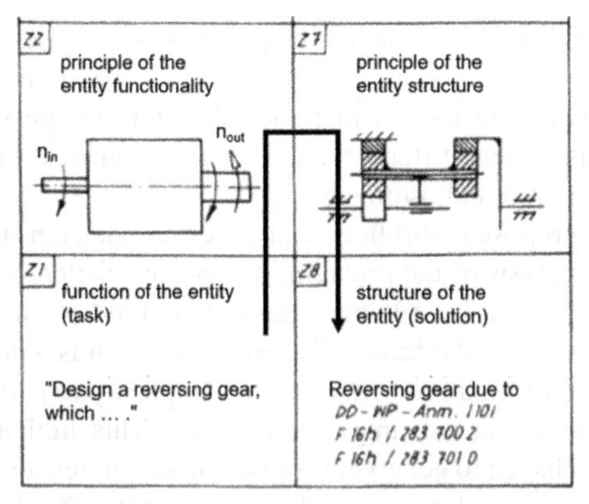

Fig. 4. Shortcut to find solutions with pictographic analogy by moving from Z2 to Z7 performing a picture search and try to get association from the search results [8, edited by the author]

the specific drawing is optional because more often it could be easier more time efficient making an abstract drawing right from the start.

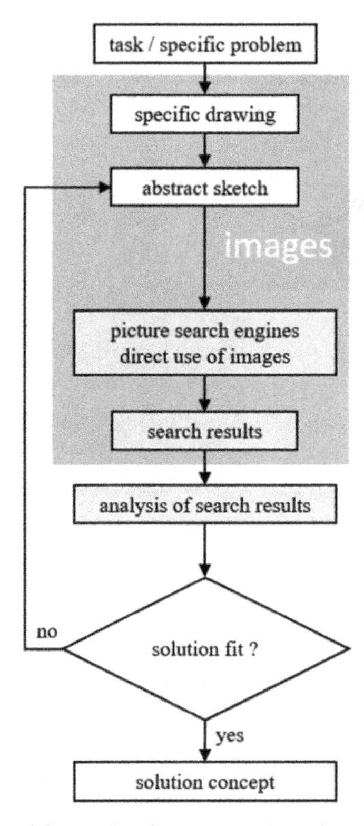

Fig. 5. Proposed workflow using image search engine to find solution ideas

The basic idea of moving from a specific situation to a general sketch of the problem is now further detailed with a case study. This study deals with improving the design of a tea press during the operating time of pressing the knob of the plunger down. This results in capturing the tealeaves between the sieve and the plunger after seeping.

Figure 6 shows the tea press with its components, the operating time (moving the knob down) and the final state which is reached after the complete movement. Figure 7 shows a function model of the addressed system, in this case the press assembly, during this operating time and Fig. 8 displays the situation with the help of first sketches of the problem situation. The problem is perceived to be the bending of the rod due to the applied force for moving. Now the first two steps of the workflow have been carried out, the problem analysis and the drawing of the problem sketch.

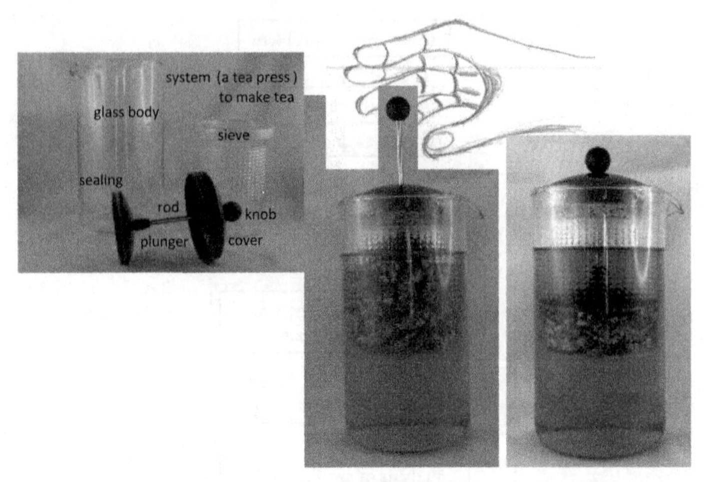

Fig. 6. A household tea press and its components (left), tea press filled with tealeaves and hot water (middle) and tea press after the knob has been moved down [9]

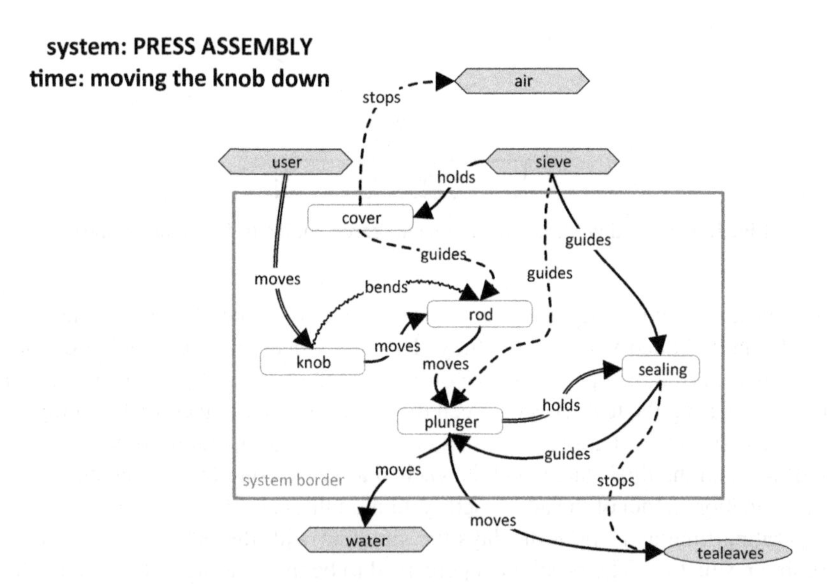

Fig. 7. Function model of the tea press [9]

The sketches still contain specific information which are not suitable for an image search. To make them more general or abstract the loads are removed and the deformed state is left out. The cover is model as a fixed mounting for the rod and all other details are removed because there are no issues there. Now only two parts are left, the knob and the rod. The sketch is shown as image number two in Fig. 9. This sketch is used to run a search with internet search engines that allow image search with a picture or a sketch. In the example the following search engines have been considered (in alphabetical order):

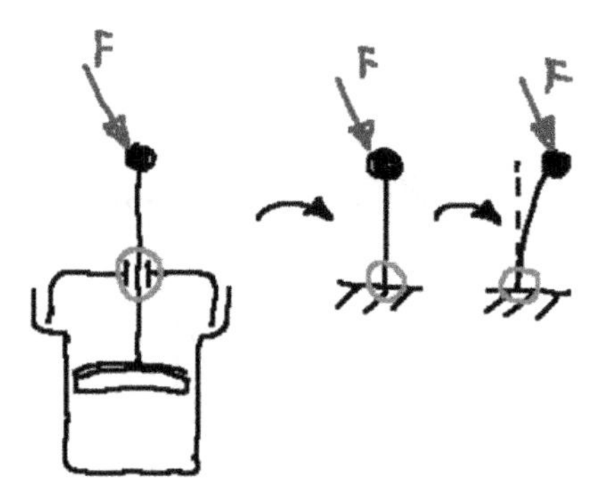

Fig. 8. Reduction of the amount of detail for the problem – sketch of the time tea and the acting loads on the rod and its support

- bing - https://www.bing.com/?scope=images&nr=1&FORM=NOFORM, last accessed 22.06.2021
- gettyimages - https://www.gettyimages.de/, last accessed 22.06.2021
- google - https://images.google.com/, last accessed 22.06.2021
- nypl (New York Public Library) - https://digitalcollections.nypl.org/, last accessed 22.06.2021
- picsearch - https://www.picsearch.com/index.cgi?q = , last accessed 22.06.2021
- shutterstock - https://www.shutterstock.com/de/, last accessed 22.06.2021
- TinEye - https://tineye.com/, last accessed 22.06.2021
- yahoo - https://images.search.yahoo.com/, last accessed 22.06.2021
- yandex - https://yandex.ru/images/?rdrnd=662869&redircnt=1625165951.1, last accessed 22.06.2021

Some of the above listed search engines only allow a search for images with the help of search terms, these could not be used.

As hinted earlier, the picture quality and level of abstraction plays an important role for getting useful hits. To get an impression of the impact of these two parameters the search was carried out with eight pictures.

These search images are shown in Fig. 9. Starting from left to right and from top to down the level of abstraction is reduced and in some case the parameter of perspective is added. The last two images represent a combination of a photo with a sketch and a section of a photo.

The number of obtained pictures were huge. The list below shows associations which were assumed useful together with the search engine that produced the result:

- Bing

  1) no direct association
  2) knot path, tree
  3) no results
  4) two legs, bone, chisel, streetlamp
  5) tower, flask, tuba, test tube

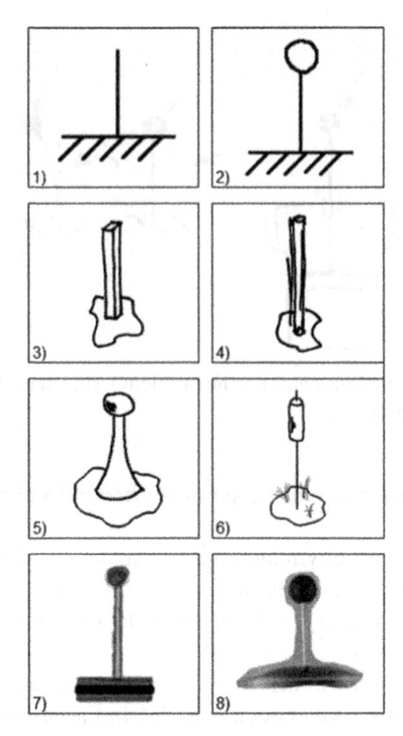

Fig. 9. Images used for the direct picture search using standard search engines

6) buoy, wire suspension of a lamp, syringe with needle
7) tool handle, support, tripod
8) office stool, handle

- Gettyimages

1) pyramid, dust bin
2) no direct association
3) no results
4) to 6) no direct association
7) tool, syringe with needle, test tube, fork,
8) no results

- Google:

1) chandelier
2) rail for sliding gate
3) test tube, lighthouse, exhaust pipe, flask, grid or drilled wires, syringe with needle, tripod, tower, tuba, nail, paperclip, stub + additional info: cylinder (with first ideas for variation)
4) tree, bamboo, cactus, wind wheel

5) flask, tower, barrel, lighthouse, moka, water tap
6) flask, test tube, syringe with needle, tree + additional info: cylinder
7) u-shaped metal part, damper, handle of floor wiper, tool handle, two or more rods in parallel, cylinder head valves, telescope supports, tripods + additional info: cylinder
8) funnel, spoon, pop-up stopper, fence top

- Nypl:
  - No direct use of images
- Picsearch:
  - No direct use of images
- Shutterstock:

  1) handle of tool, sword, syringe with needle
  2) tree, umbrella
  3) cactus
  4) flask, test tube, bone
  5) streetlamp, tower
  6) streetlamp
  7) handle
  8) bone
- Tineye
  - No direct use of images
- Yahoo
  - No direct use of images
- Yandex

  1) tool handle, rake
  2) rake
  3) bone, two or more supports in parallel
  4) hollow cylinder, bone, nested tubes, tree, two or more supports in parallel, pillar, grid, chisel, plant, cactus, rolled metal tool, needle
  5) flask, bone, tower, valve
  6) flask, tower, needle, thermometer
  7) tool handle, support, tower, valve
  8) tool handle, railway track, valve, flask

Possible associations for solving the problem are:

- Chandelier – use multiple rods in parallel (increase the number of rods while possibly decreasing the diameter of the individual rod)
- Rail for sliding gate, spoon – higher dimension (moving material to a greater diameter and introduction of hollow centre)
- Test tube, lighthouse and tower – moving the material to the outside without changing the mass of the rod (see above)
- Stub, tree – local quality (rigid close to the surface and ductile in the near the centre) (heat treatment)
- Cactus, pop-up stopper, fence top – higher dimension (star shaped cross-section)
- Damper – nested doll (rod with multiple sections, telescope)
- Funnel – higher dimension, hollow (change of outer diameter)
- Buoy – the other way round (could the plunger be pulled with a rope rather than pushed)

This list of free associations is built from the result analysis and the comparison to the problem situation. All ideas are perceived to have a fit to the solution and could be used in the next step to create a solution concept. Combinations of solution ideas and minimal changes of the system could indicate the ideality of the sketched concept.

If the concepts are not suitable the picture search could be repeated with a slightly different sketch. The level of abstraction of the sketches should decrease for the following iterations, e.g., moving from sketch 1 to sketch 3 [see Fig. 9]. This could be applied to a certain detail or the complete sketch.

3 Results

First of all, it could be said that the obtained results show a suitable and sufficient analogy to the problem situation. It seems that even a solely pictorial analogy could be used to search for solution ideas. A comparison of the search results shows the amount, and the quality of the hits depend strongly on the quality of the created sketch. It appears that 3d pictures are more likely to come across good and numerous results. The pixel size also plays an important role. Due to the size some search engines have not returned any results at all. The more artistic or organic the search image the more closely the results images are to nature. Straight lines and perpendicular angles yield results, which bear a more technical area. If photos are used there might be the danger of limiting the search due to the metal surface texture. Though in the tea press example this was not the case.

For details the reader is asked to carry out the described searches for himself. Pictures of the result hits have been omitted because the number would be quite large. Figure 10 shows the layout of the hits of a very useful search engine. This engine also provides a further and additional information in a column on the right-hand side. This information contains an associated symbolic image and first suggestions of a result variation.

The search engine also provides additional keywords. These could be taken for an image search using the keyword search instead of the images.

The searches have been carried out using the English language and search engines of English-speaking countries except for yandex. The navigation in the later search engine was perceived very difficult by the author due to the language barrier. The advantage of a picture search is that it is not limited to a certain language. Nevertheless, language specifics should be taken into account.

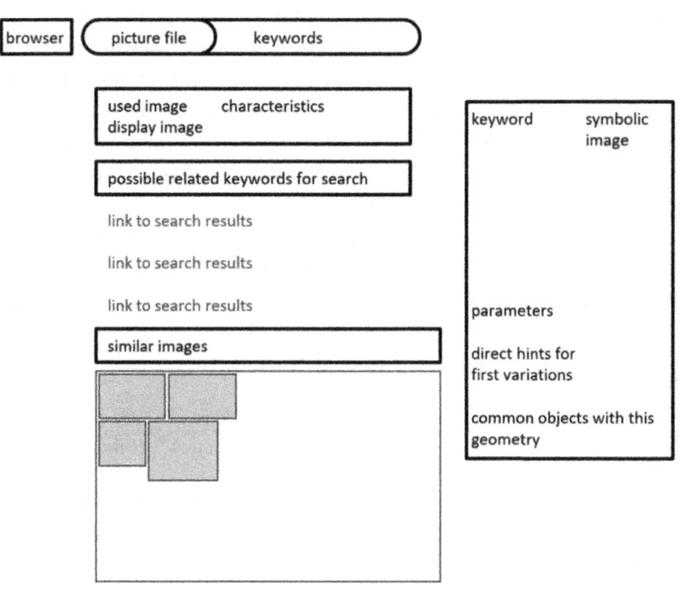

Fig. 10. Result depiction of the search engine google

The described parameter variation contains some elements of the acronym "galfmobs", which was later extended to "galfmorbus" [10]. This German acronym stands for:

- Größe – size
- Anzahl – amount/number
- Lage – position/orientation
- Form – shape/geometry
- Material – material
- Oberfläche- surface/surface finish
- Reihenfolge – order
- Bewegungsart – type of movement
- Umgebung – environment
- Schlussart – type of connection/coupling

Overall is it noticeable, that the quality of the result images gives a good impression of possible fields of industry where the sketched situation plays a vital role for function fulfilment. Usually, the image search also supplies new keywords with which a reverse search could be carried out.

The validity of the proposed workflow has to be substantiated with further case studies. Nevertheless, the case study was carried out with eight sample sketches and returned useful images to come up with associations for solution ideas.

4 Conclusions

The assumption was that the single application of pictorial analogy could be sufficient for a search to find useful solution ideas. To carry out this search in a systematic and directed way a workflow with an iteration loop is proposed. This workflow is illustrated by means of an example to facilitate the application. Due to the application of images the workflow possesses the advantage that it is not restricted to language specifics. The obtained results of the example support the assumption that solely pictorial analogy could be enough. Furthermore, no special software is needed, the functionality of present software tools and internet browsers seem to be adequate.

However, the retrieved data has to be transformed into relevant information to deploy it for a problem solution. This could be done by looking at the required amount of changes of the system. The smaller the changes the more ideal the solution seems to be.

References

1. Hayward, A., Sparkes, J.: The Concise English Dictionary, 4th edition, pp. 284, 604, 966. Orbis Verlag (1990). ISBN 3-572-01024-1
2. Hill, B.: Innovationsquelle Natur, p. 48. Shaker Verlag (1997). ISBN 3-8265-2887-5
3. Litvin, S.: Training Material Level 3, GEN TRIZ Training Material (2019)
4. Savelli, S., Feygenson, O.: Algorithm of Enhanced Function-Oriented Search, pp. 23–34. International TRIZ Association – MATRIZ (2019). ISSN 2374-2275
5. Schneider, J.: Konstruktionskataloge in der Antriebstechnik, p. 9. Hoppenstedt (1986). ISBN 3-8203-0127-5
6. Gross, D., Wriggers P., Reese S.: Aufgabensammlung Technische Mechanik III, p. 18. Institut für Mechanik (1991)
7. Herrig, D.: Programmpaket HEUREKA – Hilfsmittel zum Erfinden unter Rechner- und Karteinutzung, pp. 25, 39. Bauakademie der DDR (1988)
8. Herrig, D.: Rechnerunterstütztes Erfinden – eine Einführung, p. 23. Freies Wort Suhl (1986)
9. Wessner, J.: Time and FTA with Function Analysis, pp. 392, 395. International TRIZ Association – MATRIZ (2019). ISSN 2374-2275
10. Linde, H., Hill, B.: Erfolgreich erfinden, pp. 187–195. Hoppenstedt Technik Tabellen Verlag (1993). ISBN 3-87807-174-4.

Open Inventive Design Method (OIDM-Triz) Approach for the Modeling of Complex Systems and the Resolution of Multidisciplinary Contradictions. Application to the Exploration of Innovative Solutions to Deal with the Climate Change Impacts

Amadou Coulibaly[1](✉), Florence Rudolf[2], Murielle Ory[2], Denis Cavallucci[1], Lucas Bastian[3], and Julie Gobert[4]

[1] University of Strasbourg, Icube, Strasbourg, France
amadou.coulibaly@insa-strasbourg.fr
[2] INSA, 7309 Strasbourg, UR, France
[3] CESI Informatics School, Strasbourg, France
[4] Ecole des Ponts ParisTech, LEESU, Champs-sur-Marne, France

Abstract. This paper aims to propose an approach for Open IDM-Triz oriented analysis for complex systems.

We propose an approach to model a complex system involving multiple points of view. In this way, a complex system is modelled by a meta-graph that is built using partial knowledge graphs based on the viewpoints of the different stakeholders and users around the complex system.

This work is also a contribution for resolving non-conventional pluridisciplinary contradictions by extending the limits of classic Triz patents databases which is mainly focused on technical aspects.

Thereby, we propose an approach to extend IDM-Triz solutions concepts exploration area by using web-scraping and AI reasoning to search for similar contradictions. These approaches are implemented in a TrizAlerts software that scans various databases and websites to find out ideas matching given contradictions.

Our Open IDM-Triz approach is applied in the Interreg Clim'Ability Design project, to analyze the climate change impacts on the activities that usually manifest themselves in multiple ways.

By applying the Open IDM-Triz approach to the "Low Water" issue, we are moving away from the classical application domain of Triz because of the context extended to a multi-scalar territory corresponding to a complex system of actors impacted by climate change.

The study is carried out by a multidisciplinary team made up of researchers from engineering and social sciences backed by a network of companies involved in logistics on the Rhine (institutional, shippers, operators, etc.).

Keywords: Open IDM-Triz · Open innovation · Complex system modeling · Pluridisciplinary contradiction resolution · Climate change impacts analysis · TrizAlerts artificial intelligence

Published by Springer Nature Switzerland AG 2022
R. Nowak et al. (Eds.): TFC 2022, IFIP AICT 655, pp. 121–134, 2022.
https://doi.org/10.1007/978-3-031-17288-5_11

1 Introduction

The traditional approach of IDM TRIZ is mainly based on the use of generic design parameters and does not take into account some aspects needed in the analysis of a complex system which, in addition to the technical dimension, includes other dimensions related to the difference in perception and appreciation if we consider several groups of actors. The main motivation of our contribution is to propose an extended approach to these non-technical dimensions.

Indeed, the successful feedback of the IDM-Triz contributes to the extension of its fields of application, from the industrial products to the most complex system involving multiple stakeholders. Our article is part of this dynamic initiated by the request of social science colleagues who work on the scale of complex territories where actors are intertwined in space and time. The colleagues wished to extend their usual survey approach to digitally assisted forms. Their demand arose in the context of the 2018 low water crisis on the Rhine in an Interreg V project. It has allowed an experimentation of the method to complex systems. After presenting the classical way to perform IDM-Triz study, we propose an approach for complex system modeling by taking into account structural and functional viewpoints.

In Sect. 2 the classical way to perform IDM-Triz study is presented [2], then Sect. 3 outlines an approach for complex system modeling by taking into account structural and functional viewpoints. Section 4 presents a Social Sciences methodology for problem analysis. Our Open IDM-Triz approach is described in Sect. 5. Then in Sect. 6 we propose a method for resolution of pluridisciplinary contradiction. Our TrizAlerts software is presented in Sect. 7. And finally, Sect. 8 outlines an application of the proposed Open IDM-Triz approach to analyze the low water problematic and its impacts [11].

2 Inventive Design Method Based on Triz (IDM-Triz)

The Inventive Design Method (IDM) derives from TRIZ, the theory of inventive problem solving whose goal is to theorize on the act of inventing using the bases of patent's analysis [7].

2.1 Solving Classical Inventive Problems

The conduct of inventive design classic study proceeds from a linear approach going from the analysis of the initial situation to the generation of solution concepts through the identification of contradictions and their resolution (Fig. 1).

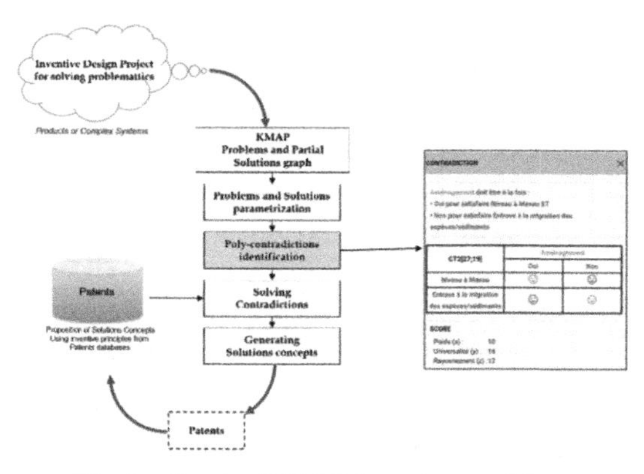

Fig. 1. Classical IDM-Triz project workflow

2.2 Semantic Limitations in Problem Graph

In the problem graph, a problem is formulated according to the syntax:

$$Subject + verb + complement \tag{1}$$

and is characterized by Evaluating Parameters.

And a partial solution is expressed in the syntax:

$$Verb + complement \tag{2}$$

This formalism is insufficient to analyze a complex system. Indeed, in this formalism, the evaluation parameters are generally defined in relation to the 39 generic design parameters whereas in a complex system the semantic aspects must be taken into account to evaluate a problem. Similarly, for a partial solution is generally formulated by deterministic action verb. Therefore, such parameters do not take into account situations that require the search for consensus with non-deterministic actions in advance. In the next section, we propose an approach for complex system modeling.

3 Complex Systems Modeling

3.1 Object-Oriented Modeling

The complexity of a system is characterized according to different points: structural, functional, static and dynamic [8]. It is also necessary to take into account the level of perception of the system, which implies the multi-scale dimension. Finally, we must take into account the viewpoints of the actors involved in the implementation of the system [3, 4]. Figure 2 gives a general perception of different aspects of a complex system.

Taking into account these aspects, a complex system can be structurally modeled in terms of conceptual classes using UML language formalism (Fig. 3).

The different activities of the stakeholders around a complex system can also be represented with a UML use case diagram (Fig. 4). This allows to identify activities and actors and to structure this knowledge.

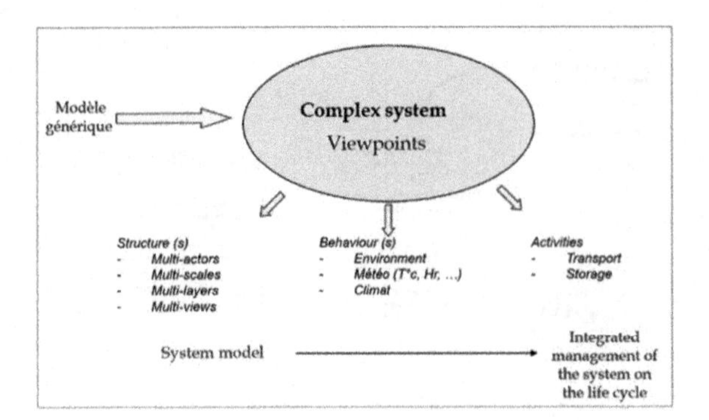

Fig. 2. A complex system modeling levels

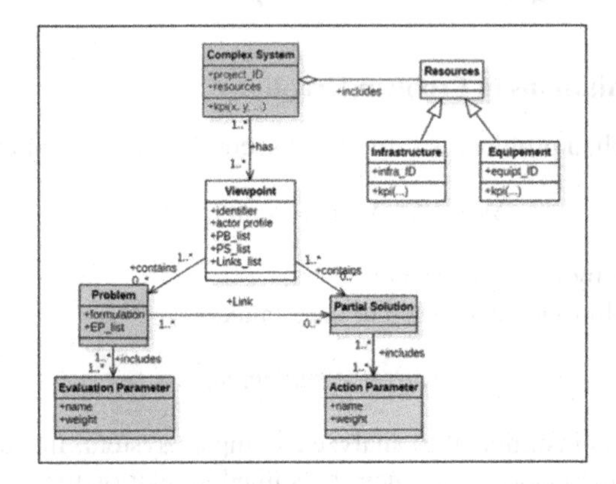

Fig. 3. A complex system UML class model

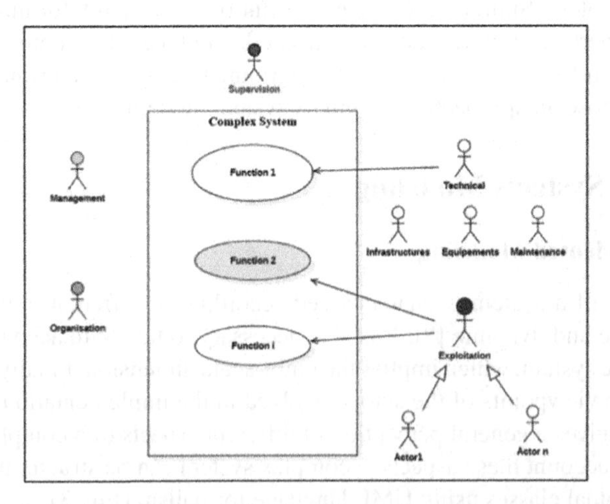

Fig. 4. Multi-scale and multi-actors' perception of a complex system

3.2 Multi-views Problems Graph

The multi-actors' perceptions are modelled as viewpoints that describe the different actors' profiles [5]. Each viewpoint has its own problems graph with specific sets of problems and sets of partial solutions. Thus, a global problems graph of a complex system is made by the superposition of these specific problem graph related to the different actors' profiles. The identification of the interactions between the different evaluation and action parameters follows from this superposition. As hypothesis, we define the following concepts used to elaborate the multi-views problems graph of a complex system:

– *Graph*: a graph is specified as:

$$VP\_Graph(PB, PS, Links) \tag{3}$$

where:

PB: is a set of problems (PB_1, PB_2, $...PB_n$);

PS: is a set of partial solutions (PS_1, PS_2, $...PS_n$);

Links : are relationships between problems *PB* and partial solutions *PS*.

VP_Graph: is a graph that describes a set of problems corresponding to a specific viewpoint;

CS_{graph}: represents the multi-views graph of the complex system (Fig. 5).

$$CS_{graph} = \sum_{vp=1}^{k} \{VP\_Graph\}$$

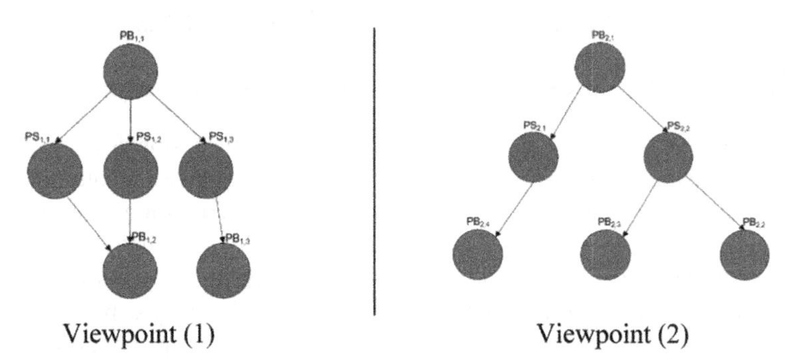

Fig. 5. Multi-views problem graph of a complex system

Each viewpoint chart can be selected separately or the charts can be combined. In the latter case, the analysis of the complex system must merge the 2 viewpoint graphs. To do this, it is necessary to make a semantic analysis of the problems and partial solutions in order to identify the identities and the differences. This analysis must be extended to the evaluation parameters (EP) of the problems and to the action parameters (AP) of the partial solutions.

The problems and partial solutions interrelationships may be represented using a semantic matrix. Such matrix allows a more comprehensive and readable representation.

Table 1 is a representation in a matrix form of the relationships between the Problems PBi,j and the Partial Solutions PSn,m identified by the different groups of project stakeholders who generally have different points of view on the analysis of the project.

Existence and absence of a relationship between a problem and a partial solution are respectively indicated in a binary way by 1 (one) and 0 (zero).

$EP_{i,j,k}$: represents an Evaluation Parameter of the problem $PB_{i,j,k}$: where i designates the group of actors, k is an evaluation parameter numberer, k represents the number of partial solutions associated with the problem $PB_{i,j}$.

Analogously, $AP_{i,j,k}$: represents an Action Parameter of the partial solution $PS_{i,j,k}$: where (i) designates the group of actors, (j) is an action parameter numberer, (k) represents the number of problems associated with the Partial Solution $PS_{i,j}$.

Table 1 Semantic matrix representation of a complex system

	$PB_{1,1}$	$PB_{1,2}$	---	---	$PB_{k,1}$	$PB_{k,2}$	
$PS_{1,1}$	1	0	1	0	0	1	$Ap_{1,1,k}$
$PS_{1,2}$	0	1	0	1	1		
…	0	0	1	0	1	0	
$PS_{n,1}$	1	1	1	0	0	1	
$PS_{n,2}$	1	1	0	1	0	0	$Ap_{n,2,n}$
	$Ep_{1,1,1}$				$Ep_{k,1,1}$		

4 Social Sciences Methodology for Problem Analysis

4.1 Individual and Semi-structured Interviews and SWOT Analysis

In-depth qualitative interviews favour the identification of the rationality into action of an actor. Each actor we met testimonies about concrete practices situated in specific contexts. Compared to quantitative survey methods based on questionnaires and to collective interviews in form of focus groups, the added value of the individual interview is to precise the respective understanding of the situation which condition their modus of action or agentivity, according to the organisational models they expect to act. Each interview is analysed through the SWOT grid which offers a compact view of the main stakes for an actor. These compact representations make it easier to extract the similitudes and differences between the actors [13].

4.2 The Gap Between SWOT and Focus Groups as Indicator for the Dialectic Between Individual and Society

Compared to the individual interview method, the IDM-Triz approach aims rather to capture common visions of a shared issue by the members of a group. The focus groups aspire to a common vison instead of many different ones. The consensus won from the focus groups can be compared with the SWOT deriving from the individual interviews to identify the gap between the individual position and the collective one [12].

5 Open IDM-Triz Approach

5.1 Parameters Semantic Formulation

The attribution of state and action parameters constitutes a key moment of inter-disciplinary and thus of the interest of crossing engineering sciences with social sciences.

In the IDM-Triz approach, the parameterisation step is very useful for placing problems on a scale of intensity or severity and specifying possible solutions to them. However, when this approach is applied to a complex system, it encounters an important coding question because of the distinction between two kinds of variables. Coding is easy when it refers to measurable and quantifiable elements in euros, metres, tonnes or days. This is not the case when it comes to determining the social acceptability of implementing a partial solution, for example, or the extent of its ecological impact on ecosystems. These questions involve the subjectivity and sensitivity of the respondent and can hardly be reduced to a binary code (high/low social acceptability; high/low impact on ecosystems). The coding of these parameters requires the search for consensus. In the application of the IDM-Triz, the step of parametrization of the problems and partial solutions privileges all the quantitative dimensions of the problem and marginalises, or even evacuates, all the elements of a qualitative nature, which are more difficult to weigh up but which are nonetheless fundamental for the identification of solutions that are acceptable to all the stakeholders.

5.2 Weighting Parameters

When it comes at least to the weighting of the solutions drawn by the stakeholders, the distinction between the variables with calibrated devices and the one without it may discriminate the latter one. Unlike physicians, social scientists are used to encounter variables which are intertwined with esthetical, ethical, political and normative values. This situation obliges them to construct some indicators without established universal equivalents. This consideration put the question of objectivation in the forefront. The objectivation can be win by inquiries which provide some statistics. The statistics may discriminate between some considerations without calibrated measuring device. These situations happen many times in multi-scalar, multi-actor and open systems. The result is a loss of objectivity that can be problematic especially when it comes into competition with variables well equipped. This distinction may minimize the social and ecological acceptance of the solutions when it comes, in the last step, to weight solutions [6].

6 Pluridisciplinary Contradictions Formulation and Resolution

6.1 Contradiction Formulation

A pluridisciplinary contradiction requires more than a simple relationship between an given Action Parameter (AP_x) and its effects on a couple of Evaluation Parameters (Epi,

EPj). To overcome this limitation, we propose a more general formulation in order to take into account simultaneously the good, harmful or non-influence effects on various evaluation parameters. These cases are marked respectively by semantic parameters Va and Va^{-} with values in $\{+1, -1, 0\}$ as shown in the following table (Table 2).

Table 2. Pluridisciplinary contradiction representation

	AP_X	
	Va	Va^{-}
EP_1	+1	0
EP_2	0	−1
-	−1	0
-	−1	0
-	0	+1
EP_{1k}	+1	0

6.2 Contradictions Resolution

A multidisciplinary contradiction translates the differences in perception of the problems and partial solutions according to the different profiles of project actors. Therefore, the same action parameter (AI) could induce contradictory effects depending on the profiles. In such situations, a consensus-building process is needed to validate the action.

7 Extending Solutions Concepts Area

7.1 Triz Patents Databases Scope and Limitations

The Triz reference patent database [10] is generally inspired by industrial patents. Also, it is not necessarily suited to the study of complex systems with non-technical dimensions such as organizational, sociological, environmental or even political aspects [18].

7.2 Web Scraping and Artificial Intelligence for Contradictions Resolution

We implement a TrizAlerts software which uses the Web scraping technique for extracting alerts from the content of websites with the aim of identifying articles that match a contradiction formulated by the user. To evaluate the correspondence between the articles and a contradiction, this software implements an algorithm based on Artificial Intelligence which evaluates the coverage rate of the inventive principles eligible to resolve the considered contradiction [1] (Fig. 6).

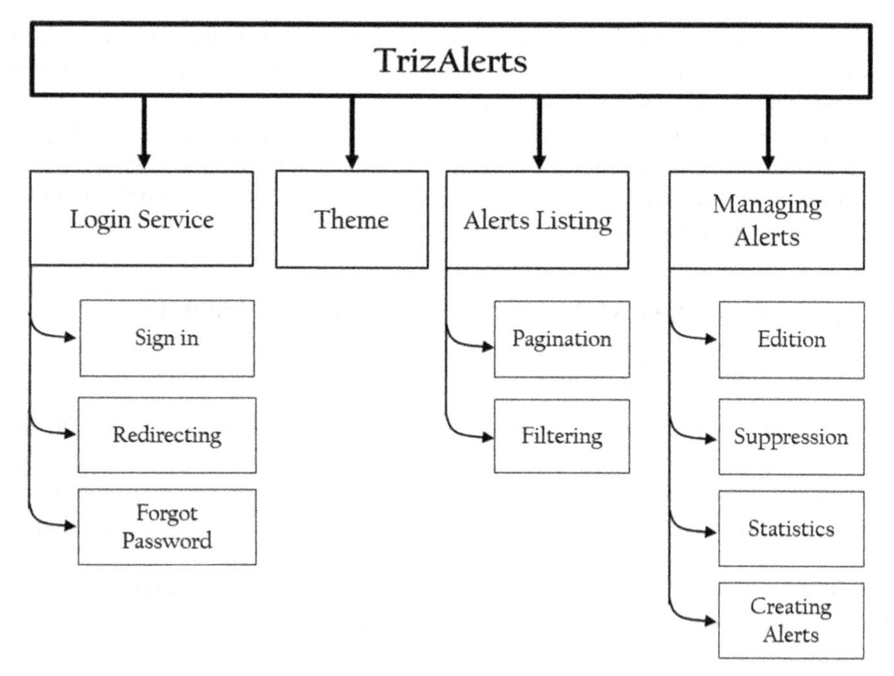

Fig. 6. TrizAlerts functional structure

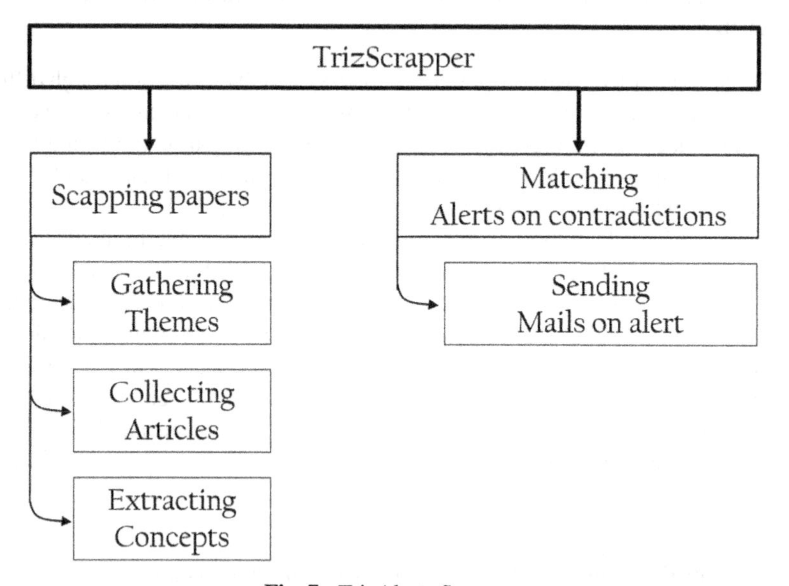

Fig. 7. TrizAlerts Scrapper

As shown in Fig. 7, the TrizAlerts is a web service dedicated to the creation of alerts based on the Triz contradictions to detect the publication of research articles on scientific sites and to inform by sending regular emails.

Based on the different CSIP's services, particularly in artificial intelligence, the performance of TrizAlerts is mainly based on the reliability of this software ecosystem. With it's friendly user interface, TrizAlerts allows an easy way to formulate multidisciplinary contradictions and adjust the emails sending frequency.

8 Application Open IDM-Triz to Low Water Problematic

8.1 CLim'Ability Design Project

In the continuity of Clim'Ability (2016–2019), the Interreg Clim'Ability Design (2019–2022) project offers companies in the Upper Rhine region devices to outline solutions to deal with the climate crisis. The project is engaged with companies willing to contribute to in-depth studies on the scale of specific branches or clusters to improve adaptation strategies [14].

In this context, we apply our Open IDM-Triz approach to study the problematic of low water and their impacts on various activities related to the navigability of the Upper Rhine.

8.2 Freight Transport on Inland Waterways and the Rhine Low Water Crisis in 2018

The low water crisis on the Rhine draws a complex system by its linear through different countries and their different national jurisdictions. These national stakes are redoubled by economic stakes depending on the importance of navigation on the Rhine for each country. Dependence on the Rhine is greater for Germany than for France, and is crucial for Switzerland, whose port is both the only sea and river port [16]. However, the severity of this crisis, during which large cargo vessels were no longer able to navigate certain sections of the river (CIPR 2018, 2019), was a wake-up call for the inland navigation sector. Transport operators and infrastructure managers have been forced to question their practices and organisational models by reintegrating them into a context of tensions and strong uncertainties linked to the future climate and its consequences on the water levels of the Rhine. Conducted as part of the Interreg Clim'Ability Design project, our study aimed at outlining, together with river transport stakeholders, sustainable adaptation paths for both the river and the economic activities associated with it and for the territories [17–19].

8.3 Process of the Study and the Multi-actors' Problem Graph

From July to October 2020, social scientists conducted a series of individual interviews with French river operators regarding the impacts of the 2018 low water crisis on their activities. They met with river and port infrastructure managers, shippers, container carriers and dry and liquid bulk carriers. The study provided a clear picture of the internal and external factors that affect the ability of companies to cope with low water. In order to structure a common vocabulary and avoid misunderstandings, the implementation of the inventive design method took place from September to March 2021.

Figure 8 gives an overview of the multi-actors' problem graph co-constructed by the different teams of actors made up of carriers (containers and bulk) and managers.

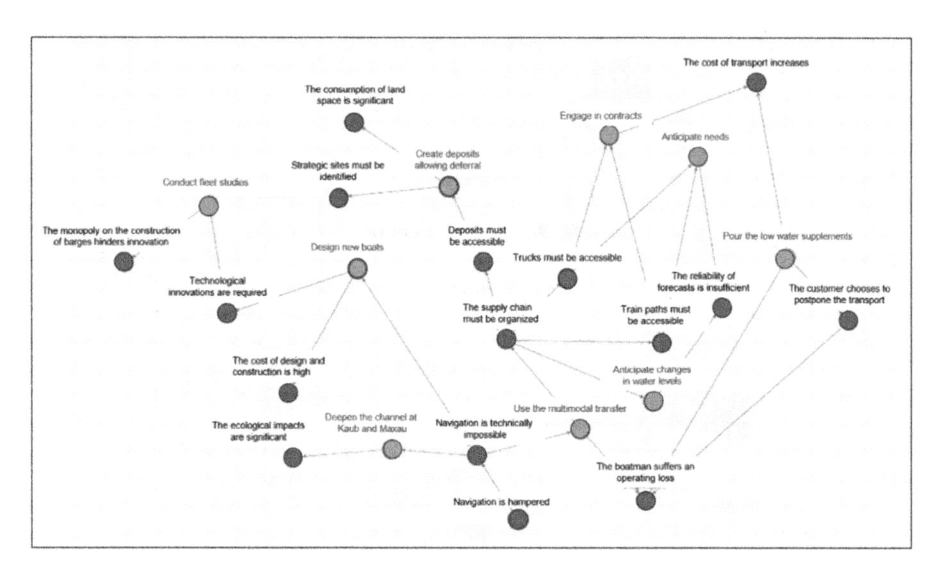

Fig. 8. Co-constructed multi-actors' problem graph

8.4 Results

To solve the contradictions identified from this problem graph, we analysed the system by combining the sociological inquiry and the inventive design method. This permitted to identify 3 types of solutions: The infrastructural one, the organisation one and the technical one. The transformative infrastructural adaptation is the kind of solution which convinced the most stakeholders involved.

The organisation one which is mainly based on the inter- and multimodality attract some attention. The technical solutions which focus on the vessel didn't win much attention because of previous progresses made in this sector.

8.5 Using TrizAlerts for Solutions Concepts Exploration

In order to extend the search domain of solution concepts, we use the TrizAlerts software to explore open databases containing scientific articles or documents dealing with contradictions similar to those we wish to resolve.

Figure 9 and Fig. 10 show the editor for contradiction formulation and submission.

Search tests in the documentary databases were carried out over 3 months and made it possible to identify 31 documents, including 6 articles dealing with issues similar to our project on low water.

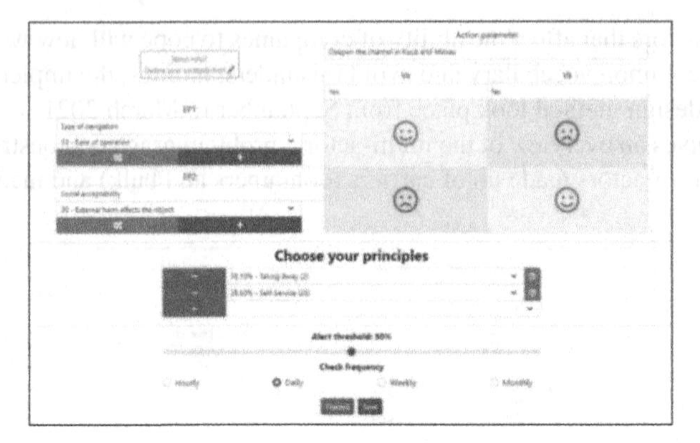

Fig. 9. TrizAlerts editor

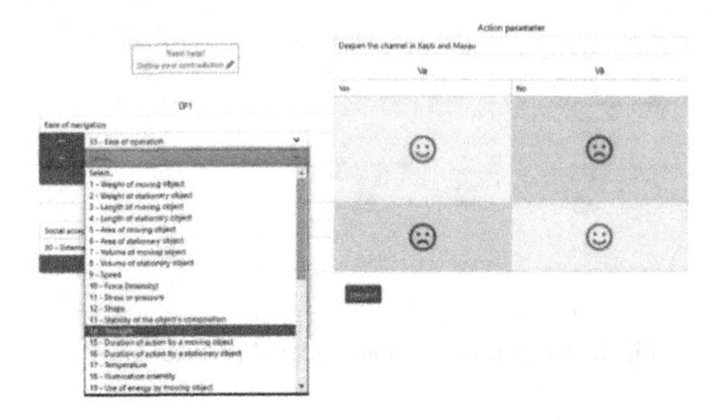

Fig. 10. TrizAlerts editor (Inventive principles)

9 Conclusions

In this article, we have proposed an approach to model a complex system in order to apply the inventive design method to it. To achieve this task, we combine different scientific cultures. The parametrization and the weighting of the solutions imagined by the stakeholders was a stimulating momentum of this interdisciplinary experience which conducted to a method for the resolution of multidisciplinary contradictions. The differences between the individual priorities and the one drawn by the focus groups was another important result. We observe that the delegation's solutions where largely elected instead of the options which necessitate to reorganize their production collective. Finally, in order to strengthen the concept of solutions, we have developed TrizAlerts, an internet search software based on the formulation of contradictions resulting from the analysis of the effects of planned actions.

References

1. Hanifi, M., et al.: Problem formulation in inventive design using Doc2vec and cosine similarity as artificial intelligence methods and scientific papers. Eng. Appl. Artifi. Intellig. **109**, 104661 (2022)
2. Chou, J.-R.: A TRIZ-based product-service design approach for developing innovative products. Computers & Industrial Engineering **161**, 107608 (2021)
3. Zhou, J.-H., et al.: Recognizing and coordinating multidimensional dynamic stakeholder value conflicts for sustainability-oriented Construction Land Reduction projects in Shanghai, China: An integrated SA-SNA-TRIZ approach. Journal of Cleaner Production **348**, 131343 (2022)
4. Tao, L., Helen, L.: Craig lawson using requirement-functional-logical-physical models to support early assembly process planning for complex aircraft systems integration. Journal of Manufacturing Systems **54**, 242–257 (2020)
5. Jordan, J., et al.: Graph-based modeling and simulation of complex systems. Computers and Chemical Engineering **125**, 134–154 (2019)
6. Muhammad Irshad Yehya, H.C., Houssin, R., Coulibaly, A., State of the art for Evaluation of Inventive Design Solution Concepts. [Online]. Available: http://www.jcm2020ct.com/en/programme/ataglance/j1/
7. Yehya, M.I., Houssin, R., Coulibaly, A., Chibane, H.: Towards evaluation of solutions concepts in inventive design. ACTA Tech. NAPOCENSIS-Series Appl. Math. Mech. Eng. **63**(3S) (2020)
8. Gero, J.S., Kannengiesser, U.: The function-behaviour-structure ontology of design. In: An anthology of theories and models of design, Springer, pp. 263–283 (2014). https://doi.org/10.1007/978-1-4471-6338-1_13
9. Sheu, D.D., Chen, C.-H., Yu, P.-Y.: Invention principles and contradiction matrix for semiconductor manufacturing industry: chemical mechanical polishing. J. Intell. Manuf. **23**(5), 1637–1648 (2012)
10. Souili, A., Cavallucci, D., Rousselot, F.: A lexico-syntactic pattern matching method to extract IDM- TRIZ knowledge from on-line patent databases. Procedia Eng. **131**, 418–425 (2015). https://doi.org/10.1016/j.proeng.2015.12.437
11. Pasimeni, F., et al.: International landscape of the inventive activity on climate change mitigation technologies. A patent analysis. Energy Strategy Reviews **36**, 100677 (2021)
12. Averbeck, P., Frör, O.: SWOT-Analyse der Klimawandel-Anpassungsbereitschaft von Unternehmen. Revue d'Allemagne et des pays de langue allemande **50**(2), 319–323 (2018)
13. Averbeck, P., Frör, O., Gartizer, N., Lützel, N., Rudolf, F.: Climate change preparedness of enterprises in the Upper Rhine region from a business perspective- A multidisciplinary, transboundary analysis, Nachhaltigkeits Management. Forum **27**, 83–93 (2019)
14. Gobert, J., Rudolf, F., Kudriavtsev, A., Averbeck, P. : L'adaptation des entreprises au changement climatique. Questionnements théoriques et opérationnels. Revue d'Allemagne et des pays de langue allemande **49**(2), 491–504 (2017)
15. Rudolf, F.: (éds.) Revue d'Allemagne et des pays de langue allemande, Dossier: Humanités environnementales – Quoi de neuf du côté des méthodes ? **50**(2) (2018)
16. Kriedel, N.: La navigation fluviale sur le Rhin au XIXe siècle – avec un regard sur le trafic rhénan actuel et son influence sur la métropolisation rhénane. In: Rudolf, F. (ed.) Revue d'Allemagne et des pays de langue allemande, T. **47**, pp. 307–320 (2 2015)
17. Commission Internationale pour la Protection du Rhin (CIPR): Inventaire des conditions et des situations d'étiage sur le Rhin, Rapport n°248 (2018)

18. Commission Internationale pour la Protection du Rhin (CIPR) : Surveillance des étiages du Rhin et de son bassin par la CIPR, Rapport n°261 (2019)
19. Rudolf, F., Grino, C.: The Nature-Society Controversy in France: Epistemological and Political Implications. In: Erasga, D. (ed.) Sological Landscape. Theories, Realities and Trends, Croatia: InTech (free online editions of InTech), pp. 41–54 (2012)

Market Impact Chain Analysis - MICA, New TRIZ Tool

Jerzy Obojski(✉)

Novismo Sp. Z O.O., Warsaw, Poland
jerzy.obojski@novismo.com

Abstract. In case when we have a key disadvantage that stops us from achieving the goal, we know what to do. But in situation when we have a key positive feature, this is not obvious how to proceed. This work is trying to start a discussion about this direction using a TRIZ toolbox. From many possibilities it looks that none of them is not a perfect choice.

In TRIZ there are many tools that use the opposite state as a source of new, creative ideas. The author just builds a hypothesis that something should exist as a mirror tool compared to CECA (Cause and Effect Chain Analysis). The base, starting point will be not a disadvantage but a unique, positive feature that will be used for creation of marketing message. All those results should pay off in the future and make a stronger position company on the market.

The present work is only a proposal on how to deal and fulfill the gap in the very narrow area – transfer features into benefit language.

Better understanding purpose decision of buying from perspective specific features and parameters of existing products or services will be another strong advantage for using TRIZ by wilder spectrum on clients.

Keywords: CECA · MICA · Feature · Benefit · Marketing · Value · Marketing · TRIZ

1 Market Background

1.1 Why do Great Products Fail Sometimes?

In the current market, where we have many products to choose from, each producer wants to know how to distinguish. The easiest way to do this is to provide a feature that the competition does not have. Many companies spend millions to find this unique feature and finally are disappointed with the result of the sale. Two examples that I will present now will be an illustration of situations where people responsible for providing unique features may struggle with it in a different way. Finally, I will try to make life easier and show how we can use new tools that will help us achieve this goal.

Developing an innovative product is only half of the success to achieve a good sales result. In the past many products have landed on the shelf due to the fact that the target market did not appreciate their potential. In Sweden, there is even a dedicated place

Published by Springer Nature Switzerland AG 2022
R. Nowak et al. (Eds.): TFC 2022, IFIP AICT 655, pp. 135–144, 2022.
https://doi.org/10.1007/978-3-031-17288-5_12

where such products are presented to the warning of other entrepreneurs. The name of this place is Museum of Failures [1]. Among the anonymous, failed products, we can find surprise cases that are not associated with a business failure at first glance such as: sticky notes, WD40, bubble wrap, Super Glue or corn flakes. So what happened is that despite the initial failures, we see these products on the market and are successful today.

The answer to this question will be my article and thesis that: we need a strong marketing message that will emphasize the key feature of the device/service and meet the basic needs of customers.

1.2 Can TRIZ Help Here?

TRIZ is a set of tools that evolve over time and must deliver the expected results with changes in markets, products and customer expectations. Integration of TRIZ with tools from other fields (e.g. planning, analysis, marketing) is a must nowadays.

Let's start with the first example. A device that will help us build a marketing message is a bottle opener. The bottle opener is a simple and handy tool that can help you open a bottle of beer or a soda. But, these days, there are so many different types of bottle openers. It can be hard to find the right one. In this article, I am going to show you how create an unique bottle opener and what kind of features can be equipped. Along with that, I will also show you how to create the suitable marketing content to convince customers to buy it!

There are many bottle openers on the market and this market segment has been very saturated for years, so how can we provide customers with something that will stand out from the competition? A bottle opener is a must-have tool for any camping trip. It's perfect for opening beer cans, wine bottles, and even soda cans. There are a lot of different types of bottle openers available, but they all have one thing in common. They're all shaped differently and they all do a different job. To make the right choice, you should first consider the type of camping trip you're going on. If you're going on a backpacking trip, you should get a small, lightweight bottle opener. If you're going on a car camping trip, you should get a larger and heavier bottle opener. You should also consider what kind of camping you're going on. If you're going on a camping trip with your family, you should get a larger and heavier bottle opener. If you're going on a camping trip with friends, you should get a smaller and light-weight bottle opener.

Let's add to this simple device a unique function that is not available in any of the products available on the market. Make it the loudest opener on the market.

We have an idea for a new product with a comma; knowing the principles of TRIZ, we can develop such a device based on the feature transfer hybridization or solving technical contradictions. This part of the work can be done relatively quickly. How-ever, uncertainty remains as to how to persuade a new customer to buy.

The client will not know what value to expect from the new product as there are no benchmarks against the known solution. If there are no such tools, it means that they need to be created. The inspiration to create such a tool was TRIZ itself.

Many of the tools that provoke us to look at a given situation from a different point of view are based on using extremes. The SIZE - TIME - COST operator, or innovative rules such as: dividing or combining, adding or subtracting, introducing or avoiding symmetry, are examples of such an approach.

Now that we know how we deal with problems, defects or unwanted functions, the question is how can we make effective use of the positive qualities we have?

The starting point was the CECA tool, which allows you to reach the deep scales hidden behind the obvious sources of problems. Then, according to the extreme approach, the question arises as to whether we can obtain something surprising by doing the opposite. Let the starting point be a positive feature, and our goal should be a deeper and deeper understanding of what we can gain. By showing what we can get from the fact that we have a unique feature in our device or service, we can more easily convince the customer to make a purchase decision. Our message is no longer just the delivery of a technical specification, but above all a language of values. This change simply looks so natural, and actually shows us what we can gain.

1.3 How to Connect a Feature with a Benefit?

Let's go back to our bottle opener. What would be unique about this product? It would be the loudest bottle opener in the world.

Let's think what the user of this device will be able to achieve after receiving this product. First, he will be able to open his beer loudly in the bar. Loud opening will attract the attention of others. The next step may be to get them interested and start a conversation after receiving the answer that the sound was the result of using the device. You will probably be asked to show how it works. Then the user can ask for a new beer bottle for demonstration, and after a few moments, become the owner of the newly opened beer bottle. After all, there will be some free beers around it. What more can the owner of this device get? We can go to the social level, i.e. making contacts, raising social position, recognition and respect, all respect or celebrity status.

Each of these deeper answers can be used to persuade a person to buy using the language of values.

It is difficult to connect worlds as distant as technical parameters and social aspects. This difficulty can be overcome thanks to successive attempts to understand what the client will get when the first, obvious need is satisfied [2].

2 The Tool

2.1 The Logic of MICA

So far we have chain of events, logically connected as a result of using bottle opener. How to connect desired situation: celebrity status with specific feature of device. This connection is not obvious, and similar to CECA there is a need to make a few iterations to see answers on deeper levels of analysis. The procedure for searching for what kind of strong, desired effect can be connected with a starting positive parameter I named MICA – Marketing Impact Chain Analysis. This process is shown in the pic. 1 (Fig. 1).

2.2 Marketing Mindset

Marketing is the activity, communication and management used to attract and keep a target audience's interest. It can be done in the context of a commercial enterprise, either

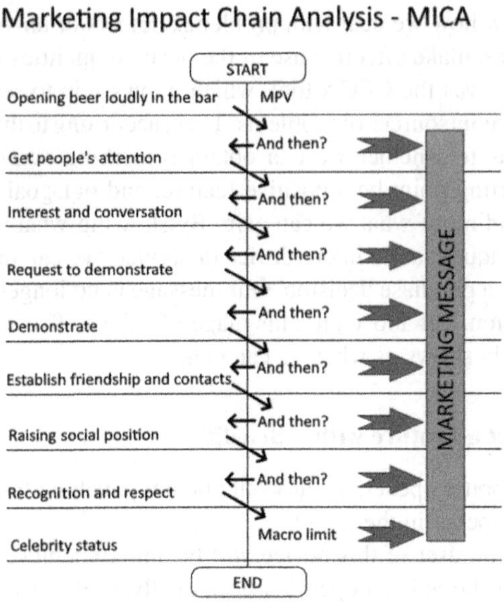

Fig. 1. Graphical illustration of a chain of sequel after acquiring a bottle opener. Further questions: "And then?" provide us with more and more new information.

to simply inform or to persuade a prospective customer or client to purchase products and services. The term developed from its original meaning, which meant the spread of gossip (as usually going from person to person) in the 1550s, took on its current meaning in the 1900s to distinguish it from other forms such as advertising [3].

Marketing is all about getting your message in front of the right people. And if you want to be successful, that message needs to be compelling. It needs to be some-thing that people want to hear, something that makes them curious and inspires them to take action. It starts with a clear purpose of identifying the audience and their needs. It also helps in coming up with a good strategy on how to address those needs. This can be achieved by conducting research on the target market and understanding their needs, wants, and values. Once the research is complete, a company can then create a marketing message that talks directly to those needs, wants, and values. Some companies even go as far as creating personas for each type of client they have so that they can tailor their messages accordingly.

Your marketing message is what sets you apart from the competition. It's what makes people want to do business with you instead of someone else. So it's important to make sure that your message is clear, concise, and memorable. If you want to get a client, you need to know how to do it. That's why there is the need to under-stand how to find a client and convince them that they need your products. What are the key elements of a compelling marketing message? When crafting a marketing message, it's important to keep in mind the key elements that make it compelling. Here are four essential tips to get you started:

- Make it personal. Your customers want to feel like they're special, so make sure your marketing messages address them directly.
- Keep it relevant. Your messages should be relevant to your customers' interests and needs.
- Make it interesting. Nobody wants to read or hear a dry, boring message. Be sure to add some personality and excitement!
- Make it easy to act on marketing message

The way to catch a client's eye in a sea of thousands competitors? You need to use marketing messages that speaks for you, sells your product and presents it as the best choice for consumer. We do not only want the recipient to click on our link, we want them to purchase the product [4].

Everybody wants to know how to catch a client in order to sell a product. It is not easy, because actually you are selling a future benefit of the "product". You are selling the things the person will have or the state he or she could be if he or she used your product.

Research says, we buy from people we like, on the principle of reciprocity, be-cause they are much better communicators and often more knowledgeable than others. In marketing slogan you will find not only promises but also specific recommendations how to catch client's attention, how to grab his preferences toward your product or service and how to convince him to buy [5].

So let's back to our story and check how can looks like the marketing message for the loudest beer bottle opener in the world. (Pic 2) (Fig. 2).

Fig. 2. An exemplary form of a marketing message for the loudest bottle cap opener on the word. Logical connections, between "BOOM" and free beers for cap opener owners there are not so obvious.

On the above example, attempts were made to highlight one feature of the product in such a way as to justify the decision to purchase it. In addition to the features of the device itself, the focus is on what the end user can gain from using it, what he can experience and how it can affect his life.

Bottle openers are an essential item for any kitchen. They are used for opening beer bottles, wine bottles, and other types of bottles. Some people don't like to use a knife to open a bottle, because they are afraid of cutting their hand. However, there are many different types of bottle openers that can be used to open a bottle without cutting your hand. One of these types of bottle openers is called a bottle opener. They come in many different shapes and sizes, but are always made of metal. The above example can be an inspiration to offering something we well known in a new form, with an interesting consumer story in the background. That way of presenting can be a good source for not conventional dialog with customer [6].

2.3 Case Study

Another story concerns a tender on the market for energy devices, and more specifically, transformers. In every tender bid, we have to convince the clients that our product is the best for their needs. Many times, it all came down to offering the lowest price, and we should do that by convincing our team that it's possible to do more with less. We've always prided ourselves on being cost-effective, but this tender required us to do even more than what we thought was possible, and everyone really stepped up. In the procedure, the distribution of points for meeting the requirements was as follows (Fig. 3).

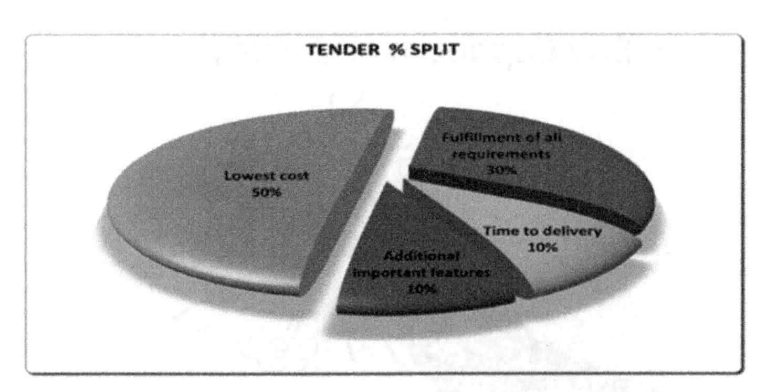

Fig. 3. The proportions of the distribution of points for meeting the requirements of each of the tender categories.

We were going through a round of tenders—a process where potential clients put out bids for what they're looking for, and interested companies submitted proposals and were competing against three other firms, with big brands and established reputations. I knew we had a great product, but I wasn't sure if we could beat those bigger companies. To avoid product placement or advertising, let's use the simple symbols: A, B, C, D to differentiate participants. As usual before the tender, it was some speculation about who was the closer to win, and who should be looser, according to previous results.

One of the smaller companies offering the lowest unit price was the clear candidate to win this tender.

One of the competitors (company B) organized an internal meeting to simulate what the situation was looking like so far. After calculating theoretical points for each category, it resulted that, based on predictions, the result of 88 was a certain victory for company D. (See Table 1).

The only open question was, what about the category: Additional important features? In that category engineers decided to find and deliver something that would bring victory. And here the MICA tool was used to get to what we already have and what could be shown as an added advantage in this particular case. The biggest problem itself was getting started. Nobody believed that something else could be found, after all, so many times have tried different methods (Table 2).

Table 1. Tender settlement simulation made by Company B.

	Criterion	A	B	C	D
50%	Lowest cost	48	45	40	50
30%	Fulfillment of requirements	30	30	30	30
10%	Time to delivery	9	9	10	8
10%	Additional important features	?	?	?	?
	TOTAL	87	84	80	88

Table 2. Final result of tender.

	Criterion	A	B	C	D
50%	Lowest cost	48	43	45	50
30%	Fulfillment of requirements	30	30	30	29
10%	Time to delivery	9	10	9	8
10%	Additional important features	1	10	5	0
	TOTAL	88	93	89	87

After submitting the bids and announcing the results, the victory went to B. What kind of additional important features were delivered, and why other companies, don't get such a high score?

The secret was creating a catching marketing message with usage of existing features of existing components.

All transformers have sand inside steel structures for noise dumping. The company B, spends time underlining other features that can be realized by using sand grains. For this case it was used the MICA approach.

The fact of using sand inside the transformer structure gives us the effect of filling the space, which in turn increases the weight of the device. By itself, there is an additional weight that stiffens the structure while dampening waves and vibrations. Thanks to this, we stop all disturbances of objects, even heavy or moving at high speed. And in this way, we can understand what we have at the product level, smoothly moving it to the point where we get a clear picture of the end-user benefit from the perspective of safety, reliability or user-friendliness.

The initial positive features of the product are clearly associated with what we get after using it. This step is obvious and seems trivial. The beauty of the MICA tool lies in repeating the question several times what will we gain when we achieve the expected results. The following hints brought us closer to understanding whether what we obtained may be an additional important feature for the client. Usually the first answers did not give satisfactory results.

On the other hand, the deeper we went into the issue, the more interesting the results were. What at first seemed to be an insignificant feature became more and more important with the passage of time.

It was also interesting to see that the team itself felt a growing excitement as the analysis developed. Ideas of how to use the received response to offer the client special value became more and more creative. The "Aha !" moment that we got to the end, it was really priceless.

As in the example with the bottle opener, the significant value for the user, which was emphasized by sand, was a surprise for the participants of the meeting. The results of this analysis are presented in the picture below (Fig. 4).

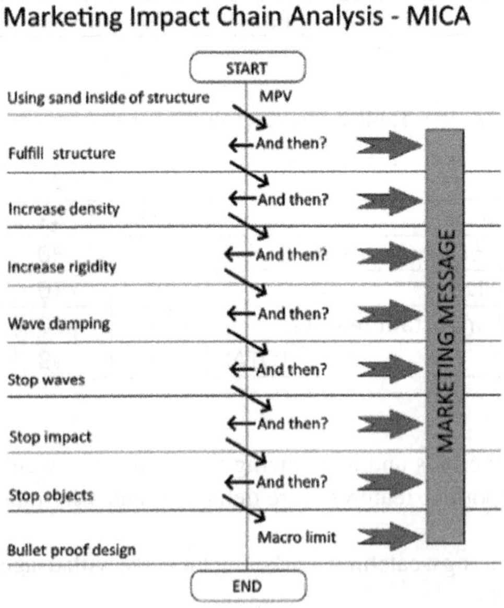

Fig. 4. Graphical illustration of a chain of sequel after using a sand inside the structure. Further questions: "And then?" provide us with more and more new possibilities

Suddenly, they find other uses for the sand. The key was to create a document that would be interesting, memorable and easy to read. We divided the text into different parts, making the design simple and without unnecessary elements. Also, to make our content more lively, we created several illustrations for it. The final marketing message was that they delivered: Bulletproof design! (Fig. 5)

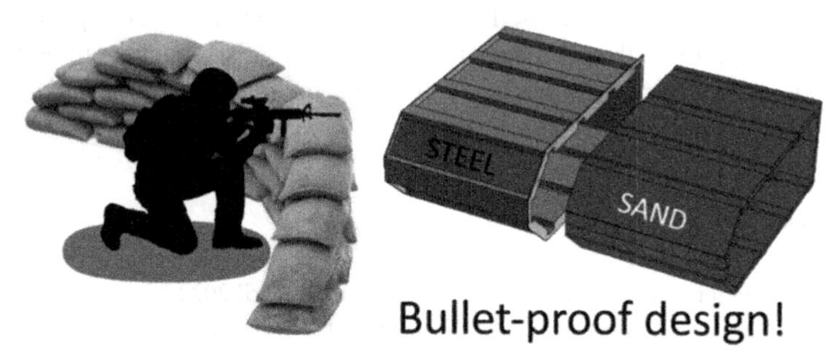

Fig. 5. A soldier, wanting to protect himself from enemy fire, builds a shooting position for himself, secured with sandbags. Similar function can offer sand inside the structure.

That unique feature brings for Company B an additional 10 points, and the final winning tender. All other competitors they have already sand inside, but the differentiator it was clear marketing message and of course certificate from bullet proof test.

The key thing I learned from this experience is that you have to translate your strengths into benefits for your clients. It's not enough just to be good at what you do. You have to show your clients why your product or service is better than all the others out there.

3 Summary

A marketing message communicates from a company to its customers or potential customers. The purpose of a marketing message is to persuade customers to take some desired action, such as buying a product or service. Marketing messages can be delivered through a variety of channels, including print advertisements, television commercials, radio spots, online ads, and email campaigns.

All of the previous points are important, but the key is the content of the marketing message. As it was presented the new TRIZ tool MICA can be a first try to make this process much more efficient, to translate features to the benefits. That way, as a result, the most catching and wanted effect of using a product or service can be used to create a strong marketing message. Promotion and marketing are critical in getting a job to the level of success that is intended. To begin with, it can serve you in connecting with your target audience and informing them about the position and its benefits. To get the desired results, marketing strategies must be specific and obvious. Marketing is the art of persuading people to do something. Purchasing a product, donating money, posting a link on social media, and signing up for an email list are all common marketing actions. This definition is shared by the word "persuasion," which means that outstanding marketers are masters of persuasion to a large extent [7].

Advertising slogans have been around for a long time. They are used to create a memorable message that will stick with the potential customer and give him an idea of what the company is. There are many benefits to having an advertising slogan. It makes it easy for people to remember your company and also can be used in marketing campaigns. Advertising slogans can also be used to show your company's value, personality or creativity. Why not, to do that in more professional way? The conclusion: nothing can replace good, relevant content. But when you have something good to say and you want others to believe it, you have to present it well.

References

1. https://collection.museumoffailure.com. Last accessed 30 April 2022
2. Obojski, J., Kaplan, L., Młynarski, M.: New TRIZ business in times of global crisis. In: TRIZCon 2020 Proceedings of virtual event
3. American Marketing Association: Definitions of Marketing, approved 2017. Accessed 24 January 2021. Author, F., Author, S., Author, T.: Book title, 2nd edn. Publisher, Location (1999)
4. Lamb, C., Hair, J., McDaniel, C.: Principles of Marketing. Cengage Learning, Boston, MA (2016). ISBN 978-1-285-86014-5
5. DeMers, J.: The Four Elements of any Action, and How to Use them in Your Online Marketing Initiative (Jun. 2013)
6. Kaplan, L.: Strategic Innovation. How to Address Unsolvable Challenges. OutCompete (2009). https://www.lulu.com/content/paperback-book/strategic-innovation/5923939?page=1&pageSize=4
7. Kaplan, L.: The Mind of The OutCompete Strategist OutCompete (2009). https://www.lulu.com/content/paperback-book/the-mind-of-the-outcompete-strategist/5865145?page=1&pageSize=4

Modular Ideality for Systematic Segmentation

Marek Mysior(✉)

Faculty of Mechanical Engineering, Department of Machine and Vehicle Design and Research, Wroclaw University of Science and Technology, ul. Łukasiewicza 7/9, 50-371 Wroclaw, Poland
marek.mysior@pwr.edu.pl

Abstract. In modern developing world, there is a growing significance of the contradiction between versatility and ease of manufacturing. Products should exist in many user-centered variants, that are manufactured with the least resources possible. The purpose of this work was to formulate a novel method of modular design and assessment to improve design of multivariant products that undergo evolution of Main Parameters of Value. The proposed method represents a systematic approach in technical system segmentation, that is one of the 40 inventive principles in TRIZ. It is based on identification of system evolution and optimized selection of modules and connections between them. As a part of the method, a novel modular ideality parameter was defined that quantitatively describes segmentation principle in the scope of increase of the useful function, increasing versatility of a product with no to little harm to ease of manufacturing. This approach makes it possible to design multi-variant products that represent a systematic and structured application of a segmentation principle. The application of the methods was shown on an example of a Mobile Biogas Station. It was shown in the study, that modular ideality has an influence on ease of manufacturing and resources usage of multivariant product design.

Keywords: Modularity · TRIZ · Segmentation · Ideality

1 Introduction

Rapid development of customers' needs makes it necessary for engineers to constantly develop new products that will satisfy them. In order to meet those evolving needs, technical systems are required to undergo numerous changes that affect its ideality. According to [1–3], development of technical systems is based on elimination of system contradictions that exist within that system. This leads to increase of a technical system ideality, that is defined as the ratio of useful to harmful functions. In the context of ever-changing supersystem, engineers struggle to deliver high-quality products at minimum costs that are capable to satisfy wide variety of customers' needs. Growing significance of evolutionary based design and design through analogy [4], developing solutions have to be compatible to previous and future generations, increasing versatility and adaptivity. Taking into consideration a traditional approach in design, decisions about technical system or technological process upgrade are rarely correlated with detailed forecasts,

Published by Springer Nature Switzerland AG 2022
R. Nowak et al. (Eds.): TFC 2022, IFIP AICT 655, pp. 145–153, 2022.
https://doi.org/10.1007/978-3-031-17288-5_13

leading to much lower than desired ideality levels (Traditional Design line on Fig. 1). A desired approach is to take into consideration evolving needs as soon as possible, tailoring system and process development to the desired outcome. Because such system or process dynamically adapts to changes originating in supersystem, such an approach is referred to as Adaptive Design for variable customers' needs (Fig. 1).

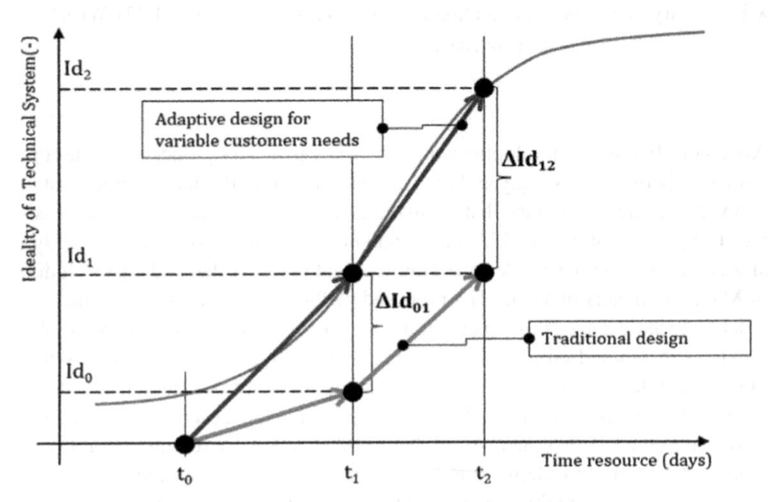

Fig. 1. Ideality in traditional and adaptive design.

Instead of having several technical systems that perform different functions, engineers tend to develop an adaptive technical system that is capable of changing its functionality in exploitation, reducing significantly manufacturing costs. Technical systems that are dedicated to satisfy dynamically changing needs can be often characterized by existence of a pair of conflicting system parameters, that form a system contradiction in the form of (Fig. 2).

Two key parameters of a technical system that describe possibility for a change to be implemented are Adaptability and Manufacturability. There are numerous works on how to measure adaptability that are directly connected with identification of MPV's and their variability [5] as well as ideality and innovativeness of a technical system at a given time [6, 7]. Apart from that, technical system deteriorates during their lifecycle, which makes it necessary monitor their critical to quality parameters to make sure it performs as intended [8]. Adaptability can be understood as a change of innovativeness in time or space, depending on if the change is evolution-based (time) or market-based (space). Manufacturability of a technical system is related to its ability to be manufactured, assembled and modified during its lifecycle. Having in mind, that customers' needs often change within a single lifecycle of a technical system, ability of a technical system to adopt to those changes is of a great importance. Using TRIZ Contradiction Matrix to solve the problem, segmentation principle is suggested.

Segmentation in its purest form is the process of division of an object into independent parts. There are many possible methods of such a division, and more efforts are dedicated towards development of novel methods on systematic segmentation. One of

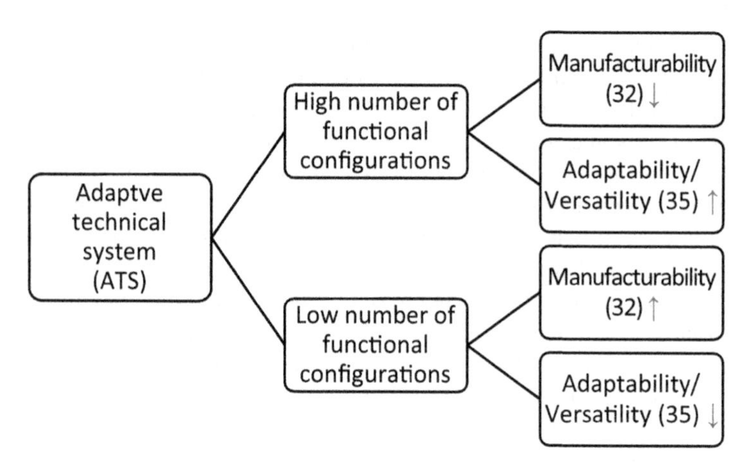

Fig. 2. System contradiction of an adaptive technical system

the possible solutions for systematic segmentation is to use size ranges, that provide adaptivity of useful function within a given range of size, mass, power, etc. [9]. This approach is effective if the change of needs is related to one single function, changing its intensity. In a different situation, where evolving needs require addition, removal or change in functional model of a technical system, a different rationalization can be implemented, that is referred to as modularity. The concept of modularity is referenced by many researchers all around the globe [10–15] mostly because it provides tools and rationalization techniques to develop highly customizable products with the highest possible manufacturability.

There are many studies on how to measure manufacturability in the context of expected changes in the structure of a technical system. This measure is often referred to as modularity, which is defined as "the relationship between a product's functional and physical structures such that (1) there is a one-to-one correspondence between the functional and physical structures, and (2) unintended interactions between modules are minimized" [16]. Modularity is often quantified by means of modules similarity or module-independence, taking into account a number of inter- and intra-module connections. Proposed in literature metrics that makes it possible to assess modularity (and hence manufacturability) are possible to be used during deployment phase of a technical system and are not possible to be implemented at an early design stage, when the only information available is related to components and functions of a technical system. This creates an urgent need to develop a novel metrics that will make it possible to assess manufacturability at an early design stage using available at this point information.

2 Materials and Methods

The purpose of this work was to formulate a novel method of modular design and assessment to improve design of multivariant products that undergo evolution of Main Parameters of Value as a part of design-for-change rationalization. Ideal technical system in the context of design-for-change is defined as a system that at any given time is fulfilling customers' needs without a need of redesign and reassembly, thus for all of its configurations is sharing the same set of components and its occurrences. In real technical systems this situation is very often impossible to be achieved which means, that there is always some adaptation work needed to change useful behavior of a technical system. This work is defined as manufacturability, which means the bigger the adaptability of the technical system, the smaller the manufacturability, which is a result of existence of custom-made modules and components that have to be incorporated in product architecture. In this article, a novel measure of manufacturability is proposed that takes into account expected changes in physical structure of a to-be-modified technical system at an early design stage. Similar to [15], proposed metric is dependent on number of non-standard components that are introduced to a technical system as a result of its adaptation to a different need. What is new in this approach is, that apart from change in component itself, a proposed metric takes into account a number of occurrences of standard and unique components, which was not distinguished in earlier works.

For a given set of configurations of an adaptive technical system (hereafter referred to as ATS), a number of changes in components and its occurrences have to be implemented between various configurations of ATS. For a given technical system, one can distinguish a set of components (C1–C7) and its occurrences that constituents it. This set is presented on Fig. 3.

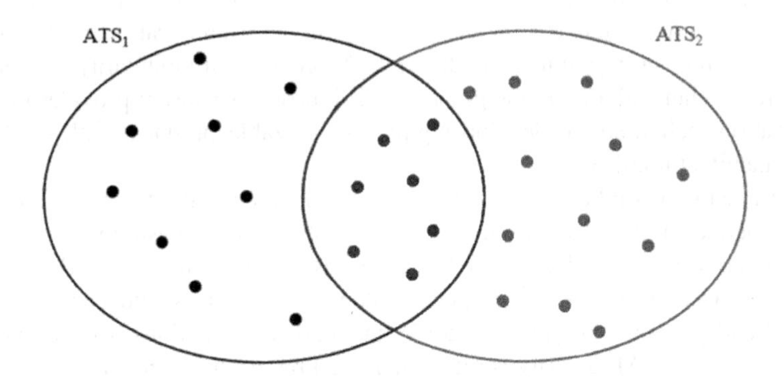

Fig. 3. Scheme of sets of components and its occurrences for two configurations of ATS.

As a result of implemented change in functionality of the ATS, a second configuration was derived (ATS_2) that uses some existing components of ATS_1, referred as common components (CC) and a set of new components, referred to as configuration-specific components (CSC). Ideally adaptive technical system is such, that uses all of its components to provide required functionality, hence for which all configurations contain only common components. Opposite to that, a non-adaptive technical system is such in which all of its configurations contain only configuration-specific-components (CSC). For the purpose of this work it was assumed, that manufacturability decreases with an increase of number of occurrences of configuration-specific components, here and thereafter referred to as n_{CSC}. Increase of manufacturability metric is related to increase of number of occurrences of common components, here and thereafter referred to as n_{CC}. In mathematical form, decrease in manufacturability can be described as:

$$\Delta M = -r\left(\frac{n_{csc}}{n_{cc}}\right)M\,\Delta\left(\frac{n_{csc}}{n_{cc}}\right) \tag{1}$$

The rate of manufacturability decrease is dependent on the value of manufacturability at given n_{CSC}/n_{CC} ratio and a proportionality coefficient r, defined as:

$$r = \frac{N_{CSC}}{N_{cc}} \tag{2}$$

In the Eq. (2), N_{CSC} denotes number of configuration-specific components in ATS and N_{CC} denotes number of common components in ATS. Integrating Eq. (1) one may obtain the equation for manufacturability metrics of a given ATS:

$$M_{MM} = e^{-\frac{1}{2}\cdot\frac{N_{CSC}}{N_{CC}}\cdot\left(\frac{n_{CSC}}{n_{CC}}\right)^2} \tag{3}$$

Change in M_{MM} metric is represented on graph:

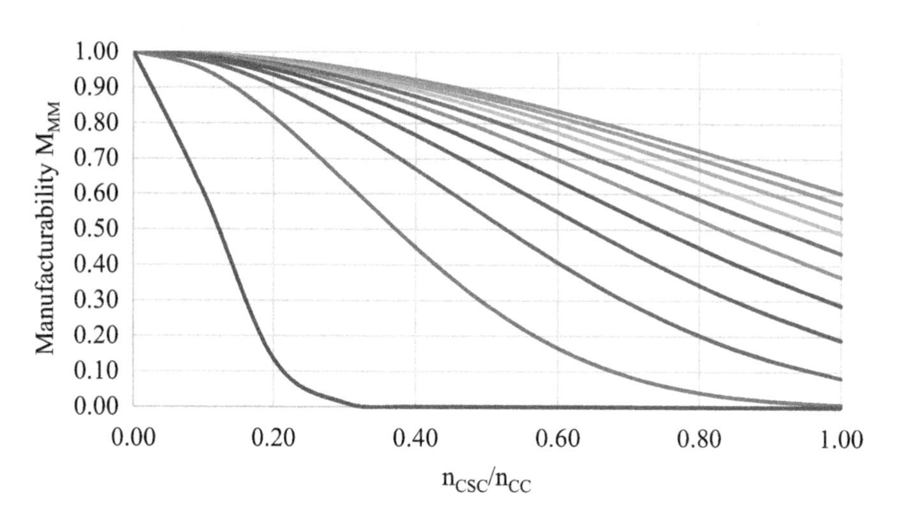

Fig. 4. Manufacturability in a function of component occurrences.

Maximum value of M_{MM} metrics is equal to one and can be achieved only if there are no occurrences of configuration-specific components in a given adaptive technical system (ATS). This represents a situation, when change in functionality is achieved without and need for redesign and/or reassembly of a technical system. Minimum value of M_{MM} tends to zero in the situation where there are only configuration-specific components in a given ATS. This is equivalent to the situation, when for obtaining a new functionality of the product, an entirely new technical system is being created. The latter scenario is the most common if changes of technical system are not anticipated at an early design phase.

When there are n configuration of an adaptive technical system (ATS_1, ATS_2, ..., ATS_i, ..., ATS_n), manufacturability measure of the entire product family consisting of all those adaptive technical systems can be described as:

$$M_{PF} = \frac{\sum_{i=1}^{n} M_{MMi}}{n} \tag{4}$$

The above equation means, that manufacturability of a product family can be equal to one if and only if each configuration uses only and all of common components, providing different functionality.

The proposed metric of manufacturability was applied to evaluate two concepts of gas storage unit in order to identify, which solution is more beneficial to be implemented based on customers' needs evolution. Detailed description of the study is presented in the next chapter.

3 Segmentation and Modular Ideality of Gas Delivery Systems

Development of gas delivery systems are of huge interest currently. One of the solutions to meet growing demand on energy obtained from environmentally-friendly sources is connected with utilization of biogas for the purpose of green energy generation [17, 18]. Biogas production is constantly growing, and so are perspectives to use other gases (i.e. Syngas) for energy generation. The evolution of gas supply system from the perspective of a source gas is being shown along with corresponding research in the field (Fig. 5). The above means, that technical systems in the area of gas delivery systems should be able to adopt to those changes, and one of the possible solutions to make this process less resource-consuming is to perform its systematic segmentation by modularization.

Fig. 5. Evolution of gas energy systems in time

Gas can be transported in dedicated gas storage units, that can store compressed gas at elevated pressure. Transitioning from one gas to another, it is often required to adopt the containers and/or piping of the system to meet conditions and behavior of compressed biogas or syngas. Additionally, due to different energy content of gas taken from various sources, there is a demand for size variability of such systems. Derived in Sect. 2 metric was used to assess manufacturability of compressed gas transport container. Based on a required gas capacity, there are two types of gas container needed that differ in size. As a result of TRIZ oriented design [19], two concepts of gas container were developed. The first concept comprises of a frame and 60 gas containers placed horizontally in the transverse direction. The second concept differs in container orientation. Both concepts are shown on Fig. 6.

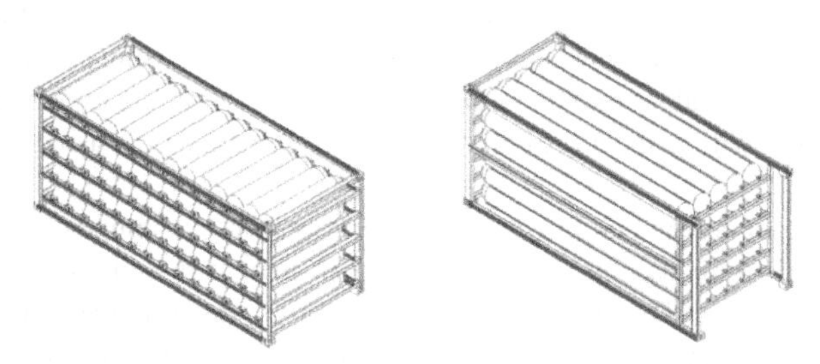

Fig. 6. Gas transport concepts, first concept on the left, second on the right.

Each of two concepts was evaluated in terms of manufacturability using proposed metrics in the scope of size change.

3.1 Concept 1

Basic configuration of gas transport container shown on Fig. 4 (concept 1) comprises of 13 components and 100 of their occurrences. Changing size of the container (understood as change in length from 20 ft to 40 ft) results in 14 occurrences of 2 configuration-specific-components, hence for ATF_{40ft}, $N_{CSC} = 2$, $n_{CSC} = 14$, $N_{CC} = 11$, $n_{CC} = 156$. Substituting those values to Eq. (2) one can obtain:

$$M_{20ft} = M_{40ft} = e^{-\frac{1}{2}\cdot\frac{2}{11}\cdot\left(\frac{14}{156}\right)^2} \cong 0,99$$

For concept 1, manufacturability M_{PF} (Eq. 4) equals to 0,499.

3.2 Concept 2

Basic configuration of gas transport container shown on Fig. 4 (concept 2) comprises of 13 components and 52 of their occurrences. Changing size of the container results 36 occurrences of 3 configuration-specific-components, hence for ATF_{40ft}, $N_{CSC} = 3$, $n_{CSC} = 36$, $N_{CC} = 10$, $n_{CC} = 16$. Substituting those values to Eq. (2) one can obtain:

$$M_{20ft} = M_{40ft} = e^{-\frac{1}{2}\cdot\frac{3}{10}\cdot\left(\frac{36}{16}\right)^2} = 0,47$$

For concept 1, manufacturability M_{PF} (Eq. 4) equals to 0,235.

4 Discussion and Summary

In this article, a novel metrics of technical system manufacturability was proposed and demonstrated on an example of gas transport container. It was shown, that manufacturability of a technical system depends on number of components and their occurrences, which can be estimated at an early design phase. This makes it possible to assess manufacturability of an adaptive technical system and all its configurations at the time, when implementation of change is of least expense. This means, that using proposed method of identification of technical system manufacturability makes it easier and more effective to increase its ideality with respect to evolving customers' needs, increasing innovativeness of the product.

The influence of product architecture and the way physical connections and components are arranged within technical system are proven to be sufficient to describe potential of the product to change. This makes it possible to choose an optimal design with respect to planned changes in early development stages, reducing costs and time-to-market.

Acknowledgements. The presented research results were carried out as part of a research task "Forecasting development of sandwich panels in modular technical systems" financed by a pro-quality subsidy for the development of the research potential of the Faculty of Mechanical Engineering of the Wroclaw University of Science and Technology in 2022.

References

1. Bukhman, I.: TRIZ Technology for Innovation. Cubic Creativity Company, Tulsa (2012)
2. Gadd, K.: TRIZ for Engineers: Enabling Inventive Problem Solving. Wiley, Hoboken (2011)
3. Altshuller, G.S.: The Innovation Algorithm: TRIZ, Systematic Innovation and Technical Creativity. Technical Innovation Center, Inc., Worcester (2007)
4. Koziołek, S.: Design by analogy: synectics and knowledge acquisition network. In: Rusiński, E., Pietrusiak, D. (eds.) RESRB 2016. LNME, pp. 259–273. Springer, Cham (2017). https://doi.org/10.1007/978-3-319-50938-9_27
5. Lok, A.: Combination of Hidden Customer Needs tools and MPV to generate product concept (2016)
6. Pryda, B., Mysior, M., Koziołek, S.: Method of innovation assessment of products and processes in the initial design phase. In: Cavallucci, D., De Guio, R., Koziołek, S. (eds.) TFC 2018. IAICT, vol. 541, pp. 75–83. Springer, Cham (2018). https://doi.org/10.1007/978-3-030-02456-7_7
7. Koziołek, S.: Inżynieria Wynalazczości. Metodologia projektowania innowacyjnych systemów technicznych. Oficyna Wydawnicza Politechniki Wrocławskiej, Wrocław (2019)
8. Koziolek, S., Rusinski, E., Jamroziak, K.: Critical to quality factors of engineering design process of armoured vehicles. In: Marcinkevicius, A.H., Valiulis, A.V. (eds.) Mechatronic Systems and Materials: Materials Production Technologies, pp. 280–284 (2010)
9. Pahl, G., Beitz, W., Feldhusen, J., Grote, K.-H.: Engineering Design, 2nd edn. Springer, London (1996). https://doi.org/10.1007/978-1-4471-3581-4
10. Gershenson, J.K., Prasad, G.J., Zhang, Y.: Product modularity: measures and design methods. J. Eng. Des. **15**, 33–51 (2004). https://doi.org/10.1080/0954482032000101731
11. Kashkoush, M., ElMaraghy, H.: Designing modular product architecture for optimal overall product modularity. J. Eng. Des. **28**, 293–316 (2017). https://doi.org/10.1080/09544828.2017.1307949

12. Sanaei, R., Otto, K.N., Hölttä-Otto, K., Wood, K.L.: An algorithmic approach to system modularization under constraints. In: Understand, Innovate, and Manage your Complex System! - Proceedings of the 19th International DSM Conference, pp. 15–24 (2017)
13. Hölttä-Otto, K., de Weck, O.: Degree of modularity in engineering systems and products with technical and business constraints. Concurr. Eng. **15**, 113–126 (2007). https://doi.org/10.1177/1063293X07078931
14. Mattson, C.A., Magleby, S.P.: The influence of product modularity during concept selection of consumer products. In: Proceedings of the ASME Design Engineering Technical Conference (2001)
15. Mikkola, J.H., Gassmann, O.: Managing modularity of product architecture: towards and integrated theory. Trans. Eng. Manag. **50**, 204–218 (2003). https://doi.org/10.1109/TEM.2003.810826
16. Ulrich, K.: The role of product architecture in the manufacturing firm. Res. Policy **24**, 419–440 (1995). https://doi.org/10.1016/0048-7333(94)00775-3
17. Ptak, M., Koziołek, S., Derlukiewicz, D., Słupiński, M., Mysior, M.: Analysis of the use of biogas as fuel for internal combustion engines. In: Rusiński, E., Pietrusiak, D. (eds.) RESRB 2016. LNME, pp. 441–450. Springer, Cham (2017). https://doi.org/10.1007/978-3-319-50938-9_46
18. Debska, A., Koziolek, S., Bieniek, J., Bialowiec, A.: The biogas production potential from Wroclaw zoological garden. Rocznik Ochrona Srodowiska **18**, 337–351 (2016)
19. Ptak, M., Koziołek, S., Derlukiewicz, D., Mysior, M., Słupiński, M.: Mobile biogas station design: the TRIZ approach. In: Koziołek, S., Chechurin, L., Collan, M. (eds.) Advances and Impacts of the Theory of Inventive Problem Solving, pp. 113–120. Springer, Cham (2018). https://doi.org/10.1007/978-3-319-96532-1_11

Hypotheses Analysis as a Development of the System Operator Method Used in TRIZ

Marek Mysior(✉) and Sebastian Koziołek

Faculty of Mechanical Engineering, Department of Machine and Vehicle Design and Research, Wroclaw University of Science and Technology, ul. Łukasiewicza 7/9, 50-371 Wroclaw, Poland
{marek.mysior,sebastian.koziolek}@pwr.edu.pl

Abstract. The purpose of this work was to formulate a novel method of problem definition and to apply this method as an extended version of the System Operator. In TRIZ, System Operator is used to describing context of the problem in time and space, which is of vital importance in many problem-solving approaches and forecasting applications. However, this known method concentrates on a single technical system and its alternatives and does not take into account other systems, that directly interact with each other. The proposed approach takes into consideration not only sub- and supersystems in the past and future but also the outcome of the problem and phenomena occurring in relation to the identified problem. The method is organized in a systematic way through a dedicated diagram and also utilizes a heuristic approach to identify and describe the context of a problem. The proposed method was applied to identify drivers and barriers to the development of vehicles and biogas distribution solutions. It was shown, that the proposed approach represents a significant and novel development of a System Operator that improves problem definition in the problem-solving approach.

Keywords: System Operator · TRIZ · Problem-solving · Inventive Engineering

1 Introduction

In modern design problems, there is an increasing importance of proper problem identification and its description, which is often regarded as the most important part of conceptual design [1, 2]. In an ever-changing world around us, the capability to cope with the evolution of technical systems and technological processes often determines the ability of companies to remain on the market and compete with each other [3]. Concentrating our efforts on solving a problem that is not significant consumes available resources of time, information, material, energy, and space, giving no to little improvement. It was shown, that lack of understanding of customers' needs is one of the key issues in modern design approaches [4].

Proper identification of the outcome of the problem as well as its causes is thus of vital importance in an effective problem-solving approach. There are already existing approaches available that concentrate on problem modeling and identification of the root

Published by Springer Nature Switzerland AG 2022
R. Nowak et al. (Eds.): TFC 2022, IFIP AICT 655, pp. 154–162, 2022.
https://doi.org/10.1007/978-3-031-17288-5_14

cause of the problem. In TRIZ, such tools as Contradiction Matrix [5], including analysis of the contradiction clouds [6], Cause-Effect Chain Analysis [7] and System Operator (also known as the 9 boxes) [8, 9] are commonly used. Those approaches are effective in defining the correspondence between different components of the problem but are not directed toward the quantitative description of the problem neither in the scope of its outcome, nor its cause.

Contradiction Matrix is an effective tool used to find potential solutions to the problem by modeling contradictions existing within a technical system or a technological process. But without a proper understanding of parameters describing the system or a process, there is a risk of solving a contradiction, that does not affect the outcome of the analyzed problem. Such a situation contributes to increased resources usage and limited usefulness of the outcome, decreasing the ideality of the design process.

In Cause-Effect Chain Analysis, a problem is analyzed in the form of the creation of a cause and effect relationship. This analysis is usually performed heuristically, without an emphasis on the design of a measuring system that can verify the potential dependence between various layers of the problem.

A more broad approach is used in the System Operator Analysis, commonly known as the 9 windows analysis, in which a problem is being analyzed in the time domain (before-at-after the moment of problem occurrence) as well as the space domain (subsystem, system, super-system level). This is a very efficient approach in technology forecasting since it combines past, present, and future descriptions of technical systems and technological processes at different contextual levels [10, 11]. Although by taking into account space and time dimensions, there is still a lack of a systematic approach in the identification of parameters that directly describe the surrounding of the problem to be solved at different levels with the possible examination of their mutual dependence.

Having roots in Inventive Engineering by combining several TRIZ concepts into a single problem modeling tool, Hypotheses Analysis makes it possible to formulate hypothetical dependencies between various parameters describing the problem directly and indirectly. Each of those dependencies can possibly indicate the quantitative influence of the source parameter on the one resulting in the outcome of the problem. This in turn allows establishing distinct limits for the parameters to be designed to solve the problem, leading to the formulation of the technical specification of a technical system or a technological process. In the next chapter, a detailed description of the process of creation of the Hypotheses Map is shown. The following chapters present examples of Hypotheses Maps for various examples and a selected example of the application of the analysis as an expansion tool for the Contradiction Matrix.

2 Materials and Methods

Hypotheses analysis is primarily used to establish a set of parameters, that directly and indirectly contribute to the outcome of the problem. This leads to the formation of the measurement system, in which the mutual influence of parameters is examined, determining causal relations. Knowing this influence, one can identify the source of the problem and define quantitatively the expected outcome of the problem-solving approach. The first step in the creation of the hypotheses map is to identify parameters that describe the outcome of the problem. If there is more than one key parameter describing the outcome of the problem, hypotheses analysis should be performed for each one of them separately. Each problem refers to the particular object, that is being directly affected by the problem. The presented approach is based on the assumption, that the problem to be solved exists due to the existence of a strong relationship between the subject and the object. The subject is a technical system or technological process that is directly connected with the problem to be solved. This stands in agreement with the principles of functional analysis used in TRIZ [12, 13]. From the point of view of the subject, there is an ambiguity in its selection, because at this stage it is unknown, which system directly influences the outcome of the analyzed problem. One can select the subject at different levels, depending on the area in which the problem-solving approach is being conducted. The subject is being analyzed from the point of view of its function (*what does it do?*), supersystem (*what is its environment?*), subsystem (*what does it have?*), and outcome of the function it performs (*what does it make?*). Let us consider the exemplary problem that is the bicycle accident. In this case, there will be a strong relationship between the Bicycle (*Subject*) and the Cyclist (*Object*), since the Cyclist is directly affected by the outcome of the problem. The selection of the bicycle as a subject creates a hypothesis, that the problem can be solved within that Technical System. The scheme of the hypotheses map at this stage is shown in Fig. 1.

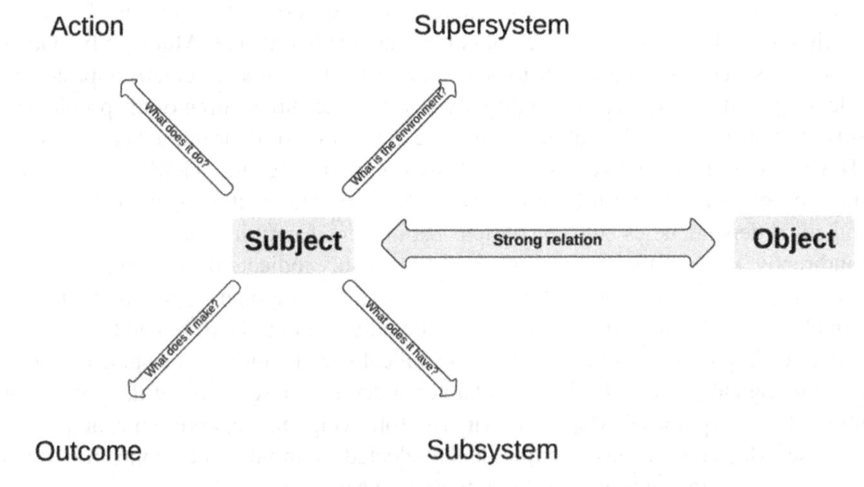

Fig. 1. Scheme of the hypotheses map, strong relation

The following process involves the definition of 6 relations in the form of physical phenomena that exist in conjunction between each of the four elements of the hypotheses map, as shown in Fig. 2. Those relations, existing between:

- Action and Supersystem,
- Supersystem and Subsystem,
- Subsystem and Outcome,
- Outcome and Action,

- Supersystem and Outcome,
- Subsystem and Action,

are regarded as indirect relations, thus not directly corresponding to the subject and outcome of the problem to be solved. Each of those relations can be described by a set of measurable parameters, that can potentially form a direct cause and effect relation with the outcome of the problem, reinforcing or weakening it.

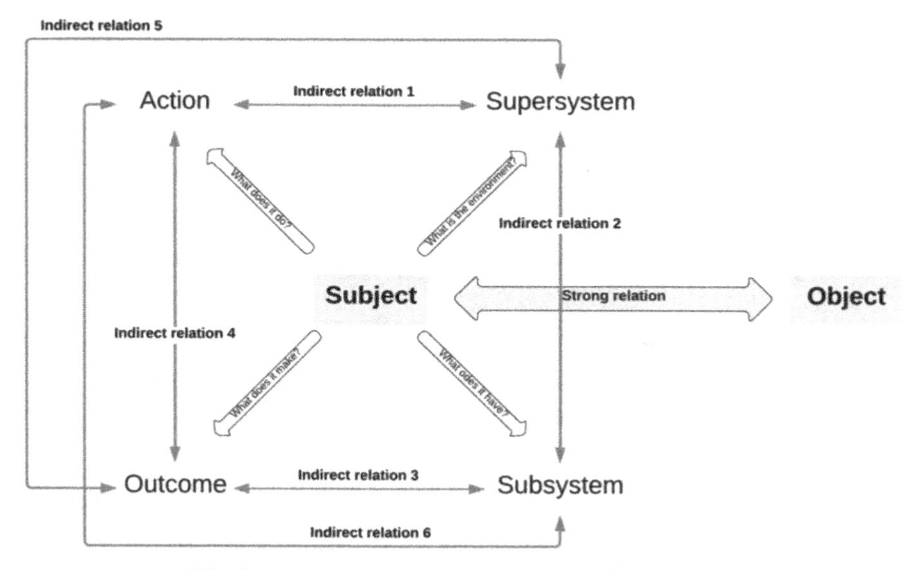

Fig. 2. Scheme of the hypotheses map including relations

At the final stage of the process, 4 direct relations are being defined between the object and the Action, Supersystem, Subsystem, and the Outcome. Those are the relations that are directly related to the outcome of the analyzed problem and can represent a set of measures describing it in a quantitative way. Usually, those relations are described in a similar way to those indirect relations, being physical phenomena that occur in conjunction with relating elements. A measure of those direct parameters contributes directly to the outcome of the problem and may represent a measure of the significance of the problem for the customer, A complete scheme of the hypotheses map is shown in Fig. 3.

Performing Hypotheses Analysis at different levels makes it possible to identify many potential solutions to the problem, forming a set of potential solutions. Each one of them is then analyzed from the point of view of ideality or innovativeness, leading to the selection of the ideal/most innovative solutions.

3 Examples of Hypotheses Maps

In this chapter, several exemplary diagrams are presented to demonstrate the idea behind the formation of the Hypotheses Map. The first problem to be analyzed is connected with

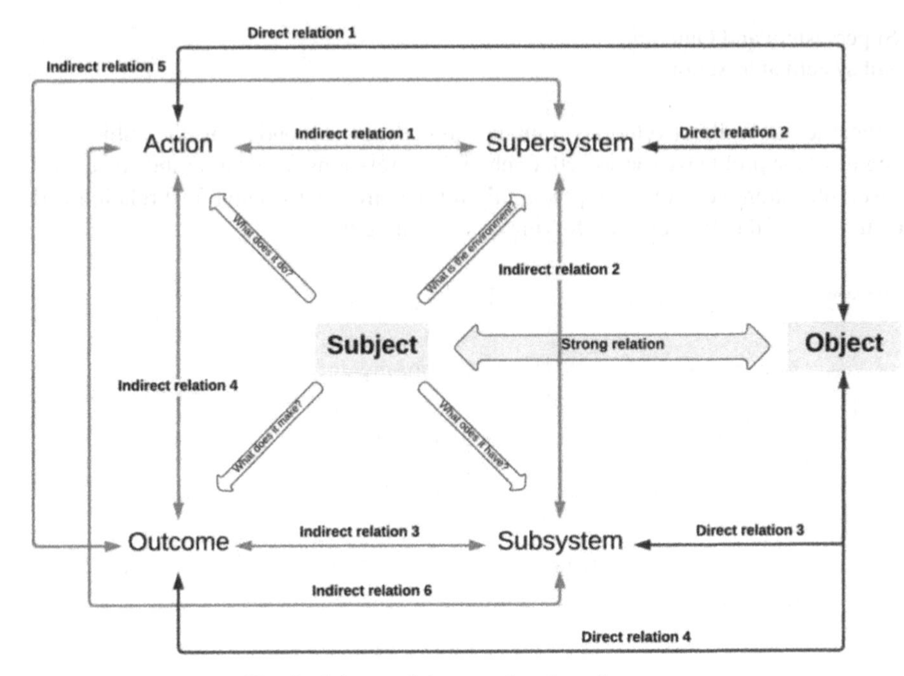

Fig. 3. Scheme of the complete hypotheses map

preventing of accidents occurring on motorbikes. The outcome of the problem, which can be measured by the number of accidents every year, is directly affecting motorcyclists. The area, in which this problem can be solved can be oriented towards the motorcycle or the protective equipment he/she is wearing. This forms two relations:

1. Motorcycle – Motorcyclist
2. Protective equipment – Motorcycle

On Fig. 4, a hypotheses map of the first relation is shown, and on Fig. 5 of the latter one.

4 Applications of the Method in Problem Solving

Development of gas delivery systems are of huge interest currently. One of the solutions to meet growing demand on energy obtained from green, environmentally friendly sources is located in Biogas. This opportunity is of growing concern worldwide [9, 14–16]. Applying a Hypotheses Analysis to this problem results in the following map, as shown in Fig. 6. The strong relation is defined between the biogas (as a potential source of energy) and the user that will consume it.

For each relation shown on the map, a set of parameters can be defined that directly (shown in blue) or indirectly contributes to the outcome of the problem. Examining those parameters can uncover contradictions existing within the analyzed Technical System or

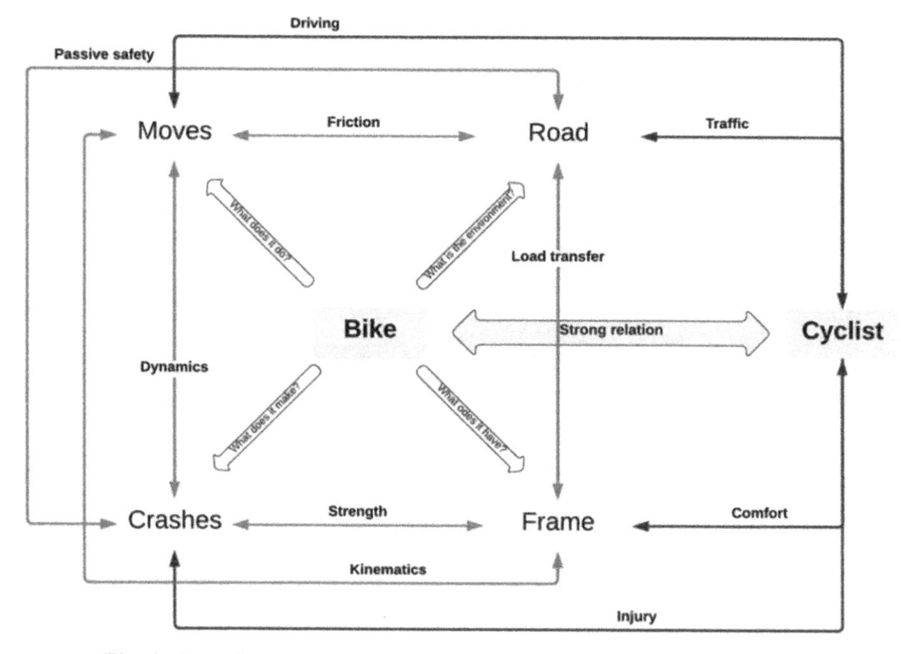

Fig. 4. Hypotheses map of a bicycle crash problem, point of view of the bike.

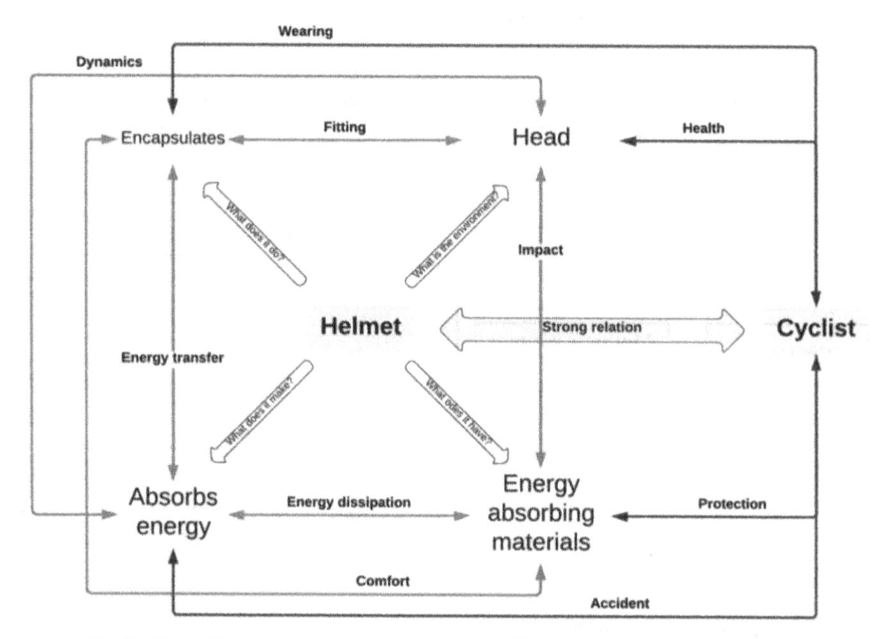

Fig. 5. Hypotheses map of a bicycle crash problem, point of view of the helmet.

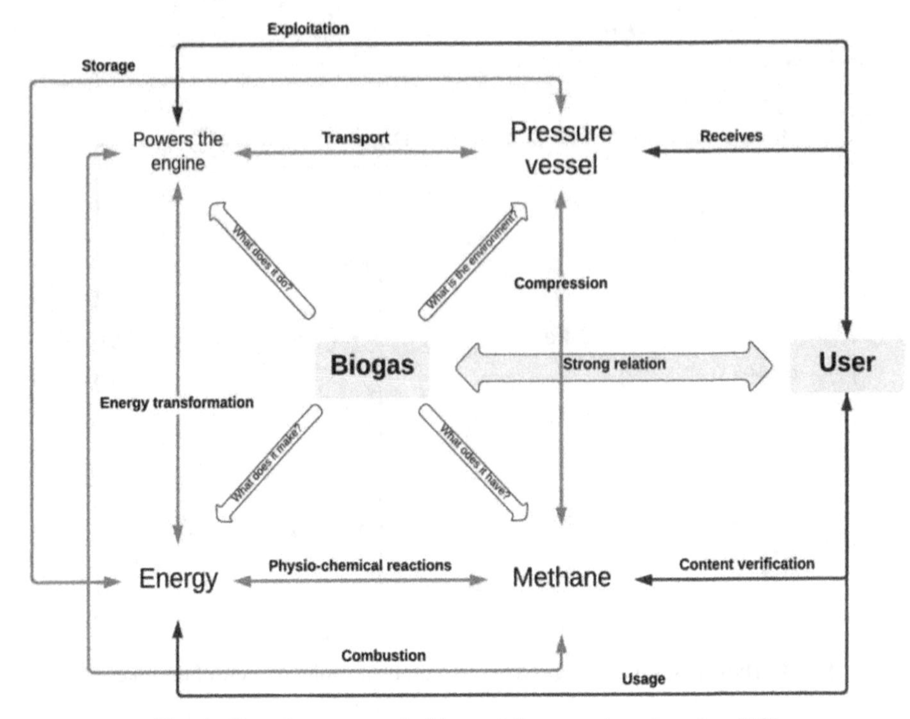

Fig. 6. Hypotheses map of a biogas delivery system, based on [17].

Technological Process. An example of such contradiction for "combustion" that exists in conjunction with Methane and powering of an engine is described in Fig. 7.

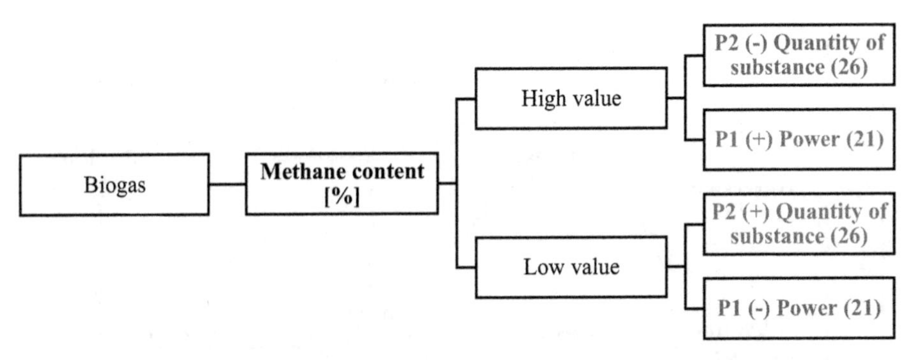

Fig. 7. Contradiction model based on the parameter describing relation of "combustion"

From this point, there is a possibility to use Contradiction Matrix and 40 Inventive Principles to find potential solutions to the problem, in this case:

- (4) assymetry;
- () discarding and recovering;

- () periodic action.

5 Discussion and Summary

The proposed approach represents a significant development in the problem-solving approach and expands the application of the System Operator known in TRIZ to be utilized as a tool for measuring system design. Working with conceptual design is difficult due to very limited access to measurable that are of vital importance in innovativeness assessment [18]. Hypotheses analysis makes it possible to identify a broad set of parameters related to many different Technical Systems or Technological Processes associated with the analyzed problem. The goal of this approach is to identify those parameters, that affect the outcome of the problem the most and concentrate the problem-solving approach on alternating the value of the said parameter, improving it. Additionally, identification of parameters influencing the outcome of the problem, it is very common to define technical and physical contradictions, which makes it possible to use other TRIZ tools like Contradiction Matrix and 40 Principles to seek the solution. As compared to the System Operator, Hypotheses Analysis concentrates not only on the context of the Technical System or Technological Process from the point of view of scale and time but also manifests causal relations within the surrounding of the problem. By applying this approach to the problem of green energy generation from methane, it was possible to identify the set of solutions based on solving contradictions between the Volume of the stationary object (8) and Strength (14). The quantitative feature of this tool is of great importance in the formulation of contradictions.

Acknowledgements. The presented research results were carried out as part of a research task "Forecasting development of sandwich panels in modular technical systems" financed by a pro-quality subsidy for the development of the research potential of the Faculty of Mechanical Engineering of the Wroclaw University of Science and Technology in 2022.

References

1. Koziołek, S.: Inżynieria Wynalazczości. Metodologia projektowania innowacyjnych systemów technicznych. Oficyna Wydawnicza Politechniki Wrocławskiej, Wrocław (2019)
2. Koziołek, S.: Design by analogy: synectics and knowledge acquisition network. In: Rusiński, E., Pietrusiak, D. (eds.) RESRB 2016. LNME, pp. 259–273. Springer, Cham (2017). https://doi.org/10.1007/978-3-319-50938-9_27
3. Chybowska, D., Chybowski, L., Souchkov, V.: R&D in Poland: is the country close to a knowledge-driven economy? Manag. Syst. Prod. Eng. **26**, 99–105 (2018). https://doi.org/10.1515/mspe-2018-0016
4. Livotov, P.: Estimation of new-product success by company's internal experts in the early phases of innovation process. Proc. CIRP **39**, 150–155 (2016)
5. Gadd, K.: TRIZ for Engineers: Enabling Inventive Problem Solving. Wiley, Hoboken (2011)
6. Cavallucci, D., Rousselot, F., Zanni, C.: On contradiction clouds. Proc. Eng. **9**, 368–378 (2011). https://doi.org/10.1016/j.proeng.2011.03.126
7. Dobrusskin, C.: On the identification of contradictions using cause effect chain analysis. Proc. CIRP **39**, 221–224 (2016). https://doi.org/10.1016/J.PROCIR.2016.01.192

8. Altshuller, G.S.: The Innovation Algorithm: TRIZ, Systematic Innovation and Technical Creativity. Technical Innovation Center, Inc. (2007)
9. Ptak, M., Koziołek, S., Derlukiewicz, D., Mysior, M., Słupiński, M.: Mobile biogas station design: the TRIZ approach. In: Koziołek, S., Chechurin, L., Collan, M. (eds.) Advances and Impacts of the Theory of Inventive Problem Solving, pp. 113–120. Springer, Cham (2018). https://doi.org/10.1007/978-3-319-96532-1_11
10. Koziołek, S., Mysior, M., Pryda, B., et al.: Forecasting of product and technology development using heuristic-systematic approach. J. Eur. TRIZ Assoc. **04**, (2017). https://etria.eu/portal/index.php/innovator-etria-official-journal/120-etria-journal-02-201704
11. Cascini, G., Becattini, N., Kaikov, I., et al.: FORMAT – building an original methodology for technology forecasting through researchers exchanges between industry and academia. Proc. Eng. **131**, 1084–1093 (2015). https://doi.org/10.1016/J.PROENG.2015.12.426
12. Litvin, S., Feygenson, N., Feygenson, O.: Advanced function approach. Proc. Eng. **9**, 92–102 (2011). https://doi.org/10.1016/J.PROENG.2011.03.103
13. Makino, K., Sawaguchi, M., Miyata, N.: Research on functional analysis useful for utilizing TRIZ. Proc. Eng. **131**, 1021–1030 (2015). https://doi.org/10.1016/J.PROENG.2015.12.420
14. Scarlat, N., Dallemand, J.-F., Fahl, F.: Biogas: developments and perspectives in Europe. Renew. Energy **129**, 457–472 (2018). https://doi.org/10.1016/j.renene.2018.03.006
15. Mysior, M., Stępień, P., Koziołek, S.: Modeling and experimental validation of compression and storage of raw biogas. Processes **8**, 1556 (2020). https://doi.org/10.3390/pr8121556
16. Jerzak, W., Sikora, J., Łyko, P., Kuźnia, M.: Analysis of the combustion products of biogas produced from organic municipal waste. J. Power Technol. **95**, 158–165 (2015)
17. Koziołek, S., Białowiec, A., Mysior, M., et al.: Rozproszone systemy dystrybucji biogazu: badania, projektowanie i rozwój. Oficyna Wydawnicza Politechniki Wrocławskiej, Wrocław (2017)
18. Pryda, B., Mysior, M., Koziołek, S.: Method of innovation assessment of products and processes in the initial design phase. In: Cavallucci, D., De Guio, R., Koziołek, S. (eds.) TFC 2018. IAICT, vol. 541, pp. 75–83. Springer, Cham (2018). https://doi.org/10.1007/978-3-030-02456-7_7

AI in Systematic Innovation

An Interactive Artificial Intelligence System for Inventive Problem-Solving

Stelian Brad[1](✉) and Emil Ștetco[2]

[1] Technical University of Cluj-Napoca, Memorandumului 28, 400445 Cluj-Napoca, Romania
stelian.brad@staff.utcluj.ro
[2] Zetta Cloud, Govora 16A, 400535 Cluj-Napoca, Romania
emil.stetco@zettacloud.ro

Abstract. There is a vast space of potentiality for inspiration in the design and engineering of technical systems that are poorly valorized; the cyberspace that stores and daily adds high volumes of global collective intelligence. This space could be more productively tackled with the assistance of Artificial Intelligence algorithms led by Natural Language Processing (NLP) models. We investigate the application of Structured Activation Vertex Entropy (SAVE) method in combination with Question Answering Machine (QAM) algorithms to explore information that is stored in big datasets, accessible within unstructured dataspaces. The SAVE method is transformed with the assistance of TRIZ into a set of searching meta-terms or meta-concepts. Taking off from a clear description of the problem, target results, and the current (eco)system, meta-terms, and concepts are incorporated into a spiral searching-answering process called 'D-SIT-SIT-C', driven by a Retrieval Augmented Generation (RAG) model to create an "intelligent" Natural Language Processing pipeline, with inserting the human in the loop at each iteration. We have found that the proposed pipeline based on a RAG model brings new valences to the creative thinking process and unleashes new dimensions of investigations that lead to higher quality solutions than those formulated with limited resources.

Keywords: Artificial Intelligence (AI) · Natural Language Processing (NLP) · Question Answering Machine (QAM) · Deep Learning (DL) · Theory of Inventive Problem Solving (TRIZ) · Structured Activation of Vertex Entropy (SAVE) · Retrieval Augmented Generation (RAG) · Humans in the Loop (HL) · Interactive AI systems

1 Introduction

With the development of algorithms for text mining that use regular expressions [1], as well as considering the improvements in natural language processing models (NLP) [2], researchers, scientists and engineers benefit a lot for searching in a smarter way within dataspaces for specific information. This is more than using search engines and key words to display a list of sources, such as the case of Google or other similar web search

Published by Springer Nature Switzerland AG 2022
R. Nowak et al. (Eds.): TFC 2022, IFIP AICT 655, pp. 165–177, 2022.
https://doi.org/10.1007/978-3-031-17288-5_15

engines [3]. In inventive-problem solving frameworks we count on previous knowledge, discoveries, and inventions (e.g., patents, solutions published in scientific papers, etc.) to inspire us to solve a new problem. TRIZ explicitly encourages engineers to investigate knowledge bases at a given step in the problem-solving process [4], mostly databases with patents, where information is well-structured.

Case-based reasoning (CBR) is another approach to search for inspiration in problem-solving [5]. It is a model from the field of artificial intelligence (AI) and cognitive science that defines reasoning patterns to solve new problems by retrieving 'cases' that are stored in knowledge bases. In CBR, previous problem-solving experiences are extracted and adapted to fit new needs. CBR incorporates in the searching process a model of human reasoning as a mechanism to constructing intelligent systems [6].

Combinations of semantic models (which go beyond simple ontologies) and TRIZ have been considered to develop expert systems to search for ideas in structured patent databases and to make various analytics in the process of idea generation for a new problem. A representative technological development in this direction is PatentInspiration [7]. Semantic models with TRIZ inventive principles are embedded in another technology called InnovationQ Plus [8] that is designed to search in massive databases of patents and IEEE indexed scientific papers (over 100 million documents in all).

Matching and extraction of relevant knowledge from patent documents to be integrated within Inventive Design Method is another research contribution to support creative design [9]. In the same line, there are some other researches. Thus, our investigation in the mainstream databases with scientific publications (e.g., Clarivate Analytics, SCOPUS) revealed a research work about Artificial Intelligence (AI) driven inventions [10]. The authors cited in [10] articulate the idea that automating inventions look interesting, but they consider that AI will rather complement the intelligence of engineers and scientists, rather than replace it. They also present a semantic model that can form the basis of future AI approaches. As part of their work, over eight million patents and scientific publications have been screened with NLP techniques to extract semantic concepts. Another relevant research is about reproducing TRIZ reasoning through the Deep Learning (DL) on a large number of transdisciplinary patent sets [11]. The investigation of online data sources for pairing engineering problems with knowledge of physics for classification have also recently been published [12]. The last reference with relevance in relation to the topic considered in this paper is about the use of DL NLP-related models (Doc2vec and Cosine Similarity) for automatic information retrieval and introduction into an Inverse Problem Graph (IPG) process (note: IPG is a lean-based method for defining problems in the initial analysis phase of the inventive design process) [13].

The above-mentioned researches indicate a growing interest of the scientific community in introducing AI algorithms, mostly NLP DL-related ones, for smarter searching in databases on information and knowledge to support the inventive design process. However, our inquiry in mainstream databases of scientific papers does not indicate relevant research in the field of automating idea proposals for problem solving. This niche is going to be further elaborated on in the next sections of this paper. We consider the latest developments in artificial intelligence, specifically Question Answering Machine

(QAM) algorithms [14] in combination with the inventive principles promoted by Structured Activation of Vertex Entropy (SAVE) method [15] to assist the ideation process for engineering problems. This is about neuro-symbolic models of AI.

2 Methodology

Our aim is to investigate the application of Structured Activation Vertex Entropy (SAVE) method in combination with Question Answering Machine (QAM) algorithms to explore information that is stored in big knowledge/information/datasets (in a large number of documents of various formats), accessible within structured or unstructured dataspaces (e.g., open Internet) and formulate an inventive solution to a problem.

The SAVE method is transformed with the assistance of TRIZ into a set of searching meta-terms or meta-concepts. Thus, by applying TRIZ contradiction matrix on the SAVE method in conjunction with QAM, the conflict is: how to transform SAVE for an easy operation (parameter 33 in TRIZ) without losing essential information (parameter 24 in TRIZ). This leads to the following set of TRIZ inventive principles (IP): asymmetry (IP 4); preliminary action (IP 10); cheap short-living objects (IP 27); turn harm into a benefit (IP 22). IP 27 inspired us to covert the ten vertexes of SAVE into key words, key actions, and target-related key words. IP 4 inspired us to include as many as possible synonyms to the key words. IP 10 encourages organizing the key words into some semantics. IP 22 was not considered in this research. With these indications, the results are as follows:

VERTEX 1: Activation of resonance [capable to resonate, work at the same frequency]. *Vertex 1 associated key words*: {resonance, resonant, reverberate, reverberation, harmony, harmonize, agreement, agreed, accord, consensus, unity, vibrate, vibrant, resonate, alignment, placement, configuration, positioning, pact, congruence, converge, convergence, synchronize, synchronization, pulsate, pulsating}. *Vertex 1 related key actions*: {capable, able} {work, operate, act, perform} + {resonant, same, similar, harmonic, natural} + {intervention, opinion, frequency, vibration}.
VERTEX 2: Introduction of neutral elements [capable to annihilate; activate a new path]. *Vertex 2 associated key words*: {annihilate, defeat, crush, overwhelm, overpower}; {path, trail, track, way, trajectory, direction, route, road, pathway}. *Vertex 2 related key actions*: {capable, able} + {annihilate, cancel, overwhelm, overpass, overcome}; {activate, work, create, generate, make} + {(new) + path, trajectory, direction, way, route, pathway}.
VERTEX 3: Action against the wolf-pack spirit [operate and reach a target with no support from other systems; operate and reach a target with a fully volunteer support from other systems]. *Vertex 3 associated key words*: {operate, reach, target, aim, objective, volunteer, support, system, structure, field, domain, no, without, other, external, single, alone, against}. *Vertex 3 related key actions*: {operate, activate, reach}; {target, goal, aim, desire, dream}; {(no, against, without) + support, assistance, help, guidance}; {alone, single}; {(full, entire, complete, big, large) + (volunteer, open, disinterest) + support, assistance, help, guidance}; {(other, outer, external) + system(s), domain(s), structure(s), field(s)}.
VERTEX 4: Activation of centrifugal forces [benefit from the dynamics of individual elements]. *Vertex 4 associated key words*: {benefit, advantage, activation, generation};

{speed, rotational, centrifugal, centripetal, dynamics}. *Vertex 4 related key actions*: {activate, generate, create, combine, produce} + {dynamics, element(s), force(s), field(s), motivation, (new) + (interest(s), attractor(s), influence(s))}.
VERTEX 5: Application of multi-level connections [act in alignment and synergy with other systems]. *Vertex 5 associated key words*: {action}; {synergy, alignment}; {multiple, external, other}; {system(s), module(s)}. *Vertex 5 related key actions*: {act, do, perform, execute, operate, align} + {synergy} + {system(s), unit(s), element(s), person(s), people, object(s), part(s), module(s)}.
VERTEX 6: Application of asymmetry [counterbalance a much bigger system]. *Vertex 6 associated key words*: {asymmetry, counterbalance}; {super-system, system}; {larger, bigger, wider, higher, longer, greater, comprehensive}. *Vertex 6 related key actions*: {equilibrate, leveraging, counterbalancing} + {activity, system, part, module, unit}.
VERTEX 7: Harmonization of individual goals with collective goals [aligned to a higher-level target]. *Vertex 7 associated key words*: {harmonization, alignment}; {performance, goal}; {improvement, radical}. *Vertex 7 related key actions*: {harmonize, aggregate, align} + {target, performance, goal} + {individual, collective}.
VERTEX 8: Transformation for value-added [can provide more outputs than before, using the same inputs]. *Vertex 8 associated key words*: {value-added, efficiency}; {input-output, transformation}. *Vertex 8 related key actions*: {transform, generate, create} + {efficiency, value-added}.
VERTEX 9: Application of prisoner paradox [use only the existing local resources; rearrangement, and utilization of local resources]. *Vertex 9 associated key words*: {reconfiguration, combination, rearrangement, reorganization}; {tool(s), material(s), element(s), resource(s)}; {local, limited}. *Vertex 9 related key actions*: {reconfigure, rearrange, combine, reorganize} + {local, limited, existent} + {resource(s), mean(s), tool(s), element(s), material(s), system(s)}.
VERTEX 10: Application of shipwrecked paradox [transform some local negative factors into positive factors, identify hidden resources]. *Vertex 10 associated key words*: {transformation, identification, disclosure, revealing}; {negative, positive}; {local, limited, existent}; {resource(s), factor(s)}. *Vertex 10 related key actions*: {dig, mining, discover, explore, investigate, search, identify, unhide, disclose, reveal, display, show}; {convert, transform, modify} + {negative, positive} + {factor(s), resource(s)}.

In addition, we consider the target-related key words representative for the particular use case. For the use case introduced in the Sect. 4 of this paper, the *particular target-related key words* are: {common, same, aligned} + {goal(s), interest(s)}; {no, low, missing, absent, irrelevant, little} + {obstacle(s), barrier(s), conflict(s), stress(es), tension(s), struggle(s), fight(s), impediment(s), problem(s), difficulty, complication(s), barricade(s), blockage(s)}.

To combine SAVE with QAM, we propose an algorithm that uses the problem, the system, and the target result in a series of transformative processes led by the SAVE vertexes and embedded in the QAM model. The algorithm is named D-SIT-SIT-C, meaning:

DESCRIBE

1. Describe the problem, including synonyms for the keywords.

2. Describe the system, including synonyms for the keywords.
3. Describe the target result, including synonyms for the key target words.

PROCESS FOR EACH VERTEX

STEP 1: SEARCH

1. Search using the mix "system-problem"
2. Search using the mix "problem – target result"

STEP 2: INVESTIGATE (WITH HUMAN-IN-THE-LOOP)

1. Investigate the extracted information – if it does not deliver acceptable solutions move to the next step

STEP 3: TRANSFORM

1. Add to the system vertex-associated keywords
2. Add to the problem key actions

STEP 4: SEARCH

1. Search using the mix "transformed system- transformed problem"
2. Search using the mix "transformed problem – target result"

STEP 5: INVESTIGATE (WITH HUMAN-IN-THE-LOOP)

1. Investigate the extracted information – if it does not deliver acceptable solutions move to the next step

STEP 6: TRANSFORM (WITH HUMAN-IN-THE-LOOP)

1. Create an intermediary solution by best possible use of the information collected in the previous steps
2. Make a new description of the TRANSFORMED system (intermediary solution)

CONTINUE OR END

1. Take the next vertex and continue the process

As one can see, in the proposed algorithm human expert is present in the loop to analyze the results and to introduce additional inputs that would facilitate the progress in problem resolution. Human-in-the-loop (HITL) is met in NLP, including QAM algorithms [16]. HITL in this paper is inspired from traditional online HITL, but it follows a different path. We do not consider HITL for improving the training dataset. We make corrections to the given answer and introduce the corrected answer in the QAM's loop without altering the training and testing datasets, and the trained model.

3 Results and Discussions

For our experiments we decided to use the recent advances in NLP domain with a pre-trained long form question answering system [17], which engages a question (engineering problem) and transformed with the assistance of SAVE method, fetches couple of relevant passages from the dataspace (e.g., in our case from Wikipedia snapshot), and writes a generated multi-sentence answer based on the question and retrieved passages. In particular, for our experiments we used the pre-trained ELI5 model from Hugging Face Model Repository [18], a model that was trained using the ELI5 dataset described in [19], and the Wiki Snippets Indexes [20] generated using the Wiki-40B: Multilingual Language Model Dataset [21] loaded from Hugging Face Datasets [22].

To implement the algorithm, we wrote the code in Python, using eli5_utils.py, lfqa_utils.py specific modules, as well as some general libraries such as faiss_gpu, nlp, transformers, and torch. The running environment was Jupyter Notebook, and for faster computation we used GPU resources (note: import torch; torch.cuda.is_available()).

As example, we selected a situation from the field of oil industry. Oil is deposited in huge reservoirs (e.g., 6 m height, 10 m radius or even bigger). Residuals settle over time to the bottom of these reservoirs and cover the heating pipes. To clean tanks from these residuals, oil must be extracted and deposited into new reservoirs. In this context, we can formulate an engineering challenge as follows:

The problem: how to clean the oil tanks from residuals collected on the tank's floor without extracting the oil from the tank, and without working manually to complete this kind of job.
The system at start: oil tank, hole on the top of the tank, height of the tank, area of the floor, area of the hole, hole on the bottom side of the tank.
The target result: An automatic installation capable to enter a small hole on the top or bottom of the oil tank and cover a large surface on the floor of the tank to extract residuals.

First, we will use ARIZ [23] to find a solution to this problem. After that we will apply the proposed AI-driven algorithm to see the results. Discussions around results with the two approaches (a traditional one and one driven by AI for searching within the space of collective intelligence) will be also considered.

ARIZ recommends separating the opposite properties over time. The characteristics to be considered are: "12. Shape" (seen as the robot configuration in this case), which conflicts with: "8. The volume of the static object". The following inventive principles result: "7. Nest-in-nest: an object is placed inside another object, etc.; through a cavity, an object passes into another object"; "2. Extraction: extracts, removes, or separates a part or property from the object that is bothering it"; "35. Transforming object properties: changing the degree of flexibility". Results are shown in Fig. 1.

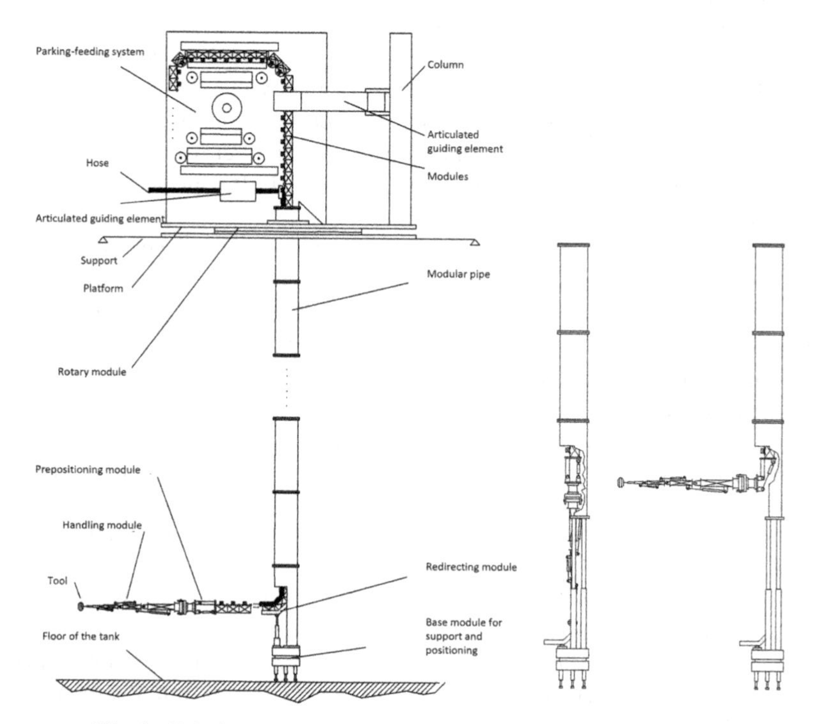

Fig. 1. Solution generated with indications from ARIZ application.

Now, we consider the D-SIT-SIT-C algorithm for the same problem. Intentionally, we will not consider all resources in system description to see how the D-SIT-SIT-C algorithm is dealing with. This means, we will intentionally exclude from system description the hole from the bottom side of the tank just to be somehow aligned with the solution proposed in Fig. 1, with the help of ARIZ method. We expect that the AI model will search in the database for other resources, including those excluded by us.

DESCRIBE

Describe the *problem*, including synonyms for the *key words*: how to [clean, suction, extraction] the [oil, petroleum, gasoline] [tanks, reservoir] from [residuals, waste] [collected, deposited] on the [floor, bottom] without [extracting, emptying] the oil from the tank, and without working manually.

Describe the *system*, including synonyms for the *key words*: height of the tank, area of the floor, area of the hole, robot structure, robot joints, robot arm.

Describe the *target result*, including synonyms for the *target key words*: A [robot, technology, mechanism, device] capable to enter a [small, tiny] [hole, hollow, cavity] on the [top] of an oil tank and cover a large surface on the floor of the tank.

STEP 1: SEARCH

Search using the mix "system-problem". *Question*: the text from "Describe the problem" + the text from "Describe the system".

The Wiki top 10 support documents retrieved by the system are not included here because of space considerations, but we can reveal the areas indicated in these documents: mud agitator, desander solids control, triplex-style plunger pumps & compressed-air-powered double-diaphragm pumps, industrial extraction with centrifugal systems, artificial lift with rod less pumping & continuous belt transportation, electrostatic coalescer, automated tank cleaning machine from chemical industry, Petrosix oil extraction process, artificial lift with hybrid gas and rod pump, artificial lift with electric submersible pumps (ESP).

[eli5 + Wiki] *automatic answer 1*: "You don't need to suction the oil out of the tank, you just need to remove the oil from the tank. You can do this by using a pump to suck up the oil, and then you can pump the oil back into the tank with a pump attachment. The pump attachment is a bit more complicated, but the basic idea is the same."

We want to highlight the clear, human-like answer automatically formulated by the AI-driven model. This is in our opinion a step forward in helping the creative problem solving of untrained people in structured innovation.

Search using the mix "problem – target result". *Question*: the text from "Describe the problem" + the text from "Describe the target result".

The Wiki top 10 support documents reveal two new indications: multi-stage gas extractor, and enhanced stripping methods from glycol dehydration.

[eli5 + Wiki] *automatic answer 2*: "There are a lot of different ways to do it, but the most common is to pump the oil out of the tank or reservoir, and then pump it back into the tank/reservoir. You can also pump it into the reservoir, but that's more expensive than just pumping it out. There are also ways to remove the oil from the reservoir and pump it out, but those are more expensive."

STEP 2: INVESTIGATE (WITH HUMAN EXPERTS)

Investigate the extracted information – if it does not deliver acceptable solutions move to the next step. In our case we decided to move to the next step just to explore the space of potentiality, even if relevant information has been revealed at STEP 1.

STEP 3: TRANSFORM

Add to the system *SAVE vertex-associated key words* (in this step VERTEX 1). *Question*: [eli5 + Wiki] automatic answer 1 + [resonance, harmony, vibrate, configuration, positioning, converge].

The Wiki top 10 support documents retrieved in addition to the previous information other new directions of investigation: fluid coupling with scoop control, multi-stage centrifugal pumps & centrifugal pumps for solids control & magnetically coupled pumps, hit-and-miss engine, velocity triangle & vertical centrifugal pumps & froth pumps, hot-bulb engine, fluidyne engine.

[eli5 + Wiki] *automatic answer 3*: The answer retrieved is identical with answer 1. It looks that the system is not capable to provide additional insights.

Add key actions in conjunction to problem. *Question*: the text from "Describe the problem" + [activate, generate, create, combine, produce] + [dynamics, element(s), force(s), field(s), (new) + (interest(s), attractor(s), influence(s))].

The Wiki top 10 support documents retrieved by the system indicates other new directions of investigation: Lurgi–Ruhrgas process, RDF-power station, petroleum microbiology, Carabobo field.

[eli5 + Wiki] *automatic answer 4*: "You don't need to do any of those things. You just need to be able to get the oil out of the tank or reservoir. It's not like you have to do anything to get it out, *it's just a matter of getting it out fast enough* that it doesn't build up to the point where it's a problem."

With the indications from the last answer, we made some transformations to the level of the answer and the problem. This process is required by the SAVE algorithm, in order to move forward with the investigation.

Transformed answer: It's just a matter of getting the residuals fast enough that it doesn't build up to the point where it's a problem.

Transformed problem: Is there a robot that cleans an oil tank from residuals without extracting the oil?

STEP 4: SEARCH

Search using the mix "transformed system- transformed problem". *Question*: "Is there a robot that cleans an oil tank from residuals without extracting the oil? It's just a matter of getting the residuals fast enough that it doesn't build up to the point where it's a problem."

The Wiki top 10 support documents retrieved by the system indicates other new directions of investigation: hydropneumatic suspension, bioclogging, green nanotechnology for cleaning up oil spills, dry milling and fractionation of grain, oily water separator (marine), Alberta Taciuk process.

It looks like the deeper we investigate the problem indications move towards chemical processes rather than mechanical systems. This is a paradigm shift from the initial way of setting up the target result. This means, the AI system was not bordered by the limited perception (phycological inertia) of the team. It is also interesting the answer generated by the system.

[eli5 + Wiki] *automatic answer 5*: "Yes, there is (about the robot). *It's called a scrubber*, and it's used to remove the residuals from the oil tank. The problem is that scrubbers don't remove all of the oil. Some of it is still in the tank, and the scrubber can't get rid of all of it, so it builds up to the point where it's a problem."

Thus, the system was able to "judge" the relevance of having a pure mechanical solution and to "decide" that it is not necessarily the optimal one.

Search using the mix "transformed problem – target result". *Question*: "It's just a matter of getting the residuals fast enough that it doesn't build up to the point where it's a problem [robot, technology, mechanism, device] capable to enter a [small, tiny] [hole, hollow] on the top of an oil tank and cover a large surface on the floor of the tank".

The Wiki top 10 support documents retrieved by the system indicates other new directions of investigation: Nokomis 3 process, rebreather with CO_2 scrubber, artificial pressure, annular fluidized bed with fine particulates, Finnish pollution control vessel Louhi.

[eli5 + Wiki] *automatic answer 6*: "It's not just a matter of getting the residuals fast enough that it doesn't build up to the point where it's a problem. The problem is that there's no way to get rid of it fast enough for it to be a problem, so it builds up until it's too big to be contained." The answer shows that the system did not advance in "judgment". This somehow indicates that the searched resources have been covered.

STEP 5: INVESTIGATE (WITH HUMAN-IN-THE-LOOP)
Investigate the extracted information – if it does not deliver acceptable solutions move to the next step. In principle, for the problem exemplified here, the system was able to reveal sufficient advices and directions to think to a novel solution. However, we can investigate further, sometimes just for the sick of curiosity and consolidation of the decision.

STEP 6: TRANSFORM (WITH HUMAN-IN-THE-LOOP)
Create an intermediary solution by best possible use of the information collected in the previous steps. In this case, the intermediary formulated solution is: "Pump the oil out of the bottom of the tank, filter it and then pump it back as quickly as possible into the tank from the top of it. The installation that cleans an oil tank from residuals without extracting the oil is called the scrubber [24]."

Make a new description of the *transformed system* (intermediary solution). The new system includes scrubber, pump, handling unit, chemical reactions, filters, units that generate fields or forces.

CONTINUE OR END
Take the next SAVE vertex and *continue* the process. In the second cycle we have considered VERTEX 2. The detailed steps are not further described in the paper. In continuation we limit ourselves to present only the last relevant answer generated by the system.

[eli5 + Wiki] *automatic answer 7*: "Pumping the oil out of the bottom of the tank, filtering it, and then pumping it back as quickly as possible into the tank from the top of it. The robot that cleans an oil tank is called the Crude Oil Washing (COW). The COW washing system, or COW system, is part of the cleaning process."

Now it's the time to use the list of information and indication to ideate the solution to the problem. Actually, the last answer is a solution to the problem, automatically indicated by the system. We have also considered additional indications from the list retrieved by the system. It cannot be disclosed here because of an ongoing research project with inventive mission.

Comparing the solution generated with the help of ARIZ method and the one proposed by the system we can conclude that the first one is novel in the series of solutions destinated to petroleum tank cleaning, whereas the second is extracted from the pool of knowledge already created by humans. However, the first solution was generated with a huge creative contribution of human experts, and it is a pure mechanical one. In the case of the AI-driven approach we did not want to develop a new tool for ideation, but rather a tool that helps engineers in the application of ARIZ or similar algorithms for inventive problem. It is a more powerful form of CBR because it moves beyond just displaying existent solutions, meaning it suggests vectors of investigation, as actually the TRIZ does, but with much more practical indications. These vectors of investigation are extracted from the documents retrieved in various steps of the proposed algorithm, as one already saw in the previous paragraphs of this section. They can be combined using morphological charts to help engineers thinking to various options for ideation.

4 Conclusions

This research introduces a possibility to transform a human-like nonlinear pattern of deep thinking into a format that is suitable for implementation into AI NLP-driven algorithms that search throughout big information/data/knowledge sets and automatically generate "intelligent" machine answers to a given problem. SAVE method is embedded in the algorithm to increase the search effectiveness and provoke the system in answer formulation. But there are no limitations to replace SAVE vertexes with TRIZ inventive principles or to add them to the vertexes list. The single issue is the productivity of doing the job.

The use case illustrated in the paper reveals a sign of encouragement, indicating that AI models of QAM can be a reliable assistant for experts in solving complex technical problems in shorter periods of time and at higher levels of performance. In fact, this research brings a new tool to the world of ARIZ, where the use of knowledge bases is required for documentation. Traditional software systems were limited in providing a massive pool of resources. Modern tools such as those cited in [7] and [8] better fulfill this job. In the case of our model, we claim that it brings something more, by also indicating basic solutions or clues towards the solution in a very natural language. Our research also indicates that there are still huge steps until the automatic systems will be able to replace human mind and become sufficient creative. We see the current systems rather being more explorative and highly productive in doing this job than people, and less creative in the sense of proposing out-of-the-box ideas. But they can help in indicating disruptive vectors of investigation, which is a relevant finding of this research. This means, we should not be limited in the short list of 40 generic inventive principles proposed by traditional TRIZ, and expand its pool of suggestions, moving towards the modern TRIZ.

It is important to highlight the fact that the capacity of the AI system is limited by the pool of training data, and from this angle one area to support inventive problem solving is to work on a dataspace of information that collects as many as possible information created by humankind (e.g., patents, open Internet, scientific papers, images, videos). Thus, new spaces for AI are open in this regard. Further effort is needed to expand these experiments on various knowledge bases and online resources, including scientific web databases or the unstructured open Internet, with making sure that not only the English language content is targeted but also other languages, as well. For this we would need to train our own dense models from within the knowledge base and also sequence-to-sequence models which are built upon a BART architecture [25] domain and language specific. Our future researches in the direction opened by this paper are also to train new models capable to deal with managerial problems, as well as to refine the algorithm in terms of formulating questions, etc. We also see a space of opportunity in testing some other tools of creativity in conjunction to QAM. Of course, the challenge to create a smarter system, capable to play with creative patterns is also in view for our future researches.

References

1. Friedl, J.: Mastering Regular Expressions, 3rd edn. O'Reilly, Sebastopol (2006)

2. Lane, H., Hapke, H., Howard, C.: Natural Language Processing in Action. Manning, New York (2019)
3. Dwyer, D.: Top 12 Best Search Engines in the World (2016). https://www.inspire.scot/blog/2016/11/11/top-12-best-search-engines-in-the-world238. Accessed 20 June 2022
4. Gadd, K.: TRIZ for Engineers. Wiley, Chichester (2011)
5. Schmidt, R., Montani, S., Bellazzi, R., Portinale, L., Gierl, L.: Cased-based reasoning for medical knowledge-based systems. Int. J. Med. Inform. **64**, 355–367 (2001)
6. Lee, C.H., Chen, C.H., Li, F., Shie, A.J.: Customized and knowledge-centric service design model integrating case-based reasoning and TRIZ. Expert Syst. Appl. **143**, 13062, 14 pp. (2020)
7. Dewulf, S., Childs, P.R.N.: Patent data driven innovation logic: textual pattern exploration to identify innovation logic data. In: Borgianni, Y., Brad, S., Cavallucci, D., Livotov, P. (eds.) TFC 2021. IAICT, vol. 635, pp. 170–181. Springer, Cham (2021). https://doi.org/10.1007/978-3-030-86614-3_14
8. Ip.com: Why Non-Patent Literature Can Make or Break Your Business. https://ip.com/wp-content/uploads/2020/09/IQ_NPL_ebook_P2.pdf. Accessed 02 June 2022
9. Souilia, A., Cavallucci, D., Rousselot, F.: Natural Language Processing (NLP) - a solution for knowledge extraction from patent unstructured data. Proc. Eng. **131**, 635–643 (2015)
10. Kaliteevskii, V., Deder, A., Peric, N., Chechurin, L.: Concept extraction based on semantic models using big amount of patents and scientific publications data. In: Borgianni, Y., Brad, S., Cavallucci, D., Livotov, P. (eds.) TFC 2021. IAICT, vol. 635, pp. 141–149. Springer, Cham (2021). https://doi.org/10.1007/978-3-030-86614-3_11
11. Guarino, G., Samet, A., Cavallucci, D.: Patent specialization for deep learning information retrieval algorithms. In: Borgianni, Y., Brad, S., Cavallucci, D., Livotov, P. (eds.) TFC 2021. IAICT, vol. 635, pp. 162–169. Springer, Cham (2021). https://doi.org/10.1007/978-3-030-86614-3_13
12. Boufeloussen, O., Cavallucci, D.: Bringing together engineering problems and basic science knowledge, one step closer to systematic invention. In: Borgianni, Y., Brad, S., Cavallucci, D., Livotov, P. (eds.) TFC 2021. IAICT, vol. 635, pp. 340–351. Springer, Cham (2021). https://doi.org/10.1007/978-3-030-86614-3_27
13. Hanifi, M., Chibane, H., Houssin, R., Cavallucci, D.: Problem formulation in inventive design using Doc2vec and cosine similarity as artificial intelligence methods and scientific papers. Eng. Appl. Artif. Intell. **109**, 104661 (2022)
14. Hugging Face: What is Question Answering?. https://huggingface.co/tasks/question-answering. Accessed 04 June 2022
15. Brad, S.: Domain analysis with TRIZ to define an effective "Design for Excellence" framework. In: Borgianni, Y., Brad, S., Cavallucci, D., Livotov, P. (eds.) TFC 2021. IAICT, vol. 635, pp. 426–444. Springer, Cham (2021). https://doi.org/10.1007/978-3-030-86614-3_34
16. Wang, Z.J., Choi, D., Xu, S., Yang, D.: Putting humans in the natural language processing loop: a survey. https://arxiv.org/abs/2103.04044 (2021). Accessed 04 May 2022
17. Roy, A.: Progress and Challenges in Long-Form Open-Domain Question Answering. https://ai.googleblog.com/2021/03/progress-and-challenges-in-long-form.html. Accessed 03 Apr 2022
18. Jernite, Y.: ELI5 Model from Hugging Face Model Repository. https://huggingface.co/yjernite. Accessed 02 Feb 2022
19. Fan, A., Jernite, Y., Perez, E., Grangier, D., Weston, J., Auli, M.: ELI5: Long Form Question Answering. https://arxiv.org/abs/1907.09190 (2019). Accessed 20 Jan 2022
20. Wikipedia: User scripts/Snippets. https://en.wikipedia.org/wiki/Wikipedia:User_scripts/Snippets. Accessed 05 Apr 2022

21. Guo, M., Dai, Z., Vrandečić, D., Al-Rfou, R.: Wiki-40B: multilingual language model dataset. In: Proceedings of the 12th Language Resources and Evaluation Conference, pp. 2440–2452. European Language Resources Association, Marseille, France (2020)
22. Hugging Face Data Sets. https://github.com/huggingface/datasets. Accessed 05 Apr 2022
23. Cameron, G.: ARIZ Explored: A Step-by-Step Guide to ARIZ, the Algorithm for Solving Inventive Problems. Create Space, Scotts Valley (2015)
24. Wikipedia. Scrubber: https://en.wikipedia.org/wiki/Scrubber. Accessed 02 June 2022
25. Lewis, M., et al.: BART: denoising sequence-to-sequence pre-training for natural language generation, translation, and comprehension. https://arxiv.org/abs/1910.13461 (2019). Accessed 04 May 2022

Inventive Principles Extraction in Inventive Design Using Artificial Intelligence Methods

Masih Hanifi[1,2,3(✉)], Hicham Chibane[2,3], Remy Houssin[1,3], and Denis Cavallucci[2,3]

[1] Strasbourg University, 4 Rue Blaise Pascal, 67081 Strasbourg, France
masih.hanifi@insa-strasbourg.fr
[2] INSA of Strasbourg, 24 Boulevard de la Victoire, 67000 Strasbourg, France
[3] ICUBE/University of Strasbourg, 4 Street Blaise Pascal, 67081 Strasbourg, France

Abstract. Today, companies are seeking effective approaches to improve their innovation cycle time. Among them, it is possible to mention Inventive Design Methodology (IDM) as a TRIZ-based systematic inventive design process. However, the application of this approach is time-consuming due to requesting a complete map of a problem situation at the initial phase of the inventive design process. To solve this drawback, the Inverse Problem Graph (IPG) method has been developed to increase the agility of the process. Nevertheless, authors of IPG did not mention how the designers could achieve the innovative solutions by using the formulated problems. The purpose of the research presented in this article is to integrate the doc2vec method and machine learning text classification algorithms as Artificial Intelligence methods into the IPG process. This integration helps introduce an automatic approach for the inventive design process, helping to formulate the contradictions among TRIZ parameters in the contradiction matrix and extract the inventive principles in their intersection. The capability of the proposed methodology is finally tested through its application in a case study.

Keywords: Inventive design · TRIZ · Text classification · Document embedding · Artificial Intelligence

1 Introduction

In recent decades, many enterprises have been competing on decreasing the innovation cycle time due to its importance in their success [1]. To do so, these enterprises can apply TRIZ-based systematic approaches such as Inventive Design Methodology (IDM). This approach has been developed to complete the TRIZ body of knowledge with other theories such as graph theory [2]. Nevertheless, one of the criticisms often leveled is that Inventive Design Methodology lacks the essential agility [3]. This is mainly due to constructing a complete map of a problem situation at the start of the project, regardless of how effective it is in solving the problem. Consequently, the Inverse Problem Graph (IPG) method was developed by several authors to formulate a problem situation.

The IPG method was introduced to formulate the contradictions related to a problem situation [4, 5]. This method adds to the IDM process the characteristics such as the

Published by Springer Nature Switzerland AG 2022
R. Nowak et al. (Eds.): TFC 2022, IFIP AICT 655, pp. 178–186, 2022.
https://doi.org/10.1007/978-3-031-17288-5_16

capability for iterative development and the capacity to generate rapid response to change to increase its agility [6]. Nevertheless, the manual extraction of the TRIZ parameters and TRIZ inventive principles within the IDM process reduces its agility [7, 8]. Therefore, it is necessary to utilize automatic information retrieval methods to solve this drawback. Among them, we can mention the machine learning algorithms and the doc2vec model. The main objective of this paper is to integrate machine learning algorithms and the doc2vec model into the process of Inventive Design Methodology. This integration can facilitate and accelerate the extraction of TRIZ parameters and inventive principles in the process.

The rest of the paper is organized as follows. In Sect. 2, we present the relevant literature review. Section 3 shows the evaluation of the reviewed machine learning algorithms. Section 4 describes the structure of the proposed method and its step. In Sect. 5, a case study is presented in which the proposed method is used to extract TRIZ parameters and the inventive principles related to the lattice structures. We present the discussion and conclusion in the last section.

2 Literature Review and Background

2.1 Inventive Design Methodology (IDM)

IDM framework has been developed for several years to solve the limitations of TRIZ [2]. This framework includes the phases below [9, 10]:

1. Initial Analysis phase: In this phase, the relevant knowledge from patents, internal documents, tacit expert's know-how, and other documents is gathered. Then, the Problem Graph method is used to transform the knowledge into a graphical model to facilitate decision-making [2].
2. Contradictions Formulation phase: The physical and technical contradictions are formulated by applying the methods such as poly-contradiction template in the second phase of IDM framework [11].
3. Solution Concept Synthesis phase: In the third phase, various TRIZ tools are used to solve the formulated contradictions in the previous phase [12].
4. Solution Concept Selection phase: In the last phase, an evaluation grid is applied to measure the impact of each concept and select the most relevant one [2, 13].

In the next sub-section, we will review several methods that has been introduced to automate information extraction in the Inventive Design Methodology framework.

2.2 Automatic Extraction Methods to Assist Designers

In the literature, various methods have been introduced to support the designers in the framework of Inventive Design Methodology. For example, [14] introduced an approach to extract problems and partial solutions from the patent. Besides, in order to improve the final output of the IDM-related information extraction tools, the authors in [15] proposed to use claims hierarchical structure. In addition, [16] developed a tool within the IDM framework to extract the parameters, problems and partial solutions from scientific articles. However, these approaches retrieve the information beyond the needs of designers. This makes analysis of the extracted information time-consuming and quite laborious.

2.3 Document Embedding Techniques

The authors in [17] have proposed the word2vec model to create a word embedding representation. This model includes the two model architectures that are Skip-gram and Continuous bag-of-words [18]. Word2vec permits deriving similar words semantically. However, this model loses the word order information [19]. To overcome this drawback, the researchers proposed an unsupervised method called doc2vec.

The doc2vec method is an extension of word2vec, which has been proposed by Le et al. [20]. This method includes the following architectures [21]: 1) Distributed Memory (PV-DM) and 2) Distributed Bag of Words (PV-DBOW). Doc2vec can extract the word order information from the text [22]. Besides, it can express variable-length text, ranging from sentences to large documents, as a vector [23]. Furthermore, doc2vec is used to calculate document similarity and classify the given texts. To extract the TRIZ parameters from the sentences, it is necessary to use also the machine learning text classification algorithms, reviewed in the next section.

2.4 Text Classification Algorithms

The task of classifying a text into predefined classes is defined as Text classification [24]. There are two main groups of machine learning algorithms to automate text classification, as the following [25]: 1) Supervised learning and 2) Unsupervised learning. In our proposed method, we used the algorithms related to the first group. Below, we brought some of these algorithms:

Logistic regression (LR) is a supervised machine learning classification technique developed by David Cox in 1958 [25]. There are two types of LR, as the following [26]: 1) Binominal logistic model and 2) Multinomial logistic regression. This technique serves well to anticipate categorical results [27].

The Multilayer Perceptron (MLP) algorithm is a feed-forward neural network proposed by Rosenblatt in 1958 [28, 29]. MLP maps input data to the outputs. This algorithm is constructed of an output layer, one or several hidden layers, and an input layer [30]. In this algorithm, the input layer obtains an external activation vector including the values {X1, X2, ..., Xn}. Subsequently, these values are transferred via weighted connections to the nodes of the first hidden layer. Then, the layer calculates their activation and spreads them to the nodes in the subsequent layer [31]. MLP is used to predict and classify problems [32].

3 Experiment and Evaluation

In this section, we evaluate the accuracy of the reviewed machine learning algorithms for the "TRIZ Parameters" data set.

TRIZ Parameters data set is used for the second evaluation of machine learning algorithms. This data set consists of 3607 sentences and 2 following groups of parameters: 1) The "Positive TRIZ Parameters" group includes 40 labels (39 TRIZ parameters labels + nan label), and 2) The "Negative TRIZ Parameters" group consists of 40 labels (39 TRIZ parameters + nan label). To evaluate the reviewed machine learning algorithms by the TRIZ parameters data set, we first selected 4 highest labels in the first and second group as the following:

1) "Positive TRIZ Parameters" group (469 sentences, Parameters: 15.77% Strength, 14.07% Object-generated harmful factors, 13.21% Temperature, 12.36% Use of energy by stationary object, ...).
2) "Negative TRIZ Parameter" group (410 sentences, Parameters: 13.90% Strength, 13.41% Loss of energy, 12.92% Temperature, 12.68% Loss of substance, ...).

Then, we perform a train-test split using the 80-20 rule, where 80% of data is used for training, and the remaining 20% is used for testing. Subsequently, the reviewed machine learning algorithms in the literature were trained using the training data and tested on the test set for their accuracies. Table 1 display the accuracy for each machine learning algorithm related to the TRIZ parameter dataset. We used the algorithms with the highest accuracy to train the model in our case study section.

Table 1. Evaluation of machine learning algorithms for the "Positive TRIZ Parameter" and the "Negative TRIZ Parameter" groups

Algorithms	F1 scores	Precision scores	Recall scores
Positive TRIZ Parameter			
Logistic regression	83.35%	**85.24%**	83.67%
MLP	78.62%	81.79%	79.59%
Negative TRIZ Parameter			
Logistic regression	80.60%	**84.09%**	80%
MLP	80.66%	83.42%	80%

4 Proposed Method

In this section, we used doc2vec and machine learning algorithms to extract TRIZ parameters from the sentences. This enabled us to introduce a new method for the inventive design process. Figure 1 shows different steps of our proposal. These steps are as the following.

Step 1: **Determine the initial problem of the project**: In the first step, designers determine the initial problem of the project by considering its objective.

Step 2: **Extract the cause sentences from the corpus**: The cause sentences are extracted from the corpus in the second step of the process. To do this, designers formulate the what-cause question at the beginning of this step. Then, they use it to extract the cause sentences from the corpus.
Step 3: **Extract TRIZ parameters to formulate the contradictions:** In the third step of the process, the system extracts the positive and negative TRIZ parameters from the cause sentences to formulate their related contradictions. For this purpose, it is necessary to train the doc2vec model and machine learning algorithms by applying the provided data samples, including TRIZ parameters.
Step 4: **Connect the formulated contradictions to the inventive principles:** In the final step, the system connects the formulated contradictions in the previous step to the inventive principles of contradiction matrix.

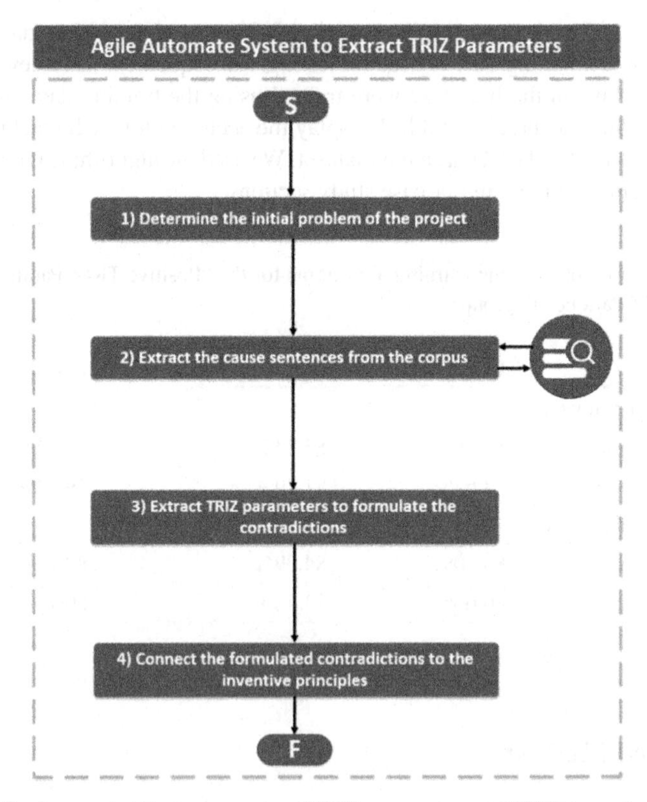

Fig. 1. Agile Automated System to extract TRIZ parameters and TRIZ inventive principles

5 Application of the Method

In this section, we used the proposed method "Agile Automated System to Extract TRIZ Parameters" to the Lattice Structure (LS) case study. This application enables us to evaluate its applicability. The energy absorption of LS has always been an interesting subject for engineers and research scientist because of its wide usage in energy-absorbing application [33, 34]. In this case study, we will determine the factors that influence the energy absorption of the lattice structure. In the following, we will describe the steps of this application.

In the first step, we determined the initial problem, which was "The energy absorption of LS is decreased", by considering the project objective.

After defining the initial problem, we had to identify the causes of this problem in the second step. To do so, we first formulated our question, which was "What reduces the energy absorption of the lattice structure?". Then, we used the most similar method and doc2vec to uncover the cause sentences nearest to our formulated question. Table 2 displays some of these sentences. We should also mention that the minimum threshold for this case study is considered to be 0.65.

We extracted the TRIZ parameters in the third step. For this purpose, we first trained our machine learning algorithm with the highest accuracy presented in the "Experiment and Evaluation" section. Next, this algorithm allowed us to extract the TRIZ parameters. Table 2 also shows the extracted parameters related to the selected causes.

In the fourth step, we connected the inventive principles of the contradiction matrix to the formulated contradiction in the previous step. Table 2 also displays the inventive principles related to formulated contradictions. In what follows, we use the principles to develop our final solution.

Table 2. The selected cause sentences related to the formulated question, their extracted TRIZ parameters, and the related inventive principles.

Sentences	TRIZ Positive Parameter	TRIZ Negative Parameter	IP
When the volume fraction is reduced from 40.4% to 12.5%, the maximum energy absorption efficiency increases from 35.20% to 46.95%, an increase of 11.75%.	Shape	nan	-
However, the increase in strength is at the expense of energy absorption, as observed experimentally.	Strength	Shape	10, 30, 35, 40
It indicated that as the volume fraction of the lattice increases, the maximum energy absorption efficiency decreases.	nan	Shape	-
But lower energy absorption due to the increase in stiffness.	Strength	Shape	10, 30, 35, 40

6 Discussion and Conclusion

In this study, we proposed a new method for extracting TRIZ parameters in the inventive design process. To propose our method, we first reviewed some of the leading document embedding techniques to highlight their advantages and limitations. Next, we reviewed several machine learning algorithms. In the end, we proposed to apply doc2vec and the machine learning algorithms to develop a new approach, helping to extract TRIZ parameters from the sentences. We then tested the capacity of our proposal by applying it to the case study of Lattice Structure. Based on this application, we realized that our proposed method could facilitate and accelerate the formulation of contradictions through automatic extraction of TRIZ parameters.

The contribution of this work to the inventive design process is reflected in several aspects. First, the system can retrieve TRIZ parameters from the cause sentences through the application of machine learning algorithms. The existence of this capability can facilitate and accelerate the formulation of contradictions in the inventive design process. As the second, the formulated contradictions can be automatically connected to the inventive principles on the contradiction matrix. This can reduce the amount of time to develop inventive solutions for the designers. The analysis of the results related to the case study reveals some limitations. As the most important of them, we can mention that providing a complete data sample for our supervised machine learning algorithms is a time-consuming task.

Further research is necessary to appreciate our proposal. To solve the drawback related to providing data samples for our used machine learning algorithms, it is possible to use transfer learning. This can decrease the amount of resources and labelled to train the models.

References

1. Cohen, M.A., Eliasberg, J., Ho, T.-H.: New product development: the performance and time-to-market tradeoff. Manag. Sci. **42**(2), 173–186 (1996). https://doi.org/10.1287/mnsc.42.2.173
2. Cavallucci, D., Strasbourg, I.: From TRIZ to Inventive Design Method (IDM): towards a formalization of Inventive Practices in R&D Departments. Innovation **18**, 2 (2009)
3. Cavallucci, D., Rousselot, F., Zanni, C.: Assisting R&D activities definition through problem mapping. CIRP J. Manuf. Sci. Technol. **1**(3), 131–136 (2009). https://doi.org/10.1016/j.cirpj.2008.09.014
4. Hanifi, M., Chibane, H., Houssin, R., Cavallucci, D.: IPG as a new method to improve the agility of the initial analysis of the inventive design. FME Trans. **49**(3), 549–562 (2021). https://doi.org/10.5937/fme2103549H
5. Hanifi, M., Chibane, H., Houssin, R., Cavallucci, D.: Problem formulation in inventive design using Doc2vec and Cosine Similarity as Artificial Intelligence methods and Scientific Papers. Eng. Appl. Artif. Intell. **109**, 104661 (2022). https://doi.org/10.1016/j.engappai.2022.104661
6. Kumar, G., Bhatia, P.K.: Impact of Agile methodology on software development process. Int. J. Comput. Technol. Electron. Eng. IJCTEE **2**(4), 46–50 (2012)
7. Hanifi, M., Chibane, H., Houssin, R., Cavallucci, D.: A Method to formulate problem in initial analysis of inventive design. In: Nyffenegger, F., Ríos, J., Rivest, L., Bouras, A. (eds.) PLM 2020. IAICT, vol. 594, pp. 311–323. Springer, Cham (2020). https://doi.org/10.1007/978-3-030-62807-9_25

8. Hanifi, M., Chibane, H., Remy, H., Denis, C., Ghannad, N.: Artificial intelligence methods for improving the inventive design process, application in lattice structure case study, vol. 36 (2022)
9. Hanifi, M., Chibane, H., Houssin, R., Cavallucci, D.: Improving inventive design methodology's agility. In: Benmoussa, R., De Guio, R., Dubois, S., Koziołek, S. (eds.) TFC 2019. IAICT, vol. 572, pp. 216–227. Springer, Cham (2019). https://doi.org/10.1007/978-3-030-32497-1_18
10. Slim, R., Houssin, R., Coulibaly, A., Hanifi, M., Chibane, H.: Framework for resolving problems resulting from lean integration from the early design phases of production 3D printing machine. FME Trans. **49**(2), 279–290 (2021). https://doi.org/10.5937/fme2102279S
11. Cavallucci, D.: Designing the inventive way in the innovation era. In: Chakrabarti, A., Blessing, L.T.M. (eds.) An Anthology of Theories and Models of Design, pp. 237–262. Springer, London (2014). https://doi.org/10.1007/978-1-4471-6338-1_12
12. Da Silva, R.H., Kaminski, P.C., Armellini, F.: Improving new product development innovation effectiveness by using problem solving tools during the conceptual development phase: integrating Design Thinking and TRIZ. Creat. Innov. Manag. **29**(4), 685–700 (2020). https://doi.org/10.1111/caim.12399
13. Hanifi, M., Chibane, H., Houssin, R., Cavallucci, D.: Contribution to TRIZ in combining lean and inventive design method. In: Cavallucci, D., Brad, S., Livotov, P. (eds.) TFC 2020. IAICT, vol. 597, pp. 280–291. Springer, Cham (2020). https://doi.org/10.1007/978-3-030-61295-5_23
14. Souili, A., Cavallucci, D., Rousselot, F., Zanni, C.: Starting from patents to find inputs to the problem graph model of IDM-TRIZ. Procedia Eng. **131**, 150–161 (2015). https://doi.org/10.1016/j.proeng.2015.12.365
15. Berduygina, D., Cavallucci, D.: Improvement of automatic extraction of inventive information with patent claims structure recognition. In: Arai, K., Kapoor, S., Bhatia, R. (eds.) SAI 2020. AISC, vol. 1229, pp. 625–637. Springer, Cham (2020). https://doi.org/10.1007/978-3-030-52246-9_46
16. Nédey, O., Souili, A., Cavallucci, D.: Automatic extraction of IDM-related information in scientific articles and online science news websites. In: Cavallucci, D., De Guio, R., Koziołek, S. (eds.) TFC 2018. IAICT, vol. 541, pp. 213–224. Springer, Cham (2018). https://doi.org/10.1007/978-3-030-02456-7_18
17. Naili, M., Chaibi, A.H., Ben Ghezala, H.H.: Comparative study of word embedding methods in topic segmentation. Procedia Comput. Sci. **112**, 340–349 (2017). https://doi.org/10.1016/j.procs.2017.08.009
18. Karvelis, P., Gavrilis, D., Georgoulas, G., Stylios, C.: Topic recommendation using Doc2Vec. In: 2018 International Joint Conference on Neural Networks (IJCNN), Rio de Janeiro, pp. 1–6 (2018). https://doi.org/10.1109/IJCNN.2018.8489513
19. Mimura, M., Tanaka, H.: Leaving all proxy server logs to paragraph vector. J. Inf. Process. **26**, 804–812 (2018). https://doi.org/10.2197/ipsjjip.26.804
20. Mikolov, T., Sutskever, I., Chen, K., Corrado, G.S., Dean, J.: Distributed representations of words and phrases and their compositionality. Adv. Neural Inf. Process. Syst. **26**, 3111–3119 (2013)
21. Le, Q., Mikolov, T.: Distributed representations of sentences and documents. In: International Conference on Machine Learning, pp. 1188–1196 (2014)
22. Zhang, H., Zhou, L.: Similarity judgment of civil aviation regulations based on Doc2Vec deep learning algorithm. In: 2019 12th International Congress on Image and Signal Processing, BioMedical Engineering and Informatics (CISP-BMEI), Suzhou, China, pp. 1–8 (2019). https://doi.org/10.1109/CISP-BMEI48845.2019.8965709

23. Aman, H., Amasaki, S., Yokogawa, T., Kawahara, M.: A Doc2Vec-based assessment of comments and its application to change-prone method analysis. In: 2018 25th Asia-Pacific Software Engineering Conference (APSEC), Nara, Japan, pp. 643–647 (2018). https://doi.org/10.1109/APSEC.2018.00082
24. Dalal, M.K., Zaveri, M.A.: Automatic text classification: a technical review. Int. J. Comput. Appl. **28**(2), 37–40 (2011). https://doi.org/10.5120/3358-4633
25. Sarkar, D.: Text Analytics with Python: A Practitioner's Guide to Natural Language Processing. Apress, Berkeley (2019). https://doi.org/10.1007/978-1-4842-4354-1
26. Park, H.-A.: An introduction to logistic regression: from basic concepts to interpretation with particular attention to nursing domain. J. Korean Acad. Nurs. **43**(2), 154–164 (2013). https://doi.org/10.4040/jkan.2013.43.2.154
27. Kowsari, K., Jafari Meimandi, K., Heidarysafa, M., Mendu, S., Barnes, L., Brown, D.: Text classification algorithms: a survey. Information **10**(4), 150 (2019). https://doi.org/10.3390/info10040150
28. Gulia, A., Vohra, D.R., Rani, P.: Liver patient classification using intelligent techniques. Int. J. Comput. Sci. Inf. Technol. **5**(4), 5110–5115 (2014)
29. Panchal, G., Ganatra, A., Kosta, Y.P., Panchal, D.: Behaviour analysis of multilayer perceptrons with multiple hidden neurons and hidden layers. Int. J. Comput. Theory Eng. **3**(2), 332–337 (2011). https://doi.org/10.7763/IJCTE.2011.V3.328
30. Ramchoun, H., Amine, M., Idrissi, J., Ghanou, Y., Ettaouil, M.: Multilayer perceptron: architecture optimization and training. Int. J. Interact. Multimed. Artif. Intell. **4**(1), 26–30 (2016). https://doi.org/10.9781/ijimai.2016.415
31. Riedmiller, M.: Advanced supervised learning in multi-layer perceptrons—From backpropagation to adaptive learning algorithms. Comput. Stand. Interfaces **16**(3), 265–278 (1994). https://doi.org/10.1016/0920-5489(94)90017-5
32. Adwan, O., Faris, H., Jaradat, K., Harfoushi, O., Ghatasheh, N.: Predicting customer churn in telecom industry using multilayer preceptron neural networks. Life Sci. J. **11**(3), 75–81 (2014)
33. Edouard, R., Chibane, H., Cavallucci, D.: New characterizing method of a 3D parametric lattice structure. FME Trans. **49**(4), 894–895 (2021). https://doi.org/10.5937/fme2104894E
34. Li, D., Liao, W., Dai, N., Xie, Y.M.: Comparison of mechanical properties and energy absorption of sheet-based and strut-based Gyroid cellular structures with graded densities. Materials **12**(13), 2183 (2019). https://doi.org/10.3390/ma12132183

AI Based Patent Analyzer for Suggesting Solutive Actions and Graphical Triggers During Problem Solving

Davide Russo[1(✉)] and David Gervasoni[2(✉)]

[1] University of Bergamo, Viale Marconi 5, 24044 Dalmine, Italy
davide.russo@unibg.it
[2] TRIX SRL, Viale Mazzini 11, Correggio, Reggio Emilia, Italy
david.gervasoni@trix.ai

Abstract. This paper proposes an idea for developing a computational model of creative processes in design. This model facilitates and accelerates idea generation in the inventive design, increasing the solution space definition by suggesting technical actions and graphical triggers.

The problem solver has to state the required design objective using any verbal action, then an automatic system generates an appropriate set of triggering actions indicating different ways of accomplishing that goal. In addition, for each verb is associated a list of evocative images indicating how that action can be implemented in space/time and through specific physical effects. The system is capable of handling the huge number of verbs that the English language offers. To select all functional verbs of the technical lexicon, the patent database has been processed using the most advanced text mining techniques. Among them, a customized version of Word2Vec model has been exploited to learn word/actions associations from a large corpus of patents.

The article explains how the libraries have been created, the progress the software prototype and the results of a first validation campaign.

Keywords: CAI - Computer aided inventing · Cosine similarity · Problem solving · Creative trigger · TRIZ · AI

1 Introduction

Creative processes in computer-aided design are identified as those that introduce new design variables into the design process. In TRIZ, variables are usually associated to new resources helping designer to go out of routine design towards a design space extended.

But how can these resources be suggested during the problem-solving phase? TRIZ experts are used to cross-fertilize different tools to accurately scan the problem space: they use abstraction, continuous problem reformulations in functional/temporal/spatial terms, decomposition in sub-problems at different levels of detail up to the identification of the technical parameters that control the contradictions.

Published by Springer Nature Switzerland AG 2022
R. Nowak et al. (Eds.): TFC 2022, IFIP AICT 655, pp. 187–197, 2022.
https://doi.org/10.1007/978-3-031-17288-5_17

Attempts to replicate this model using CAI (computer aided inventing) system are very topical, just think of semantic engines for functional search and physical effects [1–3], proposals for integrating semantics and case-based reasoning CBR [4], morphological matrix [5, 6], Information retrieval by artificial intelligence [7], up to the latest thinking machines for writing patents [8]. But to date we can say that we are still far from replicating human reasoning. Although very different all of them have in common the choice of patents as knowledge sources and the use of text mining tools to extract key technical concepts from them. The advent of new natural language processing algorithms, word and sentence embeddings, and pre-trained language models such as BERT (Bidirectional Encoder Representations from Transformers) [9] has accelerated a race toward new paradigms of CAI computer aided inventing that has origins far back in time.

This article is part of this series of attempts. Its purpose is suggesting ideas, in form of a combination of actions and related images helping to adapt the verb to the specific context. This simplified approach, although it cannot replicate the complexity of a methodological path such as TRIZ, has the advantage that it can also be used by non-experts in problem solving. Section 2 shows the state of the art of tools for CAI development. In Sect. 3, the construction of the language library is explained to suggest other ways to functionally address the high-level problem. The graphical triggering mechanism is instead introduced in Sect. 4. This is followed by the results of a validation campaign and a final chapter on the progress of the work and future developments.

2 State of Art

Solving a problem by suggesting the good action is not a novelty. The literature counts numerous methodological approaches based on the systematic use of verbal forms, pointer to effects, Inventive Principles, Function Oriented Search, Matchem-ib [10], Scamper [11], functional basis by Nist [12] and other forms of cataloguing typical of linguists studying the English verb lexicon.

Theoretical methods are supported by information technology to make the most of the technical knowledge base, especially patent knowledge that is widely used in the TRIZ community. The pioneer of all CAI software was Goldfire Innovator [1], that in the early 2000s showed the potential of a SAO-based (subject–action–object) text mining techniques for implementing FOS approach. Goldfire's most important limitations were that it worked almost exclusively on SAO triads, and that it had problems handling the Object if it appeared as complex multiwords. Today we know that the use of SAOs alone is very limiting [13], in spite of other linguistic forms that are more common in the English language (e.g. for + …ing, to + …inf).

Luckily, research on natural language processing has made big leaps forward in the last years, thanks to projects such as IBM Watson [14], GPT-3 by OpenAI [15], BERT by Google [9], and open-source word embedding models like Word2Vec, GloVe, FastAI. These advancements, matched to a cloud computing environment, have allowed the processing of huge corpora in a reasonable computing time.

The only drawback of these models is the fact that they are trained on general-purpose datasets (Wikipedia, BookCorpus, etc.), and they lack knowledge of technical and patent lexical terms.

Although important steps have been made [16], automatic technology features recognition is still an open problem in order to well understand technical content.

Automatic technology features recognition is still an open problem in order to well understand technical content.

In this work, the ability to automatically recognize the function of a component described within a patent text is crucial. The software system has to work real time on a corpus of many millions of documents without any form of limitation.

Thanks to this innovative technology developed by the start-up TRIX it was possible to develop the verb libraries described in more detail in the next chapter.

3 How to Create a Problem-Solving Actions Library

The English language has more than one million terms, of which about 10% are verbal predicates, approximately just over 100,000 terms. Among these, an important number that is impossible to estimate is represented by functional verbs, i.e. useful in problem solving to suggest a solution.

To identify this set of relevant verbs there are no ready-made text mining tools in the literature. It was therefore necessary to start from a manual classification of verb forms. Two different models of classification were taken as reference: the Functional basis of NIST and the classification proposed by Beth Levin [17].

3.1 Functional Basis

Functional basis has been conceived to support functional basis of an artifact. It is a collection of functional verbs hierarchically organized into categories according to their level of generality. Inspired by the modelling of Paul & Beitz, Functional basis has been subject to improvements by many authors until its version "Functional basis reconciled", which includes the NIST taxonomy [12]. The final result is a model characterized by 3 levels of classification and a list of correlated verbs as shown in figure below. In each category the list of verbs is exclusive to eliminate ambiguity (Fig. 1).

Class (Primary)	*Secondary*	*Tertiary*	*Correspondents*
Branch	Separate		Isolate, sever, disjoin
		Divide	Detach, *isolate*, release, sort, split, disconnect, subtract
		Extract	Refine, filter, purify, percolate, strain, *clear*
		Remove	Cut, drill, lathe, polish, sand
	Distribute		Diffuse, dispel, disperse, dissipate, diverge, scatter
Channel	Import		Form entrance, *allow*, input, *capture*
	Export		Dispose, eject, *emit*, empty, *remove*, destroy, eliminate
	Transfer		Carry, deliver
		Transport	Advance, lift, move
		Transmit	Conduct, convey
	Guide		Direct, shift, steer, straighten, switch
		Translate	Move, relocate
		Rotate	Spin, turn
		Allow DOF	*Constrain*, unfasten, unlock

Fig. 1. Functional basis reconciled by Nist.

The total number of verbs belonging to this classification is about 50 divided into 8 macro categories. The corresponding verbs that can be associated with these categories unfortunately do not exceed 200.

3.2 Beth Levin Classification

Beth Levin is an American linguist and Professor in the Humanities at Stanford University who conducts her research on the English language and its lexicon [17]. Her recent work investigates the representation of events and the ways in which events and their participants are expressed in English. Those studies necessitate the development of verbal meaning models, since verbs are the main element used to describe events.

She proposes a classification of over 3 thousand verbs, grouped according to their "diathesis alternations" and into classes of actions with similar meanings. It counts 58 classes, typically with about five subclasses. The guiding principle is semantics first, syntax second; so, for example, a class of near-synonyms may be divided according to whether or not it allows a certain diathesis alternation. Verbal forms have no connection to the more technical world of engineering and problem solving, but they are a great starting point for excluding non-functional ones and exploring the others.

Unlike functional basis, a single class can contain many more verbs, and individual verbs can repeat in multiple categories since they can have different meanings.

An example of a class is "funnel" verbs, consisting of: bang, channel, dip, dump, hammer, ladle, pound, push, rake, ram, scoop, scrape, shake, shovel, siphon, spoon, squash, squeeze, squish, sweep, tuck, wad, wedge, wipe, wring.

Again, the greatest limitation of this approach is the limited sample compared to the total number of verbs present in the English language. The verbs already classified and related to the technical sphere are little more than a hundred.

3.3 Building Our Own Patent Based Library

In order to implement the system to generate suggestions, it was necessary to create a new library of functional verbs, much richer than those seen so far in the literature. The first step was to put together the work already described in the previous paragraphs. The system architecturis outlined in the Fig. 2 and is composed of multiple actors like cloud functions and operations, virtual machines and elasticsearch databases. The cloud infrastructure leverages the Google Cloud Platform (GCP) system and the python language has been used in all stages of development.

The verbs found were organized according to a very simple 2-level hierarchy: a generic verb represented by a whole category and a list of verbs associated with it. Generic verbs are unique, while verbs included in the second level can appear several times in different categories. Compared to previous work, the functional verbs associated with the second level are much more numerous, to date just over 2 thousand terms have already been validated. Each of these verbs was then expanded in turn semantically using its own network of lexical resources by using synonyms, troponyms, iperonyms.

This approach allowed us an initial level of pool expansion that was notable greater than previous approaches, but still limited in comparison to the ultimate goal. The biggest

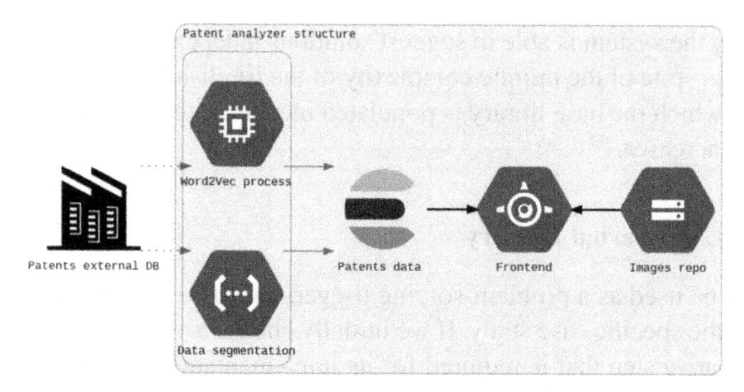

Fig. 2. Overall architecture

problem is the limitations in the available semantic network libraries, which are designed for general purpose languages and almost never for the technical lexicon.

In order to complete the work and to guarantee the total coverage of the verbal forms compatible with the problem solving, it was decided to extract the most specific verbal forms directly from the patent corpus. Currently, our patent libraries contain over 120 million documents, all digitized and translated into English. These documents were obtained from one of the largest accurate patent database through the use of a commercial api; the api provides the data in a structured json format with logically segmented fields like textual sections, images, bibliographical information.

Each patent necessarily contains a description of the starting problem, the solution and a description of how it should work. The patent pool is semantically very rich and contains the technical vocabulary typical of every technological area known to date without exception.

To identify the verbal forms contained in the corpus we used an algorithm that combines part-of-speech tagging and word embeddings. Initially, all the words in the text have been tagged with their part-of-speech function (verb, noun, etc.); then, a word embedding model (Word2Vec) has been trained using the text of the words and their part-of-speech as a training feature.

Word embeddings model are a type of neural network that maps words to a vector space of fixed dimension (in this case 1024), they learn vector positions through training on big textual corpus. Once the training is done, it's possible to find related words by computing the distance among their vector. The most common technique to compute those distances is cosine similarity [18].

In the specific case presented in this paper, as the model has been trained on the patent domain, it allows for better precision and reliability in a technical problem-solving. Furthermore, using the part-of-speech as a feature allows precise searches for similar or correlated actions.

At this point, for each functional verb contained in the corpus it is possible to calculate a distance with respect to the reference verbs already identified in the library; moreover, for each suggested verb there is a list of further verbs that are close to it in meaning, therefore potentially useful as further suggestions.

In this way the system is able to suggest solutions independently from the proposed verbal target, in spite of the infinite complexity of the English language. As the number of verbs with which the base library is populated increases, the level of precision of the system itself increases.

3.4 How to Use a Verbal Library

Any verb can be used as a problem-solving trigger, but to be effective it needs contextualization to the specific case study. If we initially choose a very specific verb, e.g., "*to weld*", the creative step that is required for its implementation will be more restrained than with a verb that is broader in meaning, such as the verb "*to join*".

There are in fact many different ways in which I can join an object, not only s*oldering* but for example also *gluing or embedding.*

Greater generality tends to create more associations between alternative verbs, conversely, specificity of meaning reduces associations but facilitates its application to the problem. The most specific situation is when we choose a verb that already contains a physical effect in its meaning (e.g. *freezing*).

One last case that might occur is the extreme case of a too generic verb (e.g. *ameliorate, improve, increase, decrease, maintain,* etc.); these verbs alone are not useful to indicate a solution direction but if coupled with a technical parameter (e.g. increase the thickness) they are really effective. The way to treat this last category of verbs is postponed to future publications.

The proposed method requires the user to make a minimal effort: only to indicate a verb describing the objective to be achieved or the action that is thought to be useful for solving the problem (for example I want to *remove* limestone or I already decided to mechanically *break* it). The task of the library is to broaden the concept and allow navigation by functional verbs, used in patents in similar ways. Each suggested verb, in order to be used appropriately, needs its contextualization to the specific case study. The way the authors have chosen to aid in its implementation is to pair it with imagery.

4 How to Create a Graphical Triggers Library

For each verbal form described in the previous chapter, the system must return a predefined set of images. The images have been manually collected. They come from a direct interpretation of TRIZ approach and are conceived to work also with those who have no experience in problem solving.

Not being able to create images for all the thousands of verbs potentially useful for problem solving, a subset of preferential verbs was chosen, covering the first hierarchical level of the classification and over. Verbs with a similar meaning will refer to the same set of images.

The main difficulty in the compilation of this library consists in providing images capable of stimulating creativity. Therefore, the choice of images cannot be random but follows precise rules.

4.1 Rules for Creating Images

Although the library of images was created by hand, the process by which they were identified followed strict rules:

1. *By Physical effects*. Each action can be carried out through different physical effects. The first step will be to find images of applications of that function associated with a physical principle that is well recognizable. If the image alone is not enough evocative, or the effect is too hidden, it is possible to associate it with a very brief description at the bottom. Considering the verb *to shake* for example, I can find images about sound waves that shatter glass (Acoustic effect), shaker of olive trees (mechanical effect) or wind that makes bridges oscillate (resonance effect).
2. *By Space-mode*. A further selection criterion concerns the ways in which action can be implemented by considering the operating space. These include Inside/Outside, Above/Below, High/Low, Independent/Part of a process.
3. *By Time-mode*. Similarly, the operational time of the action is considered: full time/partial, slow/fast, absence or presence, dynamic or static, repetitive or not, involuntary/voluntary effect, autonomous or with intervention, ordered or random.
4. *By detail level*. The size scale is then considered: macro/micro, oversized or undersized, changes of state (liquid, solid, gaseous), quantity.
5. *By other criteria*. Whether involves other actions or if it is an isolated, level of precision, multifunction.

4.2 Prototype - SW Implementation

The technology that allows the navigation of functional verbs with the possibility of associating any verb with a set of images can have many applications. Even just in the field of problem solving, it can be implemented in different ways. To date, the implementation phase of this technology has included the creation of a user interface with the sole objective of supporting an experimental campaign to test the effectiveness of the methodology. Given a problem, the user enters a verb describing the goal to be achieved and the system automatically get back a list of related actions. From here the user can choose one of the suggested actions or enter a new one resulting from the verbal stimulus action. We can go on iteratively until we come across a new verb that convinces us of its effectiveness. When the desired action has been chosen, the system will generate a set of images and the short list of related actions from which the starting one was drawn.

If an image evokes a solution, it can be easily uploaded to the system by clicking on the triggering image and writing a free-form text in the form created for it. Being able to record every user action allows developers to make analytics, monitor which paths statistically (graphical and verbal) are most followed, how many iterations are made before fixing a response (Fig. 3).

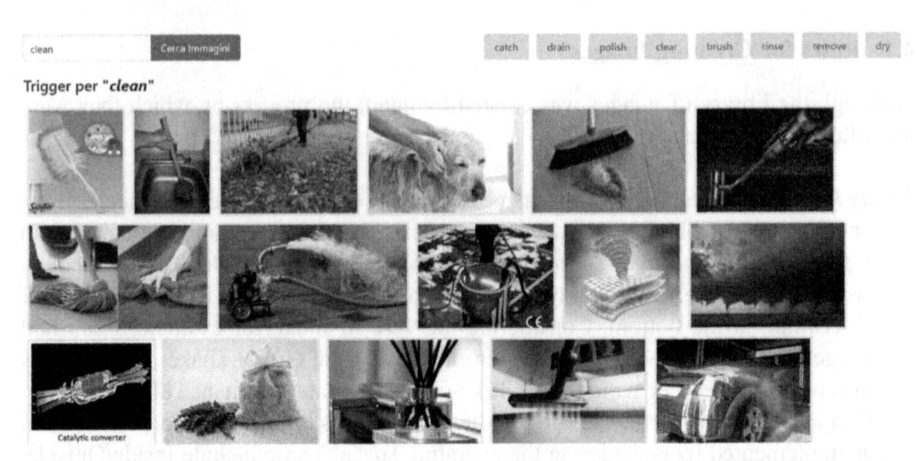

Fig. 3. User interface of the software prototype. In the picture a partial list of the images for "Cleaning" actions and in the right top of the page the actions list of near-actions.

5 Evaluation

The stated objective of this method is to give hints to enlarge the number of the practicable ways and increase the variety of solutions in terms of physical domains. The effectiveness of graphic activators has been tested with over than 100 first year engineering management students and 40 students at the last year of academic course in mechanical engineering with a mechatronic specialization (who followed the TRIZ course). Both first-year students and a group of engineering PhD students were used for the control sample.

5.1 Validation Criteria

The objectives of the tests were to:

a. estimate the *number of solutions* suggested for each problem
b. define *breadth*, i.e. the number of solutions attributable to different physical effects.
c. Compare results with those of a international test promoted by Iouri Belski using Matchemib [10].

5.2 Case Studies

In order to test the efficacy of the method eight problems were considered: crankshaft breakage, killing lice in the scalp, removing ice from the wings of airplanes, cleaning the windows of skyscrapers, order box matches, rotate a solar panel to ensure its efficiency, stop leaks in oil pipelines. A final problem was about deposition of limestone residues in pipes, the same problem already proposed in a previous test on Matchemib. For each problem students had 16 min maximum for record all solutions on the software platform.

5.3 Main Results

A first control group consisting of first-year students and a second control group consisting of PhD students was set up to evaluate the test. In addition, we had available the results of the International Matchemib Test Control Group [10]. The test showed that the mean of the solutions of the control sample was rather low, compared to the international sample. This can be explained by the fact that the current evaluation of the solutions was stricter, the solutions that resembled each other were all eliminated.

In contrast, there were no substantial differences between the control group consisting of first-year and doctoral students.

As shown in Table 1, the results obtained by applying the proposed method to first-year students are slightly lower than the results conducted with the international Matchemib test, while when it is applied by students skilled in TRIZ, it gives the best results, higher even than those of the international test (quite 7 solutions each vs 6, 4 of international test students).

If we compare results with and without method, the first-year students improve the average from 2.5 to 6 (+255%), while students of international test made a growth of rate about 144%. Unfortunately, it was not possible to calculate the growth of rate for TRIZ students due to the lack of the corresponding control group.

For breadth the results are similar, but with some differences: the percentage growth for first year students using the method compared to their control group is +152%, very similar to what happened in the international Matchemib group, with +167%. On an absolute level, the average breadth for the first-year group is about 2.8 different physical effects, while 3.05 for the TRIZ group, and 4.7 for Matchemib (which is a method designed precisely to expand the breadth).

Table 1. Test results.

	Control group			Results		
	1 year-Eng	Phd-Eng	Intern. test	1 year-Eng	TRIZ group	Intern. test
n° solutions	2,3	2,9	4,4	6	6,95	6,4
Rate of growth				+255%		+144%
Breadth	1,85	1,9	2,8	2,8	3,05	4,7
Rate of growth				+152%		+167%

6 Conclusion and Future Developments

The method proposed in this paper is not an alternative to the traditional TRIZ approach, but it wants to be a contribute to systematize the research of the variants in the early phases of the process. It does not help in the reformulation of complex problems, for example to better frame the problem and to reformulate it in simpler form. In fact, to use this tool it is assumed to have already defined the specific problem on which to intervene. From here the system can give hints to enlarge the number of the practicable ways. Initial results from a test of nearly 200 engineering students have shown that the proposed method can help triple the number of solutions to simple problems, while also increasing the variety of solutions in terms of physical effects. It can also help students who are TRIZ experts to further diversify their skills.

The results obtained are due to an impressive software infrastructure able to eliminate certain technological barriers that affected syntactic parsers in the past: due to the cost-effective availability of computing resources it is now possible to process large volumes of data in relatively short periods of time.

This system processes a corpus of many millions of documents without any form of limitation and it is able to manage all functional verbs from the over 100,000 verbs in the English language. Processed data are then made real-time searchable by keyword and also by mean of relationships between them.

Further developments will involve the use of BERT models, more precisely Google's Bert-for-patents [19], a BERT model trained on more than 100 million patent documents, trained on all parts of a patent (abstract, claims, description).

Finally, the latest innovation lies in graphic activators. At the moment it has not been possible to create a set of verbs for each functional verb, but the extension work is in progress and already in this way an important number have been covered.

References

1. Tsourikov, V.M.: Inventive machine: second generation. AI & Soc. **7**(1), 62–77 (1993)
2. Kucharavy, D.: Thoughts about history of Inventive Machine Projects. Presentation at LICIA/LGECO-Design Engineering Laboratory INSA Strasbourg (2011). http://www.seecore.org/d/20110923(2).pdf
3. Montecchi, T., Russo, D.: FBOS: function/behaviour–oriented search. Procedia Eng. **131**, 140–149 (2015)
4. Zhang, P., Cavallucci, D., Bai, Z., Zanni-Merk, C.: Facilitating engineers abilities to solve inventive problems using CBR and semantic similarity. In: Cavallucci, D., De Guio, R., Koziołek, S. (eds.) TFC 2018. IFIP Advances in Information and Communication Technology, vol. 541, pp. 204–212. Springer, Cham (2018). https://doi.org/10.1007/978-3-030-02456-7_17
5. Rakov, D.: Okkam-advanced morphological approach as method for computer aided innovation (CAI). MATEC Web Conf. **298**, 00120, 1–9 (2019)
6. Korobkin, D., Fomenkov, S., Vereschak, G., Kolesnikov, S., Tolokin, D., Kravets, A.G.: The formation of morphological matrix based on an ontology "patent representation of technical systems" for the search of innovative technical solutions. In: Kravets, A.G., Bolshakov, A.A., Shcherbakov, M.V. (eds.) Cyber-Physical Systems. Studies in Systems, Decision and Control, vol. 350, pp. 149–160. Springer, Cham (2021). https://doi.org/10.1007/978-3-030-67892-0_13
7. Hanifi, M., et al.: Problem formulation in inventive design using Doc2vec and Cosine Similarity as Artificial Intelligence methods and Scientific Papers. Eng. Appl. Artif. Intell. **109**, 104661 (2022)
8. Wendrich, R.E.: Computer aided creative thinking machines (CaXTus). Comput.-Aided Des. Appl. **18**(6), 1390–1409 (2021)
9. Devlin, J., et al.: BERT: pre-training of deep bidirectional transformers for language understanding. arXiv preprint arXiv:1810.04805 (2018)

10. Belski, I., Skiadopoulos, A., Aranda-Mena, G., Cascini, G., Russo, D.: Engineering creativity: the influence of general knowledge and thinking heuristics. In: Chechurin, L., Collan, M. (eds.) Advances in Systematic Creativity, pp. 245–263. Palgrave Macmillan, Cham (2019). https://doi.org/10.1007/978-3-319-78075-7_15
11. Eberle, B.: Scamper on: Games for Imagination Development. Prufrock Press Inc., Austin (1996)
12. Hirtz, J., et al.: A functional basis for engineering design: reconciling and evolving previous efforts. Res. Eng. Des. **13**(2), 65–82 (2002)
13. Russo, D., Spreafico, M., Precorvi, A.: Discovering new business opportunities with dependent semantic parsers. Comput. Ind. **123**, 103330 (2020)
14. High, R.: The era of cognitive systems: an inside look at IBM Watson and how it works. IBM Corporation, Redbooks, vol. 1, p. 16 (2012)
15. Floridi, L., Chiriatti, M.: GPT-3: its nature, scope, limits, and consequences. Minds Mach. **30**(4), 681–694 (2020)
16. Lee, J.-S., Hsiang, J.: Patentbert: patent classification with fine-tuning a pre-trained BERT model. arXiv preprint arXiv:1906.02124 (2019)
17. Levin, B.: English Verb Classes and Alternations a Preliminary Investigation. The University of Chicago Press, Chicago (1993)
18. Mikolov, T., et al.: Efficient estimation of word representations in vector space. arXiv preprint arXiv:1301.3781 (2013)
19. Srebrovic, R., Yonamine, J.: Leveraging the BERT algorithm for Patents with TensorFlow and BigQuery. https://services.google.com/fh/files/blogs/bert_for_patents_white_paper.pdf

Automated TRIZ Domain Mapping

Guillaume Guarino(✉) and Denis Cavallucci

ICUBE/CSIP Team (UMR CNRS 7357)-INSA Strasbourg, Strasbourg, France
{guillaume.guarino,denis.cavallucci}@insa-strasbourg.fr

Abstract. The automatic analysis of patents is still one of the main challenges in R&D, particularly in terms of establishing automatic states of the art. Indeed, this is still mostly done manually, which is very time-consuming. The progress of artificial intelligence allows us to go a step further in the understanding of patents and in particular of the issues they address. In this paper we present an end-to-end tool that allows us to map the main trends in term of research directions in a sector in a few minutes from a simple keyword search. To do so, we will rely on TRIZ formalization with contradictions and evaluation parameters.

Keywords: Domain mapping · TRIZ · Deep learning

1 Introduction

The identification of key issues in a field is very important to guide the research work of companies, especially when a company enters a new field of activity. With the technological challenges that lie ahead due to the energy transition, many companies will have to innovate in new areas. Reading patents does not provide a reliable mapping of a field, particularly because it is impossible for a team of engineers to read hundreds or even thousands of patents in a few days. In parallel, TRIZ theory allows problems to be formulated in a uniform manner across domains, in the form of contradictions. Progress in AI, particularly with deep neural networks, provides extremely powerful tools for analyzing textual content. Indeed, in an increasing number of NLP tasks, algorithms are actually becoming better than humans. It therefore seems possible to create a tool for fined-grained analyses of patents contents, regardless of the field, to provide a mapping of it in terms of research priorities. This is what we are going to present in this paper. From a simple search by domain, by keyword or even by applicant names we are able to provide a map in the form of a matrix highlighting the contradictions that are most dealt with by the patents. To build this map, an automatic extraction of contradictions resolved by each of the analyzed patents had to be set up. We will therefore start by presenting the contradiction extraction mechanism, then the construction of the map. Finally, we will show concrete examples of mappings.

Published by Springer Nature Switzerland AG 2022
R. Nowak et al. (Eds.): TFC 2022, IFIP AICT 655, pp. 198–205, 2022.
https://doi.org/10.1007/978-3-031-17288-5_18

2 Related Work

So far, mapping has been proposed for patents mainly to represent the technologies used in certain fields or companies [3–5]. [1] proposes to visualize conflicts between patents, [2] proposes to identify TRIZ trends for given technologies.

Contradiction mining is well known theme in TRIZ domain but not much approaches tackled this problem as understanding the content of a patent remains a very difficult task. Despite the importance of this challenge and the NLP techniques that have been developed, very few have tackled this problem. However, we find classification of inventive principles [6–8], extraction of parameters [9, 10], reconstruction of TRIZ matrices for targeted domains [11].

However, these methods often use simplifications such as the reduction of the number of inventive principles [6, 7], the use of keywords or key phrases [8, 9] or assumptions on the structure of patents [10] which makes these approaches unusable in practice on "new" content.

3 Contradiction Mining

Understanding which contradiction a patent addresses is a very difficult task. Indeed, many patents are, in fact, not inventive (as Altshuler himself concluded) and therefore do not resolve a contradiction. Moreover, the drafting of a patent depends very much on who wrote it and the contradiction that the patent resolves, if there is one, is not always explicitly cited. All these limitations mean that the analysis process must be fine-tuned and have a validation mechanism.

3.1 Identification of the Areas of Interest

The preliminary step to identifying contradictions is to identify the areas that are likely to be most interesting. Patents have the advantage of being structured content. Some parts are indispensable, such as the abstract, the description or the claims. Other parts are not always present or are sub-parts of the description, such as the summary or the state of the art. A preliminary manual study convinced us to use the state of the art as a source of information concerning the contradictions that the patent seeks to resolve. Indeed, one often finds structures such as "To improve (EP 1), Patent... Proposes to.... However, this implies that (EP 2) is degraded". We found that the contradictions encountered by the state of the art were in fact the contradictions that the patents in question sought to overcome. The textual analysis will therefore only be carried out on the part of the state of the art of the patents which is previously extracted.

3.2 Mining Process

The extraction process is separated into three distinct phases. The first phase is an automatic summary phase with the selection of sentences with the highest probability of containing one or more parameters related to the contradiction that the patent allows to solve. For example, in the US5316377 patent two distinct sentences can be selected:

"For limited use or lightweight applications, such as with barbecue carts, lawn mowers, trash containers and many other devices, plastic wheels can serve the same purpose, but at relatively lower costs."

"Inherent negatives of such wheels, however, are that the core of the wheels are hollow and thus the wheels tend to be noisy."

The first of these two sentences isolates a "lightweight" evaluation parameter, while the second isolates "noisy". We are therefore dealing here with a weight/noise contradiction. The first of these sentences will be called the "first part of the contradiction" while the second will be called the "second part of the contradiction". Automatic summarization is a common task in language processing. Deep neural networks such as transformers [15] are used. These neural networks have become popular in recent years in all language processing tasks thanks to their ability to be pre-trained. The best known of these is BERT (Fig. 1).

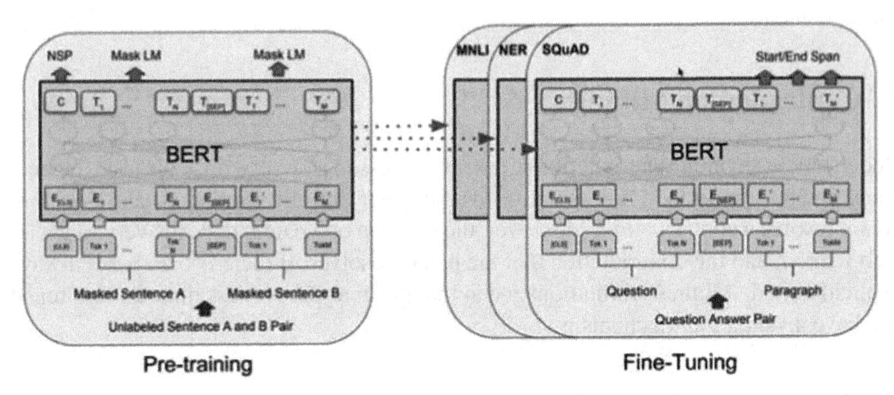

Fig. 1. BERT [12], a pretrained deep neural network

BERT [12] was introduced by Google in 2019 and allows to generate contextual representations of tokens for textual content. These vector representations can then be used to classify tokens, documents, etc. These tools have replaced non-contextual representations such as Word2Vec [16] which are less efficient. Indeed, with contextual representations, the same word can have different representations depending on the words that are close to it in the sentence or paragraph. This makes it possible to integrate better quality information on the meaning of words. It is relatively easy to use this type of network for automatic summarisation (Fig. 2). An additional Transformer layer on top of the encoder is used to combine the special CLS token representations intended to be the sentence representations [14]. The decision is then made by two (binary) sentence classifiers which will select the best first part of the contradiction and the best second part of the contradiction.

A document classifier is associated to the summarization model to validate the extraction. It is supposed to predict whether there is a contradiction to mine. If it predicts that the patent indeed contains a contradiction the sentences considered as first and second part of contradictions are assumed to be correct and the process continues. If not, the results are ignored.

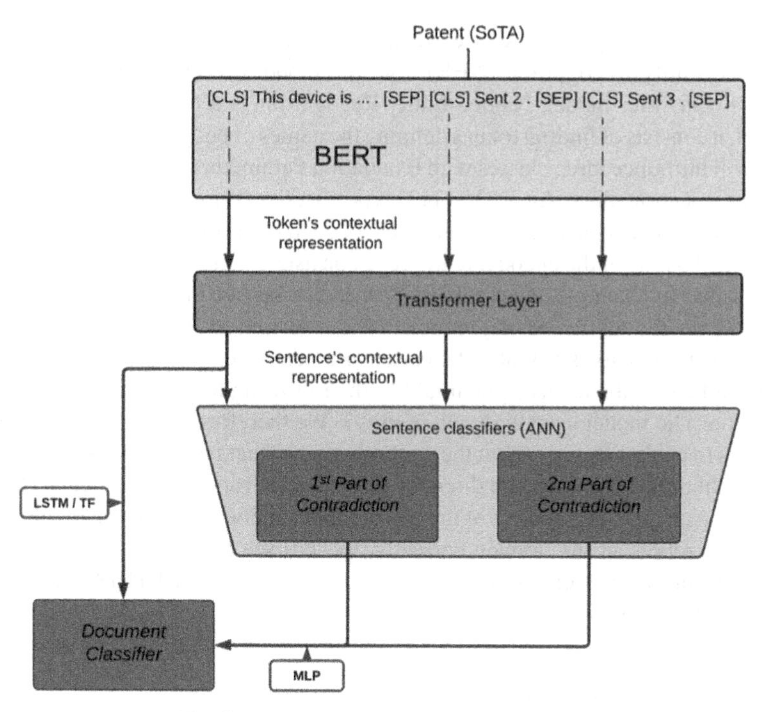

Fig. 2. BERT for extractive summarization

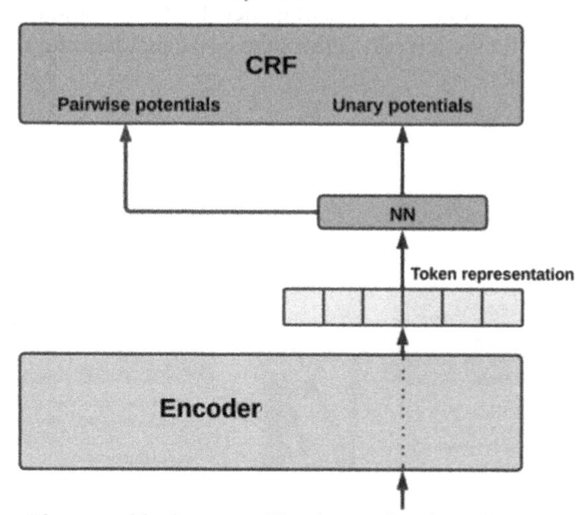

Fig. 3. Parameters mining model

The last step of the process is the extraction of parameters. As this step is very similar to a Named Entity Recognition (NER) task, the model used will be a classical NER Encoder + Classifier model. Named Entity Recognition is a token-wise classification. In general, it consists of finding tokens defining the names of people or locations but in our case we will introduce three classes with Evaluation Parameters (EP), Action Parameters (AP) and a rejection class. An XLNet [13] type encoder is used as it is one of the most efficient currently. It is also a pre-trained encoder like BERT. A Conditional Random Field is added on top of the encoder to improve consistency in predictions. Indeed, with a linear classifier for example, the predictions for each token are independent of each other, i.e. if a token is thought to belong to an EP, the following token will not be influenced by this prediction. This is actually the case as PEs are often formed by several words, so when a token is labelled as a potential PE, the following tokens are more likely to be part of it too. The model used is shown in Fig. 3. We therefore have a complete process allowing us to extract from a patent the contradiction(s) that it seeks to resolve. This will allow us to find the main research directions in the field from a set of patents. However, if keeping the original parameters of the patents (thus applied to the domain) also makes it possible to represent the domain correctly, we will see that in the optics of keeping a constant form of representation we fall back on the generalized TRIZ contradictions with the 39 original parameters.

4 Mapping Construction

From a multi-criteria search, our mapping tool allows to select a subset of patents that will be used to map the domain. After running the mining process, a selection of best phrases appears for each patent. The left part gathers the best candidates for the first part of the contradiction while the left part gathers the best candidates for the second part of the contradiction (Fig. 4.)

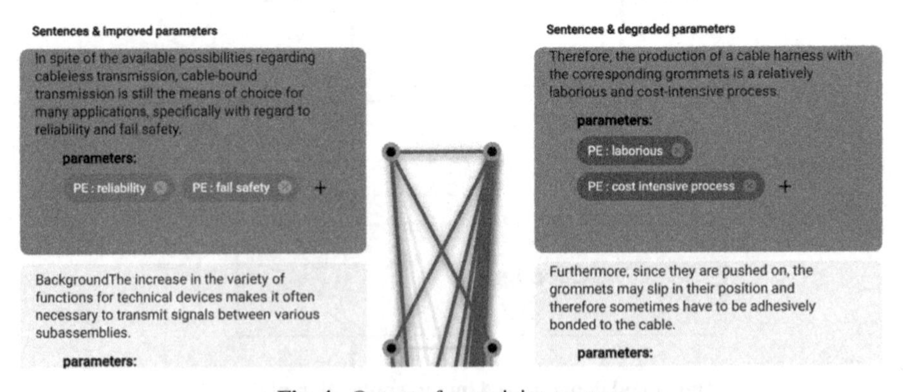

Fig. 4. Output of our mining process

The parameters of the first sentence of each part are then translated into the original TRIZ parameters before being displayed. This translation is done by similarity. An embedding is constructed for each parameter and this is then compared by cosine similarity with those of the TRIZ parameters. Experimentally, we have shown that building the embedding from a sentence Parameter + must be studied allows to build better embeddings than by taking only the few tokens contained in the parameters. Indeed, the model used for the construction of the embeddings is a model learned on sentence similarity datasets and it is therefore logical that it works best with sentences.

5 Visualization

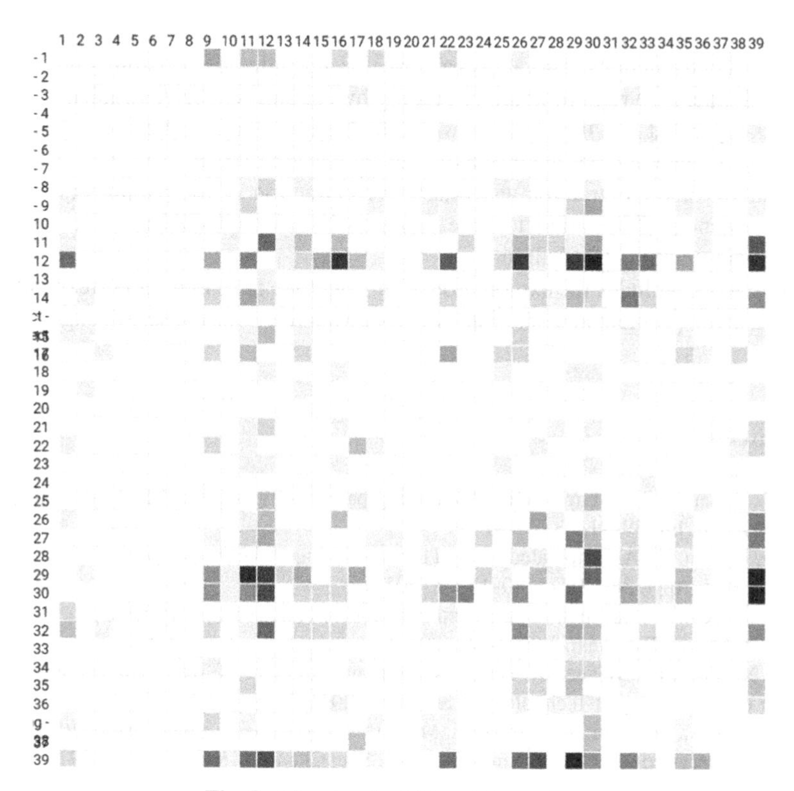

Fig. 5. Mapping for 100 molding patents

The map (Fig. 5) has the form of a 39 × 39 matrix. The first part of the contradiction is on the ordinate while the second part of the contradiction is on the abscissa. We only deal here with contradictions between two parameters. In the quite frequent case where several evaluation parameters are in contradiction with several evaluation parameters (if there are several evaluation parameters in one or more of the selected sentences in the patent). In this case, the contradictions are decoupled and it is considered that each of the parameters in the first part of the contradiction will be in contradiction with each of the parameters in the second part of the contradiction. For example, in Fig. 4, reliability is considered to be in contradiction with "laborious" and "cost intensive process", and the same is true for "safety". The colour scale is then adjusted with a maximum intensity for the most common contradiction. In this example, patents with "molding" in the title were used. It can be seen that the most common contradiction is between parameters 29 and 39 i.e. manufacturing precision and productivity. This would of course require each patent to be checked by an expert in the field but at first sight the results do not seem to be outliers.

6 Conclusion and Perspectives

In this paper we have presented an algorithm to build a multi-domain or single-domain mapping of contradictions. This allows to have, in a few minutes, a visibility on a domain that would otherwise require hours of reading by engineers. The contradiction mining process is separated into three phases with a first phase of automatic summarization to isolate the sentences containing the parameters of the contradiction resolved by a patent, the validation of the extraction and finally the extraction of the parameters from the selected sentences. The parameters are then translated into TRIZ parameters to allow multi-domain mapping if required.

Future work includes the exploitation of this pipeline to search for correspondences between a given problem and solutions. In this way, a real solution search engine can be set up.

References

1. Li, Z., Atherton, M., Harrison, D.: Identifying patent conflicts: TRIZ-Led patent mapping. World Patent Inf. **39**, 11–23 (2014)
2. Yoon, J., Kim, K.: An automated method for identifying TRIZ evolution trends from patents. Expert Syst. Appl. **38**(12), 15540–15548 (2011)
3. Russo, D., et al.: A new patent-based approach for technology mapping in the pharmaceutical domain. Pharm. Pat. Anal. **2**, 611–627 (2013)
4. Guo, X., Park, H., Magee, C.: Decomposition and Analysis of Technological domains for better understanding of Technological Structure (2016)
5. Suominen, A., Toivanen, H., Seppänen, M.: Firms' knowledge profiles: mapping patent data with unsupervised learning. Technol. Forecast. Soc. Chang. **115**, 131–142 (2017)
6. Loh, H., He, C., Shen, L.: Automatic classification of patent documents for TRIZ users. World Patent Inf. **28**, 6–13 (2006). https://doi.org/10.1016/j.wpi.2005.07.007
7. Cong, H., Tong, L.H.: Grouping of TRIZ inventive principles to facilitate automatic patent classification. Expert Syst. Appl. **34**, 788–795 (2008)
8. Liang, Y., et al.: Computer-aided classification of patents oriented to TRIZ. In: 2009 IEEE International Conference on Industrial Engineering and Engineering Management, pp. 2389–2393 (2009)
9. Chang, H.-T., Chang, C.-Y., Wu, W.-K.: Computerized innovation inspired by existing patents, pp. 1134–1137 (2017)
10. Cascini, G., Russo, D.: Computer-aided analysis of patents and search for TRIZ contradictions. Int. J. Prod. Dev. **4**, 52–67 (2007)

11. Berdyugina, D., Cavallucci, D.: Setting up context-sensitive real-time contradiction matrix of a given field using unstructured texts of patent contents and natural language processing. In: Cavallucci, D., Brad, S., Livotov, P. (eds.) TFC 2020. IAICT, vol. 597, pp. 30–39. Springer, Cham (2020). https://doi.org/10.1007/978-3-030-61295-5_3
12. Devlin, J., et al.: BERT: pre-training of deep bidirectional transformers for language understanding. In: NAACL-HLT (2019)
13. Yang, Z., et al.: XLNet: generalized autoregressive pretraining for language understanding. In: Wallach, H., et al. (eds.) Advances in Neural Information Processing Systems 32, pp. 5754–5764. Curran Associates, Inc. (2019)
14. Liu, Y., Lapata, M.: Text summarization with pretrained encoders. In: EMNLP/IJCNLP (2019)
15. Vaswani, A., et al.: Attention is all you need. In: Guyon, I., et al. (eds.) Advances in Neural Information Processing Systems 30, pp. 5998–6008. Curran Associates, Inc. (2017)
16. Mikolov, T., Chen, K., Corrado, G., Dean, J.: Efficient estimation of word representations in vector space. In: Proceedings of Workshop at ICLR (2013)

Systematic Innovations Supporting IT and AI

Using NLP to Detect Tradeoffs in Employee Reviews

Justus Schollmeyer(✉)

University of Bremen, 28359 Bremen, Germany
justusschollmeyer@gmail.com

Abstract. This paper presents a methodology to identify industry-specific tradeoffs using natural language processing (NLP) to analyze employee reviews from Glassdoor. The analysis is based on 400,000+ reviews from employees working in the financial sector between 2008 and 2020. For each review, the pros and cons sections are classified in a one-to-many approach in terms of the most prevalent topics in the sector. The most prevalent noun chunks within a representative sample of reviews are used as topics. The classification of reviews is based on the cosine similarity between the sentence embeddings of these topics and the sentence embeddings of the comment sections using a sentence-transformer model. Based on this classification, the count of pro-con topic pairs is tested for statistical significance against a control of randomly generated pairings from the same sample of classifications. The process is repeated 10,000 times and only pairs with a p-value $< .05$ (after Bonferroni-Holm correction) are considered. If both, a pro-con pair of topics and its opposite pairing are significantly frequent, this combination of topics qualifies as a tradeoff. Depending on whether the same pairing constitutes a significantly frequent pro-pro pair, it can be determined whether a solution for the tradeoff exists in the sector (at least within the limitations of this approach). Using 13 topic labels, 2 tradeoffs – *work-life balance vs. opportunity* (without solution in the sector) and *work-life balance vs. management* (with solution in the sector) – have been identified. In addition to that, 8 more pro-con pairs were identified as significant.

Keywords: Tradeoffs · Glassdoor · Employee reviews · Natural Language Processing · Machine learning

1 Introduction

The goal of this paper is to demonstrate that the free text in employee reviews from online platforms such as Glassdoor [1] or Indeed [2] can be used to identify sector-specific tradeoffs that organizations face from the perspective of their employees. For this purpose, 439,599 Glassdoor reviews written between 2008 and 2020 by employees from the financial industry in the US will be processed using NLP.

The paper starts with a discussion of why tradeoffs matter to innovation (Sect. 2.1) followed by brief introductions to employee reviews on Glassdoor (Sect. 2.2) and the

Published by Springer Nature Switzerland AG 2022
R. Nowak et al. (Eds.): TFC 2022, IFIP AICT 655, pp. 209–219, 2022.
https://doi.org/10.1007/978-3-031-17288-5_19

use of NLP for their analysis (Sect. 2.3). Section 3 will introduce the method of using the pros and cons sections to identify significantly frequent pro-con pairs and their opposites. Section 4 presents the results followed by a brief discussion of the findings and how to improve the approach further (Sect. 5).

2 Background

2.1 Why Tradeoffs Matter to Innovation

Tradeoffs are problems with conflicting targets. Improving one results in the deterioration of the other and vice versa. Problems that present themselves in this form bear high potential for innovation [3]. This is because to improve the values for both conflicting parameters, it is required to rethink the structure of the system (or some aspects of it).

A simple example: A container to deposit trash is 300 m away from someone's house. The trash gets collected in a bin in the kitchen. The larger the bin, the fewer meters must be walked to deposit the trash over a given period. Increasing the bin size comes at the cost of a heavier bin once filled with trash. "Meters walked to the container in a month" can be seen as a negatively sloped function of the filled bin's mass if everything else remains stable (see left side of Fig. 1). The function defines the options if only the size of the bin can be changed.

If, however, someone decided to go beyond these artificial limitations, they might come up with a solution that improves both parameters to the better, which can come in two forms: The system is modified such that.

(i) the entire function shifts to the better for both parameters, or
(ii) the tradeoff disappears entirely.

Type (i) is a more general version of Schumpeter's definition of innovation: a shift of the production function [4]. Type (ii) is even preferable (at least as far as these parameters are concerned) and lies on the vector of the ideal solution defined in [3] as *function without a system*.

For both types of solutions, creative thinking is required since mental limitations – such as only allowing ourselves to change the size of the bin – must be abandoned. One could e.g., decide to produce less trash. This would shift the function to the better in the sense of (i) (see right side of Fig. 1). One could also decide to integrate the walk to the container into one's daily routines, such as going to work, grocery shopping, jogging, etc. The problem becomes obsolete (ii) since no extra time is being spent on walking to the container if it is on the way of other errands. Unsurprisingly, most people apply this strategy to deal with the tradeoff.

We cannot know for certain that a tradeoff is not resolvable in principle (because we cannot be certain that all our assumptions are true and exhaustive). However, we can check whether a solution to what looks like a tradeoff already exists. If none exists and we find a type (i) or (ii) solution, we have made a discovery or invention that could result in innovation if implemented at scale [4].

The approach of analyzing employee reviews from the financial sector presented in this paper will show how to distinguish tradeoffs that have a solution in the sector from those that seem to have no solution yet.

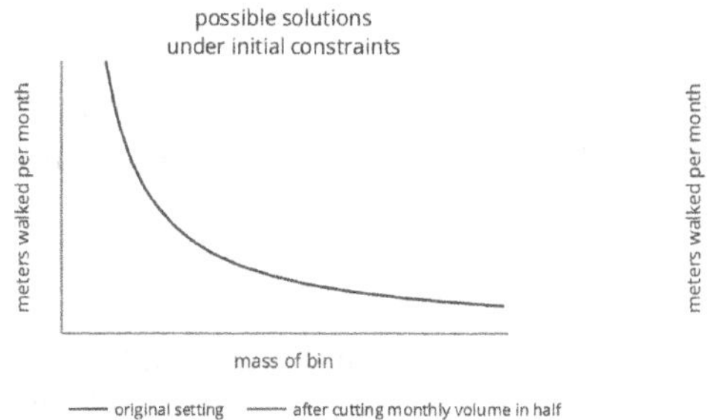

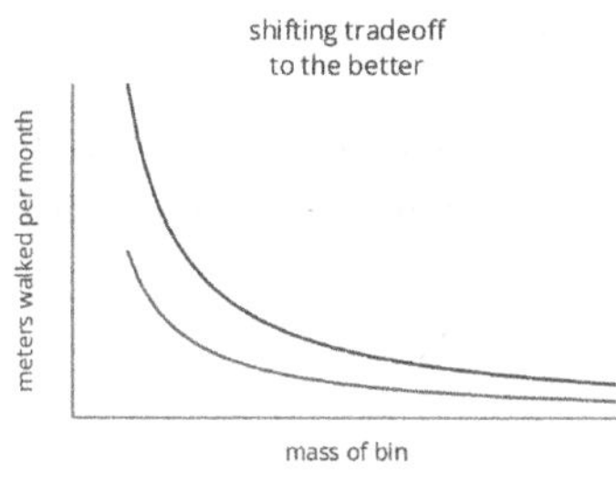

Fig. 1. The left side figure shows the meters walked per month as a function of the mass of the bin. The graph assumes that other variables such as type of trash, volume of trash, distance to the container, etc. remain stable. The graph to the right shows how the curve changes if the volume of trash is cut in half.

2.2 Employee Reviews on Glassdoor

Glassdoor [1] is a website that allows current and former employees to review their companies by leaving ratings for *career opportunities*, *compensation & benefits*, *work-life balance*, *senior management*, and *culture & values*. In addition to that, employees can decide whether they approve of the CEO, whether they would recommend the company to a friend, and how they feel about the business outlook for the next six months. Finally, they are asked to write about the pros and the cons and to provide feedback to management. The analysis in this paper makes use of the pros and cons structure in the free text section. For a detailed description of the method see the methods section (Sect. 3) and the code on GitHub [5].

2.3 NLP for Analyzing Glassdoor Reviews

Recent advancements in natural language processing (NLP) have proven useful in *human resource management* allowing for the analysis of qualitative data without the need for laboriously coding text into categories [6]. Both, bag-of-words approaches (e.g., [7]) and approaches based on vector-representations (e.g., [8]) have been used successfully on Glassdoor data. This article relies on both of them: a bag-of-words approach combined with part of speech tagging to identify the most prevalent noun chunks that will serve as categories, and sentence-embeddings to compute the cosine-similarity between the categories and the comments using a sentence-transformer model [9], which maps paragraphs and sentences to a 768 dimensional dense vector space [10].[1]

3 Methods

For more details see the implementation of the approach on GitHub [5].

[1] This model is also used to reduce the number of categories by removing semantic duplicates (see Sect. 3).

3.1 The Data

The analysis is based on 439,599 comments from the financial sector featuring 12,868 companies[2] submitted between April 23rd, 2008, and September 8th, 2020, to Glassdoor.

3.2 The Labels

To identify the most prevalent categories, 4,000 comments have been sampled from the *pros* and the *cons* sections, respectively. The resulting 8,000 comments were used to identify the 15 most prevalent noun chunks using spaCy [11] (ignoring stop words and removing adjectives by means of spaCy's parts of speech tags). Their number has been further reduced by removing semantic duplicates. Duplicates were determined in terms of the cosine similarity between the categories' semantic vector representations. The vector representations were obtained by means of the *paraphrase-distilroberta-base-v1* transformer model from the Huggingface library [10]. Categories with cosine similarities above .7 were considered as duplicates. To cluster the similarity scores, the similarity matrix is transformed to a distance matrix (i.e., a matrix of dissimilarity scores). Similar topics are identified by means of hierarchical clustering and a dissimilarity threshold of .3, which corresponds to a similarity score of .7. Among such semantic duplicates, the more frequent category was kept unless it contained lemmata appearing in other topics. In the case of the latter, the less frequent category was kept. For example, the topics "work" and "job" are duplicates according to this definition. "Work" is more frequent than "job". However, "work" also appears in the topic "work life balance". "Job" is thus kept and "work" discarded. This procedure reduced the 15 categories down to 13.

3.3 Mapping Comments to Labels

To classify the comments, the same model used for label reduction (see Sect. 2.2) is used to generate vector representations [10]: The vector representation of each comment is compared to the vector representation of each label in terms of their cosine similarity. If the similarity score is larger than .35, the label is assigned to the comment.

Three more labels are added to the existing list: *nothing negative to say, nothing positive to say, none*. Their only role is to identify comments that are to be removed because they do not carry relevant information for the purpose of identifying tradeoffs.

3.4 Identifying Tradeoffs

The general idea behind the methodology to identify tradeoffs is described in Fig. 2. Each review has a *pros* and a *cons* section. These sections are assigned to category names (i.e., labels) in a one-to-many approach (see Sects. 2.2 and 2.3). The procedure results in a list of pro-con pairs (the maximal count of pairs is equal to the number of

[2] The companies comprise organizations from the following industries: Financial Transaction Processing, Investment Banking & Asset Management, Banks & Credit Unions, Lending, Brokerage Services, Financial Analytics & Research, Stock Exchanges, Venture Capital & Private Equity.

labels squared; in this case, 169 pairs). Pairs with the same label in the pros and the cons section are discarded (leaving maximally 156 pairs).

Statistical significance of the count of pairs is tested against a control obtained by randomly reshuffling the pairings 10,000 times. A pro-con pair counts as significant, if the Holm-Bonferroni adjusted p-value [12] is below <.05 when compared to the randomly created distribution.[3] If both, a pro-con pair and its opposite meet the significance test, the combination qualifies as a tradeoff (see Fig. 2).

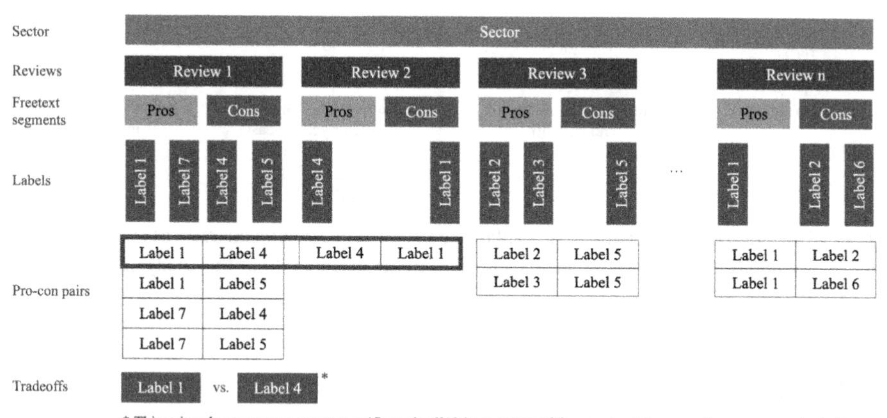

Fig. 2. The pros and cons sections of employee reviews are classified in terms of categories (labels) relevant to the sector. Only pairs of labels that are significantly frequent qualify as pro-con pairs. If the opposite of the pro-con pair constitutes another significantly frequent pro-con pair, the combination counts as a tradeoff.

To evaluate whether a sector has a solution for a tradeoff, the same procedure is carried out for the labels within the *pros* and the *cons* sections, respectively. The goal is to check whether the *either-or* combination also exists in the form of *both-both*. For example, if significantly many reviewers report *good work-life balance* and *bad management*, and another group reports the opposite, it could still be the case that another group of reviewers reports both *good work-life balance* and *good management*. The latter is called a *pro-pro pair*. Tradeoffs whose labels also appear significantly often as pro-pro pairs count as tradeoffs that have a solution within the sector. In contrast, if a pro-pro combination of the labels constituting a tradeoff is significantly infrequent, it

[3] The more pairs tested for statistical significance, the higher the probability of obtaining type I errors (false positives). To control for this without unnecessarily increasing the risk of type II errors (false negatives), the Holm-Bonferroni method is used to adjust the family-wise error rates (FWER) for each pair. The p-values are sorted and ranked from lowest to highest starting with rank 0. For each pair that is to be tested, the FWER (in this case .05) is then divided by the number of tests (in this case 156) minus the pair's rank. This results in an array of adjusted FWERs. If an adjusted FWER is smaller than its predecessor, it is set equal to the rate of the previous pair. A pair is significant if its p-value is lower than its adjusted FWER.

can be assumed to have no obvious solution within the sector – at least not within the constraints of this approach (see Fig. 3).

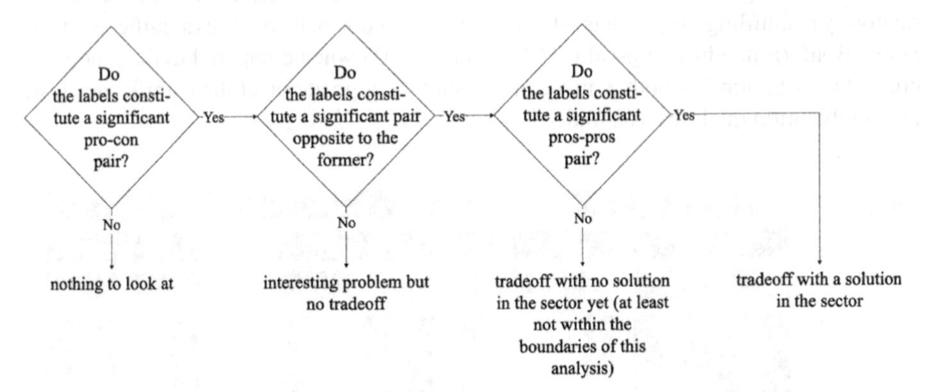

Fig. 3. Illustration of the decision logic explaining the difference between tradeoffs with a solution in the sector and tradeoffs without such a solution (in the boundaries of this analysis).

4 Results

4.1 Tradeoffs

Filtering for the most prevalent nouns (noun chunks included) results in the following list of 13 categories used as labels: *company, people, management, employee, benefit, time, job, pay, opportunity, culture, hour, training*, and *work life balance.*

Running the significance test on the total counts of all pairings reveals two tradeoffs (see Table 1): *management vs. work life balance* (870, $P < .0001$ and 118, $P < .0001$), and *opportunity vs. work life balance* (47, $P < .0001$ and 54, $P < .0001$). The labels *management & work-life balance* appear significantly frequently together in the pros section as a *pro-pro* pair, which suggests that a solution exists within the sector (145, $P < .0001$). In contrast, the pair *opportunity & work life balance* is significantly infrequent (11, $P < .0001$) and thus has no obvious solution in the sector (at least not within the boundaries of this analysis).

Table 1. Summary of the tradeoffs.

Con	Pro	Count	p-val (adj.)	Cons pairing	Pros pairing
Management	Work life balance	870	0	Insignificant	Frequent
Work life balance	Management	118	0		
Opportunity	Work life balance	47	0	Infrequent	Infrequent
Work life balance	Opportunity	54	0		

For an illustration of the tradeoff ratios within firms, see Fig. 4.

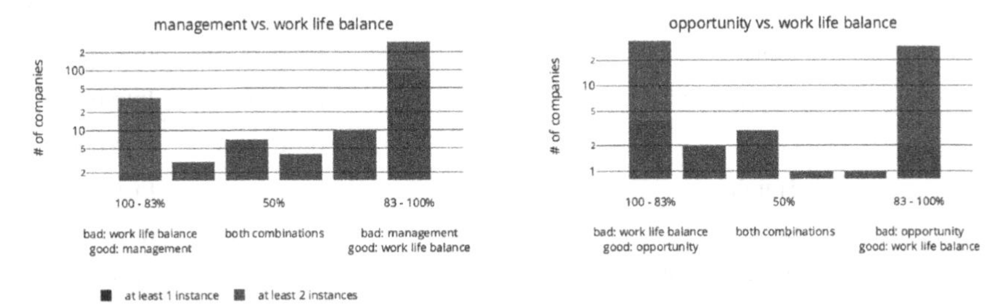

Fig. 4. Illustration of the ratios between the two sides of the respective tradeoffs within companies. The y-axis (logarithmic scale) shows the number of companies that fall into the respective ratio bins.

For examples of reviews illustrating the tradeoffs, see Table 2.

Table 2. Examples of reviews illustrating the tradeoffs.

Label cons	Label pros	Cons	Pros
Work life balance	Opportunity	Not the best work life balance	Opportunity to make a lot of money
		Work/Life balance could be better	Opportunities for growth and advancement
Opportunity	Work life balance	Opportunity for growth, financial compensation, location	Work-life balance, job security, culture, tuition reimbursement
		Opportunity for advancement and lack of growth	Family oriented, balanced life, encourage learning
Work life balance	Management	Long hours and not enough work life balance	Great people and management
		Work/life balance quite difficult	Excellent management at all levels
Management	Work life balance	Sometimes management is unorganized	The work/Life balance is good, they understand that you have many things going on with your life
		Management problems at the middle levels	Culture, benefits, compensation, work life balance

Table 3. Summary of frequent pro-con pairs.

Con	Pro	Count	p-val (adj.)	Cons pairing	Pros pairing
Hour	Culture	86	0	Insignificant	Infrequent
	Management	132	0	Infrequent	Infrequent
	Pay	177	0	Infrequent	Infrequent
	Training	94	0	Infrequent	Insignificant
Management	Benefit	523	0.0001	Infrequent	Infrequent
Pay	Work life balance	280	0	Infrequent	Infrequent
Work life balance	Culture	77	0,0001	Infrequent	Frequent
	Training	85	0	Infrequent	Insignificant

4.2 Frequent Pro-con Pairings

In addition to the 4 pairs constituting the 2 tradeoffs, 8 more pairs have been identified as significantly frequent (see Table 3). None of them constitutes a significantly frequent pro-pro pair except for con: work life balance & pro: culture. Examples can be found in Table 4.

Table 4. Examples of reviews illustrating significantly frequent pro-con pairs.

Label cons	Label pros	Cons	Pros
Work life balance	Training	Work/life balance. A lot of information to be responsible for	Good benefits and good training
		No work life balance and promise big bonuses but do everything they can to cut bonuses and salary increases at year end	Gives solid training when starting out
	Culture	Work life balance not the best. Needs work on diversity of employees	Great culture and good people
		Worklife balance especially for families	Great culture and great people
Pay	Work life balance	Pay on the low end	Stress is low and work/life balance is good

(continued)

Table 4. (*continued*)

Label cons	Label pros	Cons	Pros
		Pay is well below industry average	Great work-life balance, knowledgeable people, benefits
Management	Benefit	Horrible management. HR management worse	The benefits are good and not too costly
		The management is really horrible	The benefits are really great
Hour	Training	Hours, a lot of turnover	Training, no other pros to mention
		Long hours and stressful work environment	Great training and team environment
	Pay	Long hours and constant pressure	Pay is good hard to beat
		Long hours on the job	Good pay for work done
	Management	Hours and ongoing changes.	Solid management and great location
		Very long hours raise very small	Great work environment management hand off remote work
	Culture	Long hours and over time	Like the culture and people
		A lot of hours worked	Great pay and good culture

5 Discussion

5.1 Tradeoffs

The approach identified two tradeoffs: *management vs. work life balance*, and *opportunity vs. work life balance*. A tradeoff between management and work-life balance means that a significant number of people complains about bad management but praises the good work life balance (870, $P < .0001$), while another group of people complains about poor work-life balance but praises the company for good management (118, $P < .0001$). That good work-life balance and good management are not incompatible can be seen from the fact that both labels appear significantly frequently together as a *pro-pro* pair (145, $P < .0001$). This makes intuitive sense: A manager can be good at managing people and care about their work-life balance if circumstances allow for it. However, more often (870, $P < .0001$) employees in finance seem to think that they have bad managers. These managers are potentially less organized, which might result in less pressure at work and thus better work-life balance. For example, one employee wrote: *Cons: Sometimes management is unorganized. Pros: The work/Life balance is good, they understand that you have many things going on with your life* (see the examples in Table 2). Less (but significantly) often, the work-life balance is considered bad, but the management is given credit for doing a good job, for example: *Cons: Long hours and not enough work life balance. Pros: Great people and management.*

The second tradeoff between opportunity and work-life balance showed up in fewer reviews but marks a tradeoff with less compatible labels (significantly infrequent pro-pro pairing: 11, $P < .0001$). The tradeoff suggests that working less comes with fewer opportunities for career development in the financial industry (47, $P < .0001$), while having less time outside of work – and therefore probably working more – creates additional opportunities (54, $P < .0001$). Intuitively, it makes sense and is probably due to time being a scarce resource.

It is striking that work-life balance is involved in both tradeoffs. The concept describes in and of itself a tradeoff evolving around how to allocate one's time. To see whether a solution could be found for *work life balance vs. opportunity* in the sense of Fig. 1, and how to explain *work life balance vs. management*, more analysis would be required, but this is outside of the scope of this paper. For inspiration regarding the former, we could investigate other sectors or use the generalized version of the *algorithm of inventive problem solving* that was designed for dealing with tradeoffs [13].

5.2 Pro-con Pairs

The theme underlying the two tradeoffs, i.e., the concern of employees with work-life balance and thus their *time*, is reinforced by the list of significantly frequent pro-con pairs (Table 2). The time-related label, *hours*, appears in the cons section of 4 pairs, and *work life balance* twice in *cons* sections and once in the *pros* section of another pair. 7 out of 8 pairs are thus time related (for a few examples, see Table 4). On the most general level, the explanation seems to be straight forward: employees sell their time to their employer for a salary. If a resource is scarce and different systems exploit the same shared resource, tradeoffs emerge around this resource [14]. While each of the conflicts would deserve further investigation (especially the less obvious ones such as *con: work life balance & pro: culture* (77, P .0001), or *con: management & pro: benefits* (85, $P < .0001$), it is outside of the scope of this paper.

5.3 Possible Future Improvements on the Approach

Four main aspects stand out as potential improvements for the approach. First, the classifier to assign labels to the comments requires further improvement. Currently, it is based on a model that has not been fine-tuned for the task of classifying free text on Glassdoor in terms of the identified labels. Second, it is desirable to use more than 13 labels to identify less obvious combinations. Third, it might be worthwhile to do a comparison between sectors. Fourth, the scope of applicability of the approach could be broadened by making it independent of the pros-cons structure that is immanent to reviews on Glassdoor and Indeed. This might be achievable by adding a layer of sentiment analysis. Currently, the approach piggybacks on the pre-categorization of the reviews themselves.

Acknowledgment. I am grateful to Glassdoor for making this research possible by providing access to their proprietary data, to Marco Meyer for the discussions of this topic and his advice, to Georg Müller-Christ and his colloquium at the university of Bremen for the feedback and guidance, and to Isabel Estevez for proof-reading the paper.

References

1. Glassdoor. https://www.glassdoor.com. Accessed 29 Apr 2022
2. Indeed. https://www.indeed.com/. Accessed 29 Apr 2022
3. Altshuller, G.S.: Creativity as an Exact Science: The Theory of the Solution of Inventive Problems. Gordon and Breach Science Publishers, London (1984)
4. Schumpeter, J.A.: Business Cycles: A Theoretical, Historical and Statistical Analysis of the Capitalist Process. Porcupine Press, Randburg (1939)
5. Tradeoff-detection. https://github.com/datamaunz/tradeoff-detection. Accessed 29 Apr 2022
6. Guo, F., Gallagher, C.M., Sun, T., Tavoosi, S., Min, H.: Smarter people analytics with organizational text data: demonstrations using classic and advanced NLP models. Hum. Resour. Manag. J. (2021). https://doi.org/10.1111/1748-8583.12426
7. Luo, N., Zhou, Y., Shon, J.: Employee Satisfaction and Corporate Performance: Mining Employee Reviews on Glassdoor.com. 16 (2016)
8. Smith, D., Choudhury, P., Chen, G., Agarwal, R.: Weathering the COVID storm: the effect of employee engagement on firm performance during the COVID pandemic (2021). https://papers.ssrn.com/abstract=3841779. https://doi.org/10.2139/ssrn.3841779
9. Reimers, N., Gurevych, I.: Sentence-BERT: sentence embeddings using siamese BERT-networks. arXiv:1908.10084 [cs] (2019)
10. sentence-transformers/paraphrase-distilroberta-base-v1. Hugging Face. https://huggingface.co/sentence-transformers/paraphrase-distilroberta-base-v1. Accessed 29 Apr 2022
11. spaCy. Industrial-strength natural language processing in python. https://spacy.io/. Accessed 29 Apr 2022
12. Holm, S.: A simple sequentially rejective multiple test procedure. Scand. J. Stat. **6**, 65–70 (1979)
13. Schollmeyer, J., Tamuzs, V.: Deducing Altshuller's laws of evolution of technical systems. In: Benmoussa, R., De Guio, R., Dubois, S., Koziołek, S. (eds.) TFC 2019. IAICT, vol. 572, pp. 55–69. Springer, Cham (2019). https://doi.org/10.1007/978-3-030-32497-1_6
14. Schollmeyer, J., Tamuzs, V.: Discovery on purpose? Toward the unification of paradigm theory and the theory of inventive problem solving (TRIZ). In: Cavallucci, D., De Guio, R., Koziołek, S. (eds.) TFC 2018. IAICT, vol. 541, pp. 94–109. Springer, Cham (2018). https://doi.org/10.1007/978-3-030-02456-7_9

TRIZ-Based Approach in Capturing and Managing Indigenous Innovation and Knowledge

Timothy George Mintu[1](✉), Narayanan Kulathuramaiyer[1](✉), Franklin George[2](✉), and John Phoa Chui Leong[2](✉)

[1] Institute of Social Informatics and Technological Innovations, University of Malaysia Sarawak (UNIMAS), 94300 Kuching, Sarawak, Malaysia
timothygeorgemintu@gmail.com, nara@unimas.my

[2] People's Association for Development and Education of Penan Sarawak (PADE), Miri, Sarawak, Malaysia
fg36093@gmail.com, jphoacl@gmail.com

Abstract. Indigenous people who are still connected to traditional lifestyles and are living closely in touch with nature's patents, remain custodians to vast treasures of knowledge. The ability to tap on indigenous inventions can be useful within a contemporary context in providing insights to help solve emergent problems such as global warming and climate change. This research focuses on exploring ways to capture this implicit and tacitly held knowledge among these remote indigenous communities of Sarawak, Borneo. Engaging with the local community in exploring the immense challenge requires a participatory model for eliciting innovative expressions across time and space boundaries. Mechanisms to associate such discovered knowledge within the context of current scenarios requires a standard framework for achieving the alignment. In this paper, a TRIZ-based framework for connecting to and mapping these past innovations has been proposed. The 40 inventive principles of Genrich Altshuler has been adopted as a means of bridging knowledge gaps and connecting the diverse knowledge forms. The collection of customized TRIZ instruments served as a collaborative visual knowledge mapping framework for acquiring and organizing knowledge for local indigenous communities. This study has demonstrated the ability to unlock tacit knowledge amongst community knowledge-custodians living in remote and isolated communities. The 40 Inventive Principles served not only as an index for innovative expressions but also as a good platform for these communities to make systems innovation as a way of life, and also to acquire expertise from external sources. The continuing efforts in knowledge-based activities has a potential for expansion to be used by other communities. Despite the initial challenges where there was a need to address language and intergenerational gaps, the proposed model has also demonstrated interest amongst youths to connect to their roots and share the past inventive moments with community elders.

Keywords: TRIZ · Indigenous inventions · Knowledge-based innovation · Knowledge acquisition and knowledge representation

R. Nowak et al. (Eds.): TFC 2022, IFIP AICT 655, pp. 220–229, 2022.
https://doi.org/10.1007/978-3-031-17288-5_20

1 Introduction

1.1 Knowledge and Innovation

Societies where people have learned to live together in a civilized manner demonstrates the long pathways undertaken to achieve this state. Knowledge is the thread in that can be passed down to successively to bring cumulative benefits [17].

However, there are well known types of knowledge such as the explicit and tacit knowledge [3]. Explicit knowledge is something that can be easily passed down to another person i.e. how to do a certain task. Meanwhile, tacit knowledge is something that is learned in an intangible way which is usually harder to express as it is a sort of knowledge by 'feel'.

The sharing of knowledge brings about innovations [7] that can help a community to improve their methods of doing things, improving their sustainability [6]. Genrich Altshuler, the father of TRIZ, has highlighted innovation is repeated, when a systematic approach of transfer is in place.

In relation, this paper aims to exploit TRIZ based approach for capturing and managing traditional knowledge can be adopted by the indigenous communities in Sarawak.

The TRIZ instruments co-created with the community serves as a collaborative visual knowledge mapping framework for acquiring and organizing knowledge for local indigenous communities.

1.2 The Indigenous People of Sarawak

According to Mamo [9], the International Work Group for Indigenous People assessed 13.8% (4,369,177) of all Malaysians as indigenous in 2017. In the Peninsular of Malaysia, 18 Orang Asli tribes, namely Semang, Senoi, and Aboriginal-Malay, are referred to as indigenous people, accounting for 0.7% (182,000) of the total 26,000,000 Peninsular Malaysians. Meanwhile, the indigenous people of Sarawak are known as Dayak or Orang Ulu, and they are made up of ethnicities such as Berawan, Bidayuh, Bisaya, Iban, Kejaman, Kenyah, Kayan, Kedayan, Kelabit, Lun Bawang, Melanau, Penan, Punan, Sekapan, and Ukit. They account for 70.5% (1,932,600) of Sarawak's total population of 2,707,600. We therefore see an importance of indigenous knowledge being passed down across generations.These communities have rich history in terms of their story, innovation spirit in their capacity to overcome hardships and difficulties in their quest for resilience.

1.3 Problem Statement

In this study, we explore the tacit knowledge capture and management for use in a particular community as a case study. After several discussions with community members in building on our past engagements [Franklin 2022] with the nomadic Penen community of Long Lamai, we have come up with a cause-and-effect chain analysis which illustrates that indigenous innovations requires a dedicated framework together with an ecosystem for it to be captured and managed properly (Fig. 1).

Major part of the problem has to do with:

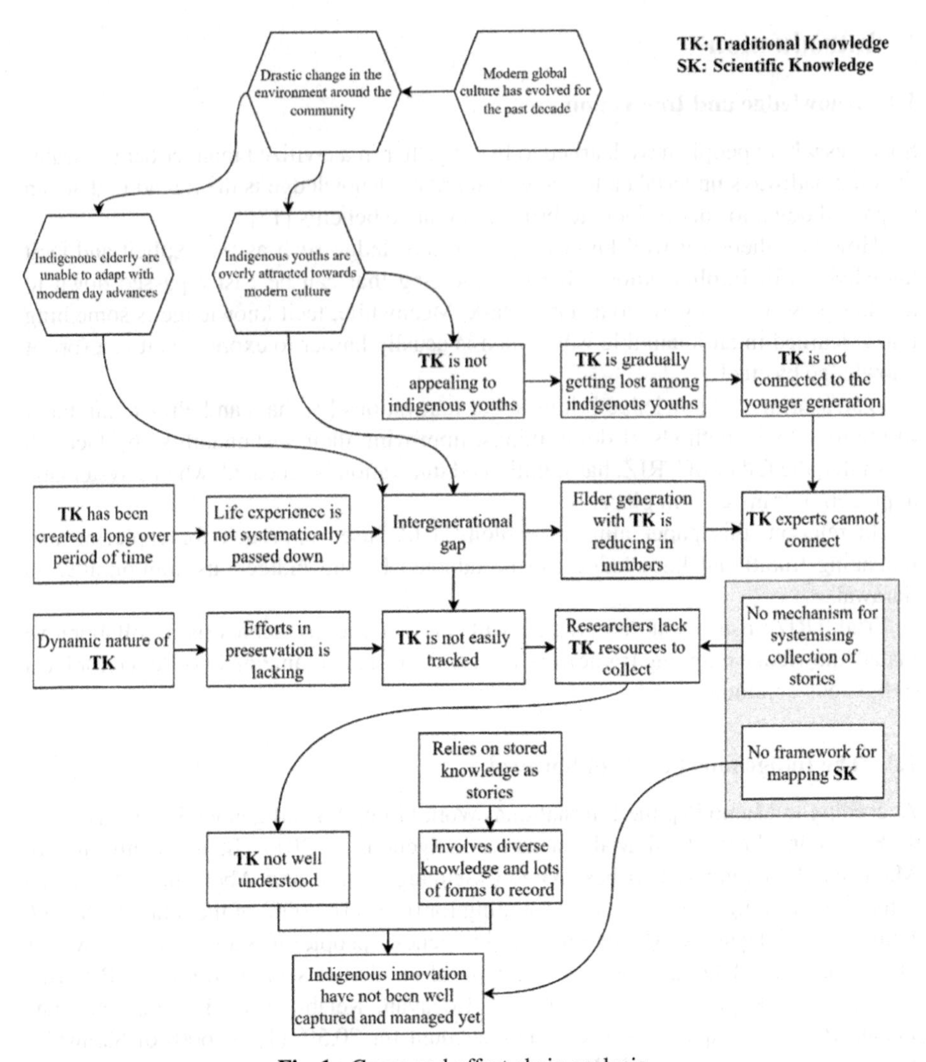

Fig. 1. Cause-and-effect chain analysis

(i) no available mechanisms in use for the capture the indigenous knowledge.
(ii) TRIZ (other forms of SK) tends to be too complex to be well understood by rural indigenous communities.
(iii) integration of TRIZ concepts with indigenous creative expressions is not well-founded

It seems that the traditional knowledge has faced several implications due to the lack of capturing methods and it is difficult to map the traditional knowledge with scientific knowledge. That is partly because traditional knowledge is cumulated over time making it harder to trace.

2 Case Study and Related Works

2.1 Innovation in Indigenous Community

The Penans are an indigenous group that may be found across central Borneo Island. Their communities are mostly concentrated near the Baram River in Sarawak, Malaysia. Historically, the Penans were divided into two groups known as the Western Penan and the Eastern Penan [11]. The livelihoods of these two Penan groups are virtually the same. According to Survival International [15], the Penan people live a stable existence in contrast to their nomadic lifestyle before to the colonial era in Sarawak.

According to several sources [10, 12] and [8], the Penans' major source of income is still hunting wild animals, scavenging for wild foods or plants, and recently expanding into agriculture and fishing. They are also observed to perform arts and handicrafts like as basketry and musical instruments in terms of creativity [2, 13].

Even though they have formed settlements, a few of the Penans continue to live as nomads. In essence, there are marginalized people who are really alienated from the modern world devoting their majority of their time in nature. However, just because they have been living with nature for so long does not imply that they should be subjected to being left behind in terms of innovation capacity for advancement.

2.2 Co-creation and Participatory Practices

Both of co-creation and participatory practices are the methods used to get communities to involve themselves in a development. Based on a work done by Grant [5], the identification of hierarchy within the co-creation and participatory model is well defined.

The purpose of the hierarchy shows that the importance of each approach has its own specific purpose but still aiming to make community organisers to work together with the stakeholders. Refer to Table 1 for a complete view of both methods with the highlighted ones as the as our main practices for this study.

2.3 Preliminary Works in Integration of Indigenous Sign Language (Oroo') with TRIZ

The works as described in this research was started by our research team with an ethnographic study by the co-author, Franklin George. As a Penan who learnt the traditional language form as a youth, initiated the mapping of the Penan forest twigs-and-leaves language symbols into the appropriate local meanings. The approach evolved through co-creation and participatory practices the joint understanding of indigenous knowledge and TRIZ concepts from the Penan indigenous perspective. The training on TRIZ principles was facilitated by team members undergoing Level 1 and Level 2 MAyasian TRIZ syllabus. The use of oral interviews and focus group discussions with community experts and the research team led by the second author who is a International TRIZ L3 certified, ensured the correctness in mapping with TRIZ inventive principles. [4]. The oral interviews involve local champions that consisted of Penan elderly and formulated a way of integrating TRIZ inventive 40 principles and Oroo' (a visual communication methods as used by the Penan). Over 50 Penan Oroo' symbols were mapped to the 40

Table 1. Different approach between co-creation practices and participatory practices

Co-creation	Building connections	Connectivity	Actions by community member and organizers
		Collaboration ethos	
	Collaborative project design	Community consultations	
		Oral history interviews	
		Storytelling and knowledge sharing	
Participatory Practices	Participatory methods for developing collections	Informal description practices	
		Informal archival or preservation practices	
		Post-custodial practices	
		Social media as an online forum	
	Collecting multiple narratives or perspectives	Content organization	
		Narratives in description	

TRIZ Inventive principles. The list below represents the steps of mapping process of his Oroo' to 40 Principles.

1. A collection of Indigenous Principles with their own name and meaning/message.
2. A collection of localized 40 inventive principles
3. Categorization of 40 inventive principles based on its physical state, function and meaning.
4. Mapping of the Inventive principles and 40 inventive principles
5. A collection of indigenised inventions as a community knowledge bank

2.4 Indigenous Innovation with Function-Behaviour-Structure (FBS)

Based on the work of Altshuller [1], the compilation of 40 Inventive Principles is an instrument based on an intensive study of existing patents serving as an index for mapping traditional inventive knowledge to help solve emergent complex problems such as climate change and global warming.

However, with just the 40 Inventive Principles, we can use the characteristics of each of it in with the proposed framework. In relation with the principles, there are other ways for us to go in-depth with the principle by applying the *Function-Behaviour-Structure (FBS)* with each principle [14]. See Table 2 for a simplified connotation.

Table 2. Simple FBS connotation based on Russo & Spreafico [14].

	FBS connotation	Proposed FBS connotation
Function	Purpose and intention of the design [14]	What response do we expect from an indigenous innovation expression?
Behaviour	Expected or obtainable characteristic from structure [14]	What kind of activity that needs to be done to fulfill the function?
Structure	The relationship of the elements in the design [14]	Definition of ideas to some intermediate structure to scientific knowledge

3 Findings and Discussions

3.1 TRIZ Based Approach to Capture Tacit Knowledge

In the first run, we collect relevant stories from their indigenous experience and perspective on solving a problem the traditional way and about their traditional artifacts.

Similarly, these stories will undergo a co-organisation with the community to get an in-depth opinion through oral exchange.

Then, we will conduct a consultation for the TRIZ experts to brainstorm the optimal way to link the traditional knowledge to structural knowledge from TRIZ.

After consulting with the experts, we referred to the community to validate the proposed framework or model. Making the model also required us to translate it into a local language to be understood with the indigenous community.

The proposed mechanism shows a full view of the process in a bigger picture. The application of TRIZ inventive principle will also be discussed later.

3.2 40 Inventive Principles in Capturing Indigenous Knowledge

In this paper, 40 Inventive Principles will be the basis of knowledge capture. This will be an index to classify various indigenous processes and products. This index will also assist the experts and the indigenous community to map the indigenous products.

Based on *Function-Behaviour-Structure* (FBS) ontology by Russo & Spreafico [14], there is an emphasis of function and behaviour in the structure. As FBS requires a thorough understanding of the concepts, a model has developed to support the TRIZ experts and indigenous community experts to map their indigenous inventive moments. Table 4 shows the elements in the capturing mechanism meanwhile, Table 3 and 4 shows how the way the indigenous innovation is mapped with TRIZ 40 Inventive Principles.

Table 3. *Telikit* in FBS and TRIZ Form

<table>
<tr><td>Indigenous Innovation</td><td colspan="2">Telikit</td></tr>
<tr><td colspan="3">Structure</td></tr>
<tr><td colspan="2">Application
How can we hone an apprentice's hunting-skills without risking their life or their safety?</td><td>Method
We use a training in simulated gamified environment to hone the skills until they are ready for the real hunt.</td></tr>
<tr><td colspan="3">Innovation Description

Telikit is a game used to train Penan youngsters to hunt using bamboo hoops and a long stick. Bamboo hoops represents the prey, and the long stick represents the spear used to stab the prey. When the hoops are thrown to a player holding the stick, the player must catch the fast-moving hoops by swinging the stick into the center of the hoop, simulating stabbing the prey into their vital organs.
The game is used to evaluate the readiness of the Penan youngster to be a hunter. The difficulty of the game also can be increased by throwing a smaller bamboo hoop with increased speed and intensity.</td></tr>
<tr><td>Function</td><td colspan="2">Inform Penan youngsters on basic hunting skills and knowledge</td></tr>
<tr><td>Behaviour</td><td colspan="2">Training using a game to simulate hunting in a safe environment</td></tr>
<tr><td>Function predicates</td><td colspan="2">inform(youths), empower(apprentice), imbibe(core values)</td></tr>
<tr><td colspan="3">Inventive Principles as Structural Elements</td></tr>
<tr><td colspan="2">10. Preliminary Action</td><td>Early preparation for optimal result</td></tr>
<tr><td colspan="2">26. Copying</td><td>Use simple and easily available model-copies</td></tr>
</table>

As seen above, the proposed framework based on FBS can aid in the collecting and mapping process significantly. This process should be co-created with the community and mentored by TRIZ experts for interpreting and extracting the innovations from the stories and interviews done with the Penan community.

Table 4. Water logistics in FBS and TRIZ Form

Indigenous Innovation	Logistics using water vehicle
Structure	
Application How to move large objects using small water vehicles?	**Method** Change the physical capability of water vehicle to float large objects or alter the objects to fit in water vehicle.
Innovation Description Transporting objects in the waterways are a challenge for the Penan community in Long Lamai because the only way to get there is through the river. Based on the situation, the object can be too large to carry in a single boat, thus a way of transporting it by the boat must be tackled.	
Function	Moving large object through the waters
Behaviour	• Use a bigger carrier • Disassemble the object • Compact the object • Combine smaller boats to increase surface area
Function predicates	move(large-object), supports(movement), merge(boats), configure(boat-structure)
Inventive Principles as Structural Elements	
1. Segmentation	Divide object into smaller parts
3. Local Quality	Make objects to operate in its ideal condition
5. Merging	Merge objects or operations

3.3 Managing Captured Indigenous Innovations

In managing the indigenous innovations, it is more suitable to store the indigenous innovations using technology which can handle a large volume of information. The template for knowledge capture has been discussed in the previous chapter. However, some terms have to be simplified (Fig. 2).

3.4 Future Works

The potential of this preliminary work is for the mechanism to have a sustainable model where every community can benefit from this. The next version of this model will introduce us with a standalone software which can be utilized by using any mobile device with ability to store the multimedia contents of the products in a cloud-based storage.

The knowledge-base of past inventions will be conserved, and the communities can take advantage of a potential software to be developed by using the stored knowledge to foster innovation as a culture. Future works will explore the modelling of Altshuler's matrix as a knowledge-base of inventive patterns.

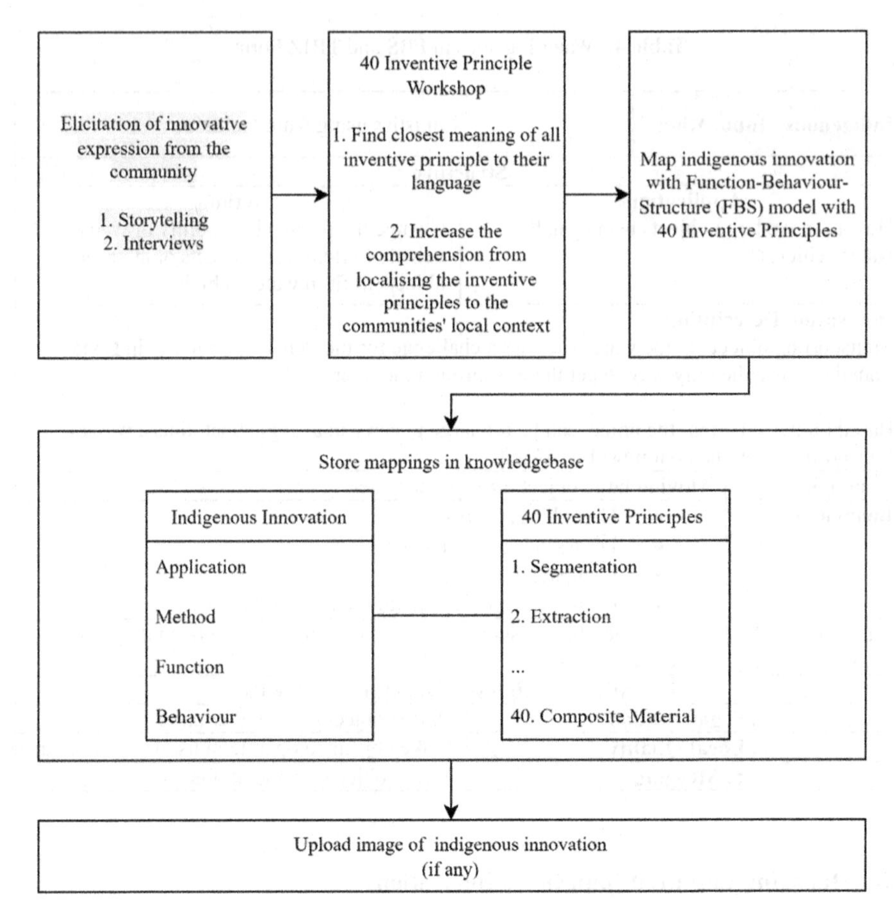

Fig. 2. Methods of managing indigenous innovations using Google services.

3.5 Benefits of Implementing Co-creation and Participatory Practices

The importance to know how to work the community has been rising because there are a lot of protocols that needs to be followed. By providing the community a clear course of action that needs to be done in the project will help the organisers and experts to work flawlessly.

In the Sarawakian indigenous communities, the researchers must follow the community protocols. Since the communities are the ones to manage their own resources, we will have to learn the community protocols to avoid against legal actions [16].

Co-creation and participatory practices also must be shown to the community before proceeding to do any further steps. If implemented properly, the community will be more than welcome for us to do research in their native land.

Acknowledgements. The authors are greatly thankful for the project funding from Ministry of Higher Education (*Kementerian Pengajian Tinggi*) and *Formulation of TRIZ-based Sustainability-Oriented Innovation Model for Indigenous Knowledge Management* (I03/FRGS/2009/2020). The authors would also like to thank the community leaders, local champions, and villagers for sharing their indigenous knowledge.

References

1. Altshuller, G.: 40 principles: TRIZ keys to innovation, vol. 1. Technical Innovation Center, Inc. (2002)
2. BorneoTalk: Beautiful Sounds of Bamboo. BorneoTalk, vol. 52, pp. 76–77 (APR–JUN 2019) 1 April 2019
3. Davies, M.: Knowledge-explicit, implicit and tacit: philosophical aspects. Int. Encycl. Soc. Behav. Sci. **13**, 74–90 (2015)
4. George, F., Kulathuramaiyer, N., Bala, P.: Fostering a TRIZ-based grassroots innovation among penans. In: MYTRIZ Conference, vol. 6, no. 7, p. 185 (2020)
5. Grant, K.A.: Affective Collections: Exploring Care Practices in Digital Community Heritage Projects (2020)
6. Järvensivu, T., et al: Helping local innovation ecosystems to become custodians of global sustainability. In: The XXIV ISPIM Conference-Innovating in Global Markets: Challenges for Sustainable Growth Conference (2013)
7. Johannessen, J.A., Olsen, B., Olaisen, J.: Aspects of innovation theory based on knowledge-management. Int. J. Inf. Manag. **19**(2), 121–139 (1999)
8. Koizumi, M., Mamung, D., Levang, P.: Hunter-gatherers' culture, a major hindrance to a settled agricultural life: the case of the Penan Benalui of East Kalimantan. Forests, Trees Livelihoods **21**(1), 1–15 (2012)
9. Mamo, D. (ed.) The Indigenous World 2021, 35 th edn. International Work Group for Indigenous Affair (2021)
10. Needham, R.: The system of teknonyms and death-names of the Penan. Southwest. J. Anthropol. **10**(4), 416–431 (1954)
11. Needham, R., Beidelman, T.O.: Penan friendship-names. Transl. Cult.: Essays EE Evans-Pritchard **117**, 203 (1971)
12. Puri, R.K.: Hunting knowledge of the Penan Benalui of East Kalimantan, Indonesia. University of Hawai'i at Manoa (1997)
13. Puri, R.K.: Transmitting Penan basketry knowledge and practice. In: Understanding Cultural Transmission: A Critical Anthropological Synthesis, pp. 266–299 (2013)
14. Russo, D., Spreafico, C.: TRIZ 40 inventive principles classification through FBS ontology. Procedia Eng. **131**, 737–746 (2015)
15. Survival International: The Penan. Survival International. https://www.survivalinternational.org/tribes/penan. Accessed Jan 2022
16. Swiderska, K.: Consent and conservation: getting the most from community protocols. In: IIED Briefing Paper-International Institute for Environment and Development, no. 17137 (2012)
17. Yu, P.K.: Traditional knowledge, intellectual property, and Indigenous culture: an introduction. Cardozo J. Int. Comp. Law **11**(2), 239 (2003)

The Potential of Creative Methods for IT Project Management

Monika Woźniak[1](✉) and Anna Boratyńska-Sala[2]

[1] University of Gdansk, ul. Piaskowa 9, 81-864 Sopot, Poland
monika.wozniak@ug.edu.pl
[2] Cracow University of Technology, al. Jana Pawła II 37, 31-864 Cracow, Poland
anna.boratynska-sala@pk.edu.pl

Abstract. IT support is now one of the key elements in the competitive advantage of almost every organization. Hence, the IT project management has started to play an increasingly important role. Therefore of concern is the observed low percentage of successful IT projects and use of IT systems functionality in organizations. Business expects more and more flexibility and innovation from the IT sector. Solutions not supported by creative techniques will not be satisfactory for the recipient. Although there is a proven positive correlation of creativity with effective project management, overall organization performance and client satisfaction, there is no detailed research in the literature on the actual use of creative methods by IT teams in project management. The research undertaken combines the area of IT project management with creative methods. The article presents the results of the study of IT teams in the Polish SME sector in the field of their knowledge and experience in the application of creative methods in IT project management. Experimental research allows an assessment of the potential and effectiveness of selected creative methods in the IT project management and an indication of the areas of their application in the project life cycle.

Keywords: IT project management · Creativity · TRIZ · Problem solving · Innovation

1 Introduction

The current twenty-first century has changed the role of creativity to the dominant one, making it sometimes more important than the technical knowledge [1]. Creativity has become an important factor in all team work, including projects. Creativity finds its application more and more often in organizations implementing long-term large-scale projects related to complex products. Multifunctional teams created for such projects often experience a crisis in communication as early as at the stage of designing. At the same time, this crisis is often transformed into a chance to stimulate different creativity techniques and generate new technical knowledge, thus enabling implementation of the main goal of the project [2].

Published by Springer Nature Switzerland AG 2022
R. Nowak et al. (Eds.): TFC 2022, IFIP AICT 655, pp. 230–247, 2022.
https://doi.org/10.1007/978-3-031-17288-5_21

On the other hand, in the IT environment, there is a constant focus on technical aspects and the methodology of project implementation [3]. Project management standards are improved, but they do not relate in any way to creative methods. Meanwhile, there is still a low percentage of IT projects completed successfully with additional problem of alignment IT - business and low percentage of functional IT systems use [4–6].

Although a positive correlation between creativity and effective project management has been proved [7], there is no detailed research in the literature on the actual use of creative methods by IT teams in project management. Therefore, it was considered reasonable to undertake research and determine how various aspects of creativity are perceived by IT teams themselves. The research was supposed to answer the following questions:

1. Do IT teams know and use creative methods in IT project management?
2. Do creative methods have potential for IT project management and, if they have, in what areas?

The structure of the article is as follows. In the next section, the research background is presented. In the third section, there is the subject literature review, which allowed for the identification of research gaps. Next, attention has been focused on creative methods. In the next section, the research purposes are presented. Then, the research method and process are described. As a next step, the basic research results are disclosed and referred to a comparative research with the discussion. The article ends with conclusions and suggestions referring to the role and potential of creative methods in IT projects management.

2 Research Background

Project management in an enterprise using creative methods aimed at improving the competitiveness of enterprises is becoming a fact. This has been confirmed by the results of the research done by Plotnikova and Romanenko [8], based on which the reasons for implementing the innovative enterprise development model, as well as innovative project management and creative technologies to manage projects in modern conditions were identified. The results of research encourage a conscious, systemic and targeted impact on the creative process and activities of creative employees as a basis for project management.

In the era of high competition, companies are still looking for ways to introduce technological innovations. More and more emphasis is placed on the process of IT project management improvement. Often the question arises as to how projects can be implemented better and faster than the projects of the competitors. IT project management requires the use of various skills. These are purely IT skills but also the skills from the area of project activities. Studies conducted among software development specialists by Chan, Jiang and Klein confirm the model, which indicates that basic teamwork skills improve other specialist skills contributing to the achievement of better project results. These software developer skills are the subject of research into software development, but it turns out that in practice inadequate resources of programming staff and narrow specialist skills are one of the main reasons responsible for the failure of software development projects [9].

Studies carried out by Belski and Belski [10], who write about the harmful influence of expertise on creativity, also confirm the above statement. This harmful effect is a natural consequence of extensive professional experience and profound knowledge in a given area. A psychological barrier is formed that blocks flexibility and creative thinking. It is called "psychological inertia" or "vector of psychological inertia" and consists in the fact that previous professional experience is preferred in the process of solving problems. Psychological inertia is such a powerful mechanism that without proper techniques one cannot overcome certain patterns and habits of thought. Experts in projects quickly find solutions to problems without a great cognitive and time-consuming effort. On the one hand, it is beneficial, but on the other hand it has negative consequences. In this way, expert solutions are usually limited to knowledge specific to a given field and with no use of new ideas.

Cheng, studying the cultures and systems of East and West, developed "C theory" [11], where "C" means creativity. In this theory he included all valuable East and West management practices in the form of five mutually supporting elements, which include control, creativity, adaptability, personnel, and decision-making, Then I-Chan investigated the use of "C theory" in project management [7]. Research has shown that the higher is the use of "C theory" in project management, the better is the organizational efficiency. Additionally, among all the factors of "C theory", the creativity factor was characterized by the highest positive correlation in project management with three of the four dimensions of the Balanced Score Card, i.e. internal process, client satisfaction, finance control.

The results of the above studies have demonstrated a significant impact of creativity on project management and organizational efficiency, which can translate into an increased probability of project success. It should also be remembered that client satisfaction is one of the main elements of the modern definition of the success of an IT project. And here creativity has also demonstrated the highest positive correlation and the greatest predictive power (the influence of creativity on client satisfaction is the biggest).

Interestingly, research shows that engaging people with the use of their creativity is motivating. By encouraging people to think creatively, the leader creates at the same time an internal motivation. This also shows the importance of the project team leader, who can either effectively encourage or block the creative activities of his subordinates. This has been confirmed by research involving owners of companies [12]. In the study, three variables were used, i.e. the owner's experience in a given field, the personality of the owner and the style of leadership. From the results of the research it follows that the personality of the owner and his style of leadership have a significant correlation with the degree of creativity in the organization.

The degree to which a company needs creativity and discipline - the main elements of creative tension - depends on the speed with which the industry changes. To a certain extent, freedom of thought and flexibility are essential for productive innovation teams. The management faces the challenge of developing control mechanisms that will guide projects in the right strategic direction and monitor companies progress towards their organizational and project-related goals. Brem in his research work discusses various creativity techniques and proposes a procedure for implementing a creativity workshop and a time plan. He also discusses typical mental barriers for decision-makers to be considered when planning such workshops [13].

It has been observed that creativity workshops are slowly becoming a common way to engage employees in innovative undertakings. Nevertheless, it is difficult to find examples in the literature on the implementation of creative techniques in organizations in the IT industry. Perhaps this is due to the fact that the creative techniques are underappreciated, and if they are used by IT teams, it is only the widely known Brainstorming method [14].

3 Literature Review

The third section is the result of an earlier literature analysis, and its aim is to present the current publications on the research on the application and potential of creative methods in the field of IT project management.

The selected set of keywords included three main domains: "Creativity", "IT project" and "Software". Synonyms were identified for the domain "Creativity". The selected keywords have been connected with logical operators and a search query was created as follows: ("IT project" AND (creativity OR creative OR innovative) AND "software"). The Scopus database was chosen for the selection of publications, as the most voluminous database of reviewed scientific literature, including a wide variety of data in each publication, useful for the analysis process. By applying the above-mentioned search query, 59 results were obtained.

After analyzing 59 studies, 15 publications were qualified for detailed analysis. They discussed the importance and use of creative methods in IT projects in the era of innovative technological development and social progress. Most of the other studies concern only the issues of IT projects and software. Despite including the keywords creativity/creative/innovative in the search query, they do not refer to creative techniques and do not raise the issue of creativity. Nevertheless, their analysis allowed to identify the following contexts in which the "Creativity" domain is contained:

- innovative project,
- innovation knowledge,
- creative design task,
- creative processes,
- human creativity and innovation,
- individual creativity in virtual worlds,
- Scrum as an incentive for creativity,
- creative information systems,
- innovative cross border IT platforms.

In IT project management, rigid software development rules and attachment to processes and tools are applied, which according to Borucki "kills" creativity [15]. The current market situation is difficult and uncertain at the same time and therefore stressful. This reduces the efficiency of employees in terms of IT projects that do not fully meet client expectations. An additional aspect that IT specialists must pay attention to is ubiquitous innovation, the basis of which is thorough research and understanding of client needs [16]. Stamelos, Settas and Mallini see the need for effective and innovative education of IT specialists and IT managers in the field of various creative techniques. In their article, they proposed the use of the so-called Management Antipatterns, which consists of rediscovering bad patterns (Smoke and mirrors, Software bloat, Bullshit Management, etc.) [17]. Similar conclusions regarding the need for education in the field of creativity were presented in the study aimed at checking the effects of introducing the creative method Design Thinking and Smart education to the subject of "IT project management" during undergraduate studies. Intelligent education along with the Design Thinking method gave positive results in teaching business students how to manage IT projects successfully [18]. Some studies indicate that some IT companies also see the importance of using creative techniques in IT projects and focus on developing the creative potential of their employees [19].

Often, IT projects are called "black swans" because of the high degree of difficulty and uncertainty, and the highly innovative ones differ so much that standard ways of managing them and developing systems cannot be successful [20]. Research confirming the validity of using creative techniques in managing uncertainty in IT projects was presented by Marinho, Lima, Sampaio and Moura. They developed a project uncertainty management guide in which they proposed the use of brainstorming and teamwork in multidisciplinary teams [21].

Other authors emphasize the understanding of techniques used in IT projects, which should improve the existing project management processes used in most software development works, as well as take into account unusual problems and creative ways of solving them [20].

It turns out that most companies have IT specialists with the much-needed creative potential and knowledge necessary to implement innovative solutions. This is important due to the need to go beyond certain rigid patterns (so-called "out of the box") in IT project management, creating teams with creative IT specialists, in order to collaborate and understand what the users really require. Simonette claims in her article that it is the attitude of the management that creates an atmosphere in the team for the emergence of creative ideas and innovative solutions to current problems. He proved in his research that promoting such an approach will ensure greater efficiency and innovation in IT processes [16].

Very interesting conclusions were presented in the study by Wu, Rose and Lyytinen using the Delphi method. During sessions with panel groups of IT managers, it proved justified to use out-of-the-box thinking due to a large number of technical problems. So-called innovation points have been established as a result of this study, which are key problem areas that required innovative, unconventional solutions. Each point required a different problem solving approach and creative methods for finding a solution. According to the project managers, who are the experts, "the adoption of such practices is novel and adds value to both practice and future research". [20].

In the literature on the subject, there are studies in which the authors analyze SCRUM as a method supporting creativity by dividing a project or large tasks into smaller parts carried out incrementally. This approach stimulates the activity of the "amygdala" in the brain, activating the creative thinking of the project team members. [22].

The above literature review points out research gaps in the following areas:

- the shortage of studies showing the use of creative techniques in IT projects,
- the shortage of studies on the ways and places of applying creative methods in IT projects,
- the best creative techniques recommended at each project stages or in specific problem areas (so-called innovation points) have not been identified.

Moreover, when referring to project management standards promoted by the two most important associations, PMI and IPMA, the practices of using creative methods are not taken into account, although they are important for highly innovative projects, which include IT projects. In these projects, standard project management and systems development pathways are not sufficient for full success. Therefore, it is necessary to develop this research area not only for practice but also for supplementing and developing the theory in the IT project management field.

For this reason, the authors of the article have attempted to fill the above-mentioned gaps in the area of IT project management in the software development category.

4 Inventive Methods

The science of inventive creation is a methodology of searching for creative solutions, stimulating creative thinking in various fields. When there is a problem in the organization for which no ready solution is available, and the traditional methods and techniques used so far are not fully operative, then the situation can be solved by using inventive (creative) methods.

In the modern literature on the subject, many different divisions of the inventive methods can be found. One of the divisions proposed by Koch [23] divides the methods of creative problem solving into:

- intuitive and creative methods, involving the generation of new ideas through subconsciousness, logic is not involved in this process but rather intuition, the solution comes by launching the imagination,
- systematic and analytical methods based on logic, accompanied by a conscious process,
- mixed methods - a combination of both categories, aimed at supporting the logical method with intuitive and creative methods.

The most common intuitive and creative method, often used in companies, is Brainstorming. This is a typical quantitative method. It is presented in detail in a monograph written by its creator Alex Osborn [24]. The essence of this technique is searching for problems and solving problems in groups diverse in many respects. The method was developed by the advertising industry in the 1950s and remained popular in this industry. Group Brainstorming can be an effective technique for generating creative ideas. According to many researchers, Brainstorming helps to develop and support creative thinking, which gives grounds for the statement that it is useful as a tool to stimulate the team to think creatively in many areas [25, 26].

The second method included in the same group is Six Thinking Hats de Bono. Established in 1985, it is still being modified and refined. The leading idea in this method is a systematic group search for solutions to a given problem. The method shows 6 different viewpoints - hats marked with six different colours, which define the area of individual thinking with artificially adopted attitudes [27]. The method allows generating a larger number of ideas of better quality for given problems compared to the usual online discussion. Serrat's [28] experience allowed him to formulate the conclusion that "The difference between weak and effective teams lies not so much in their collective potential as in how well they use their ability to think together. The Six Thinking Hats technique helps to raise the team's thinking potential".

Creative methods are constantly evolving as a result of changing external conditions and the need to solve problems on time. Part of this evolution is a more flexible understanding of the process related to creative problem solving. This evolution also includes an emerging understanding of the construction of the problem solving style and the impact that the construct has on creative productivity [29].

The most powerful tool, the evolution of which began in the fifties of the last century, is the Theory of Inventive Problem Solving (abbreviated as TRIZ). It belongs to the group of systematic and analytical methods. TRIZ offers a systematic approach to problem solving, but requires creative ability to translate the recommendations proposed by the inventive principles and standard inventive solutions into a specific area of the problem [30].

The so-called System Operator, often called Nine Screens or System Analysis, was listed as the first tool in the TRIZ methodology. At the same time, it is one of the best tools for systematizing the thinking process in the category of time and space. Effective, strong thinking is primarily systemic, block thinking that allows looking at the problem as a whole. When analysing the problem in a company, it is necessary to look more broadly, not only on the problem (system), but also on its environment (supersystem system) and on the components of the problem (subsystem). Additionally, the whole is analysed in a time perspective, i.e. past, present and future. System Analysis is used in almost every problem area, from the initial definition of the problem to the assessment of the best solutions, and for all types of situations, both technical and non-technical. This method helps managers to identify basic technologies, develop product/service, adopt proper business strategy, make objective decisions, and analyse technical and organizational strategies. Owing to the use of this method, the objective decision-making process can make companies focus on the right technological strategy and take the right innovation strategy [31]. Altshuller [32] - the creator of the method - has claimed that the problem can be more effectively solved by changing the supersystem, i.e. the environment of the problem to which the given system (problem) belongs, or its components, i.e. the elements that make up the system. Thus, the qualitative difference of this method lies in the ability to see not only the system that means problem, but also the supersystem and its components in time perspective (past, present and future).

One of the recently developed TRIZ tools is the method proposed by Altshuller's student - Valerie Souchkov (TRIZ Master) - RCA+ (Root Conflict Analysis +). This is a combination of a typical cause-and-effect analysis, otherwise called 5 x Why, with a matrix of contradictions. The chain analysis leads to the main cause of the undesirable effect, while the use of matrix allows finding solutions to the problem. The method allows solving both technical and management problems from identification of the problem and its causes to finding solutions and evaluating them. RCA+ allows isolating and mapping contradictions that appear in technical or business systems and their environment, and which are the main cause of problems [33]. It often plays a major role in patent solutions, because the revealed root causes can directly suggest directions for innovative solutions.

5 Research Purposes

In the light of the above described issues of IT project management and the potential of creative methods, it has been considered reasonable to conduct research combining both these areas. Therefore the main goal of the present study is to examine if and in what areas creative methods have the potential for IT project management. The implementation of the main objective consists of the following partial objectives:

1. Evaluation of the creativity factor in IT project management in the surveyed IT teams, broken down into:

   a) knowledge and application of creative methods in IT project management,
   b) effect of psychological inertia,
   c) the potential of the proposed creative methods for IT project management.

2. Selection of application areas of the proposed creative methods in the IT project management.
3. Verification of the impact of the order in which creative methods are applied on their assessment.

6 Research Method and Process

This study is an experimental study conducted in 2018–2021. Basic research was carried out on IT teams from 52 IT organizations from the SME sector within the same category of IT projects - developing dedicated software for organizations. Then a comparative study was conducted involving another 30 IT organizations from the SME sector. Due to voices questioning the validity of the use of student samples in applied research [34], it was decided to experimental research on real IT teams in natural conditions - in their organizations while working on IT projects.

The study assumes the following stages:

- selection of IT teams involved in the comparable IT projects, the subject of which is the development of software dedicated to the organization,
- assessment of psychological inertia during the work of IT teams on problem tasks,
- lecture and workshop training on selected creative methods,
- experimental research using creative methods in natural conditions - in companies on real IT project.

IT teams were selected as a result of announcing the recruitment of IT companies for research. From among the applying organizations, the ones that met all the criteria assumed important for the study were selected:

- implementation of IT projects where the object is to develop dedicated software for organizations,
- stability of the IT team - the team has worked in the same composition at least in one project,
- the team did not specialize in the creative methods proposed in the study.

These criteria were to ensure the homogeneity of the research objects.

Before using creative methods, the effects of psychological inertia during the work of teams in IT projects were assessed. The teams were asked to solve 3 problem tasks related to IT project management in the category of custom software. All teams were given the same tasks. These tasks required, in particular, the design of new solutions and the planning of new activities. During their resolution, the work of the teams was subjected to observation. Researchers assessed in which processes and areas inertia works most strongly when solving problem tasks. The teams were then interviewed and confirmed these assessments. These interviews also made it possible to identify the most common effects of psychological inertia in IT projects.

The next stage of the research project was to introduce IT teams to the subject of inventive creation through a lecture and workshop training on selected methods. Printed methods guidelines were provided to each member of the IT team. Then IT teams went through the process of solving problem cases related to IT project management with the use of the proposed creative methods. Teams used creative methods in a specific order proposed by the researchers, i.e. Brainstorming, Six Thinking Hats, System Analysis (TRIZ), Root Conflict Analysis RCA+ (TRIZ).

One of the most important problem areas in the management of IT projects related to custom software is well-defined and organized cooperation with the client for whom the project is implemented. All IT teams have the biggest problems in this area. Therefore, working with creative methods was expected to find improvements in this area. Researchers had access to documentation from each team's previous similar project. It was used to select problem points in the above-selected area and to identify how the team dealt with them.

The problem points were repeated in most projects. Therefore, it is possible to list the problems that the teams dealt with:

- low client awareness of their IT needs,
- unspecified, generally formulated client requirements,

- lack of consistency in business expectations and needs,
- delegating incompetent people to the project by the client,
- the contractor's lack of knowledge of the client's industry specifics,
- lack of client involvement and direct users of solutions and functionalities in design works,
- lack of the project support by the client's senior management,
- lack or inadequate communication with the client,
- extensive decision path at the client,
- underestimating the duration of project tasks.

During the experiment, it was compared what new solutions the team generates for these problem points using the indicated creative methods. Both the problem solving process and the generated solutions were assessed. The method evaluation criteria were as follows: degree of difficulty, number of solutions, time to find a solution, innovation of solutions, quality of solutions. During the experiment, the researchers coded the scores in the established criteria on a 10-point scale (from 1 - lowest to 10 - highest).

In the final stage of the research, IT teams were asked to give an overall assessment of the creative methods used in the experiment in terms of the possibility of producing the best results in IT project management. Additionally, IT teams indicated in which IT project management processes the application of particular methods can be important.

Ultimately, the results of the research described above were grouped into the following areas:

- experience and knowledge of the IT team in the field of creative methods,
- effect of psychological inertia,
- assessment of creative methods,
- application of creative methods in IT project management processes.

The obtained results were subjected to statistical analysis.

In the first half of 2019, a parallel comparative study was undertaken. The aim was to verify the impact of the order in which creative methods are applied on their assessment. The comparative study was conducted on IT teams in 30 organizations from the SME sector. It differed in the order in which creative methods were applied in experimental solving of problem cases related to IT project management. The following sequence of the application of creative methods was adopted: Root Conflict Analysis RCA+ (TRIZ), System Analysis (TRIZ), Six Thinking Hats, Brainstorming. The obtained results were subjected to statistical analysis and compared with the results obtained in the basic study.

7 Research Results

In this chapter, the results of the basic research carried out on 52 SME organizations (IT teams) are presented. The next step is to compare these results with the results obtained in a comparative study (changing the sequence in which the individual methods are applied).

The aim of the first part of the study was assessment of the knowledge and application of creative methods in IT project management done by IT teams. Of the 52 teams studied,

46 teams were familiar with Brainstorming, only 4 teams were familiar with the Six Thinking Hats method, 3 teams with Design Thinking, and one with Ishikawa Diagram. 6 teams do not know any creative method. However, no team is in the habit of using any creative method in their daily work.

The second stage of the study was an experimental study of psychological inertia. As a result of the experiment, the following processes and areas were identified by observation in which inertia had the strongest effect:

- searching for the causes of problems,
- solving problems,
- designing new solutions,
- planning new activities,
- performing seemingly routine activities,
- human resources management - delegation and verification of performed activities.

Additionally, after explaining the mechanism of psychological inertia, the teams were able to identify the following effects of inertia in their IT projects:

- the habit of carrying out various projects in the same way,
- transfer/copying of solutions,
- duplication of the same schemes.

The next stage of the study was to assess the potential of the proposed creative methods for IT project management. The evaluation was made as a result of experimental studies using the established evaluation criteria. The results are shown in Fig. 1. The following abbreviations were used in the presentation of the results: BS - Brainstorming, 6H - Six Thinking Hats, SA - System Analysis, RCA+ - Root Conflict Analysis. The evaluation applied a 10-point scale, where 0 means the lowest, the smallest, the shortest, and 10 - the highest, the largest, the longest.

The most difficult method has turned out to be the RCA+ method, and the easiest – the commonly known Brainstorming method. The most numerous solutions are generated by the Brainstorming method, followed by RCA+ and System Analysis. The time of finding the solution was the shortest with the use of the System Analysis method, and the longest with the use of RCA+. As for the last two criteria: innovation of solutions and quality of solutions, the highest scores were obtained by the RCA+ method, followed by System Analysis – nearly 50% higher compared with the number of scores obtained by Brainstorming. It should be noted that these criteria are the criteria most-valued by IT teams in project management (the highest weights - see Table 1). The Six Thinking Hats method oscillates around the middle of the scale in all criteria.

Teams were asked to assign weights to each of the criteria. Table 1 shows the average results obtained in the study.

Then, the assessment of methods in individual criteria could be reduced to a single indicator, allowing for the unambiguous determination of the effectiveness of a given method in IT project management. For this purpose, the grades from individual criteria were recalculated on a 10-point rating scale in terms of 0 - the worst, 10 - the best, and

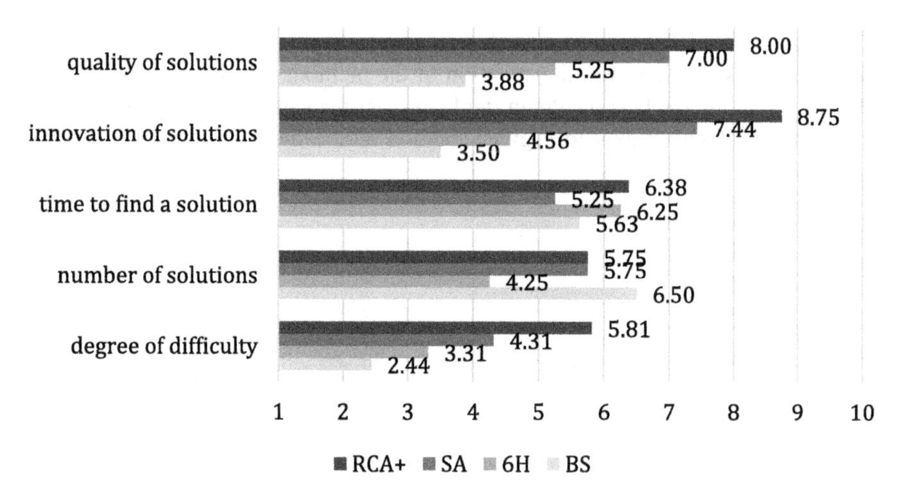

Fig. 1. Ratings of selected creative methods in individual criteria

Table 1. Average weights assigned to criteria for the evaluation of creative methods.

Criteria	Weights
Degree of difficulty	11
Number of solutions	11
Time to find a solution	14
Innovation of solutions	31
Quality of solutions	33
Total	100

the weights assigned to them were applied. Table 2 presents the effectiveness indicators for individual methods.

Table 2. Weighted effectiveness indicators for individual methods.

Method	Weighted effectiveness indicator
BS	4,57
6H	4,81
AS	6,56
RCA+	6,97

The results presented above clearly indicate that the most effective methods for IT project management from among those used in the study are RCA+ and System Analysis.

The final part of the study focused on the evaluation of creative methods by IT teams. Figure 2 shows an overall evaluation of the methods which concerned the possibility of achieving the best results in the IT project management. The graph presents the percentage of indications to a given method as the most effective in IT project management.

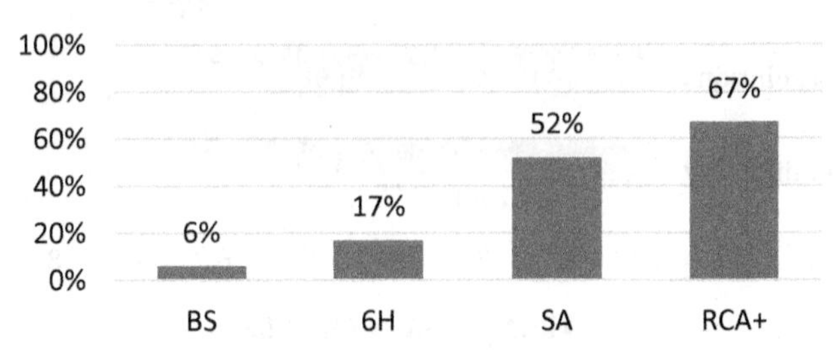

Fig. 2. Evaluation of selected creative methods in terms of the best results in IT project management

Almost half of the IT teams considered the RCA+ method as the one that could bring the best results in IT project management. System Analysis was in the second place. Brainstorming, which is the method best-known among IT teams, received the lowest percentage. Twenty two teams indicated more than one method. In most cases it was a combination of the System Analysis and RCA+ methods.

In IT project management it is also important when a given method should be used and in which phase of the project life cycle it is useful. To this end, IT teams were asked to specify in which IT project management processes the use of particular methods may be important. The obtained results are shown in Fig. 3.

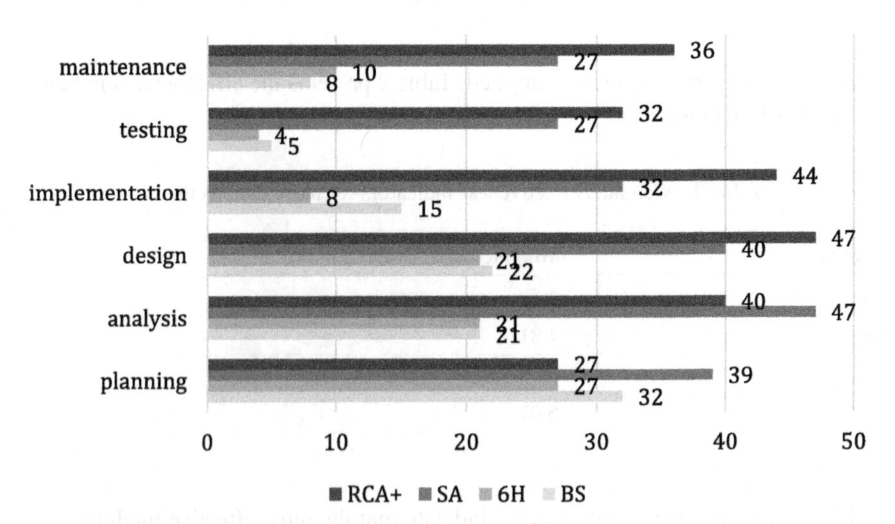

Fig. 3. Application of selected creative methods in IT project management processes.

In the planning and analysis process, System Analysis obtained the highest rating. In other processes, i.e. design, implementation, testing and maintenance, RCA+ clearly predominated, and System Analysis was in the second place receiving more than half of the scores. Brainstorming was considered among the top two only in the planning process. In the remaining processes, both Brainstorming and Six Thinking Hats were considered less important compared to the RCA+ and System Analysis methods.

An important element of the conducted research is verification of the significance of the sequence in which creative methods are applied on their assessment. For this purpose, the presented results will be compared with the results obtained in a comparative study (with the changed order of the application of individual methods).

The group participating in the comparative study had a similar structure in terms of their knowledge of creative methods, the use of these methods in work related to IT projects and the evaluation of psychological inertia. The comparison of results regarding an overall assessment of the methods in terms of the best results in IT project management did not show any significant differences (see Table 3).

Table 3. Comparison of the best results of selected creative methods in the IT project management.

	BS	6H	SA	RCA+
Basic study	6%	17%	52%	67%
Comparative study	7%	13%	53%	70%
Difference	−1%	4%	−1%	−3%

Table 4 presents the differences in the results of the evaluation of creative methods in terms of the adopted criteria. In this more detailed assessment, the differences are also insignificant and do not exceed 5%.

The last part of the research was an assessment of the applicability of individual methods in the IT project management processes. Differences in this area are presented in Table 5.

According to the data presented above, it should be stated that no significant differences were observed in this area of research.

8 Discussion

The research has shown that the knowledge of both creative methods and their use in project management by IT teams is weak. Asked about the reason for this state, IT teams in most cases answered that they could not see any possibility of applying these methods in IT project management, although their assessment of the potential of the creative methods presented to them gave completely different information. Every team has been fully aware of the fact that creative methods are of some use in the IT project management process. Only 7 teams have thought that Brainstorming is not useful in any IT project management process. The same concerned the method of Six Thinking Hats claimed to be rather useless by the 3 teams. This indicates the lack of knowledge of

Table 4. Comparison of the ratings of selected creative methods in individual criteria

Criteria		BS	6H	SA	RCA+
Degree of difficulty	Basic study	2,42	4,63	4,69	6,52
	Comparative study	2,70	4,33	4,67	6,43
	Difference	−2,77%	3,01%	0,26%	0,86%
Number of solutions	Basic study	7,10	4,46	5,40	5,92
	Comparative study	7,07	4,73	5,87	5,80
	Difference	0,29%	−2,72%	−4,63%	1,23%
Time to find a solution	Basic study	5,56	5,73	5,27	6,42
	Comparative study	5,93	5,97	5,43	6,23
	Difference	−3,76%	−2,36%	−1,64%	1,90%
Innovation of solutions	Basic study	3,44	4,88	7,44	8,63
	Comparative study	3,43	4,53	7,70	8,50
	Difference	0,09%	3,51%	−2,58%	1,35%
Quality of solutions	Basic study	3,85	4,90	7,31	8,35
	Comparative study	3,53	4,83	7,57	8,60
	Difference	3,13%	0,71%	−2,59%	−2,54%

Table 5. Comparison of indications for the use of selected creative methods in IT project management processes.

Processes	BS	6H	SA	RCA+
Planning	1,79%	4,74%	−1,67%	1,41%
Analysis	2,95%	6,28%	6,28%	9,74%
Design	−5,64%	6,28%	6,41%	−0,38%
Implementation	−5,51%	−2,05%	1,79%	−1,28%
Testing	0,38%	2,31%	−1,92%	5,13%
Maintenance	−2,05%	4,10%	1,41%	−2,56%
Average	−1,35%	3,61%	2,05%	2,01%

the creative methods and the lack of awareness of their potential among IT teams rather than the impossibility of applying these methods in IT project management. A premise for this supposition is the double number of indications regarding the use of creative methods in IT project management processes given to System Analysis and RCA+ in comparison with Brainstorming and Six Thinking Hats, remembering that, unlike the other two methods, the methods of System Analysis and RCA+ were not known to any of the IT teams.

A positive result is the IT teams' awareness of the impact of psychological inertia in their work. The most serious problem, however, is the habit of implementing various projects in the same way, while every project is, by definition, something unique. This also refers to copying ready-made solutions and applying them to situations requiring an individual approach. Creative methods have a potential here, offering the possibility of making project performer free from the psychological inertia in areas in which it interferes.

Considering the sums of applications of particular methods throughout the entire IT project management cycle, the most applications were indicated for RCA+ and System Analysis methods, more than 50% more than as for the Brainstorming or Six Thinking Hats methods.

An important element of the study was verification of the impact of the order in which creative methods were applied on their assessment. As shown by the results of the basic research and the results of a comparative study, the differences in all areas of the research were insignificant. This proves that the order in which the creative methods are applied by IT teams does not affect their assessment. This information is very important as in this way it is possible to avoid burdening the results with an error caused by the impact of the sequence in which the methods are recognized and applied on the assessment of their effectiveness and usefulness in a specific areas of IT project management.

9 Conclusions and Suggestions

The results of the conducted research suggest that in the IT environment there is a lack of sufficient knowledge about creative methods and their applicability in IT project management. This is in line with the research reports from recent years, in which doubting the potential of creative IT sector becomes more and more apparent (Gartner 2018, The Standish Group 2018). Bearing in mind the positive correlation of creativity with project management and overall organization performance, it should be considered important to undertake and deepen research in this area of the IT environment. Especially as regards the IT project management, very important is the occurrence of a high positive correlation of creativity with client satisfaction - crucial in any assessment of the IT project success [7].

The main potential of creative methods consists in the possibility to overcome psychological inertia, which causes many mistakes in IT project management and misunderstandings in relations with the clients (recipients of the project products). Nevertheless, it should be borne in mind that using creative methods does not generate random solutions but rather the most ideal solutions. This approach may prove to be crucial in the elimination of common weaknesses in IT project management, such as identification of real user needs, creeping scope, unused system functions, technological debt. Here the methods from the TRIZ group can play a special role, because TRIZ and IT projects, the aim of which is to create IT systems dedicated to a given organization, have in their assumptions the same main goal, i.e. to provide a methodology that will allow the creation of technical systems offering the best possible implementation of the functions with the least possible expenditure.

The results of the research indicate the need to build in the IT environment awareness of the role and potential of creative methods in IT project management. It would be a good practice to include workshops on creative methods into the education of IT project management. Further research should aim at locating creative methods in project management standards. This could significantly increase the success rate of IT projects.

References

1. Belski, I., Adunka, R., Mayer, O.: Educating a creative engineer: learning from engineering professionals. Procedia CIRP **39**, 79–84 (2016). https://doi.org/10.1016/j.procir.2016.01.169
2. Kazanjian, R.K., Drazin, R., Glynn, M.A.: Creativity and technological learning: the roles of organization architecture and crisis in large-scale projects. J. Eng. Technol. Manag. **17**(3–4), 273–298 (2000). https://doi.org/10.1016/S0923-4748(00)00026-6
3. Woźniak, M.: Sustainable approach in IT Project management—methodology choice vs. client satisfaction. Sustainability **13**(3), 1466 (2021). https://doi.org/10.3390/su13031466
4. Gartner Raport (2018). https://www.gartner.com/jp/products/japan-core-research-advanced/reports-list/2018-reports
5. CHAOS Report: Decision Latency Theory, Package - The Standish Group (2018). https://www.standishgroup.com/store/services/10-chaos-report-decision-latency-theory-2018-package.html
6. PMR, Market analysis and development forecasts for 2018–2023 (2018)
7. Kao, I.C.: The application and practical benefits of "C theory" in project management. In: MATEC Web of Conferences, vol. 1027, no. 119, 1–23 (2017). https://doi.org/10.1051/matecconf/201711901027
8. Plotnikova, L.I., Romanenko, M.V.: Creative methods of innovation process management as the law of competitiveness. Manag. Sci. Lett. **9**, 737–748 (2019). https://doi.org/10.5267/j.msl.2019.1.015
9. Chan, C.L., Jiang, J.J., Klein, G.: Team task skills as a facilitator for application and development skills. IEEE Trans. Eng. Manag. **55**(3), 434–441 (2008). https://doi.org/10.1109/TEM.2008.922633
10. Belski, I., Belski, I.: Application of TRIZ in improving the creativity of engineering experts. Procedia Eng. **131**, 792–797 (2015). https://doi.org/10.1016/j.proeng.2015.12.379
11. Cheng, C.Y.: C Theory - management philosophy of the book of changes. J. Chin. Philos. **19**(2), 125–153 (1992). https://doi.org/10.1111/j.1540-6253.1992.tb00115.x
12. Indriartiningtias, R., Hartono, B.: The effect of owner creativity on organizational creativity: empirical evidence from Surakarta Indonesia. In: 2018 IEEE International Conference on Industrial Engineering and Engineering Management (IEEM), pp. 473–476 (2018). https://doi.org/10.1109/IEEM.2018.8607703
13. Brem, A.: Creativity on demand: how to plan and execute successful innovation workshops. IEEE Eng. Manag. Rev. **47**(1), 94–98 (2019). https://doi.org/10.1109/EMR.2019.2896557
14. Boratyńska-Sala, A., Woźniak, M.: The state of knowledge and the use of creative methods in IT Project Management. In: Kowal, J., Keplinger, A., Klebaniuk, J., Mäkiö, J., Soja, P., Sonntag, R. (eds.) Proceedings of International Conference on ICT Management for Global Competitiveness and Economic Growth in Emerging Economies, pp. 40–52. University of Wrocław, Wroclaw (2019)
15. Borucki, A.: Ergonomic aspects of software engineering. In: Stephanidis, C., Antona, M. (eds.) UAHCI 2014. LNCS, vol. 8513, pp. 95–103. Springer, Cham (2014). https://doi.org/10.1007/978-3-319-07437-5_10
16. Simonette, M.J., Spina, E.: Enabling IT innovation through soft systems engineering. IGI Global (2014). https://doi.org/10.4018/978-1-4666-4313-0.ch005

17. Stamelos, I., Settas, D., Mallini, D.: Teaching software project management through management antipatterns. In: 2012 16th Panhellenic Conference on Informatics, Kastoria, Greece, pp. 8–12 (2011). https://doi.org/10.1109/PCI.2011.21
18. Strakhovich, E.: Using smart education together with design thinking: a case of IT product prototyping by students studying management. In: Uskov, V.L., Howlett, R.J., Jain, L.C. (eds.) Smart Education and e-Learning 2020, vol. 188, pp. 245–253. Springer, Singapore (2020). https://doi.org/10.1007/978-981-15-5584-8_21
19. Karczmarek, P., Pedrycz, W., Czerwiński, D., Kiersztyn, A.: The assessment of importance of selected issues of software engineering, IT project management, and programming paradigms based on graphical AHP and fuzzy C-means. In: 2020 IEEE International Conference on Fuzzy Systems (FUZZ-IEEE), Glasgow, UK, pp. 1–7 (2020). https://doi.org/10.1109/FUZZ48607.2020.9177591
20. Wu, W., Rose, G.M., Lyytinen, K.: Managing black swan information technology projects. In: Hawaii International Conference on System Sciences, Kauai, HI, USA, vol. 44, pp. 1–10 (2011). https://doi.org/10.1109/HICSS.2011.294
21. Marinho, M.L., Lima, T., Sampaio, S., Moura, H.: Uncertainty management in software projects - an action research. In: Proceedings of CIBSE 2015 – In XVIII Ibero-American Conference on Software Engineering, pp. 323–336 (2015)
22. AlMarar, M.S.: EPC strategies for a successful project execution. In: SPE Gas and Oil Technology Showcase and Conference, Dubai, UAE (2019). https://doi.org/10.2118/19878-MS
23. Koch, J.: O kreatywności [About creativity]. Inf. Bull. High-Tech **32**(1), 1–2 (2008)
24. Osborn, A.F.: Applied Imagination. Oxford University Press, Oxford (1953)
25. Schawel, C., Billing, F.: Brainstorming. In: Top 100 Management Tools, pp. 44–46. Gabler Verlag, Wiesbaden (2012). https://doi.org/10.1007/978-3-8349-4105-3_14
26. Dalton, J.: Brainstorming. In: Great Big Agile, pp. 139–141. Apress, Berkeley (2019). https://doi.org/10.1007/978-1-4842-4206-3_18
27. Göçmen, Ö., Coşkun, H.: The effects of the six thinking hats and speed on creativity in brainstorming. Think. Skills Creativity **31**, 284–295 (2019). https://doi.org/10.1016/j.tsc.2019.02.006
28. Serrat, O.: Wearing six thinking hats. In: Knowledge Solutions, pp. 615–618. Springer, Singapore (2017). https://doi.org/10.1007/978-981-10-0983-9_67
29. Treffinger, D.J., Selby, E.C., Isaksen, S.G.: Understanding individual problem-solving style: a key to learning and applying creative problem solving. Learn. Individ. Differ. **18**(4), 390–401 (2008). https://doi.org/10.1016/J.LINDIF.2007.11.007
30. Bertoncelli, T., Mayer, O., Lynass, M.: Creativity, learning techniques and TRIZ. Procedia CIRP **39**, 191–196 (2016). https://doi.org/10.1016/j.procir.2016.01.187
31. Zhang, F., Shen, X., He, Q.: Research on new product development planning and strategy based on TRIZ evolution theory. In: Yan, X.T., Ion, W.J., Eynard, B. (eds.) Global Design to Gain a Competitive Edge, pp. 825–834. Springer, London (2008). https://doi.org/10.1007/978-1-84800-239-5_81
32. Altshuller, H.: Развитие системного мышления – конечная цель обучения АРИЗу, (1975). https://www.altshuller.ru/triz/triz70.asp. Accessed 30 Mar 2022
33. Souchkov, V., Hoeboer, R., Zutphen, M.: TRIZ for business: application of RCA+ to analyze and solve business and management problems. TRIZ J. (2007). https://triz-journal.com/application-of-rca-to-solve-business-problems/
34. Highhouse, S., Gillespie, J.Z.: Do samples really matter that much? In: Lance, C.E., Vandenberg, R.J. (eds.) Statistical and Methodological Myths and Urban Legends: Doctrine, Verity and Fable in the Organizational and Social Sciences, pp. 247–265. Routledge, New York (2009)

Modeling IT Systems in TRIZ

Jerzy Chrząszcz[1(✉)], Tiziana Bertoncelli[2], Barbara Gronauer[3], Oliver Mayer[4], and Horst Nähler[5]

[1] Institute of Computer Science, Warsaw University of Technology, 00-665 Warsaw, Poland
jerzy.chrzaszcz@pw.edu.pl
[2] ANSYS Germany GmbH, 83624 Otterfing, Germany
[3] StrategieInnovation, 36088 Huenfeld, Germany
[4] Bayern Innovativ, 90402 Nürnberg, Germany
[5] c4pi - Center for Product Innovation, 36088 Huenfeld, Germany

Abstract. Today's world extensively uses Information Technology (IT) and heavily depends on IT systems. The evident and justified need to model IT systems in TRIZ projects faces some doubts or objections because the fundamental concepts of classic TRIZ were formulated before computer times. This paper aims to analyze terminology and guidelines regarding Function Analysis for products and Flow Analysis in the context of modeling IT systems and devising necessary adjustments so that components like hardware, software, and data may be modeled consistently.

We describe the approach taken and introduce a minimal complete IT system using the TRIZ approach, with a detailed description of components and functions supported by examples of IT systems. A conceptual perspective allowing for uniform modeling of mechanical and IT systems is intended to alleviate the problems of trainers and practitioners encountered in this area and increase TRIZ acceptance in the IT industry and IT communities.

Keywords: TRIZ · Function Analysis · IT systems · Hardware · Software · Data · Flow Analysis · Information Technology modeling · IT modeling

1 Introduction

The proliferation of Information Technology (IT) and extensive use of various IT solutions in countless application areas call for a systematic approach to modeling IT systems in TRIZ. The IT systems, as we know them today, are usually described as consisting of hardware (computers) processing data under the control of software (programs). The main difference between these building blocks is that programs and data are intangible entities.

This property is perceived as the key success factor of the computer systems, as both programs and data may be easily modified and quickly transmitted to any location. On the other hand, the intangibility raises doubts and objections in the TRIZ community because the foundations of the methodology were created before computer times.

Published by Springer Nature Switzerland AG 2022
R. Nowak et al. (Eds.): TFC 2022, IFIP AICT 655, pp. 248–260, 2022.
https://doi.org/10.1007/978-3-031-17288-5_22

In particular, the basic definitions used in Function Analysis and Substance-Field Analysis are misleading, at best, in this respect. The definitions of *Function*, *Component*, and *Parameter* taken from selected reputable sources are given in Table 1 below. As can be seen, some items are missing, and some differ significantly in scope or approach. The definitions coming from the MATRIZ-approved sources seem the most compatible with each other, but still [1, 2] require the objects to be *material*, while in [3], this adjective does not appear. Although there are also some differences between definitions of *Substance* and *Field*, the sources generally agree that a substance does have and a field does not have a rest mass.

Table 1. Selected definitions of the basic terms used in function analysis. The original capitalization of the terms has been removed for easier comparison.

Source	Function	Component	Parameter
Glossary of TRIZ and TRIZ-related terms [1]	Specification of an action performed by a material object (function carrier) that results in a change or preservation of a value of an attribute of another material object (object of the function)	A material object (substance, field, or substance-field combination) that constitutes a part of a technical system or its supersystem; a component might represent both a single object and a group of objects	A variable dimensional or dimensionless measurable factor, either specific or aggregated, that participates in the definition of an attribute of a material object of a technical system or its supersystem and determining its borders and behaviors. (…)
Selected Topics for Level 1 Training [2]	An action performed by one material object (function carrier) to change or maintain a parameter of another material object (object of the function)	A material object (substance, field, or substance-field combination) that constitutes a part of the engineering system or supersystem	A comparable value of an attribute
Level 3 training materials [3]	An action performed by one component (function carrier) to change or maintain a parameter of another component (object of function)	An object (substance, field or substance-field combination) that constitutes a part of an engineering system or supersystem	A comparable value of an attribute
A brief glossary of basic concepts and terms of TRIZ [4]	A change, stabilization or measurement of certain parameters of an object of function (or product) by the influence of the carrier of the function (tool) on it. (…)	Part of the system as an element (substance) or as a field	A quantity that characterizes a property of a process, phenomenon or system; the parameters indicate how this system (process) is different from others (…)
VDI 4521 Part 1 [5]	Effect from a system or a system component upon others which changes, eliminates, or maintains a parameter of the other component or system	n/a	n/a

A rest mass neatly corresponds with the material nature of a substance, but how a field may be perceived as a material object when it does not have a rest mass and contains no matter? Moreover, if we attempt to identify the components of an IT system, it seems reasonable to perceive hardware as a substance and software and data – as fields. Therefore, if one insists that a field must be a material object, the conclusion seems obvious: neither software nor data may be a legitimate system component (and such a statement was communicated during some TRIZ workshops, usually without hardcopy on slides).

The first viable solution to this puzzle is to remove the *material* adjective from the definitions of the component and function, just like it was done in [3], eliminating the doubts regarding a material field. The second way is less obvious, as it requires reconsidering the native meaning of this adjective. Although it is not emphasized in international literature, TRIZ originated from the materialistic and dialectic worldview [6]. Therefore, the adjective *material* in the mentioned definitions should be interpreted as *objective*, *measurable*, and *capable of interacting with other entities* rather than *made of matter*. This is an example of a notion that was literally "lost in translation" during the internationalization of TRIZ methodology. Since software and data are undoubtedly material in a sense mentioned above, we will consider them legitimate field-type components of technical systems in the following sections.

Several authors have already addressed the topic of applying TRIZ in the IT domain. A comprehensive overview of software-related works is given in [7], which also provides some forecasts for this area. Other interesting publications focus on functional modeling [8, 9], Substance-Field modeling [10], transferring the 40 Inventive Principles to the IT domain [11–14], and Laws of Evolution of information systems [15]. Nevertheless, the appropriateness of TRIZ for IT still raises doubts – see two quotes from [9]: *"Key Assumptions: Components in Function Model can be material objects only (...) Uncertainties: Uncertainty how to deal with functions in IT systems"*.

2 The Building Blocks

The first decomposition of a computer system is between hardware and software. The hardware comprises processor(s), memory, and input/output devices (i/o) required for communication with the rest of the world – in particular, sensors and effectors. There are some subcategories, as memory may appear in several layers of a computer system (processor registers, cache memory, operating memory, etc.) and several technologies (semiconductor, magnetic, optical, ferroelectric, etc.) accounting for specific characteristics of the devices – for instance, contents volatility vs. non-volatility upon power off.

Software is generally meant as programs, being sequences of instructions for the hardware units, but there are also some shades here. For instance, the system start-up (boot-up) program is typically stored in non-volatile memory, and such built-in programs are often called firmware to differentiate it from software applications, which are usually loaded from a storage device into operating memory to be executed. To configure the i/o devices, several settings must be preset by writing specific values to the internal registers of i/o controllers, so these values must also be stored in the non-volatile memory.

The data also appears in two main categories, namely the constants and the variables. As the names indicate, the difference is that variables' values may be modified during processing while the constants retain their values. Some constants are usually built in the program code, which may be stored in read-only memory. Hence, such constants should be considered as belonging to the program rather than data. Read-only memory is also used for storing data items having fixed values, such as predefined lists (e.g. month names) or error messages. The variables, on the contrary, must be stored in read-write memory, and they are used for keeping the values acquired from sensors, auxiliary data structures, intermediate results, and final results.

The memory stores programs and data for different purposes. Programs are intended to be executed by the processor, while data is intended to be manipulated (processed) by programs. It should be noted that at the physical level, memory stores just binary patterns so that data and instructions cannot be directly differentiated from each other. As a consequence, instructions stored in rewritable memory may be manipulated as data (which allows for creating self-modifying programs), and also data may be executed as instructions (by omission or as a security breach).

The processor fetches and executes instructions from subsequent memory cells unless a control instruction (e.g. jump) forces a non-sequential transition. The instruction execution cycle comprises (1) fetching instruction, (2) decoding instruction, (3) reading operands from the source locations, (4) executing respective operation, e.g. addition, and (5) writing operation result into the destination location. This cycle covers the most complex scenario, as some instructions do not need operands, some do not produce a result, etc. The processing capabilities are determined by the processor's instruction set (defining possible data manipulations) and the addressing modes (defining possible ways of accessing data items during processing).

3 Function Analysis for Products

As was mentioned before, at a high level of abstraction, a computer system is usually described as a combination of hardware and software, so let us build a function model of such a system. We should start with defining the main function and boundary of the system, and the first ambiguity appears here, as the two generic types of processing systems differ significantly.

An embedded processing system (industrial controller, signal converter, etc.) may work without direct interaction with the user, and so its main function may be described as *to process data*. An interactive processing system, in turn (a PC, smartphone, etc.), is controlled directly by the user and provides results to the user so that the main functions may be described as *to process data* and *to inform user*. Moreover, taking into account that computer systems may interact with other computer systems instead of (or in addition to) humans, we may generalize *user* as *operator*.

Another source of ambiguity is that equivalent data processing may be implemented in two distinct ways. Either dedicated, application-specific hardware (e.g. ASIC) or generic software-controlled hardware (programmable processor) may be used. In the former variant, we have only hardware. The sequence of operations is defined with the preconfigured interconnections, and the operations are determined by the functions of the execution units embedded in the structure. The latter variant comprises hardware with all necessary execution units used on-demand and software determining the sequence of operations. We will only address hardware-software systems here (Fig. 1).

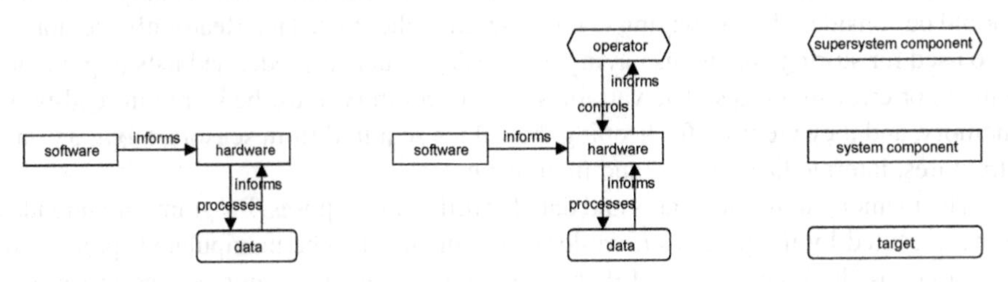

Fig. 1. Basic IT system models: embedded (left) and interactive (right).

For variant 1, the system comprises hardware (substance) and software (field), while the data (field) is a supersystem component and the functions are as follows:

- *software informs hardware* (reflects fetching instructions by the processor),
- *data informs hardware* (reflects reading values of data items),
- *hardware processes data* (reflects modifying and writing data items).

For variant 2, the system comprises hardware (substance) and software (field), while the data (field) and the operator (substance) are supersystem components. The functions *software informs hardware, data informs hardware,* and *hardware processes data* are same as above, and the new functions are:

- *operator controls hardware* (as input from the operator goes via the input devices),
- *hardware informs operator* (reflects the operation of the output devices).

Because variant 1 is a subset of variant 2, we will only consider modifications of the latter. The first extensions indicate the *parameters* as a system component (field) and distinguish *input data* (operands) from *output data* (results), as shown in Fig. 2. Parameters are meant as configuration settings e.g. coefficients in a weighted sum calculation.

- *parameters inform hardware* (reflects reading parameter values),
- *hardware processes parameters* (modifying and writing parameter values),
- *input data informs hardware* (same as *data informs hardware*),
- *hardware generates output data* (reflects modifying and writing result values).

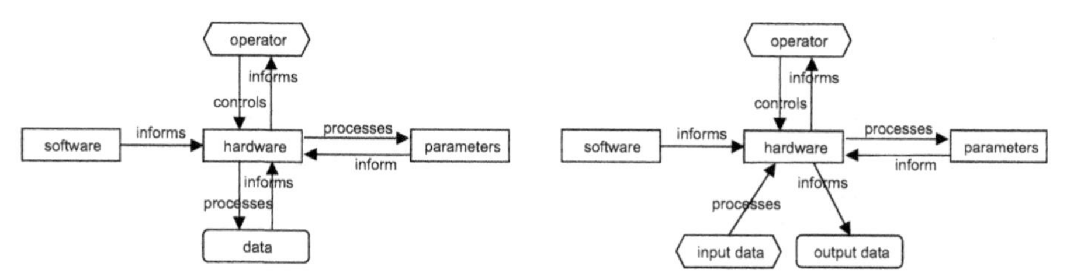

Fig. 2. A model of an IT system with indicated parameters. Data may be considered as a unified input and output component (left), or separate input and output data components (right).

A model may also be given finer granularity, like in Fig. 3, where *hardware* is decomposed into *processor*, *memory,* and *i/o devices* (substances), and memory contents are decomposed into *programs, parameters,* and *variables* (fields). The functions regarding supersystem components differ from the previous model to indicate that all interactions between the processor and the outside world go through the i/o devices. The internal functions are as follows:

- *processor controls i/o devices* (reflects indicating operations to be performed by i/o devices and, as a second function, writing data to i/o ports),
- *i/o devices inform processor* (reflects reading data from i/o ports),
- *processor controls memory* (reflects indicating operations to be performed by memory and, as a second function, writing contents to memory locations),
- *memory informs processor* (reflects reading memory contents, including program instructions, parameter values, and variable values),
- *memory holds program/parameters/variables* (reflect that memory cells are kept unchanged between write cycles).

Two additional functions are indicated in the right diagram to model the capability of direct data transfers between memory and i/o devices without consuming processor time, known in IT as Direct Memory Access functionality (*i/o devices control memory,* and *memory informs i/o devices*).

4 A Complete IT System

The concept of a complete technical system, related to the Trend of Increasing System Completeness, works well in TRIZ for mechanical systems. Surprisingly, respective definitions also vary between the sources, as shown in Table 2. The original Altshuller's description in [16] indicates four mandatory components, namely *engine, transmission, control,* and *tool.*

The same composition is indicated in [1] and in newer publications, such as [17], where the *energy source* is indicated in the supersystem. Other publications [2, 3, 18] show the energy source as a mandatory component of a complete system instead of the engine, and some even refer to the engine as an example of the energy source.

Yet another version appears in [5], where energy source and object (product) are indicated as system components in addition to the four mentioned in the original definition. Finally, a generalized composition was devised in [19], with the *source of matter,*

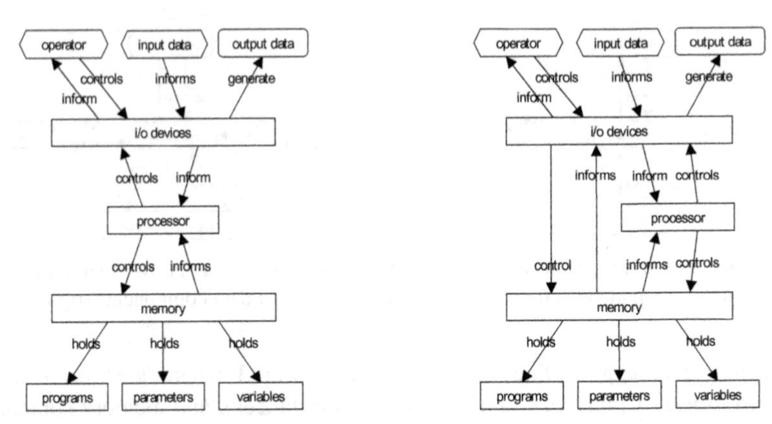

Fig. 3. An expanded model of an IT system with i/o data transfers provided by processor (left) and with Direct Memory Access capabilities (right).

energy, and information (located on the system boundary – i.e. possibly or partially in the supersystem), supposedly replacing the energy source and engine, and a *converter* (instead of transmission).

The proposed diagram of a complete IT system (see Fig. 4) has been derived from Altshuller's version. The software is perceived like fuel for the hardware engine, and this energy source may be located inside or outside the system. Data is the object of the processing, and the elementary operations indicated as the tool are determined by the instruction set of the processor. Control is provided by an operator (human or machine), and the instruction cycle is the transmission counterpart as this is how the "potential energy" stored in the software (after transformation into "kinetic energy" by hardware) is applied to data items in the form of specific instructions executed in a specific sequence.

As mentioned in Sect. 2, the instruction cycle is hardware activity supporting program execution by reading instructions and operands, performing required operations, and storing the results. The actual processing capabilities of a given processor depend on its instruction set. For example, some processors perform multiplication in a single clock cycle, others use a sequence of low-level operations, making multiplication instructions execute considerably longer than other instructions, while many processors do not have multiplication instructions at all. They still may multiply numbers, but this must be implemented in software. That is why the instruction cycle is proposed as the counterpart of transmission in a mechanical system, and the instruction set is considered a counterpart of a tool, being the component operating on the object directly. A similar diagram regarding the composition of a complete software system is presented in [20].

Table 2. Selected definitions referring to the complete technical system. The original capitalization of the terms has been removed for easier comparison.

Source	Definition
Glossary of TRIZ and TRIZ-related terms [1]	A technical system that according to the trend of technical system completeness, includes at least four components (subsystems) which provide functions of engine, transmission, control unit, and working unit
Selected Topics for Level 1 Training [2]	Control system: a common functional part of the engineering system that controls how the other parts function. For example, a thermostat in an air-conditioning system Energy source: a common functional part of the engineering system that generates energy to operate the system. For example, an engine in a car Transmission: a common functional part of an engineering system and its supersystem that transfers a field (energy) from an energy source to an operational device Operating tool: a typical function part of an engineering system that usually performs the most important basic functions
Level 3 training materials [3]	As an engineering system evolves, it acquires the following typical functions: the function of operating agent, the function of transmission, the function of energy source, the function of control system
A brief glossary of basic concepts and terms of TRIZ [4]	n/a
VDI 4521 Part 1 [5]	System consisting of the elements: energy source, drive, transmission, working means, control, object and, if applicable, further interfaces to the supersystem

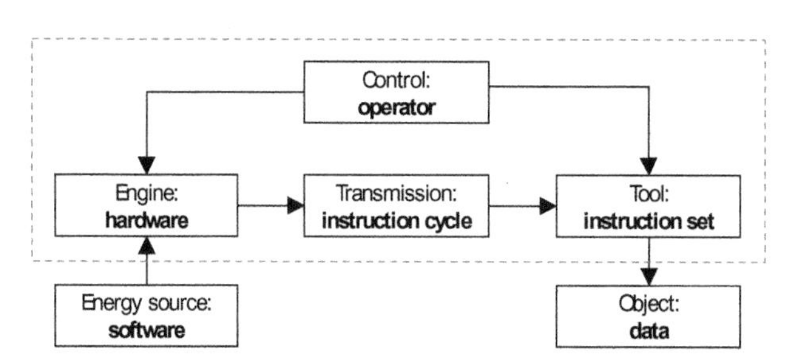

Fig. 4. The proposed composition of a complete IT system.

5 Flow Analysis

An IT system can as well be modeled using the Flow Analysis approach to describe the data manipulation process. In TRIZ, a distinction is made for the forms of flows:

- The *material flow* is about moving material objects with a mass, or around a large number of similarly movable objects, for example, cars, bottles, on a conveyor belt,

etc. Such flows are usually neglected in IT systems unless we are interested in e.g. dust transfer that may affect cooling capabilities and lead to a system failure.
- The *energy flow* is about moving nonmaterial objects, like energy, acoustic, or light (photons). High-performance IT systems consume lots of energy, while portable systems, especially those providing critical functions, need sophisticated power management so that analyzing flows of energy is of great importance in IT systems.
- And third, explicitly, the *information flow* consisting of bits and bytes, broadcasting information, and the like is stated. Here is the explicit thought to run a nonmaterial object through a process and to drive and control this process. This type of flow is the basic area of interest when analyzing IT systems, as both software (instructions) and data must be transferred between components, and the overall performance of the system is affected by stalls, delays, insufficient throughput, and other deficiencies regarding the flow of information.

In the flow analysis, there are four categories of flows:

- A useful flow is where a material object, energy, or information is performing a useful function or is subject to a useful function.
- A harmful flow is where a material object, energy, or information is performing a harmful function or is subject to a harmful function.
- A wasted flow is where a material object, energy, or information is characterized by the loss of substance, energy, or information.
- A neutral flow where a material object, energy, or information is characterized by an irrelevant or insignificant effect on the technical system.

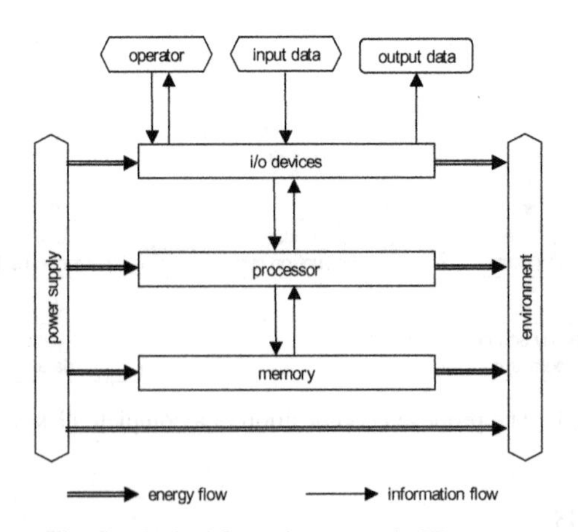

Fig. 5. Typical flows in a generic IT system.

A simplified diagram of energy and information flows in a properly operating IT system is shown in Fig. 5. Electricity from the power supply goes to the system building blocks indicated as single components (processor, memory, and i/o devices), although each of them may appear as several units, e.g. processor cores or memory modules. The mapping of the logical components onto physical devices may be much more complicated, e.g. latest generations of processors have cache memories, memory controllers, and sometimes also graphic controllers integrated on-chip.

Heat dissipation in the system is shown as energy flows sinking to the environment, which should be considered harmful. Thermal interactions between system components are neglected in the diagram for simplicity, while in reality, some elements may transfer heat to others. Finally, all information flows are shown as useful and adequate, while in a real system, there may be some bottlenecks or other flow disadvantages affecting system performance, and the flow model is capable of reflecting such situations. On the other hand, the model may also represent parasitic flows – e.g. data breaches.

As in the previous diagrams, we ended the description of the situation by generating the output data, which is typical for batch processing. In control systems, on the contrary, some input data comes from the supervised object or process through sensors, and some output data goes to the operator or to effectors to control system operation in real-time. In such cases, some additional flows should be added in the supersystem.

6 Examples

In this paragraph, an example of using the FA for products for software applications is given; the system of choice is the so-called Corona-Warn-App [21], offered and maintained by the German Authorities – in particular Robert Koch-Institut (RKI) – for disease spreading control via contact tracing, introduced during the Covid-19 Pandemic. The example is meant to be the first tentative to apply TRIZ to decentralized software systems, which is also interesting from the data security perspective. In fact, decentralization was the key technology that was claimed to assure anonymity and data privacy protection during the app roll-out. Since the system is a software application for smartphones intended to work by gathering and processing data from millions of users at the same time, the chosen template for the function model is the one depicted in Fig. 2, that is, with separate input and output components, of type interactive. With respect to the specific application example, the different components are mapped as follows (see Fig. 6):

- operator: smartphone user (only two users are shown in the example for clarity),
- hardware: smartphone + Bluetooth Low Energy module,
- software: Robert Koch-Institut Corona-Warn-App (CWA),
- input data: rolling proximity ID codes (automatically encrypted); optional: infection status, test status, vaccination status of other users,
- output data: own rolling proximity ID codes (automatically encrypted); optional: infection status, test status, vaccination status,
- parameters: time of contact, distance of contact (signal damping), time since contact, transmission risk level of the contact (evaluated on the basis of infection begin and symptom status) with respect to other smartphones with active Corona-Warn-App.

It can be observed that in this proposed function model, no function ranking has been performed. Moreover, only useful functions have been listed, while in order to identify bottlenecks and/or problematic system parts also, harmful functions should be

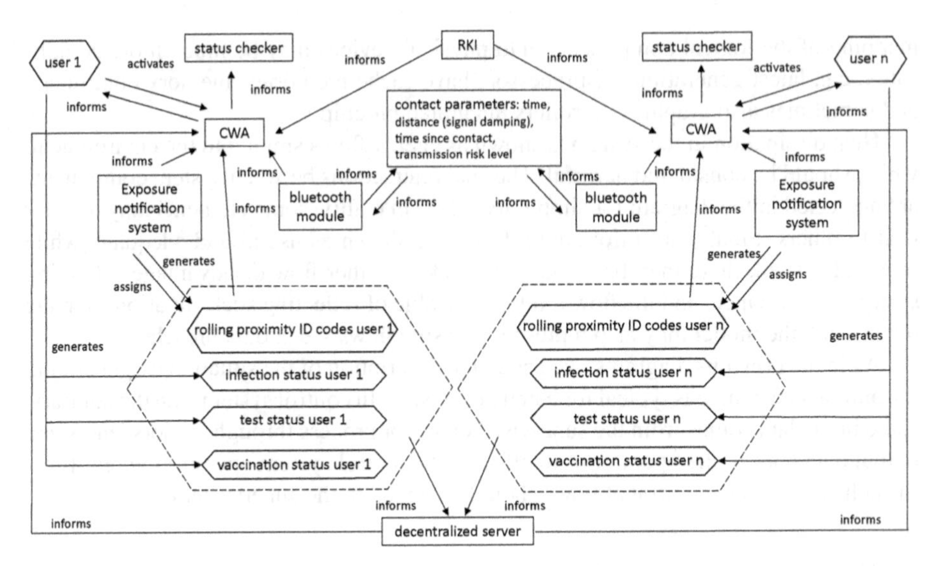

Fig. 6. Function model of a distributed IT system.

identified. Nevertheless, for this model and for the software/IT function model in general, it is advisable to rank if the useful functions are performed inefficiently or if they may represent a "weak link", i.e. if there is a danger of data loss.

The qualification of the functions (useful, performed sufficiently, insufficiently or excessively, harmful) may also vary according to the specific problem statement: for this example, the main function at the time of the App deployment was to inform user through a risk notification that was tuned to the disease understanding and level of danger to be performed satisfactorily. This changed during the pandemic evolution and resulted in being excessively performed at a later stage during spring 2022, so the contact parameters were tuned differently. An analogous model according to a Function Analysis for processes could also be built, but since most functions are performed repeatedly, possibly at the same time and not following a given sequence, depending on the movements, activation status, and input from all the users, it can be extremely cumbersome to rearrange the steps according to a process structure. For one such system, the Function Analysis for products is more convenient.

A simplified model of a recurring operation of an IT system is shown in Fig. 7. The user inputs a command onto the keyboard. The keyboard translates the keystrokes into ASCII data that is then transferred to the USB controller. The controller writes data into the memory buffer, and the processor processes the command. The result is transformed by a graphic controller, which generates the pixel map for display. Finally, the information is presented to the user on the screen, closing the interaction loop.
This is a simplified example. One can easily drill it down to further levels of abstraction or focus on specific disadvantages. For instance, a harmful flow may be indicated in the model to reflect possible crosstalks from the supersystem resulting in distorted transmission between the keyboard and USB controller. Parasitic flows may also be added to model possible eavesdropping of the keyboard data, or excessive visibility of the display

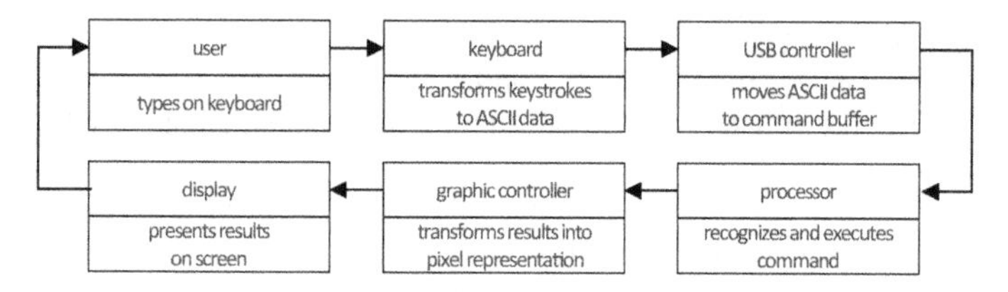

Fig. 7. Simplified information flow model of an IT system operation.

resulting in the unintended revealing of the screen contents to bystanders, gray zones or bottlenecks may be indicated on the way between USB controller and display to model performance limitations, etc.

7 Summary and Further Work

We presented a justification for considering software and data as legitimate field-type components of technical systems. This is an important change of approach since it allows to model and analyze IT-related functions of complex systems in a holistic way. It seems essential because many modern systems coming from mechanical or electrical engineering domains (ranging from headphones to cars) are invaded by Information Technology to a level that makes them vulnerable to cyberattacks like other computer systems so that hacking e.g. a Bluetooth-controlled toilet is a viable scenario.

With a whole system modeled in a unified way, one can apply seasoned TRIZ tools, such as Function Analysis and Flow Analysis, to identify functional disadvantages and flow disadvantages, including those regarding software or data. The modeling of the flow and the characterization, as well as type and degree of fulfillment, allows for a further step to specifically solve contradictions/problems in the flow. For this purpose, the tools of flow enhancement are provided by the TRIZ toolbox.

Consequently, an IT system with hardware components and the flow of information can be modeled. Using the Flow Analysis will end up in larger software and data sizes to a very big and, therefore, hard to manage model. For this reason, it is recommended to use Flow Analysis when focusing on a specific aspect and Function Analysis for the broader architectural perspective. It is comparable to Function Analysis for a system and Substance-Field modeling for a specific problem.

Acknowledgments. The authors gratefully acknowledge Dr Oleg Abramov and Mr. Dmitriy Bakhturin for their explanations regarding the philosophical roots of TRIZ.

References

1. Souchkov, V.: Glossary of TRIZ and TRIZ-related terms, version 1.2, MATRIZ 2018. https://matriz.org/wp-content/uploads/2016/11/TRIZGlossaryVersion1_2.pdf. Accessed 19 Mar 2022
2. Selected Topics for Level 1 Training, MATRIZ 2020. https://matriz.org/wp-content/uploads/2020/04/Selected-Topics-for-Level-1-Training.pdf. Accessed 19 Mar 2022

3. Ikovenko, S.: Level 3 Training Manual (2019)
4. Rubin, M., Kuryan, A., Rubina, N., Shchedrin, N., Eccardt, O.: A brief glossary of basic concepts and terms of TRIZ (in Russian). https://triz-summit.ru/onto_triz/100/. Accessed 19Mar 2022
5. VDI 4521 – Part 1. Inventive problem solving with TRIZ - fundamentals, terms and definitions, Verein Deutscher Ingenieure e.V. (2021)
6. Korolev, V.A.: On dogmatism in TRIZ. In: TRIZ Developers Summit 2017 (in Russian). https://triz-summit.ru/confer/tds-2017/article/. Accessed 19 Mar 2022
7. Govindarajan, U.H., Sheu, D.D., Mann, D.: Review of systematic software innovation using TRIZ. Int. J. Syst. Innov. **5**(3), 72–90 (2019)
8. Beckmann, H: Function analysis integrating software applications. In: TRIZ Future 2013 Proceedings, pp. 473–482 (2013)
9. Souchkov, V.: Extended Function Analysis - Overview. Presentation delivered during German Expert Day 2019, Sulzbach-Rosenberg, Germany, February, 2019
10. Petrov, V.: Using TRIZ tools in IT. In: TRIZ Developers Summit 2019. https://triz-summit.ru/file.php/id/f304862-file-original.pdf. Accessed 19 Mar 2022
11. Beckmann, H.: Method for transferring the 40 inventive principles to information technology and software. Procedia Eng. **131**, 993–1001 (2015)
12. Goethals, F.: 20 Trends in digital innovations: a TRIZ-trend-structured overview of new business information technologies and innovative applications. Amazon Digital Services LLC (2014). ASIN: B00P6GNE88
13. Lady, D.: Information Technology System Cookbook: Introducing TrizIT: TRIZ for Information Technology. AIMS Publications (2013). ISBN: 978-1478302513
14. Mishra, U.: TRIZ Principles for Information Technology. CreateSpace (2010). ISBN: 978-818465184
15. Padabed, I.: Contradictions and laws of evolution in information systems. In: TRIZ Developers Summit 2019. https://triz-summit.ru/file.php/id/f304852-file-original.pdf. Accessed 19 Mar 2022
16. Altshuller, G.: Creativity as an Exact Science: The Theory of the Solution of Inventive Problems. CRC Press, Boca Raton (1984)
17. Souchkov, V.: Development of functionality in technical and business systems. In: TRIZ Developers Summit 2020 (in Russian). https://r1.nubex.ru/s828-c8b/f3244_06/Souchkov-TDS-2020-Functionality.pdf. Accessed 19 Mar 2022
18. Lyubomirskiy, A., Litvin, S., Ikovenko, S., Thurnes, C.M., Adunka, R.: Trends of Engineering System Evolution (TESE) – TRIZ paths to innovation, TRIZ Consulting Group (2018)
19. Petrov, V.: Patterns of development of artificial systems. In: TRIZ Developers Summit 2020 (in Russian). https://r1.nubex.ru/s828-c8b/f3215_04/Petrov-TDS-2020-regularities%5b1%5d.pdf. Accessed 19 Mar 2022
20. Nähler, H., Gronauer, B., Bertoncelli, T., Beckmann, H., Chrząszcz, J., Mayer, O.: Modeling Software in TRIZ, accepted to TRIZ Future 2022 Conference
21. Corona-Warn-App. https://apps.apple.com/pl/app/corona-warn-app/id1512595757. Accessed 19 Mar 2022

Modeling Software in TRIZ

Horst Nähler[1(✉)], Barbara Gronauer[2], Tiziana Bertoncelli[3], Hartmut Beckmann[4], Jerzy Chrząszcz[5], and Oliver Mayer[6]

[1] c4pi - Center for Product-Innovation, 36088 Hünfeld, Germany
naehler@c4pi.de
[2] StrategieInnovation, 36088 Hünfeld, Germany
[3] ANSYS Germany GmbH, 83624 Otterfing, Germany
[4] Giesecke+Devrient GmbH, 81677 München, Germany
[5] Institute of Computer Science, Warsaw University of Technology, 00-665 Warsaw, Poland
[6] Bayern Innovativ GmbH, 90402 Nürnberg, Germany

Abstract. Although many papers have been published in the past on the use of TRIZ in the field of Information Technology, questions still arise as to whether the TRIZ methodology is also suitable for software as purely intangible systems.

Doubts remain because the methodology was developed at a time when patents for IT and software systems, in contrast to physical products, could not yet be considered in the underlying patent analysis.

Because of these questions, this paper examines the transferability of the TRIZ methodology to software systems on the basis of two fundamental TRIZ concepts: the Law of System Completeness and the Function Analysis for Processes.

After an introduction regarding the immaterial nature of software systems, the transferability of TRIZ concepts to the software domain is shown with the help of case studies and ends with an outlook on possible further additions to the conceptual mapping of TRIZ to the software domain.

Keywords: TRIZ · Software · Information technology · Function analysis for processes · Function modelling for software · Software development

1 Introduction

The application of TRIZ tools to the IT domain has already been extensively assessed and presented: e.g., [1] gives a comprehensive overview of software-related works, [2, 3] assess function analysis for products, [4] contains aspects of Substance-Field modelling, [5] elaborates on the Laws of Engineering System Evolution with respect to information systems. Especially the transfer of the Inventive Principles to the IT domain has been covered by [6–9] and even a dedicated view on programming under TRIZ aspects is given in [10]. Moreover, a comprehensive book with a specific contradiction matrix was published already in 2008 [11]. Despite the number of publications, the authors repeatedly are faced with questions and doubts of TRIZ students if and how TRIZ might be a useful asset in the modern world, especially when dealing with IT systems and

Published by Springer Nature Switzerland AG 2022
R. Nowak et al. (Eds.): TFC 2022, IFIP AICT 655, pp. 261–272, 2022.
https://doi.org/10.1007/978-3-031-17288-5_23

specifically with software. These doubts are mainly fueled by the origins of TRIZ in the era of mechanical engineering and the associated wording of some classical TRIZ tools or examples given in classical TRIZ literature. Basic TRIZ concepts like the model of a complete system seem to exclusively address classical material engineering systems.

While the use of TRIZ concepts within primarily hardware-related IT topics requires manageable transfer performance, e.g., see examples in [12], the perceived "immateriality" of software makes it harder to find connecting points between TRIZ concepts and the software domain. Especially topics related to high level or platform independent programming languages are very much de-coupled from the hardware level of computer systems, making the application of the TRIZ method to software more difficult.

During the work of the authors on the topic of modelling IT systems in TRIZ [13], where we assessed IT systems as a combination of software and hardware, the following question arose: *How can TRIZ concepts and modeling approaches be mapped on pure software topics?*

Consequently, the goal of this paper is to expound how basic concepts of the Theory of Inventive Problem Solving can be mapped to the software domain. For this the authors focused on the conceptual modelling aspect of Altshuller's Law of System Completeness and the Function Analysis for Processes.

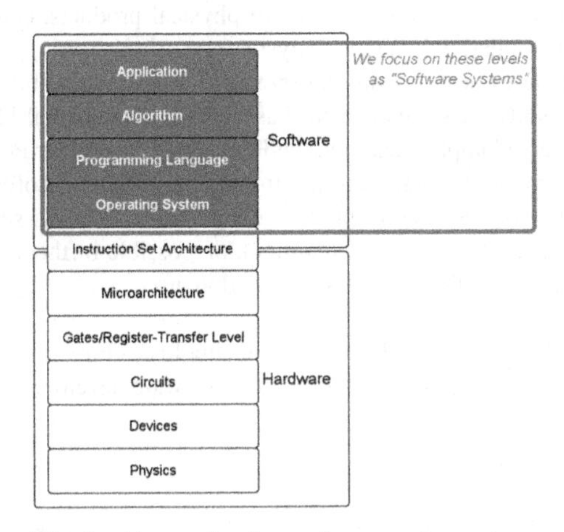

Fig. 1. Abstraction layers in computer systems

By mapping both topics to the software domain the authors want to add to a conceptual and practical foundation of using TRIZ in the software domain. In this paper, the authors strictly focus on the software level of computer systems as described in [14–16]. Within this domain, we use the term "software system" as a differentiation to "engineering system" to highlight the immaterial nature of such systems (see Fig. 1). Additionally, we provided an overview of the definitions of terms used in Table 4.

2 A Complete Software System

As a fundamental concept of TRIZ, Altshuller's Law of System Completeness represents the basic structure of an engineering system. The concept describes necessary elements of a working engineering system. Building on this concept, additional practical instruments like the sequence of acquiring necessary parts from the supersystem or transferring functionalities back to the supersystem along the evolution of engineering systems were derived [17].

To approach software systems from a TRIZ viewpoint, the Law of System Completeness can be described in an analog way. However, several modifications reflecting the specific nature of software have to be introduced (see Fig. 2).

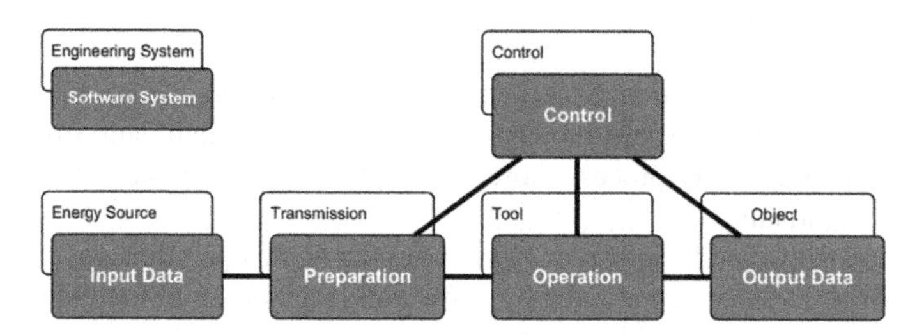

Fig. 2. Complete software system

When mapped onto a software system, the Law of System Completeness reflects the structural composition of the system and the flow and modification of input data through a software system to create output data. The parts identified as required to achieve that functionality are described in Table 1.

Table 1. Parts of a complete software system

Parts of a complete software system		RAW file editing of digital photographs
Input data	Input data is the equivalent to the energy source for a logical software system. The input data kicks off and drives the software system	RAW-file (raw image data)
Preparation	Preparation includes all necessary steps to modify data to bring it into a form and structure to be successfully processed. Preparation can include assembling, compiling, or padding of input data	Import interface/translator for camera specific RAW data
Operation	The part of the software system executing the main function, all operations which create the output data	Modification of Brightness, contrast, color,…; translation into compressed data (e.g. jpg)
Output data	Output data is the outcome of the operations and can subsequently be used for other adjacent software systems, IT systems or users as input data	JPG file (compressed image data)

The awareness of the individual parts that need to be present to form a working software system leads to conscious decisions regarding possible development directions:

1. Awareness of necessary components that form a working software system,
2. Awareness of where the component currently is located (subsystem/supersystem),
3. Exploring potential for bringing necessary components from supersystem into the software system at hand,
4. Exploring potential for bringing components currently located in the software system to the supersystem.

Examples for above mentioned mechanisms can be observed in contemporary developments. Data lakes provide vast amounts of heterogeneous input data, structured and unstructured. To efficiently process those data, software systems need to include the preparation functionality.

On the opposite, cloud applications have transferred the operations to the supersystem, which in this case is represented by software systems running on remote servers. Local software systems may just comprise interfaces for capturing input data and, if applicable, means for representing or providing output data for further use.

3 Process Function Model for Software

As indicated in the above section, software system performs specific operations on the input to generate the output. While TRIZ function analysis for products seems to be an appropriate tool for decomposition and identifying functional disadvantages regarding interactions between software modules, the authors focused on the process aspect of software systems. Therefore, the application of the TRIZ function analysis for processes in the software domain has been assessed.

In the terminology for TRIZ function analysis for processes [18], a software system can be divided into sequential or parallel operations (process steps) that perform certain functions (sub-steps). This structure represents the flow of data through the process and the modifications performed on the data. The function analysis for processes requires a conscious definition of the boundaries and the abstraction level of the system to be modelled.

One goal of TRIZ function analysis for processes is the evaluation of the functionality of each operation. This is achieved using several categories for functions inside each operation [18].

The categories listed in Table 2 were taken from the function analysis for processes for engineering systems and are applied here to qualify useful functions in software processes as well. The table reflects the importance of each function type, with productive functions being the most valuable (highest ranking value, e.g., 5) and corrective functions being the least valuable (lowest ranking value, e.g., 1).

Additionally, function disadvantages like harmful, insufficient, or excessive functions can be modelled in the context of function analysis for processes to analyze shortcomings or potential errors in a given software system.

The case study presented below demonstrate the application of the TRIZ function analysis for processes in the software domain. While the first case study focusses on

Table 2. Function categories for function analysis for processes

Function category	Explanation	Example
Productive functions Function rank 5	Productive functions irreversibly change parameters of the output, in software systems they cause irreversible and intentional changes that are "visible" in the output data	An example is Algebraic Multigrid (AMG) with which equation systems in structural analysis are solved
Providing supporting functions Function rank 4	Supporting functions only temporarily change parameters of the output. Those changes are not visible in the output data	An example is multiplying an integer number with a factor of e.g. 1000 to increase accuracy of a calculation. However, this function causes the defect of offsetting the value by the factor used. This defect needs to be corrected by a subsequent corrective function
Providing transport functions Function rank 3	Transport functions change the location of data	Exemplarily, in logical software operations this could be transfer instructions or logical shifts that move data
Providing measurement functions Function rank 2	Measurement functions provide information about parameters and properties of objects	Any function that e.g., checks the value of data (variable, parameter, input data) can be considered a measurement function, consequently conditional jumps contain measurement functions
Corrective functions Function rank 1	Corrective functions are directed towards a defect. Defects are caused by harmful, insufficient useful or excessive useful functions	As mentioned above, a corrective function can be the division of an integer number to set the value back to the original state

the function categories and identification of possible function disadvantages, the second case study explores the potential of trimming for processes in the context of software systems, leading to initial ideas for improving the given process.

3.1 Case Study 1: Multiphysics Simulation Process

The following example illustrates a possible approach to build a function model for processes for an engineering software simulation workflow from the perspective of the software user, typically a research or application engineer, in charge of a virtual prototyping project or a digital twin construction. The workflow is built as a cascade of software modules, each of them either activated by a user or by an overlay software module. Those software modules are typically CAD (Computer Aided Design) and differential equation solvers (i.e., finite element, finite difference, boundary element methods) called in a given sequence.

We can see an application example, representing the usual sequence of operations for a multi-physics software engineering workflow based on weak coupling, where a first physical domain is analyzed and the outcome serves as input for the next step; this is widespread in electrothermal applications, where the electric or electromagnetic losses computed during the first analysis serve as input for the subsequent thermal analysis. Weak coupling, where fields are solved in a sequence, is normally preferred to strong coupling, where a set of equations representing all the fields interaction are solved at the same time, since the different physical phenomena show in most application very different time constants.

Usually, losses are obtained solving for electric or magnetic field on a given geometrical domain: At first a geometry layout is produced and simplified to represent only the parts relevant to this analysis. This simplified geometry is then discretized in a number of subdomains where a numerical technique is used to solve for electric or electromagnetic fields. In the post-processing stage, the energy losses are computed from the field results; those losses are mapped to the geometry which in turn is again simplified and discretized; the subsequent numerical process to solve the thermal behavior of the system is then applied and thermal fields computed. This procedure can be again iterated if for instance a next step is introduced to analyze the effect of thermal fields on the structural behavior of the system.

The function model for this process is presented in tabular form [Table 3].

Table 3. Function model for multiphysics simulation process

Order	Operation	Function	Result	Function category
1	Initial data specification	Assign design parameters	Design parameters fixed	Providing/supporting
2	Device design	Compute initial design details	Initial design	Productive
3	Geometry generation	Translate into geometrical entity description	Geometry layout	Providing/supportive
4	Symmetry recognition	Identify model useful subsection	Minimum geometrical module	Corrective
5	Defeaturing SW	Simplify unnecessary details	Defeatured geometrical module	Corrective

(*continued*)

Table 3. (*continued*)

Order	Operation	Function	Result	Function category
6	Setup definition	Assign excitation, boundary conditions	Simulation model	Productive
7	Mesh generation	Definition of finite element shape and node positions	Geometry replaced by node assembly	Productive
8	Solution physical domain 1	Node EM field values computed	Node field values	Productive
9	Post-processing	Compute design performance data	Design performance parameters	Productive
10	New mesh generation	Definition of finite element shape and node positions	New node assembly	Productive
11	Solution physical domain 2	Node thermal field values computed	Node field values	Productive
12	Post-processing	Compute design performance data	Design performance parameters	Productive

Such a process function model can capture very well the temporal sequence of all the operations and can be as detailed as needed; it still neglects details to the specific numerical technologies adopted by each step, and this added granularity could be helpful for the developer.

Exemplarily, the following aspects highlight parts of the process where problematic situations can be identified.

Operation 7 + 12: The mesh generation stage is indeed productive, since it produces the finite element assembly upon which the field equations are solved. It must be tuned wisely since overmesh (too fine granularity of the domain discretization) will lead to an exaggerated simulation time and hardware resource requirements. Undermesh (too coarse discretization) will lead to inaccurate solution results.

Operation 5: Simplification of a geometry layout is unavoidable when working with complex data files, since drawings aimed for production are not usable for FEM (Finite Element Method) applications. In terms of an economic simulation model, the level of detail of a geometry layout is a defect, causing excessive solution time. Therefore, the simplification of this data as well as reducing the model to a subsection (Operation 4) have been categorized as corrective functions addressing defects caused by the supersystem outside the system boundary.

However, if the corrective function in operation 5 is excessive, it could bring about harmful effects which cause disadvantages: Modify the model to the point of not realistic geometries or delete features where relevant physical phenomena should be captured. This defect needs to be addressed by subsequent operations, like e.g., submodelling.

Following the function analysis for processes, the trimming rules for processes can now be used, which will be explained further in the next section. At this point it could already be mentioned that corrective functions should be avoided and can give the software developer hints to check whether this function can be changed or dispensed with. The following case study shows how this can be done.

3.2 Case Study 2: Secure Transfer and Encryption Process

The intention of this case study is to demonstrate the generation of new ideas on all solution levels by using process analysis and trimming rules [17] on pure software problems. For a well-targeted solution search, a much more detailed example would be necessary which goes far beyond the scope of this document.

The "secure transmission" of the data to the smart card means that an attacker must not be able to read or change the transmitted data without being noticed. In the world of smart cards, the data is therefore encrypted (protection against eavesdropping) and provided with a cryptographic checksum (protection against unnoticed changes). This also happens in this example.

It should also be noted that the data transfer to the smart card is still usually done using an APDU (Application Protocol Data Unit) protocol [19]. It is also common to encode the data in so-called TLVs (Type-Length-Value-Format) to save storage space [20]. This graphical version [Fig. 3] has been chosen instead of a tabular representation because is in line with the authors way of working and reflects common practice in his field.

The process shown in Fig. 3 represents the protection of data transfer to smart cards that is commonly used today. In this example, the aim is to make the data structures used simpler and more efficient. For this purpose, the step-by-step construction of the data structures is considered as a process. The trimming rules for processes are then applied to this process to find approaches and ideas for new solutions. In this paper, only two steps with ideas for solution directions are presented. The complete example can be accessed in [21].

Step 2, Supporting Function: Add length to each username and password.

- Applied trimming rule (A): The operation requiring the supporting function is trimmed.

  The operation which requires the supporting function is the operation which extracts the username/password on the smart card. Idea: Instead of using username/password, the data of the APDU or parts of it is used as a credential. Since

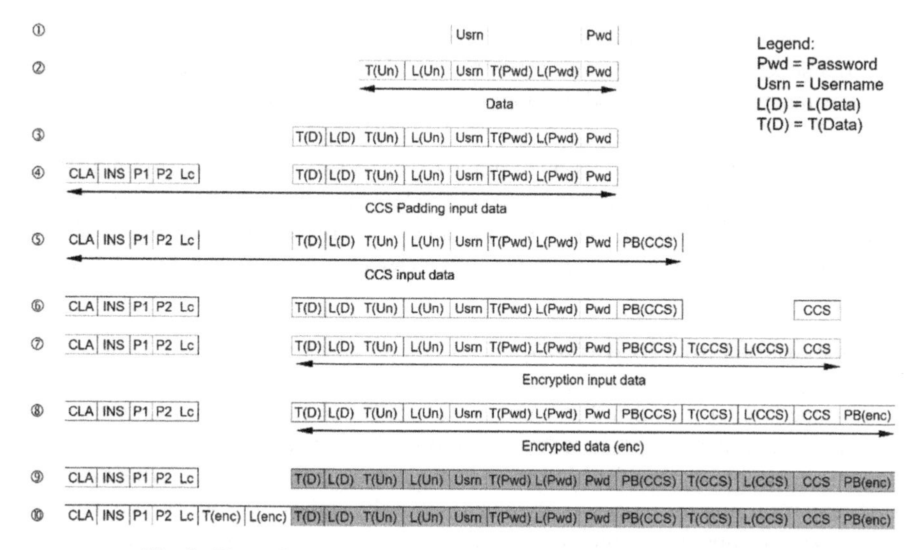

Fig. 3. Exemplary process model for secure transfer to a smart card.

this token consists of username and password, the same method can also be used when comparing username/password with the stored credential.

- Applied trimming rule (B): The supported operation is changed so that it does not require any support now.

  Thus, the data extraction does not need a length anymore. Idea: Username and password have a fixed length, maybe padded with "empty characters". This also would create great advantages regarding padding for cryptographic checksum and encryption which has a fixed size or can be omitted.
- Applied trimming rule (C): The supported function performs the supporting function itself.

  This means the data extraction method finds out the length by itself. Idea: Username and password length are coded by inserting a special "stop character" at the end instead of a length information.

  Applied trimming rule (D): The analyzed supporting function is transferred to the preceding or subsequent operation.

  Idea for preceding operation: The length is already part of the username and/or password.

  Idea for subsequent operation: The length is encoded as a part of the "tag" byte.

Step 6, Supporting Function: Calculate the cryptographic checksum and add it to the APDU.

- Applied trimming rule (C): The supported function performs the supporting function itself.

  The supported function is the message integrity check after receiving the APDU in the smart card. Idea: The integrity of the message is checked by checking the integrity of username and password. This can be done by defining a set of valid characters for username and password. If the integrity of the message is attacked, the encryption causes many bytes to be modified in an arbitrary way which can be detected easily.

This case study stops at the point of initial idea generation due to the trimming rules. Of course, during substantiation of those ideas into concepts, project requirements have to be met. This case study did not go further due to confidentiality; however, it already shows substantial potential in trimming for processes applied to software systems.

4 Conclusion and Summary

In this paper we have shown that fundamental TRIZ concepts can be mapped to software systems, even when those are strictly considered "immaterial" or "logical" systems. Modeling concepts like the function analysis for processes, together with function categories and trimming for processes are able to describe and assess software systems and can be used to elaborate on function disadvantages and spark creative ideas for further development. These prerequisites open the door for further use of TRIZ problem solving tools.

Subsequent work could include the following aspects:

- Further evaluation of the model of a complete software system. Retrospective assessment of the development of software systems could underpin or correct the presented model. Observation of the sequences of acquisition of parts from and transfer to the supersystem [17], as well as the analysis of the nature and importance of control components in software systems.
- Function model for processes for software systems. Research about possible additional function categories, which specifically could be useful for categorizing functions in software systems.
- Value analysis based on function analysis for processes. Evaluate the "cost" aspect of single operations inside a software system to be able to assess the functionality vs. cost aspect and derive a ranking for trimming of operations. The question to be answered would be, what could quantify the negative aspect of an operation in a software system? Besides memory requirements and processing time needed for the given operation, which additional "harmful" parameters can be identified and contrasted with the quantification of the function categories?

In practice TRIZ is already used in software domain. This paper shows a structured mapping of TRIZ tools and concepts to this domain to facilitate adoption while maintaining consistency with the existing theory. The extension of the systematic approach to software systems has been presented, along with two case studies.

5 Definitions

Table 4. Definitions.

Term	Definition
Algorithm	(1) A finite set of well-defined rules for the solution of a problem in a finite number of steps; for example, a complete specification of a sequence of arithmetic operations for evaluating sine x to a given precision (2) Any sequence of operations for performing a specific task [22]
Code	(1) In software engineering, computer instructions and data definitions expressed in a programming language or in a form output by an assembler, compiler, or other translator (2) To express a computer program in a programming language (3) A character or bit pattern that is assigned a particular meaning; for example, a status code [22]
Data	(1) A representation of facts, concepts, or instructions in a manner suitable for communication, interpretation, or processing by humans or by automatic means (2) Sometimes used as a synonym for documentation [22]
Information	(1) Any communication or representation of knowledge such as facts, data, or opinions in any medium or form, including textual, numerical, graphic, cartographic, narrative, or audiovisual. An instance of an information type. [23] (2) Meaningful interpretation or expression of data. [24]
Information Technology (IT) (Term used equivalently in this paper: IT-System)	Any equipment or interconnected system or subsystem of equipment that is used in the automatic acquisition, storage, manipulation, management, movement, control, display, switching, interchange, transmission, or reception of data or information by the executive agency. The term information technology includes computers, ancillary equipment, software, firmware and similar procedures, services (including support services), and related resources. [25]
Process	(1) A sequence of steps performed for a given purpose; for example, the software development process (2) An executable unit managed by an operating system scheduler (3) To perform operations on data [22]
Software	Computer programs, procedures, and possibly associated documentation and data pertaining to the operation of a computer system [22].

References

1. Govindarajan, U.H., Sheu, D.D., Mann, D.: Review of systematic software innovation using TRIZ. Int. J. Syst. Innov. **5**(3), 72–90 (2019)
2. Beckmann, H: Function analysis integrating software applications. In: TRIZ Future 2013 Proceedings, pp. 473–482 (2013)

3. Souchkov, V.: Extended function analysis - overview. Presentation delivered during German Expert Day 2019, Sulzbach-Rosenberg, Germany, February 2019
4. Petrov, V.: Using TRIZ tools in IT. In: TRIZ Developers Summit 2019. https://triz-summit.ru/file.php/id/f304862-file-original.pdf. Accessed 19 Mar 2022
5. Padabed, I.: Contradictions and laws of evolution in information systems. In: TRIZ Developers Summit 2019. https://triz-summit.ru/file.php/id/f304852-file-original.pdf. Accessed 19 Mar 2022
6. Beckmann, H.: Method for transferring the 40 inventive principles to information technology and software. Procedia Eng. **131**, 993–1001 (2015)
7. Lady, D.: Information Technology System Cookbook: Introducing TrizIT: TRIZ for Information Technology. AIMS Publications (2013). ISBN: 978-1478302513
8. Goethals, F.: 20 Trends in digital innovations: a TRIZ-trend-structured overview of new business information technologies and innovative applications. Amazon Digital Services LLC (2014). ASIN: B00P6GNE88
9. Mishra, U.: TRIZ Principles for Information Technology. CreateSpace (2010). ISBN 978-818465184
10. Nakagawa, T.: Software Engineering and TRIZ (1) Structured Programming reviewed with TRIZ, TRIZcon (2005). https://www.osaka-gu.ac.jp/php/nakagawa/TRIZ/eTRIZ/epapers/e2005Papers/eNakaTRIZCON-SE1/eTRIZCON2005-SE-050604.pdf. Accessed 25 Apr 2022
11. Mann, D.L.: Systematic (Software) Innovation. IFR Press, Frankfurt (2008)
12. Ikovenko, S., et al.: State-of-the-Art TRIZ, Theory of Inventive Problem Solving. Novismo Ltd. (2019). ISBN: 978-83-65899-05-7
13. Chrząszcz, J., Bertoncelli, T., Gronauer, B., Mayer, O., Nähler, H.: Modeling IT-systems in TRIZ. Submitted to TRIZ Future 2022 Conference
14. Asanovic, K.: CS 152 Computer Architecture and Engineering. https://slidetodoc.com/cs-152-computer-architecture-and-engineering-lecture-1-4/. Accessed 24 Apr 2022
15. https://electronics.stackexchange.com/questions/353915/what-is-the-role-of-isa-instruction-set-architecture-in-the-comp-arch-abstract. Accessed 24 Apr 2022
16. Schmalz, M.S.: Organization of Computer Systems: §1: Introductory Material, Computer Abstractions, and Technology. https://www.cise.ufl.edu/~mssz/CompOrg/CDAintro.html. Accessed 24 Apr 2022
17. Lyubomirskiy, A., Litvin, S., Ikovenko, S., Thurnes, C.M., Adunka, R.: Trends of Engineering System Evolution (TESE) – TRIZ paths to innovation. TRIZ Consulting Group (2018)
18. Ikovenko, S.: Level 3 Training Manual (2019)
19. https://en.wikipedia.org/wiki/Smart_card_application_protocol_data_unit. Accessed 25 Apr 2022
20. https://en.wikipedia.org/wiki/Type%E2%80%93length%E2%80%93value. Accessed 25 Apr 2022
21. https://triz-akademie.de/triz-software/
22. IEEE Standards Board, IEEE Std 610.121990. IEEE Standard Glossary of Software Engineering Terminology (1990)
23. NIST Special Publication 800-30 Revision 1, Guide for Conducting Risk Assessments. National Institute of Standards and Technology (2012)
24. Kissel, R., Regenscheid, A., Scholl, M., Stine, K.: NIST Special Publication 800-88 Revision 1, Guidelines for Media Sanitization. National Institute of Standards and Technology (2014)
25. FIPS PUB 200, Minimum Security Requirements for Federal Information and Information Systems, Federal Information Processing Standards Publication, National Institute of Standards and Technology (2006)

Integration of TRIZ Methodologies into the Digital Product Development Process

Vasilii Kaliteevskii[1](✉), Matvey Bryksin[2](✉), and Leonid Chechurin[1](✉)

[1] Lappeenranta-Lahti University of Technology, 53850 Lappeenranta, Finland
vkalit@gmail.com, leonid.chechurin@lut.fi
[2] Arrival UK, Hammersmith, London, UK
matvey.bryksin@gmail.com

Abstract. Iterative and waterfall software development processes such as V-model or UP (unified process) are widely used in project development. Nevertheless, modern Agile methodologies are aimed to reduce risks and achieve more predictable results in the digital world. Despite the difference between them, the software development life cycle splits into specific deliverables by breaking a project into smaller segments where the TRIZ toolkit can be effectively applied. Application of the TRIZ theory and core techniques can be valuable for new ideas generation and decision-making process at any point of product development from business model development and product vision definition to user experience design, implementation, and go-to-market strategy. As a result, the product quality grows, continuous delivery is accelerated, and new features are introduced to the final user in a more operative manner. The goal of this article is to demonstrate how TRIZ methodologies integration empowers software development for new products release and continuous improvement.

Keywords: TRIZ · Product design · Product management · Product development · Digital product

1 Introduction

1.1 Addressing the Digital Product as a Technical System

After being challenged with the dot-com bubble burst, the digital products market continued a substantial growth during the last decades with financial, technological and human resources (IT-specialists) [1]. Although, there is a substantial scientific advance in the direction of mathematical algorithms and technical solutions, there is no such a breakthrough in the direction of the digital product design and strategies. Such scientific pieces aiming user experience and digital product value [2] are rather segregated studies than made in systematic manner. At the same time the number of digital products released increased substantially as well as the number of IT-specialists involved in different digital products design, development and management [3]. In the present research

Published by Springer Nature Switzerland AG 2022
R. Nowak et al. (Eds.): TFC 2022, IFIP AICT 655, pp. 273–284, 2022.
https://doi.org/10.1007/978-3-031-17288-5_24

authors are presenting the idea of TRIZ toolset to be applied in the framework of digital product development and management.

Nowadays, TRIZ is widely used for different industries (such as automotive, telecom, aeronautics, etc.) and gives them a competitive advantage opening new opportunities for problem-solving [4, 5]. The theory provides a comprehensive set of inventive standards that engineers can apply for digital product development as well [6–8]. The powerful advantage of TRIZ tools is that they are written without precisely determining the technical system, making them applicable to any technology product [9]. Software development is a subset of modern system engineering and has the same principles for designing and developing digital products.

Thus, a digital product is considered as a technical system but higher solution flexibility because of the absence of constraints from the physical world in the IT domain. Most digital products are not tangible and can only be visible to the final user via a graphical user interface (GUI). There are different approaches to categorizing digital products, but they all contain the same examples:

- **Application software.** Such a solution provides a digital interface for the final user to deliver a value of a product. Examples: website, mobile app, e-commerce service.
- **Software platform.** Such software includes technology for data processing, event streaming, file storage, analytics, and reporting.
- **System software.** Such software includes operating systems, drivers, and utility programs.

Any digital product development process may follow the standard rules from TRIZ theory to increase performance, quality, business KPIs, and other final product parameters.

1.2 TRIZ Tools Applied for Digital Technical Systems

One of the core ideas of the TRIZ is that all its principles can be formulated in the terms of the industry where they are to be applied. The flexibility of TRIZ tools makes them independent of the digital product domains. Some of the core terms can have different meanings for software development which are described in this section [10].

Ideality. Ideality principle is a measure of the technical system which quantifies different factors of how much value a solution provides to the final user [4]. An increasing degree of ideality makes the product competitive with other market players, which causes the final product to be more profitable. Product managers can measure the same value from a different angle, such as how much the product meets business goals and KPIs.

Ideal Final Result (IFR). The principle is a formulation of the final solution that delivers the desired result to the end-user. On the one hand, IFR reaches the highest value of ideality. From another hand, IFR does not require adding extra resources or costs. The basic TRIZ principle says that an ideal system is no system, but the function is performed. Paraphrasing such a rule to the software terminology might be that the ideal application software is a zero user experience interface that provides the capability to reach expected user goals. This principle can significantly improve the quality of product strategy as a base for any new product that introduces a unique value proposition to the market. In other words, any digital product should perform a main useful function with the best-expected outcome for the end customer. Such top priority functionality defines the order for the development team how to deliver such features in the shortest time.

Su-Field. It's a model of a minimal technical system that consists of a pair of actors interacting with each other. The Su-Field requires at least two elements made of substance and a field providing interaction between the substance components. Su-Field analysis is a method of abstract modeling of a technical system. Such a model specifies substance components, fields, and interactions between them. Su-field analysis can be applied to resolve the contradiction between User and System and gain new insights.

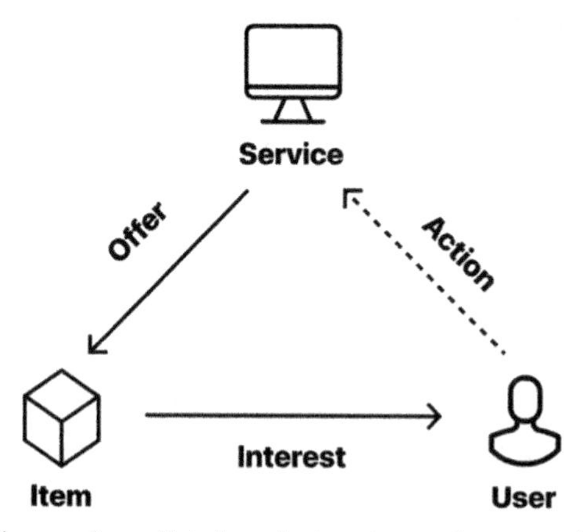

Fig. 1. The Su-field diagram for a digital service's unique value proposition. Service offers an item that interests the user to make an action on the service.

Su-Field Diagram is an outcome of the Su-Field analysis. Diagram presents a Su-field graphically as a triangle with nodes representing two substances and a field and lines between the nodes representing interactions between the Su-field components. Most software applications require a user interface (UI) for interaction between system and user. The ideal UI is designed to achieve specific user goals. The motivation field can empower the pair of User and System to create a minimal technical system that can be described using a Su-Field diagram from Fig. 1.

Mini-Problem and Maxi-Problem. It's the formulation of an inventive problem that is a result of imposing the following restrictions on this inventive situation: everything remains without any changes or becomes even simpler, but some updates are required. Every decision in software development is a part of continuous integration and improvement. Engineers decide for a specific product component on how they want to upgrade the system either keep the functionality the same or initiate a new iteration development, depending on the requirements and desired outcome.

2 Software Development Process and Product Lifecycle

2.1 Product Discovery and Delivery as Main Product Development Stages

Product Lifecycle is the process of product evolution on the entire timeline from its idea through the design and development to the product launch and further updates. The standard industry approach is to slice the lifecycle onto the short implementation iterations [11]. It helps to design and develop the new product in phases. In general, the whole lifecycle from the idea to the final product is split into two main phases – Product Discovery and Product Delivery [12]. Discovery answers how to Build the right product in order to meet the target audience needs. Delivery of how to build the product right. The interfaces between Discovery and Delivery are product backlog and feedback from the market (Fig. 2). They support the linkage between phases and make a smooth transition between them and loopback [13].

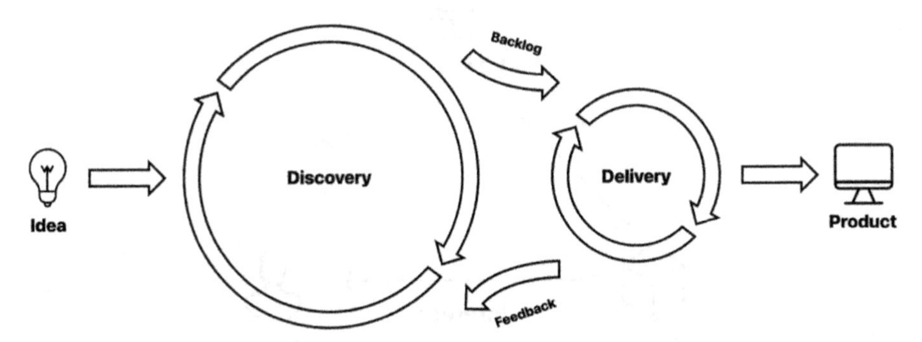

Fig. 2. Product discovery and product delivery – main phases of product continuous improvement.

Product Discovery. The very first strategic phase after a new product idea introduction is Product Discovery [12]. Most of the risks are concentrated in the ideation phase because the cost of an error in the strategic decision can lead to the whole product failure. The Discovery phase starts with generating and prioritizing existing ideas to solve the problem or achieve defined business goals. The Discovery backlog sets the requirement and right focus for the market analysis and needs research. The product manager can develop the product strategy through continuous concept improvements and hypothesis validation cycles that provide a product backlog and roadmap to guide the development team on how to deliver the right product. Such standard outputs in the software industry are required as minimum knowledge about the product to plan a transition from strategic phase to execution.

Product Delivery. The next phase after Product Discovery is the implementation and execution phase, where the requirements of the desired product are implemented in the software [14]. The main steps are developing products according to specification, running beta or pilot tests, defining a go-to-market strategy, creating a launch plan, and finally releasing the product to the market. During all these incremental steps, it is crucial to gather feedback and monitor major KPIs to adjust product strategy in short periods rapidly [15].

Even though Product Discovery and Delivery are the most critical phases, the standard industry practice is to extend the product lifecycle to the Adoption and end-of-life phases. The broader product lifecycle includes the phase of existing product evolution after introduction to the market. The growth and optimization phases are the sources of product-market fit analysis and valuable feedback for the new iteration of strategy and analysis [16]. The end-of-life phase is a decision point about whether to iterate or decline the product (Fig. 3).

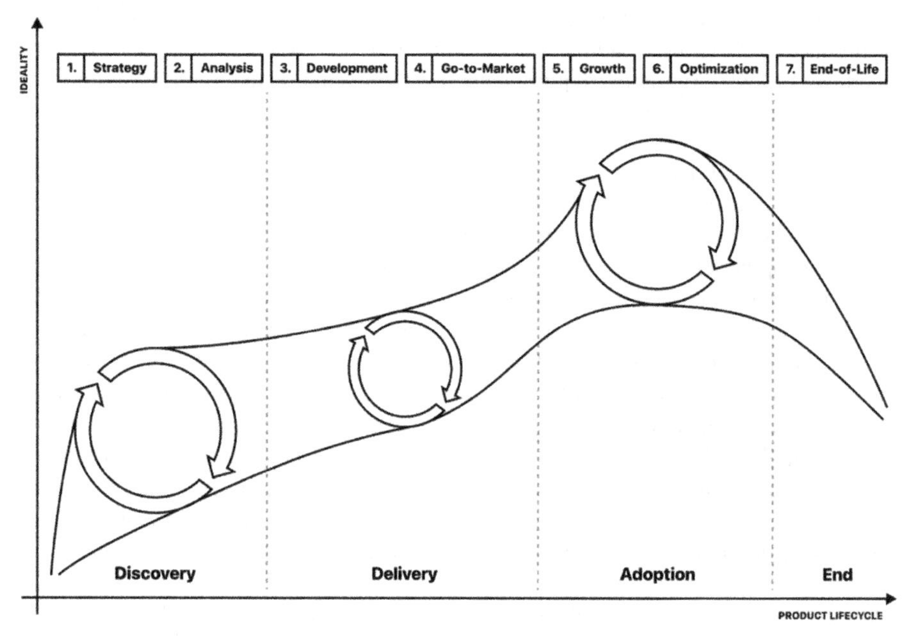

Fig. 3. Phases of product management lifecycle and feedback iterations between them.

2.2 Digital Product Lifecycle Phases Overview

Strategy. Undoubtedly the business modeling phase is the most critical starting point for any new digital product. Business goals and objectives are the main drivers of creating a long-term and mid-term product vision without implementation details. As a part of the Product Discovery phase, the major activity and focus are on creating and prioritizing ideas based on the cost to develop and value to customers or business stakeholders. Development of an initial proposition or draft product concept is required to facilitate detailed analysis and testing.

Analysis. The starting point for detailed product analysis and improvement cycles should be the defined product strategy. Product opportunity assessment requires a deep understanding of user segments and needs through market research and hypothesis analysis. During this phase, it is critical to discover the right value proposition and the problem being solved. Some rapid prototyping techniques are common techniques to validate multiple ideas through continuous feedback loops. Using paper sketches and interactive prototypes increases the quality and speed of gathering early feedback. User research methodologies help validate a market need through customer development. Research processes described in Google Design Sprint, Design Thinking, and Lean UX methodologies help to iterate quickly on the one or two-week iterations to deliver the accurate results of the analysis [17, 18].

The main objective of the Analysis phase is to understand an opportunity for development and estimate costs and risks. In case of success and the decision to move forward, the product backlog and roadmap become the bridge between strategy and delivery. Specification and formulated requirements are the results of the analysis required to create a technical backlog and product plan development.

Development. Getting product requirements delivered is the next phase of the product lifecycle after the confirmation of the product strategy. It starts with designing and delivering a solution to meet the requirements. Some Agile practices are widely used across development teams to achieve more predictable results in the middle of development. The scrum approach for release management helps to set fixed time iterations for designing, engineering, developing, and launching product updates that bring additional value to the market. In this phase of development, inaccurate sprint planning becomes the main point of failure and causes missing deadlines. If the product is delivered on time and passes verification of use cases and test scenarios, the product team ensures that requirements have been met. In that case, it becomes a ready-for-deployment artifact with a specified build date. The release plan is required to set a strategy for how a brand new product or product update will be introduced to the end-user. A soft launch strategy is the best option to pilot upcoming releases with early adopters or loyal audiences.

Go-to-Market. In the earliest stages of launching a product to the market, the launch team researches market conditions, and go-to-market strategy is planned for the new product or a service. After defining the market and target audience, the final value proposition is developed through multiple iterations of experiments and test campaigns. Then business defines the pricing strategy to fulfill company needs. Moreover, choosing the right distribution channels and distribution model followed by promotion is the most vital step before deploying a go-to-market strategy. When a competitive advantage is created, the marketing team decides how they will promote the final product and what kind of marketing campaigns are the most profitable for the go-to-live phase.

Growth. Scaling products to reach a bigger audience and increase profits is vital for any product that found its market fit and is ready for broader expansion. As the next step, value proposition and sale offers are continuously shaped with each sale by the reality of the market and feedback from customers. New functions such as customer support and customer marketing are introduced to execute marketing plans and monitor metrics of market adoption.

Optimization. As part of the natural product evolution, the maturity of the product grows with more time presented on the market. The product value proposition becomes stable, and only micro-optimizations are returned as updates of the well-developed product strategy. The product maintains saturation during the optimization phase, and scale speed goes down the same as profit growth. Most of the product enhancements may use data analysis and rely on big data to grow business metrics like ROI or LTV. Additional product features can be introduced as brand new products to support sales and extend the value proposition to the existing customer segments or new markets. Those pivots start the new iteration of the product lifecycle and can dramatically impact the future product strategy.

End-of-Life. In case of a decision to prevent keeping the product alive, stop ongoing optimizations, and decline the existing product, it goes into the end-of-life phase. Decommissioning of the product contains the procedures of planning sunset or product replacement. Performing post-mortem is the final phase of the product before its possible future reincarnation.

3 TRIZ Integration into Product Development Process

Despite some less traditional applications of TRIZ for such spheres as business and management [19], as well as for the programming (i.e. algorithmic) [20] the idea of TRIZ application for digital product development for the whole software lifecycle is rather new [6]. In the present chapter the approach to utilize TRIZ for software development looking back to product Discovery and Delivery phases is described.

Thus, as a consequence of the rapid growth of the IT industry, digital products are the most flexible systems for any market changes. The unique feature of digital products from the other technical systems is the higher speed of evolution caused by the short development cycles. Agile methodologies aim to achieve more predictable results with a smaller period and resources [21]. Modern Agile principles applied for software development make it flexible for regular updates and robust for rapid market changes. Software products follow the general Trends of Technical Systems Evolution, such as trends of static, cinematic, and dynamic [22]. The complexity of systems grows with the product scale. TRIZ techniques can be applied at any time, starting with product discovery, innovation, and analysis and finishing with go-to-market, optimization, and end-of-life.

Strategy. The strategy phase contains a high degree of uncertainty, starting from an idea and finishing with a new invention, and may utilize multiple TRIZ tools for achieving more accurate results. Su-field Analysis is a powerful tool that can be applied for business modeling or product value proposition design [23, 24]. Su-field diagram from Fig. 1 describes a general model of interaction between User and Service as a contradicted pair and Item as a field that creates interest for User to make an action such as purchase on the Service. This general model can be applied to any e-commerce service such as Amazon, Netflix, Spotify, or niche products that propose a specific value or Service to the final customer. General Su-field is a powerful tool that product managers can also apply for Social Community design, including the niche Social Networks design [25]. Ableton's sound-producing community is considered as a case to express the value of the application of the Su-Field model in digital product development (Fig. 4). It brings local artists, skilled users, and newcomers at regular intervals to locations all over the world to create an additional boost of interest in their product. Ableton creates an environment for different user groups and maintains the organizational part and sponsorship. Still, experienced users help with the basics as a part of social networking and the desire for collaboration. Moreover, almost everybody can become a community facilitator and organize their local meet-ups to grow the community. Thus, being an outcome of the heuristic approach the same model was achievable with Su-Field analysis which is a more formal tool in such a case compared with more classical but less systematic Brainstorming, Design Thinking and other rather heuristic toolsets.

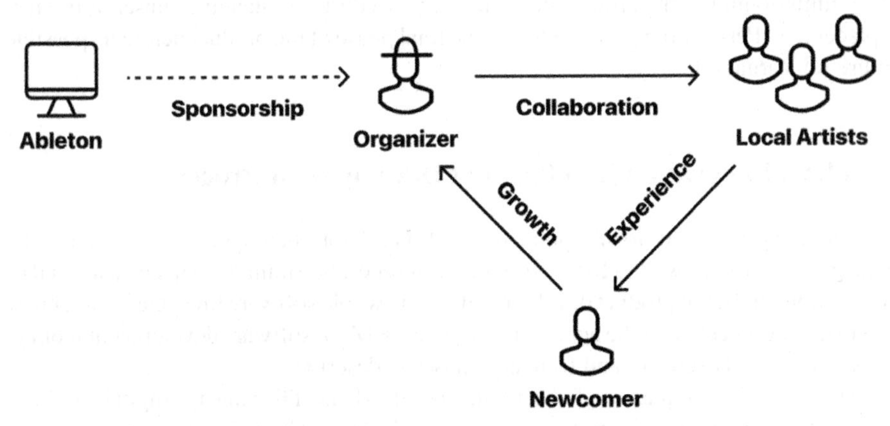

Fig. 4. The Ableton community described using the Su-Field diagram as a contradiction between organizers, local artists, and newcomers.

Analysis. Creating and prioritizing hypotheses is a challenging problem. However, digital product development is more nowadays about generating and testing hypotheses rather than development of the determined solution after continuous market and product research. This is due to the fact that market conditions are now changing promptly, and implementing a hypothesis turns out to be economically more effective than prolonged market research and subsequent development. Ability to launch digital products on big markets with a relatively small team due to the present advanced technology stack made the product development path to the short iterations starting with a set of hypotheses. And this is why it is important to find a more formal and systematic approach for hypotheses catching and formulation. Authors addressing TRIZ for the Analysis part of the product development as a toolset that have been developing for the analogous problem in classical engineering during the last half a century.

Thus, in case of wrong prioritization, the cost of failure depends on the cost of solution implementation and the amount of changes to fix the issue. That is why the integration of TRIZ tools into the Analysis phase can be considered in order to reduce risks of failure and help to increase the quality of decisions made.

One of the IFR applications for the product strategy and vision is the definition of the main function such as a user story or job story [26]. To describe the function in the user's terminology, the format of the job story proposes to describe the Situation, Motivation, and Expected Outcome. The ideal job story as a function definition is written based on the user's needs and doesn't specify any details about the technical solution (Fig. 5).

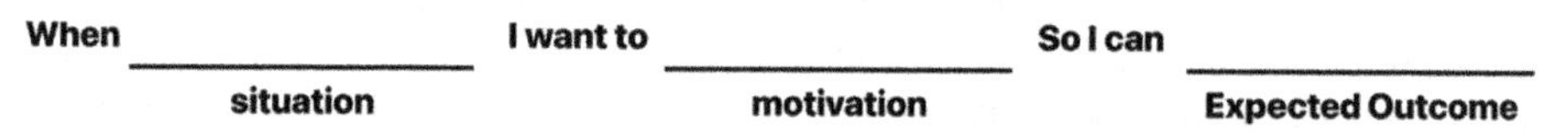

Fig. 5. Template for Job story mapping to define a user function.

IFR is a type of function that a product manager should define as the target solution and part of the product vision. After that, any iteration solution can be simplified using function-cost function and analysis. As an example, one solution can have a higher level of Ideality and higher cost, but another is lower Ideality and cost. Product managers should make such decisions and trade-offs according to available resources and constraints. The level of Ideality can be an additional parameter for function scoring and prioritization. It can be introduced into the RICE model formula that includes Reach, Impact, Confidence, and Efforts values as the main inputs for priority scoring [27]. Integration of the Ideality level to the RICE model gains an additional visionary insight for backlog prioritization which is highly important to startups with futuristic plans.

Development. The main challenge for the development phase is the design architecture that matches the product backlog requirements. Multiple architectural patterns such as microservice architecture or multi-agent systems are used as modern practices on how to build sustainable systems. The TRIZ concepts such as Subsystems or Supersystems, Monosystem, Bisystem, and Polysistem can be used as standard decision options for each technical solution. Engineers can integrate the TRIZ principles about system evolution into the system design phase to make a decision about component redesign and scalability.

Go-to-Market, Growth. All phases from the post-product launch phase have the same principles of product evolution. After introducing a product to the market, the next question to answer is what will be our next iteration after receiving real feedback from the fields. There are multiple TRIZ strategies that can be integrated into those phases. One of them is mini and maxi problem-solving. Using the result of the go-to-market deployment and monitoring KPIs helps to make the right decision on how to pivot the product forward. If profits are running low, the radical option using maxi-problem formulation can be a complete redesign and can be selected as a future improvement strategy. Another option is to solve the mini-problem, keeping the product almost the same with small optimizations.

Optimization. The standard approach for product optimization is the mini-problem formulation and solution based on micro-optimizations. The radically opposite solution to this approach is the strategy of transitioning to Macro-level or Micro-level, which is the standard TRIZ practice. Transition to the Micro-level method provides a new inventive solution by replacing the operational principle behind subsystems. In other words, a solution on the micro-level doesn't change the function itself but changes the way how the technical solution works inside.

On the other hand, transitioning to a Macro-level method provides a solution to evolving a technical system to the next stage of evolution. The first solution is increasing the number of products or interactions between existing components. It is a valid approach for the products under the umbrella brand or the microservices. The second solution is to unite all smaller product lines into a single product. One of the well-known examples is the Wechat super app that aggregates the most common functions like messages, news, and payments into one product. And finally the third solution can be bringing additional features to the existing system that increases the properties of the product and creates additional value for the final customer.

End-of-Life. At first glance, the final phase of the digital product does not have any challenges and problems to solve because of its terminal specifics. That is why in this paper, we have no examples of TRIZ integration into the End-of-Life phase. We keep this section open as an opportunity for future TRIZ integration research papers.

4 Conclusion

Being developed from the early 50s of the last century TRIZ is now saturated with different case-based knowledge and theory in the field of construction and engineering for problem solving and design problems including those challenges arising in the context of the Industry 4.0. However, due to the growth of the digital products market, more inventive challenges in the design appear in the field of such areas as Internet-of-Things, Big Data, Cloud Computing, Blockchain software and user web- and mobile- interface applications. In such areas there is a whole set of inventive problems arise in the field of customer development, product discovery and product delivery, product market-fit finding and go-to-market strategy formulation. Despite its effectiveness in problem-solving in the field of classical engineering, TRIZ has not received much popularity in the circles of product managers, software developers and UX/UI designers of digital products. Although there is a number of studies and cases suggesting the TRIZ exploitation in business, management and digital products development, the theory of TRIZ application in such areas is still on a very early stage of adoption. Thus, the approbation of such an effective tool for engineering in the area of digital product development paves the way for new innovative development approaches in a growing digital sphere opening up new challenges and opportunities. Hence, despite the efficiency of the TRIZ suggested tools there are still challenges to be overcome, such as the lack of a problem formalization model and heuristic nature of the applied steps determination compared to the more formal ARIZ approach [28]. That makes it challenging to link the inventive principles directly with model contradictions and leads to further research in the field of digital product development. One of the possible solutions for the relation identification of functional requirements with design parameters for the digital product design stage may be considered as Functional Decomposition and Morphology framework [29] and recommended to be addressed for future research to be adopted for software design and development.

Despite the mentioned challenges, the main research direction in this article of inventive toolset integration in digital makes it possible to apply proven innovation methodology to digital product delivery. The proposed toolset for the digital domain is supposed to facilitate the software development process from the early design stage to the scrum iterations for faster delivery of the key product features and faster hypotheses approbation. Being a continuation of the previous study [6], the present article uncovers the potential of TRIZ methodology at each stage of the digital product development throughout the digital product life cycle.

References

1. DeLong, J.B., Magin, K.: A short note on the size of the dot-com bubble (2006). https://www.nber.org/system/files/working_papers/w12011/w12011.pdf
2. Nylén, D., Holmström, J.: Digital innovation strategy: a framework for diagnosing and improving digital product and service innovation. Bus. Horiz. **58**(1), 57–67 (2015)
3. The number of ICT specialists in the EU grew by 50.5% from 2012 to 2021, almost 8 times as high as the increase (6.3 %) for total employment. Eurostat. ICT specialists workforce continued to grow in 2021. https://europa.eu/!Jfdpy9. Accessed 05 May 2022
4. Salamatov, Y., Souchkov, V.: TRIZ: The Right Solution at the Right Time: a Guide to Innovative Problem Solving, p. 256. Hattem, Insytec (1999)
5. Zlotin, B., Zusman, A., Altshuller, G., Philatov, V.: Tools of classical TRIZ. Ideation International, Southfield (1999)
6. Kaliteevskii, V., Bryksin, M., Chechurin, L.: TRIZ application for digital product design and management. In: Borgianni, Y., Brad, S., Cavallucci, D., Livotov, P. (eds.) Creative Solutions for a Sustainable Development, vol. 5, pp. 245–255. Springer, Cham (2021). https://doi.org/10.1007/978-3-030-86614-3_20
7. Nählera, H., Gronauerb, B.: TRIZ in the context of digitalisation and digital transformation. In: Proceedings of the MATRIZ TRIZfest-2017 International Conference, pp. 475–482 (2017)
8. Boikaa, S., Kuryana, A., Ogievicha, D.: Applications of TRIZ in business systems. In: TRIZfest-2017, vol. 52 (2017)
9. Wu, Y., Zhou, F., Kong, J.: Innovative design approach for product design based on TRIZ, AD, fuzzy and Grey relational analysis. Comput. Ind. Eng. **140**, 106276 (2020)
10. Korolyov, V.A.: Glossary of TRIZ-OTSM terms (2013). http://www.triz.org.ua/enc/. Accessed 01 Dec 2013
11. Casner, D., Souili, A., Houssin, R., Renaud, J.: Agile'TRIZ framework: towards the integration of TRIZ within the agile innovation methodology. In: Cavallucci, D., De Guio, R., Koziołek, S. (eds.) Automated Invention for Smart Industries, vol. 541, pp. 84–93. Springer, Cham (2018). https://doi.org/10.1007/978-3-030-02456-7_8
12. Blank, S.: The Four Steps to the Epiphany: Successful Strategies for Products that Win. Wiley, Hoboken (2020)
13. Patton, J.: Dual track development is not duel track (2017)
14. Greenway, A., Terrett, B., Bracken, M.: Digital transformation at scale: why the strategy is delivery. Do Sustainability (2021)

15. Sutherland, J., Sutherland, J.V.: Scrum: the art of doing twice the work in half the time. Currency (2014)
16. Dennehy, D., Kasraian, L., O'Raghallaigh, P., Conboy, K.: Product market fit frameworks for lean product development (2016)
17. Banfield, R., Lombardo, C.T., Wax, T.: Design Sprint: A Practical Guidebook for Building Great Digital Products. O'Reilly Media Inc, Sebastopol (2015)
18. Gothelf, J.: Lean UX: Applying Lean Principles to Improve User Experience. O'Reilly Media Inc, Sebastopol (2013)
19. Chechurin, L., Borgianni, Y.: Understanding TRIZ through the review of top cited publications. Comput. Ind. **82**, 119–134 (2016)
20. Chechurin, L.: Research and Practice on the Theory of Inventive Problem Solving (TRIZ). Springer, Switzerland (2016). https://doi.org/10.1007/978-3-319-31782-3
21. McCormick, M.:Waterfall vs. Agile methodology. MPCS, N/A (2012)
22. Cascini, G., Rotini, F., Russo, D.: Networks of trends: systematic definition of evolutionary scenarios. Procedia Eng. **9**, 355–367 (2011)
23. Terninko, J.: Su-field analysis. TRIZ J. **2**, 23–29 (2000)
24. Osterwalder, A., Pigneur, Y., Bernarda, G., Smith, A.: Value Proposition Design: How to Create Products and Services Customers Want. Wiley, Hoboken (2014)
25. Giuffre, K.: Communities and Networks: Using Social Network Analysis to Rethink Urban and Community Studies. Wiley, Hoboken (2013)
26. Patton, J., Economy, P.: User Story Mapping: Discover the Whole Story, Build the Right Product. O'Reilly Media Inc., Sebastopol (2014)
27. Trieflinger, S., Münch, J., Bogazköy, E., Eißler, P., Schneider, J., Roling, B.: How to prioritize your product roadmap when everything feels important: a grey literature review. In: 2021 IEEE International Conference on Engineering, Technology and Innovation (ICE/ITMC), pp. 1–9. IEEE (2021)
28. Borgianni, Y., Fiorineschi, L., Frillici, F.S., Rotini, F.: The process for individuating TRIZ inventive principles: deterministic, stochastic or domain-oriented? Des. Sci. **7** (2021)
29. Fiorineschi, L., Frillici, F.S., Rotini, F.: Enhancing functional decomposition and morphology with TRIZ: literature review. Comput. Ind. **94**, 1–15 (2018)

Lean and TRIZ for Improving the Maintenance Process

El Ghalya Laaroussi(✉), El Hassan Irhirane, and Badr Dakkak

National School of Applied Sciences, 40000 Marrakesh, Morocco
elghalya.laaroussi@ced.uca.ma, elghalya27@gmail.com, {e.irhirane, b.dakkak}@uca.ac.ma

Abstract. Maintenance plays an essential role in the progress of an organization. In order to achieve excellent performance, maintenance strategies must be linked to manufacturing strategies such as Lean. Based on the available data, this paper provides an understanding of the existing research related to the use of Lean and TRIZ within the maintenance process. In order to identify the need for further research in this area. The objective of this paper is then to answer the following question: How can Lean and TRIZ improve the maintenance process? To do this, we focus mainly on the analysis of previous work where TRIZ tools or Lean manufacturing methods have been used individually to improve the maintenance process. The results of previous studies were analyzed and then the main problems that were detected in the previous literature were identified.

Keywords: Lean manufacturing · Maintenance · TRIZ · Lean maintenance · Process

1 Introduction

Industrial maintenance plays an increasingly important role in the productivity of a company. As a result, companies can no longer neglect the maintenance of their production tools and maintenance is considered a source of optimization and even a profit factor.

On the other hand, the environment in which companies operate is constantly changing. Between more intense competition and increasingly demanding customers, knowing how to adapt quickly and improve is the key to producing better results. This is where Lean Management comes into play to ensure the sustainability of the company. This approach proposes a structured problem-solving approach to improve industrial performance in the long term. Its main strength is not to solve problems, but to recognize them [1].

On the other hand, there are more sophisticated methods for better and more innovative solutions. Innovative solutions can be found by using systematic creativity tools called TRIZ (Theory of Inventive Problem Solving).

Indeed, this article addresses the question of how Lean and TRIZ have been used to improve the maintenance process. This study conducted a literature review. The review

Published by Springer Nature Switzerland AG 2022
R. Nowak et al. (Eds.): TFC 2022, IFIP AICT 655, pp. 285–295, 2022.
https://doi.org/10.1007/978-3-031-17288-5_25

was conducted to identify and understand the existing literature on Lean maintenance and TRIZ in maintenance and to evaluate the contributions and summarize the knowledge, thereby identifying potential directions for future research. Major electronic databases were explored to gather literature on the integration of Lean and TRIZ in maintenance. A total of 31 related articles published between 2000 and 2022 were included in this study. This study is organized as follows. Section 2 demonstrates the methodology of the proposed research. Section 3 presents the research results: it gives an overview of the application of Lean and TRIZ concepts in maintenance and their developments in the literature. Section 4 concludes the study with some directions for future research.

2 Methodology

In order to achieve the research objectives and to answer the question of how TRIZ and Lean have been used individually as a means of improving maintenance process activities, a literature review was conducted.

The work of other authors provided the raw information for the research, forming the list of materials to be analyzed. However, no research process can cover the entire knowledge because documents such as theses and dissertations are difficult to collect. Therefore, knowledge integrators have to work within the limits of time and material type [2]. This study is characterized by a consultation of previous works indexed in the various scientific databases. The focus was on using the resources of the Web of Science and Scopus databases. Because there were not many publications on this topic, other research sources were included. Sources such as Google Scholar, Research gate, TRIZ Journal, direct sciences etc. Articles published between 2000 and 2022 were retrieved using the following keywords: "Lean", "maintenance", "Lean maintenance", "TRIZ" and "TRIZ and maintenance", "Lean & TRIZ" (Table 1).

Table 1. Research keywords.

Time period considered	Search engines used	keywords
2000–2022	Taylor & Francis, Emerald, IEEE, Inderscience, Sage, Science Direct, and Springer Link	Lean maintenance /Lean TPM/Lean Tools AND TPM TRIZ AND maintenance/TRIZ IN maintenance/Lean AND TRIZ AND maintenance

The publications found were studied and conclusions were drawn. The review process is illustrated in the following Fig. 1:

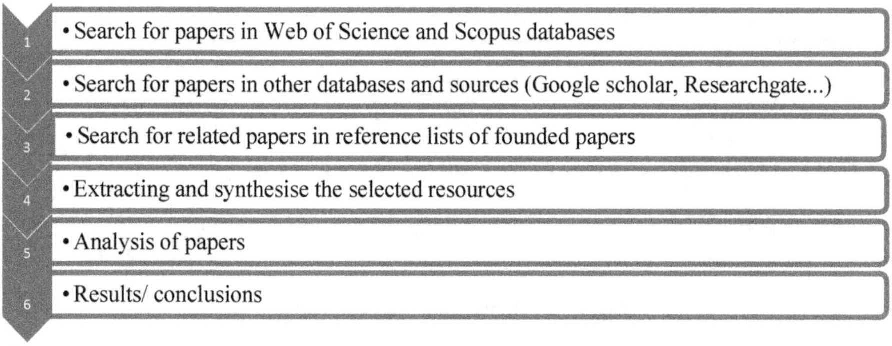

1 • Search for papers in Web of Science and Scopus databases

2 • Search for papers in other databases and sources (Google scholar, Researchgate...)

3 • Search for related papers in reference lists of founded papers

4 • Extracting and synthesise the selected resources

5 • Analysis of papers

6 • Results/ conclusions

Fig. 1. Flowchart of the literature review procedure.

3 Research Findings

This section presents the results obtained from searches of different databases, followed by a discussion and final remarks. For a better understanding, these publications can be divided into two groups. Firstly, publications that focus on the integration of Lean concepts in the maintenance function. The second group includes publications that focus on the use of TRIZ concepts to improve maintenance activities.

3.1 Lean and Maintenance

The integration of lean thinking into maintenance is known as lean maintenance (LM). Smith [3] defined lean maintenance as "a proactive maintenance operation employing planned and scheduled maintenance activities through Total Productive Maintenance (TPM) practices using maintenance strategies developed through the application of Reliability Centered Maintenance (RCM) decision logic and practiced by autonomous (self-directed) action teams…..".

Lean Maintenance is a combination of Lean and Maintenance, combining their respective approaches, methods and tools. So if Lean is more concerned with the speed of the process and anything that might slow it down, then Maintenance is essentially concerned with keeping the asset in a state that allows it to perform the functions it needs to perform and ensures its availability for production [4].

Lean Maintenance contributes to providing better quality and low cost products in the shortest possible time [5]. Lean Maintenance (LM) improves organizational profitability by identifying and eliminating maintenance-related waste [6].

Lean Maintenance Perspective. One of the main steps in improving maintenance processes is to develop a system to identify VA (Value Added) and NVA (Non-Value Added) activities and to recognize waste types [7]. To do this, Lean maintenance includes several tools and methods, such as (Fig. 2):

- 5S: as production, the aim of 5S is to eliminate clutter, tidy up tools, clean the workshop and thus keep workstations tidy. They are essential to the introduction of Lean maintenance.

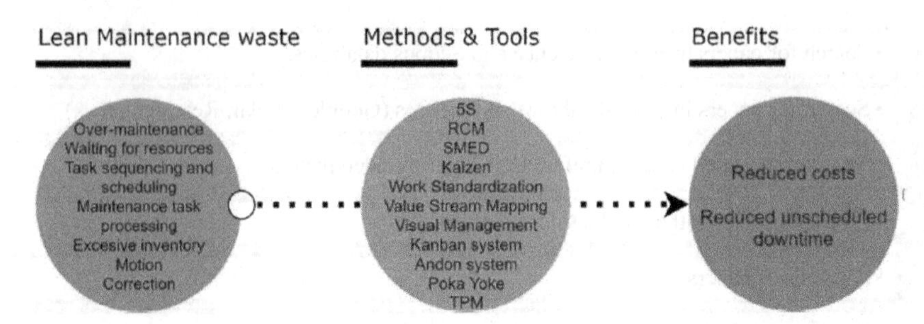

Fig. 2. Lean maintenance perspective

- Value Stream Mapping: VSM is a necessary tool at the beginning of the Lean maintenance implementation.
- Standard work: It aims to standardize the best way to carry out an operation, task or activity.
- SMED: (Single minute exchange of die) is a time analysis and reduction method. Minimize downtime for planned maintenance.
- Kanban: Kanban, i.e. cards, Task Scheduling allows for a graphical representation of tasks and maintenance operations in proportion to their priority and duration
- Kaizen: Kaizen can be used for reliability improvement.
- Jidoka: Maintenance must respect the Jidoka philosophy, which postulates that quality must be constructed a priori.
- Just In Time (JIT): Spare parts and materials are used at the time of need to avoid unnecessary stocks and thus reduce storage costs.

Development of LM in the Literature. The LM studies examined are dated from 2000 to 2022 with the total number of 20 retrievable publications as shown in the figure below (Fig. 3).

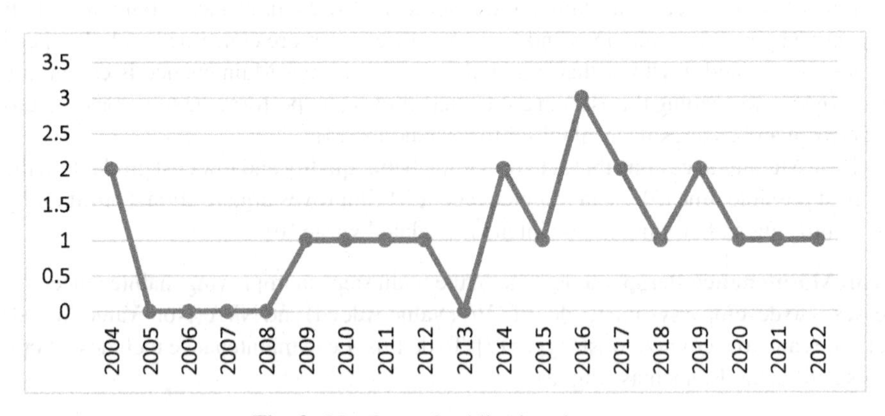

Fig. 3. Numbers of publications in years

Table 2. Review of existing literature on lean maintenance.

References	Year	Contribution
1- [3] 2 -[8]	2004	1-Definition of Lean Maintenance. Discussion of the elements of successful Lean maintenance 2-Development of a fleet repair and maintenance model based on Lean principles and current best practices
[9]	2009	Determination of the main factors and parameters of maintenance operations that are most effective in improving production towards a Lean system
[10]	2010	Formulation of a model that not only determines the minimum level of maintenance requirements, but also meets the expected level of reliability
[11]	2011	Provide an overview of the state of the art of Lean in the aviation MRO industry
[12]	2012	Analyze the challenges of Lean Maintenance to assess its impact on Lean Maintenance
[13]	2014	Present a questionnaire survey on the application of Lean engineering principles and tools in a maintenance context in the Swedish industry
[14]	2014	Optimization of maintenance of industrial systems by using Lean six Sigma bases
[15]	2015	propose a scheme for the integration of lean management in the maintenance process
1-[16] 2- [17] 3- [18]	2016	1-proposed an integration between the Tompkins audit as a ME assessment tool and the technical reverse-AMDEC as a relatively new LM tool in addition to the traditional PDCA cycle: study the effect of this model through a case study 2-development of a process to identify the resources needed for maintenance before the modularization of maintenance tasks and resources can take place. Lean is identified as an important first step before the modularization of maintenance tasks 3- Proposal of a method to measure the level of implementation of Lean in the aviation MRO industry
[19]	2017	2-presentation of the experience gathered in two projects of Lean maintenance in thermoelectric power plants
[20]	2018	Identification of the different types of maintenance losses evident in the companies and how the identified maintenance losses are reduced or eliminated by Lean tools
[21] [22]	2019	1-Implementation of Lean methodologies in the management of consumables in the maintenance workshops of an industrial company 2-proposal of a new approach: Lean maintenance based on the principles and tools of Lean manufacturing

(*continued*)

Table 2. (*continued*)

[23]	2020	Definition of a clearer scene of the implementation of the Lean philosophy in maintenance and sustainability from an empirical point of view based on two case studies
[24]	2021	1-Development of a decision support system to select the appropriate Lean maintenance methods and tools that have the greatest impact on the company's operational results
[25]	2022	Introduction of a framework to integrate sustainability criteria and develop sustainable TPM (Sus-TPM) as a complement to TPM practices

From 2000 to 2022, the publications reviewed were aimed at the integration of Lean into the maintenance process and focused on three aspects, namely the LM implementation initiative, the analysis of Lean maintenance challenges and the development of methods to measure the level of implementation of Lean in maintenance. The objectives of each publication are presented in the table below (Table 2).

3.2 TRIZ

TRIZ (Russian acronym for Theory of Inventive Problem Solving) is a heuristic approach to solving innovation problems, mainly technical. It was developed from 1946 by the Soviet engineer Genrich Altshuller [26]. TRIZ is a science that not only identifies and solves creative problems in each filed of knowledge, but also develops creative (inventive) thinking and develops the characteristics of the creative personality [27]. TRIZ leads the user to a generic and abstract formulation of his problem, and then to principles of solving the abstract problem intended to inspire inventive solutions in the real problem space. TRIZ is based on the analysis of 40,000 patents selected from 400,000 international patents. Altshuller and his team derived some basic regularities and patterns that governed the processes of problem solving, new idea creation and innovation. They have the characteristic of presenting common principles of innovation, and this in very varied domains [28].TRIZ was originally developed for technical problems. However, it has found applications in various other fields.

Development of TRIZ in Maintenance. The studies reviewed are dated from 2000 to 2022 with the total number of 11 retrievable publications as shown in the figure below (Fig. 4).

Indeed, the association of TRIZ with maintenance activities is not new. In 2003, in [29] the authors applied TRIZ methodology to solve a maintenance problem for a drinking yoghurt production system. The application of TRIZ provides both insight into how best to define the problem, and then allows for the development of potential solutions that can be used both to solve the maintenance problem and to prevent system failure. Then,in 2013 the RCA root cause analysis for identifying causes of failure is combined with TRIZ to guide the search for effective solutions [30]. The use of RCA-TRIZ allows critical defects to be found and solutions to be generated, with a lower risk of failure

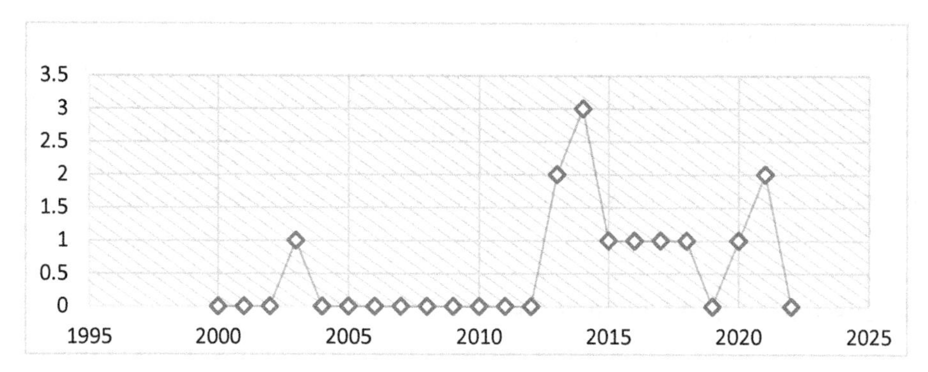

Fig. 4. Number of publications

than using trial and error methods. Similarly in [31] TRIZ was applied to maintainability design to create a set of innovative TRIZ-based aproaches to solving maintenance problems. Then in 2014, in [32] The effectiveness of TRIZ in idea generation and system innovation was demonstrated by exploiting the problem of poor quality arc welding in pipeline maintenance.This is done by combining three tools namely: FA (Function Analysis), CECA (Cause Effect Chains Analysis) are used to analyse the situation and find the key problem,then ARIZ (Algorithm of Inventive Problem Solving) is deployed to solve and search for ideal solutions. Similarly, in [33] the application of TRIZ methods to improve the maintenance of vehicles where special equipment is installed was described. The tools of "Six Sigma", especially statistical process control (SPC) and Pareto analysis, were used to identify the predominant problem. In order to overcome it, the TRIZ method was used and the new solution was analysed by matlab simulation.

Latter, in 2015, in [34] the authors formulated a roadmap to facilitate maintenance problem solving using TRIZ tools. It was proven that TRIZ tools can be used to support the 37 maintenance guidelines. In 2016 in [35] Contradiction Matrix, is used to determine a remote predictive and preventive maintenance solution. In addition Substance-Field analysis, is used in order to identify problematic situations in the existing maintenance contract and to create specific solutions. In [36] an integrated methodological proposal with a logical step-by-step sequence for solving problems related to the reliability of assets in industrial processes, using high-impact failure mode analysis and a creative proposal for the innovative use of TRIZ in reliability.

More recently, in 2020 a meta-synthesis to identify and extract the factors affecting the choice of maintenance strategy with the TRIZ approach was used [37]. Subsequently 2021, and as an extension of this work the authors identified and compared reactive maintenance strategies with TRIZ principles, developing a TRIZ contradiction matrix by explaining the dimensions and components of each of the reactive maintenance tactics [38].

3.3 Lean and TRIZ: Similarities and Differences

In [39] the author describes an overlap between Lean and TRIZ methodologies. Indeed TRIZ and Lean have many points in common. These are two ways to improve the operation of a system. TRIZ and Lean both look towards the ideal future. They both seek to optimize the use of available resources. However they have some differences. TRIZ focuses on individual elements to be optimized.while Lean considers the entire system to find potential efficiencies. Also in TRIZ, the solution to the problem often uses a resource that was previously considered a nuisance or a waste. While In Lean, the goal is to eliminate waste, because waste means there are inefficiencies and counterproductive actions in the system.

4 Discussion and Conclusions

This paper presents work initiated as part of a thesis on the potential contribution of the hybrid Lean-TRIZ approach to maintenance process improvement. The first phase of the project was devoted to researching the topic and reviewing previous research. From the literature review and the research that has been done, we can draw the following conclusions:

From this literature review, it is clear that the application of TRIZ in a maintenance environment has not yet received the necessary attention, and very little work has been done in this area. It is established that TRIZ can be used to build problem solving methods addressing maintenance issues.

Nevertheless, the review of previous studies conducted in this study revealed that the investigation of the applicability of TRIZ tools to maintenance is marginal. Existing research has been largely limited to specific case studies. The integration of TRIZ into maintenance was found in 11 publications, which mainly focused on the use of TRIZ for the development of potential solutions that can be used to solve maintenance problems according to a specific scope which is not surprising. Furthermore, few initiatives have included comprehensive frameworks or models that can integrate TRIZ into the entire maintenance process outside of specific cases. It can be observed that the vast majority of publications only provide indications of the relevance of the integration of TRIZ in solving maintenance problems. Without having given concrete examples on the sustainability of the proposed solutions. No obvious or simple reason for this limitation was provided in the articles reviewed. Thus the answer to the question of how TRIZ could be useful to improve the maintenance process will remain partial until further explorations are provided.

Lean maintenance, on the other hand, is well known and addressed in various research works as a new approach to apply Lean concepts, tools and techniques to the maintenance function to eliminate time loss and increase efficiency and quality. Many studies have looked at the application of Lean principles and tools in the maintenance environment, analyzing Lean maintenance issues to assess their impact on industrial performance. The results show the effectiveness of achieving predetermined objectives in a maintenance strategy. This is the answer to the question of how Lean has been used to improve the maintenance process.

A number of 31 publications were found. The integration of Lean or TRIZ in the maintenance process is generally beneficial, as proven by several publications. However, none of these publications provided a combination of Lean and TRIZ simultaneously to improve the maintenance process. This indicates that more research on the integration of Lean and TRIZ in maintenance is definitely needed, especially to establish a logical relationship between the two concepts and to see the possibility of their combination as a complete set of tools to improve the maintenance process. Therefore, until further exploration is provided, the answer to the initial question already posed in the introduction on how to use Lean and TRIZ methods to improve the maintenance process will remain partial. The practices resulting from this literature review do not allow any conclusions to be drawn at this stage.

Thus, in-depth work is essential. A complementary approach to the problem posed in this article is to propose a standard "LEAN-TRIZ" improvement model that simultaneously integrates the characteristics of both methods to optimize the maintenance process. Then verify to what extent this new lean-TRIZ approach could be useful. This will be the next step of our research.

References

1. Sojka, V., Lepšík, P.: Use of TRIZ, and TRIZ with other tools for process improvement: a literature review. Emerg. Sci. J. **4**, 319–335 (2020). https://doi.org/10.28991/esj-2020-01234
2. Feldman, K.: Using the work of others: some observations on reviewing and integrating (1971). https://doi.org/10.2307/2111964
3. Definitions–Lean.pdf. http://cstsolutionsllc.com/wp-content/uploads/2017/05/Definitions%E2%80%93Lean.pdf. Accessed 16 Feb 2022
4. Raddam, C., Boumane, A., Kamach, O.: Etat de l'art : optimisation de la maintenance selon une approche lean, Tanger, Morocco, December 2015. https://hal.archives-ouvertes.fr/hal-01260732. Accessed 27 May 2021
5. Hawkins, B.: The many faces of lean maintenance. Plant Eng. **59**, 63–65 (2005)
6. Gupta, S., Gupta, P., Parida, A.: Modeling lean maintenance metric using incidence matrix approach. Int. J. Syst. Assur. Eng. Manag. **8**(4), 799–816 (2017). https://doi.org/10.1007/s13198-017-0671-z
7. Shou, W., Wang, J., Wu, P., Wang, X.: Lean management framework for improving maintenance operation: development and application in the oil and gas industry. Prod. Plann. Control **32**(7), 585–602 (2021). https://doi.org/10.1080/09537287.2020.1744762
8. Verma, A., Ghadmode, A.: An integrated lean implementation model for fleet repair and maintenance. Naval Eng. J. **116**(4), 79–89 (2004). https://doi.org/10.1111/j.1559-3584.2004.tb00306.x
9. Moayed, F., Shell, R.: Comparison and evaluation of maintenance operations in lean versus non-lean production systems. J. Qual. Maint. Eng. **15**, 285–296 (2009). https://doi.org/10.1108/13552510910983224
10. Ghayebloo, S., Shahanaghi, K.: Determining maintenance system requirements by viewpoint of reliability and lean thinking: a MODM approach. J. Qual Maint. Eng. **16**(1), 89–106 (2010). https://doi.org/10.1108/13552511011030345
11. Ayeni, P., Baines, T., Lightfoot, H., Ball, P.: State-of-the-art of "Lean" in the aviation maintenance, repair, and overhaul industry. Proc. Inst. Mech. Eng. Part B: J. Eng. Manuf. **225**, 2108–2123 (2011). https://doi.org/10.1177/0954405411407122
12. Silva, N.D.: Maintainability approach for lean maintenance, Sri Lanka, p. 10 (2012)
13. Bokrantz, J., Skoogh, A., Ylipää, T.: Lean principles and engineering tools in maintenance organizations - a survey study, p. 8 (2014)
14. Youssouf, A., Rachid, C., Ion, V.: Contribution to the optimization of strategy of maintenance by lean six sigma. Phys. Procedia **55**, 512–518 (2014). https://doi.org/10.1016/j.phpro.2014.08.001

15. Mostafa, S., Lee, S.-H., Dumrak, J., Chileshe, N., Soltan, H.: Lean thinking for a maintenance process. Prod. Manuf. Res. **3**(1), 236–272 (2015). https://doi.org/10.1080/21693277.2015.1074124
16. Ebeid, A.A., El-Khouly, I.A., El-Sayed, A.E.: Lean maintenance excellence in the container handling industry: a case study. In: 2016 IEEE International Conference on Industrial Engineering and Engineering Management (IEEM), Bali, Indonesia, pp. 1646–1650, December 2016. https://doi.org/10.1109/IEEM.2016.7798156
17. Abreu, A., Calado, J., Requeijo, J.: Buildings lean maintenance implementation model. Open Eng. **6**(1) (2016). https://doi.org/10.1515/eng-2016-0055
18. de Jong, S., van Blokland, W.: Measuring lean implementation for maintenance service companies. Int. J. Lean Six Sigma **7**(1), 35–61 (2016). https://doi.org/10.1108/IJLSS-12-2014-0039
19. Duran, O., Capaldo, A., Acevedo, P.: Lean maintenance applied to improve maintenance efficiency in thermoelectric power plants. Energies **10**(10) (2017). https://doi.org/10.3390/en10101653
20. Jasiulewicz-Kaczmarek, M., Saniuk, A.: How to make maintenance processes more efficient using lean tools? In: Goossens, R.H.M. (ed.) AHFE 2017. AISC, vol. 605, pp. 9–20. Springer, Cham (2018). https://doi.org/10.1007/978-3-319-60828-0_2
21. Implementation of Lean Methodologies in the Management of Consumable.pdf
22. Mouzani, I.A., et al.: The integration of lean manufacturing and lean maintenance to improve production efficiency. IJMPERD **9**(1), 593–604 (2019). https://doi.org/10.24247/ijmperdfeb201957
23. Hammadi, S., Herrou, B.: Lean integration in maintenance logistics management: a new sustainable framework. Manag. Prod. Eng. Rev. **11**(2), 99–106 (2020). https://doi.org/10.24425/mper.2020.133732
24. Antosz, K., Jasiulewicz-Kaczmarek, M., Pasko, L., Zhang, C., Wang, S.: Application of machine learning and rough set theory in lean maintenance decision support system development. Eksploatacja I Niezawodnosc-Maint. Reliab. **23**(4), 695–708 (2021). https://doi.org/10.17531/ein.2021.4.12
25. Crosby, B., Badurdeen, F.: Integrating lean and sustainable manufacturing principles for sustainable total productive maintenance (Sus-TPM). Smart Sustainable Manuf. Syst. **6**(1) (2022). https://doi.org/10.1520/SSMS20210025
26. TRIZ - What Is TRIZ.pdf. https://skat.ihmc.us/rid=1206064509716_727387479_10719/TRIZ%20-%20What%20Is%20TRIZ.pdf. Accessed: 21 Feb 2022
27. Petrov, V.: TRIZ. Theory of Inventive Problem Solving: Level 1. Springer, Cham (2019). https://doi.org/10.1007/978-3-030-04254-7
28. Ilevbare, I.M., Probert, D., Phaal, R.: A review of TRIZ, and its benefits and challenges in practice. Technovation **33**(2), 30–37 (2013). https://doi.org/10.1016/j.technovation.2012.11.003
29. Case study - applying the TRIZ methodology to machine maintenance. Triz J. 22 Aug 2003. https://triz-journal.com/case-study-applying-triz-methodology-machine-maintenance/. Accessed 22 Feb 2022
30. Viveros, P., et al.: Enhancing maintenance scheduling and control process by using SMED and TRIZ theory, pp. 535–545 (2017). https://doi.org/10.1201/9781315210469-71
31. Lv, C., Zhang, M., Wang, M.: Application research of TRIZ in maintainability design, pp. 1971–1975 (2013). https://doi.org/10.1109/QR2MSE.2013.6625966
32. Benjaboonyazit, T.: Systematic approach to problem solving of low quality arc welding during pipeline maintenance using ARIZ (algorithm of inventive problem solving). Eng. J. **18**(4), 113–133 (2014). https://doi.org/10.4186/ej.2014.18.4.113

33. Petrovic, S., Lozanovic-Sajic, J., Knezevic, T., Pavlovic, J., Ivanov, G.: TRIZ method application for improving the special vehicles maintenance. Therm. Sci. **18**(suppl.1), 13–20 (2014). https://doi.org/10.2298/TSCI130204169P
34. Vaneker, T., van Diepen, T.: Design support for maintenance tasks using TRIZ. Procedia CIRP **39**, 67–72 (2016). https://doi.org/10.1016/j.procir.2016.01.167
35. TRIZ Methodology Applied to Maintenance Problem Solving on Industrial Steam Systems in Africa
36. Viveros, P., Nikulin, C., López-Campos, M., Villalón, R., Crespo, A.: Resolution of reliability problems based on failure mode analysis: an integrated proposal applied to a mining case study. Prod. Plann. Control **29**(15), 1225–1237 (2018). https://doi.org/10.1080/09537287.2018.1520293
37. Mortazavi, M.A., Amindoust, A., Shahin, A., Karbasian, M.: Identification and extraction of factors affecting maintenance strategy selection with 39 parameters of TRIZ approach using meta- synthesis. J. Crit. Rev. **7**(10), 20 (2020)
38. Mortazavi, M.A., Amindoust, A., Shahin, A., Karbasian, M.: Integration of the TRIZ matrix and ANP to select the reactive maintenance tactics using the meta-synthesis approach. Manag. Prod. Eng. Rev. **12**(1), 108–118 (2021). https://doi.org/10.24425/mper.2021.136876
39. Bligh, A.: The Overlap Between TRIZ and Lean (2006)

Adoption of Artificial Intelligence in Romania: Innovative Policies to Overpass Vulnerabilities with TRIZ and Deep-Thinking Tools

Marin Iuga[1] and Stelian Brad[2(✉)]

[1] Intertechnica SRL, Rodnei 21, Borsa, Maramures, Romania
marin.iuga@intertechnica.com
[2] Technical University of Cluj-Napoca, Memorandumului 28, 400445 Cluj-Napoca, Romania
stelian.brad@staff.utcluj.ro

Abstract. Adoption of artificial intelligence (AI) in all aspects of society and the economy is a major desiderate at the EU level. Nevertheless, this demarch is influenced by a series of contexts and endowments, such as talents, sophistication, and specificity of the business models, maturity of public processes, social mindset, etc. This paper analyzes opportunities and threats associated with AI adoption in Romania, possible measures to improve the predicted outcome together with lessons learned at the global level. The research methodology includes analysis of the current situation and formulation of innovative solutions to overpass various barriers and constraints. Based on TRIZ and deep-thinking approaches we formulate novel policies to overpass the status of evolution and create a new foundation for AI development and adoption in the Romanian ecosystem, with implications at the EU level. TRIZ and the deep-thinking systematic approach reveal several healthy patterns of evolution to deviate the current course of actions toward better results and outcomes. Critical findings relevant to AI at the EU level are also underlined in relation to the conclusions at the national level.

Keywords: Artificial Intelligence · Public policies · TRIZ · Deep thinking · Critical analysis

1 Introduction

Artificial Intelligence has become a disruptive factor in the global arena of economic and geopolitical competition. US, China, and Russia are all defining their own strategies and taking decisive measures to ensure their competitive dominance in this area [1]. European Union recognizes this fact and aims to become a major global player: *"The EU should be ahead of technological developments in AI and ensure they are* swiftly *taken up across its economy. This implies stepping up investments to strengthen fundamental research and make scientific breakthroughs, upgrade AI research infrastructure, develop AI applications in key sectors from health to transport, facilitate the uptake of AI and the access to data."* [2]. European Union is also aware that not fulfilling these objectives

Published by Springer Nature Switzerland AG 2022
R. Nowak et al. (Eds.): TFC 2022, IFIP AICT 655, pp. 296–311, 2022.
https://doi.org/10.1007/978-3-031-17288-5_26

will have severe long-term detrimental effects: *"Without such efforts, the EU risks losing out on the opportunities offered by AI, facing a brain-drain and being a consumer of solutions developed elsewhere."* [2]

In this context Romania must have a clear and sound strategy for Artificial Intelligence adoption, finding the relevant niche that fits its national strengths and capabilities. Failing to do so will drastically reduce Romanian's economic competitiveness and will lead to long-term underdevelopment of the Romanian society.

2 Methodology

Our approach applies deep thinking [3] to identify causes, and TRIZ [4] to propose proper directions of interventions. Deep thinking is a thinking process that reveals something beyond what we can simply see and sense of. Deep thinking includes the following steps: analytical thinking, communication, creativity, open-mindedness, problem-solving. Analytical thinking focuses on gathering and breaking down of information into small, till relevant parts. Communication gives us access to the thoughts of people and experts. Creativity helps us to make new models and associations. Open-mindedness reveals new possibilities, helping us resolve matters in a manner that doesn't hinder interested parties. Problem-solving indicates appropriate patterns of action. TRIZ and 5Why can be some of the supporting tools in problem-solving.

3 Systematic Analysis

For approaching the analytical thinking of the AI landscape in Romania, we consider a model proposed in AI Watch Index, which is a report generated by the Joint Research Center of the European Commission [5]. The core of AI Watch report is based on several high-level dimensions: global view on the AI landscape (G), industry (I), research and development (R), technology (T), society (S).

Based on this model we investigate the Romanian landscape. Results reveal several challenges that are further tackled with TRIZ and deep-thinking models. This creates the space for formulating smart public policies to accelerate AI adoption in Romania.

3.1 Core Dimensions

We will further refine the AI Watch report's set of dimensions to be better contextualized for Romania. We will consider the following dimensions:

Data: The data dimension measures the availability of data along with its quality and dynamics. Ensuring that the data is properly collected, processed, and utilized is a prerequisite for developing a relevant, high impact, AI industry.

Infrastructure: The infrastructure dimension measures the availability of infrastructure for developing AI applications. The infrastructure consists in essential elements such as data storage & processing services, computing resources or access to AI services; a good infrastructure is a very important accelerator for the AI industry.

Business: The business dimension analyses the applications of AI in the business domain and in the creation of economic value. AI needs to provide relevant business value in order to become adopted at large scale in the economy.

Society: The society dimension measures the social impact, acceptance, and adoption of AI, along with the education of society in regards with AI concepts, benefits, and threats. Acceptability of AI at social level, perception of AI as a positive factor in the workplace and in the day-by-day life will strongly accelerate the adoption of AI in society.

Talent: It reflects the availability of qualified personnel (be it developers, data scientists or AI engineers) capable to sustain the dynamics and complexity of an AI-based industry. The availability of talented people is essential for creating a sustainable AI industry.

Governance: This dimension measures how well the Artificial Intelligence is governed. A good Artificial Intelligence governance promotes ethical and beneficial use of Artificial Intelligence, along with the necessary regulations protecting the citizens and the state against malicious and unethical utilization of Artificial Intelligence.

3.2 Data Dimension

We will focus on two major concerns related to data: how easily is to access it (open data) and how meaningfully is the data utilized (collected, processed, or further utilized in the value creation). For measuring the data access, we are using the Open Data Inventory [6] from Open Data Watch which puts Romania on rank 37 out of 187, with a score of 73 out of 100 in regards with data openness. Romania scores very well on open data for population & vital statistics, national accounts, and financial data. It scores very low on food security & nutrition, crime & justice, and built environment.

Romania achieves 69% of maximum score according to the DESI 2021 country's profile, well below the 78% value representing the Europe's average [7]. A quick overview of Romania's Open Data Inventory scores in terms of data coverage and openness on several areas is presented in Fig. 1.

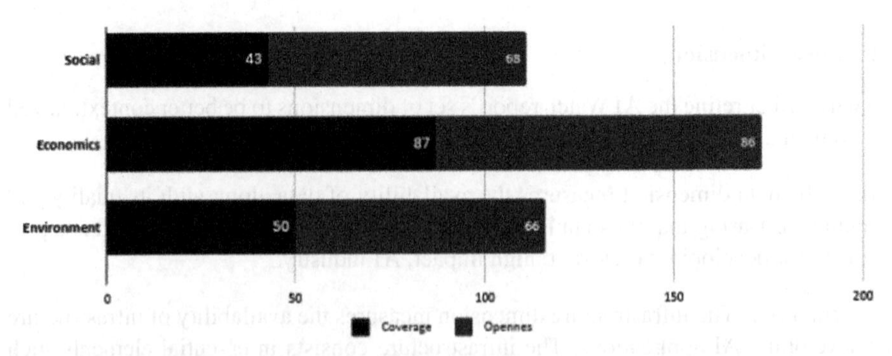

Fig. 1. Sub-scores by statistical subcategories, Romania Open Data Inventory, 2020.

Regarding data processing, Romania has a below average performance, only 5% of SMEs analyze big data (compared with an average 15% at the European level) [8]. Furthermore, according to Eurostat, only 4% of Romanian SMEs analyzed big data in 2020, well below the 12% average of EU27 zone [9]. Therefore, Romania has at best a modest performance in the area of data, especially usage of big data for analysis. On a positive note, based on the DESI 2021 Index, 31% of the Romanian enterprises are using AI – higher than the 25% European average value.

3.3 Infrastructure Dimension

The infrastructure dimension covers aspects such as connectivity, usage of cloud computing services, storage services or dedicated edge computing hardware. From the perspective of technology enablers, we can observe that in 2020 Romania outperformed the global average in the area of Broadband connectivity and underperforms in the area of Cloud, AI and IoT [10]. This is highlighted in Fig. 2.

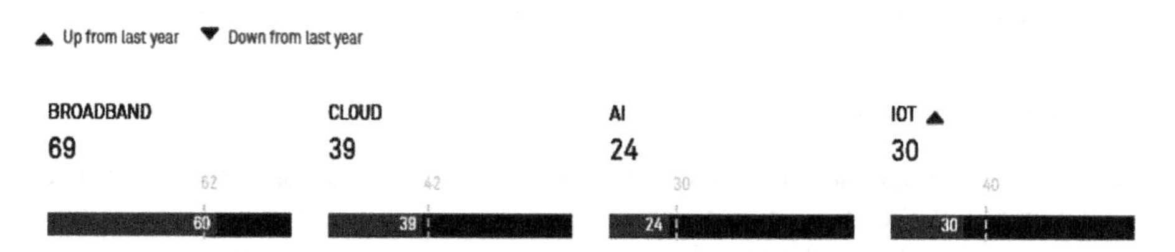

Fig. 2. Huawei GCI 2020, Romania County Profile.

Regarding the usage of cloud computing services, in 2021 only 14% of the Romanian enterprises used cloud services over the internet, compared to the EU27 average value of 41%. Compared to 2020, the trend is decreasing (in 2020 the percentage was 16%), even if the EU27 average is increasing (36% in 2020). This value is positioning Romania towards the lower end of the ranking among the EU27 countries [11].

An important aspect of the infrastructure is connectivity; here Romania outperforms the EU27 average. According to DESI 2021: "(...) while progress continued in 2020 for fixed broadband coverage, take-up of broadband services progressed at a slower pace. Nonetheless, Romania ranks 7th thanks to the high take-up of at least 100Mbps broadband (52%)" [12]. Therefore, Romania has a very good performance in terms of connectivity and a below average performance in terms of Cloud, AI and IoT infrastructure usage. A good connectivity allows Romanian companies to use global services solutions (such as Microsoft Azure or Amazon AWT) in order to compensate for the lack of competitive computing resources present at the national level.

3.4 Business Dimension

Adoption of AI by the business is a critical success factor for a competitive AI economy and market. The Romanian businesses are very open (above EU average) for adoption of Artificial Intelligence elements in their business processes [13] (see Fig. 3).

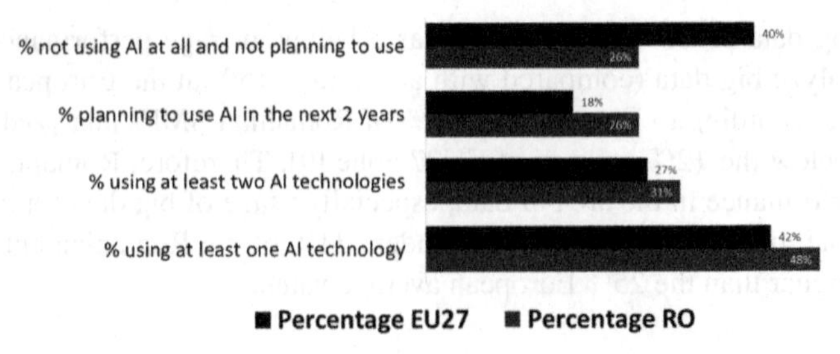

Fig. 3. AI adoption by enterprises, European Union, 2020.

Almost 50% of the Romanian companies are using at least one AI technology, the number of companies not planning to use AI are in minority (40%). From the standpoint of AI domains relative comparative advantage, Romania offers a very good performance in the area of automation [14].

As a consequence, Romania presents a very favorable climate for the adoption of AI at the business level, especially for the business oriented towards automation.

3.5 Society Dimension

Ensuring that AI is adopted at the society level creates the premises for a strong AI market backed up by a large mass of AI technology consumers. It is important to understand the social sentiment in relation to AI technology (expectations and emotions) along with the social opportunities created for AI utilization.

In a study on the Romanian generation Z and its positioning versus AI it was discovered that "*... the Romanian students' knowledge about AI is fairly limited and does not stream from formal education but rather is founded on social media finds.*" [15]

Regarding the adoption of AI at the workplace, the Romanian society displays a low level of concern [15], less than 15% of the respondents feel threatened by the changes brought by adopting AI at the workplace.

From the standpoint of the AI education offer, Romania performs very well. In terms of relative importance of AI education, approximately 14% of all the Bachelor and approximately 18% of Master places are dedicated to university education in AI [16]. With these values, Romania ranks 2nd and 1st place in AI intensity for Bachelor and Master level of studies in university.

As a conclusion, the Romanian society is relatively uninformed in regards with AI, most of the information being gathered from social media. On the other hand, the Romanian society is very open to including AI at the workplace and preparing for the knowledge necessary to do it.

3.6 Talent Dimension

Talent is crucial in any information-based economy, especially when dealing with AI. The range of AI activities that can be performed with a low degree of skills is usually very limited, mostly focused on data manual annotation.

Romanian companies are mostly at the beginning of their journey in the area of AI, and they are looking for expert skills in the area of big data management, machine learning and programming. The two most important barriers to the AI adoption are lack of skills among existing staff and difficulties to hire new staff with the right skills [17].

Romania has a low proportion of ICT specialists in total employment (2%) compared to Europe's baseline value (4%) [16]. By extrapolation, this indicates a low performance in regard to AI talent. The situation is exacerbated by the negative outlook of the productivity of the future Romanian generations because a "*(...) child born in Romania today will be 58 percent as productive when he/she grows up as he/she could be if he/she enjoyed complete education and full health. This is lower than the average for Europe & Central Asia region and high-income countries*" [18].

In conclusion, Romania has a strong deficit of talent in regards with the AI-related skillset. AI companies are identifying that lack of talent is one of the major barriers to growth, yet the projection for AI skillset demand fulfillment have a negative outlook.

3.7 Governance Dimension

A good governance will ensure that the moral, ethical, and legal criteria usage will be applied consistently for AI usage. Failing to do this will create rejection of technology, fears, algorithmic discrimination, or social bias. The Romanian legislation does not grant any legal status to, nor does it recognize the "Artificial Intelligence"-related profession. Therefore, any content produced by Artificial Intelligence is not protected by the copyright law. There is also no special liability provision for damages generated by the usage of AI. Damage and liability are covered by the provisions of the Civil Code, which in itself is poorly adapted to an environment heavily infused by AI. Also, from a strategic perspective there is no official strategy for AI yet. There is however an official initiative directed by the Authority for the Digitalization of Romania to create a first AI strategy together with the Technical University of Cluj-Napoca [19].

In conclusion, Romania is lagging behind in the area of AI Governance. While there is a general consensus in Romania that there should be a better governance (strategy, legislation, ethical framework, etc.), real results have still to be achieved.

4 Tackling with Challenges

The analysis highlighted in the previous subsections indicates several challenges for Romania in adoption of AI by both public and private sectors. It is obvious that a natural direction of intervention is to improve conditions towards creation and adoption of AI solutions and to use them for generation of public and private value. But this is not the right way of action, because it reflects a wish, not a smart, realistic strategy. This leads to the need for identifying causes of the observed situation and afterwards to the foundation of policies capable to tackle causes in an intelligent way, meaning to overpass barriers in an "out-of-the-box" way. Here we consider the other four steps of deep thinking, specifically communication, creativity, open-mindedness, and problem-solving. These steps are aggregated in a structured framework, illustrated in Fig. 4.

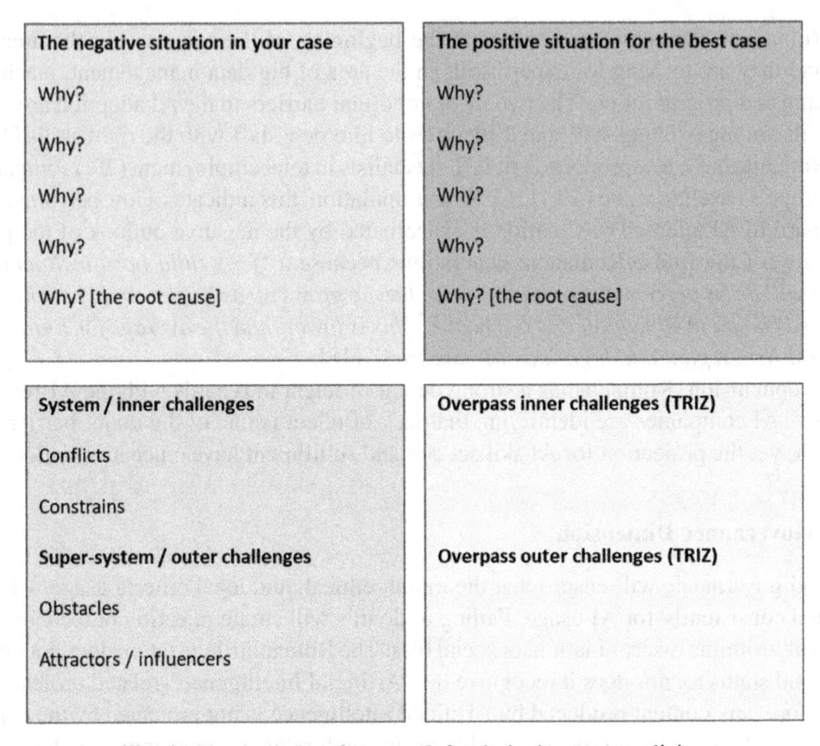

Fig. 4. The 4-window framework for designing smart policies.

4.1 Application of the 4-Window Framework

Because of the space constraints, we will introduce in the followings only one example about how the problems can be tackled with the roadmap from Fig. 4. In this respect, we have selected the situation "Romania has a strong deficit of talent in regards with the AI related skillset".

Usually, such ascertainment is reflected in a strategy under the form of "increase the number of training programs in AI" or similar policies. But this is a very wrong approach because the problem is not necessarily created by the lack of training programs in AI. The lack of sufficient training programs in AI is the effect, not the cause(s), and the lack of talent in regards with the AI related skillset is actually one of the negative impacts on data-driven innovations.

With these remarks, the subsequent part of this section will illustrate the 4-window framework application for the above-mentioned situation.

Window 1: The negative situation in our case. Romania has a strong deficit of talent in regards with the AI related skillset.

Why? – Because there are no trained personnel on the workforce market.

Why? – Because there is an imbalance between demand and supply for AI skillset. The AI skillset supply capabilities significantly lagged behind the demand curve generated by the disruption bought in by AI.

Why? – Because the mass education system (e.g., high schools or universities) cannot supply enough trained personnel to cover in due time the surge in demand for AI skilled personnel.

Why? – Because the mass education system has a significant inertia, and it is unable to adapt quickly to disruptive events.

Why? [root cause] – This is a systemic issue, mass education system – by design – takes a significant amount of time (e.g., years level of magnitude) to adapt to changes in the industry/market demand. It takes time to develop and approve specialized new learning curricula, to hire or retrain teaching personnel, to perform organizational/administrative changes or to open new educational lines. When a disruption happens, the mass education system is unable to respond quickly enough to make a difference.

Window 2: The positive situation for the best case. There will be an influx of skilled AI personnel that will cover the demand of the industry. A good example is US which "*… has a substantial lead over all other countries in top-tier AI researchers …*" [20].

Why? – Because US has the largest contingent of top AI researchers (59% of the global researchers) [20].

Why? – Because US is attracting and retaining the large majority of the AI talent (~80% of the international AI PhD students intend to stay in US after graduation) [21].

Why? – Because both the top 3 AI university programmes (Carnegie Mellon University, Massachusetts Institute of Technology and Stanford University) and the top 3 AI research centers (Google, Stanford University, Carnegie Mellon University) are in US.

Why? – Because there is a strong entrepreneurial approach towards both education and research; thus, providing the most competitive services and ensuring a viable supply chain - from education to work.

Why? [root cause] – Because the US society and economy have an entrepreneurial nature, seizing opportunities and having the necessary agility to provide the services that are in high demand on the market.

Window 3: Challenges that create difficulties to jump to the target result. Here we analyze the conclusions from window 1 (top-left in Fig. 4) in conjunction to the conclusions from window 2 (top-right in Fig. 4). Thus, the gap analysis is clearly focused.

Inner conflicts: Two major conflicts are highlighted in relation to our case.

- (CF_1) We want to increase the agility of programs in the mass education system (to include new courses, such as AI), but this requires a fast expansion of the pool of highly trained teaching staff in AI, with a balanced distribution at the national level and less inertia of the higher education programs. Thus, our challenge is how to increase the basin of highly trained teaching staff in AI at national level and how to introduce AI courses in all higher education specializations (from engineering to medicine, biology, fine art, social, etc.), without creating harmful side effects in the current study programs (e.g., additional costs, removing of current courses, etc.).
- (CF_2) We want the formation of AI skillset taking place as early as possible (e.g., high school), but this clashes with the lack of skilled teachers capable to teach AI for kids. Thus, we face with the need to motivate existent computer science/STEM teachers from schools to learn AI and teach it to young students from high school with no complicated efforts.

Inner constrains: Three major constrains are illustrated in this case.

- (CO_1) The number of students in the Romanian universities that attend and are passionate of AI is limited (even if it is growing).
- (CO_2) The skillset required to follow of AI curricula is too high for the majority of Romanian students.
- (CO_3) The availability of AI-oriented academic staff in the Romanian universities is still limited.

Outer obstacles: From the level of supersystem we extracted three obstacles.

- (OB_1) The vast majority of Romanian students trained in universities abroad (e.g., US, UK, France, Germany, the Netherland, etc.) are not returning back in Romania.
- (OB_2) Romania fails to attract foreign AI talent (e.g., see the couple of weaknesses in the Romanian ecosystem highlighted in Sect. 3).

Outer influencers/attractors: Three attractors have been seen as critical in our case.

- (AT_1) AI salaries and bonuses outside Romania remain very attractive.
- (AT_2) AI work and research opportunities are higher outside Romania (e.g., other workforce markets (such as US, Germany, France, Canada, UK) are more attractive from a professional perspective for good AI specialists).
- (AT_3) The number and quality of foreign AI startups is higher.

Window 4: Innovative solutions. At this phase, we consider the determined conflicts, constrains, obstacles, and attractors as major areas of investigation to formulate innovative solutions to our problem. For this job we use basic TRIZ tools.

TRIZ generic areas of intervention to overpass inner challenges: To tackle the conflicts CF_1 and CF_2, TRIZ contradiction matrix is very suitable. Thus, CF_1 can be expressed in TRIZ language as "Increasing < 26. Amount of substance > " vs " Reducing < 31. Harmful factors > ". CF_2 is described in TRIZ as "Increasing < 39. Capacity > " vs "Minimal < 19. Energy spent > ". Inventive principles related to CF_1 that emerge from this situation are "Local quality", "Parameter changes", "Composite structures", and "Inert environment". In the case of CF_2, inventive principles are "Preliminary actions", "Periodic actions", "Parameter changes", and "Strong oxidizers".

Constrains limit the space of action for policies. To manage constrains, we consider the ASIT method [22] and Su-Field method [23] from the TRIZ toolbox. ASIT operates with two powerful principles: the principle of "closed universe" and the principle of "qualitative change". To convert a problem into a solution, the "closed universe" principle looks for smart reconfiguration of the current system at the level of subsystems, properties, objects. The "qualitative change" principle searches for the factors causing the problem and operates in a way that changes the direction of these factors for annihilating or dramatically reducing their effects. The *list with problematic objects and properties inside the system* is ["students", "migration", "academic staff", "complexity of AI – mathematics and programming"]. The *list with objects and properties outside the system* is ["demography", "focus of local companies on AI"]. Complexity of AI reduces the long-run interest of students on this subject. Also, the limited focus of local companies on AI usage narrows the interest of many students to specialize in AI. *Unwanted effect*: low adoption of AI in economy and social spaces (note: adoption of AI increases public and private value because of optimization of systems, better predictions, acceleration of structural transformation for social and business models, etc.). *Derived action to overpass unwanted effects*: acting such that complexity of AI not representing a barrier for students, professors, and companies. *Selected objects to be transformed*: the most critical objects to be transformed are the students and professors. If they will adopt more openness to AI, the other objects can be positively altered in the chain of influences. *Main resources of the selected objects*: motivation, time, mindset, methods, and tools to learn or teach. *Application of operators*: ASIT uses five operators to act upon the system (i.e., division, elimination, unification, breaking symmetry, multiplication). We can use any of these operators to produce transformations in the system. Our analysis indicates that "breaking symmetry" is the most appropriate operator in this case. *Application of operator*: we imagine that asymmetry is applied on resources associated to students and professors (e.g., motivation, time, mindset, methods, and tools to learn or teach). *Scenarios*: (a) asymmetric effects on AI graduates; (b) asymmetric effects on AI professors; (c) non-conventional methods and tools to teach; (d) non-conventional methods and tools to learn; (e) asymmetric triggers on students; (f) asymmetric triggers on professors. *Basic idea*: certificate for AI graduates; salary bonus for professors; associate trainers from industry; online platform with video training materials accessible to everyone; video training materials with simple-terms explained AI; examples of use cases from the national ecosystems; learning by doing projects; compensate complex mathematics with libraries provided by various technologies; exercises with gradual complexity; inclusion of chapters about deployment into production.

The Su-Field method is applied on the three constrains CO_1, CO_2, CO_3, in the form of "object-function-tool". Object is associated to CO_1 (# students attending AI courses), function is associated to CO_2 (skillset required for AI courses), and tool is associated to CO_3 (#AI-oriented academic staff) (see Fig. 5).

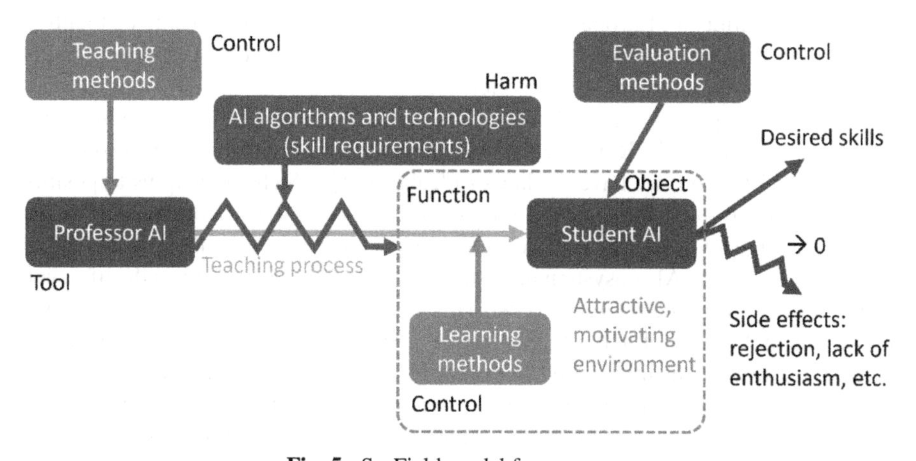

Fig. 5. Su-Field model for our case.

We can consider several standards to tackle CO_1-CO_2-CO_3 model. *Investigation of the Su-Field standards* indicates the following possible ways of action: (a) Standard 1-1-1: introduce a new substance or field in the model (see Fig. 5, teaching methods); (b) Standard 1-1-2: introduce an additive in the model to increase control (see Fig. 5, leaning methods); (c) Standard 1-1-3: introduce a new substance to an external substance (see Fig. 5, evaluation methods); (d) Standard 1-1-4: add a substance from external environment (e.g., trainers from industry); (e) Standard 1–1-7: add a substance to one of the substance to intensify the effect (e.g., prizes from industry in a graduation contest); (f) Standard 1–2-3: to minimize the undesired effects introduce an additional substance that attracts the negative effects (e.g., grants from industry for graduates in AI with high quality results).

TRIZ generic areas of intervention to overpass outer challenges: Obstacles require for more resources in the system to make it capable to move forward. Attractors generate new behavioral patterns that interfere with traditional patterns and lead to disruptions in the system, mostly with harmful effects that require reinventions in the current system to move it forward. In our case, both obstacles OB_1 and OB_2 relate to the same generic challenge: attraction of AI talents. To tackle this challenge, we selected the ARIZ method from the TRIZ toolbox. ARIZ encourages us to look first for the ideal final result (IFR). This messages us not falling into the trap of predefined actions. From this angle, we will look for the solution by investigating the four rules of ARIZ: (a) rule I (separate opposed properties in time); (b) rule II (separate opposed properties in space); (c) rule III (separate opposed properties between the system and its parts); (d) rule IV (coexistence of opposed properties in the same substance in different contexts). One idea emerging from these rules is about not necessary having these AI talents located in Romania, but rather to have them working for or running businesses located in Romania. The second idea is working in virtual companies, who have locations in more countries, thus fitting for each problem the best location (e.g., headquarter in country X, R&D in country Y, education in country Z, etc.). The third idea is to involve trainers from abroad using blended learning approaches.

All three attractors, AT_1, AT_2, and AT_3, are generically related to the same issue: other locations in Europe and Northern America have more incentives to attract talents in the field of AI than Romania. This is about clustering effect, and it cannot be compensated on short term with local alternatives.

From a TRIZ judgement, we have to analyze the S-curve of the ecosystem and to understand the place of the Romanian ecosystem on this curve and the adequate direction of evolution [24]. Here we have the law of ideality [24], which can help us to position the ecosystem on the S-curve based on the ratio between the performance of useful functions and the level of costs involved. Our analysis is graphically illustrated in Fig. 6. This indicates that the AI ecosystem in Romania is at its very early stage on the life-cycle curve, far away from the reference (US). This is mostly related to the number of local innovations in AI, level of maturity, number of adopters and effects on their businesses, as well as maturity of the innovators. One of the big barriers, as it was already highlighted is the lack of adequate data in the private sector. However, not even in the public sector there is no relevant data. For example, initiatives such as smart cities are still at their bottom level of maturity, meaning that solutions that collect massive data in cloud are only few, and in few places.

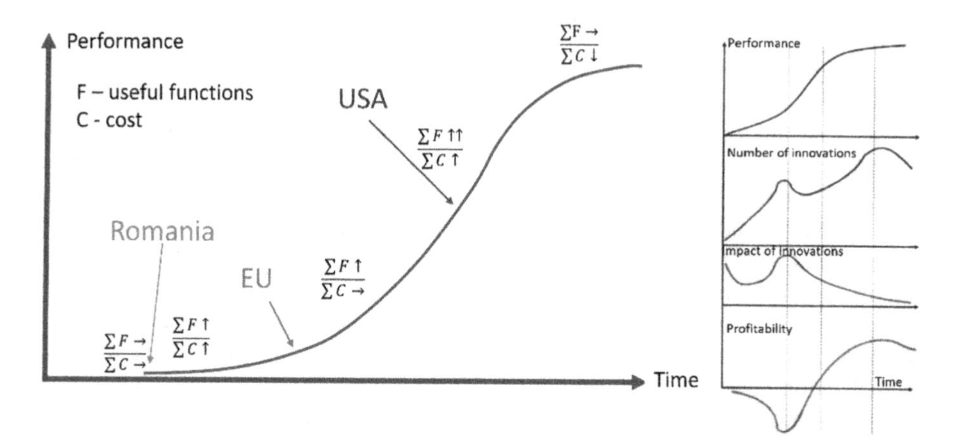

Fig. 6. S-curve of the AI ecosystem (comparative analysis).

Based on the laws of evolution and patterns of evolution [25], we can indicate that the AI ecosystem from Romania is very young. From the eight laws of evolution, it is actually marked only by static laws, still missing any kinematic or dynamic law at the level of every subsystem. The most striking law is in fact the law of the "completeness of the parts of the system". Thus, in terms of patterns of evolution, we indicate the need to evolve towards more useful functions in the ecosystem, with a careful attention paid to a balanced evolution of all subsystems, followed by the evolution of applications. Any other strategy of evolution is not effective from the perspective of TRIZ philosophy. Thus, ambitions to invest at this stage of evolution in sophisticated technologies, such that micro/nano-electronics for AI or quantum computing will be "black holes", because of many gaps currently present in the ecosystem. This kind of mistake was already done with the megalomanic project of "laser valley", which is not capable after more than 15 years to produce something and which dramatically lowered the country's funds in other research areas.

4.2 Proposed Solutions

Based on the insights from Sect. 4.1, we can formulate more argued and grounded public policies on the dimension related to "AI talents". Similar approaches can be applied for the other remaining five dimensions (see Sect. 3). These public policies will be based two strategic pillars: (a) full spectrum coverage and (b) talent retention.

Full Spectrum Coverage. AI (both as a discipline and an industry) requires a variety of skills - such as data annotation, data modelling, data engineering or machine learning model creation. The full spectrum coverage principle considers the fact that Romania should be able to provision the full spectrum of AI skills needed by the industry – both from timing and volume perspectives. Romania lacks the capabilities of supplying these skills internally, especially via mass education system; therefore, we should look for a blend of education suppliers and innovative teaching tools (see the S-curve from Fig. 6). We are proposing the following main lines of action:

Early development of generic AI skills, these skills should be taught the latest in early high school - with a focus on practical applications of AI for solving general problems.

The curricula for generic AI skills should focus on fundamental data skills (such as collection, exploration, visualization, and transformation) and black box utilization of Data Science and AI for solving practical business and industry challenges. Furthermore, the curricula for study specializations should consider applied AI as a core element for specialization's business processes augmentation [see the TRIZ principle "parameter changes" proposed in Sect. 4.1].

Externalization of specialized AI skills which are too expensive or too rare for extensive usage, thus ensuring a better cost control and a better availability of these skills. This externalization may be done towards specialized research institutes, digital innovation hubs, consulting firms or via outsourcing/near shoring. We also include here the invitation of trainers from abroad in blended learning programs [see ARIZ recommendations from Sect. 4.1].

Creation of complex educational networks, thus reducing the reliance on an inertial mass education system. This complex education ecosystem should consider both an extensive blend of education suppliers (such industry trainers, consulting companies or digital innovation hubs) and agile/innovative learning methods (online training, collaborative AI application development or learning by doing). We encourage the creation of AI courses and exercises by the best experts and professors, record and disseminate them in all universities as basic teaching materials [see from Sect. 4.1. the TRIZ results about local quality, inert environment, parameter changes, and ASIT conclusions] and renewal at every two years [see TRIZ in Sect. 4.1 about periodic action]. It is also important the creation of training programs delivered by private academies and companies, combined with tests and recognition organized in universities [see in Sect. 4.1, the TRIZ results about composite structures and Su-Field and ASIT conclusions]. In the same context, public policies should consider salary bonuses for teachers that are trained with certificate and teach AI in schools [see in Sect. 4.1. the TRIZ results about preliminary actions and strong oxidizers], as well as dedicated grants, prizes, and certificates for students that graduate AI specialization, besides their core specialization profile [see the conclusions of ASIT and Su-Field in the Sect. 4.1].

Talent Retention. Due to a lower financial and professional attractiveness, Romania should find non-conventional solutions for retaining its talent. We propose the following measures:

Incentivize the top AI performers, such as AI researchers or inventors. This incentivization should be covered by subsidies for patent applications, simplification of access to public research funding or facilitation of interaction between researchers/inventors and companies/public bodies. We are advising against indiscriminate financing (e.g., first come – first served policies) in regards with AI research [see conclusions from Sect. 4.1, ARIZ].

Ensure fiscal stimulation of AI workers via tax exemptions and deductibility of AI training and specialization expenses. This brings multiple benefits: it decreases the salary costs for companies, encourage continuous learning and motivates AI companies to relocate in Romania due to lower costs of the workforce. This measure was applied with excellent results for stimulating the development of the ICT industry. This proposal is inspired from Su-Field standard 1-2-3.

Promote the social value of AI related occupations via measures such as social recognition, state recognized certifications and deductibility of AI grants/prizes offered by the industry to high performing AI graduates. This measure has a good synergy with the esteem and self-actualization needs associated with the high level of education of top AI performers [see ARIZ indications in Sect. 4.1].

Super-competitive grant schemes for attracting international talents in the area of AI in innovation projects run by local SMEs [see ARIZ rule III from Sect. 4.1 and Su-Field standard 1-1-4].

5 Discussions and Conclusions

Given the openness of society towards using AI and the interest from business towards including AI technologies in their business process, Romania shapes itself as a potentially significant consumer of AI technology and services.

Considering the lack of AI skills, the low degree of digitalization and a relative inability to process significant amount of data, the Romanian economy does not have yet the means to sustain any relevant portion of its AI demand. Therefore, without smart interventions, Romania will be determined in import most of the AI technologies and services because of lack of local capabilities.

Furthermore, taking into consideration the fact that the level of governance and regulations is extremely low, Romania's capabilities to protect its society against misuse of AI and to execute a meaningful steering of AI adoption is basically non-existent for now. To sum it up: "*Romania's profile is that of an importer of Artificial Intelligence solutions with a low level of protection for users and consumers*" [26].

However, with a proper stewardship, Romania may find its own niche for creation of custom AI solutions and possibly for AI integration in other systems (most probably with a focus on automation). That will however require a substantial focus and investment in the area of digital literacy and for building the spectrum of skills necessary for meaningful AI application development (such as data manipulation, domain modelling or machine learning skills).

This paper introduces TRIZ tools in designing smart public policies, with a particular application in the case of AI strategies. A 4-window framework has been presented in this paper for analyzing situations and formulating adequate spaces for innovation. Various perspectives from which the problem must be seen lead to the exploration of more TRIZ tools for solution formulation, such that the contradiction matrix, Su-Field, ARIZ, ASIT, directed evolution.

This research highlights the fact that many times we are trapped in preconceptions and biases both when we formulate the problem and the related solution. The use of deep-thinking and TRIZ breaks the psychological inertia and opens new views to see and tackle a problem. One of these traps is to think in terms of national borders for something that requires a global perspective. Smart specialization (i.e., specialization of innovation policies) is one of the suitable ways of action. In the case of AI strategy, it is smarter to create niche competences at national level and cooperate internationally for other AI areas, including open polycentric innovation in virtual, distributed organizations. Polycentric sharing of resources, both in terms of talents, software and hardware at the EU level is the right approach to make a difference in AI at the global level. Therefore, Romanian AI strategy must be aligned and synchronized with the AI regulations and frameworks energized by the European Commission.

References

1. Brookings Institution, Whoever leads in artificial intelligence in 2030 will rule the world until 2100. https://www.brookings.edu/blog/future-development/2020/01/17/whoever-leads-in-artificial-intelligence-in-2030-will-rule-the-world-until-2100. Accessed 06 June 2022
2. European Commission, Artificial Intelligence for Europe, Brussels - 25.4.2018. https://eur-lex.europa.eu/legal-content/EN/TXT/HTML/?uri=CELEX:52018DC0237&from=EN. Accessed 06 June 2022
3. Byer, W.: Deep Thinking: What Mathematics can Teach us about the Mind. World Scientific, London (2015)
4. Souchkov, V.: Breakthrough Thinking with TRIZ for Business and Management: An Overview. http://www.xtriz.com/TRIZforBusinessAndManagement.pdf. Accessed 20 June 2022
5. European Commission, JRC Technical Reports, AI Watch Index 2021. https://publications.jrc.ec.europa.eu/repository/bitstream/JRC128744/JRC128744_01.pdf. Accessed 08 June 2022
6. ODIN Open data inventory, Open Data Watch, Country profile: Romania (2020). https://odin.opendatawatch.com/Report/countryProfileUpdated/ROU?year=2020. Accessed 16 June 2022
7. European Commission, The Digital Economy and Society Index - Countries' performance in digitization, Romania – Country Profile 2021, p. 16. https://ec.europa.eu/newsroom/dae/redirection/document/80496. Accessed 16 June 2022
8. European Commission, The Digital Economy and Society Index - Countries' performance in digitization, Romania – Country Profile 2021, p. 13. https://ec.europa.eu/newsroom/dae/redirection/document/80496. Accessed 16 June 2022
9. Eurostat, Big data analysis, Custom Dataset – Percentage of SMEs (without financial sectors) analyzing big data internally from any source. https://ec.europa.eu/eurostat/databrowser/bookmark/074c0cc9-ba73-4a65-80cb-c45a780a8f34?lang=en. Accessed 17 June 2022
10. Huawei, Global Connectivity Index, 2020, Country Profile – Romania. https://www.huawei.com/minisite/gci/en/country-profile-ro.html. Accessed 17 June 2022
11. Eurostat, Cloud Computing Services, Custom Dataset – All enterprises (without financial sector) buying cloud computing services used over the internet (2021). https://ec.europa.eu/eurostat/databrowser/bookmark/9abd0dac-d640-42b3-9fe5-51ca4ddcad4?lang=en. Accessed 17 June 2022
12. European Commission, The Digital Economy and Society Index - Countries' performance in digitization, Romania – Country Profile 2021, p. 3, https://ec.europa.eu/newsroom/dae/redirection/document/80496. Accessed 17 June 2022
13. Based on data from European enterprise survey on the use of technologies based on artificial intelligence, Final Report, European Commission. https://ec.europa.eu/newsroom/dae/document.cfm?doc_id=68488. Accessed 17 June 2022

14. European Commission, AI Watch Index 2021 Dataset. https://data.jrc.ec.europa.eu/dataset/e3757f41-fe54-4330-946d-ae897686164f. Accessed 17 June 2022
15. Artificial Intelligence In Education - Romanian Students' Attitudes Toward Artificial Intelligence And Its Impact On Their Career Development, Sivia Fotea, Ioan Fotea, Emanuel Tundrea (2019). https://www.researchgate.net/publication/338104741. Accessed 18 June 2022
16. European Commission, JRC Technical Reports, AI Watch Index 2021, p. 48. https://publications.jrc.ec.europa.eu/repository/bitstream/JRC128744/JRC128744_01.pdf. Accessed 18 June 2022
17. European enterprise survey on the use of technologies based on artificial intelligence, Eurostat. https://ec.europa.eu/newsroom/dae/redirection/document/68488. Accessed 18 June 2022
18. Based on data from Eurostat, Proportion of ICT specialists in total employment (2020) (%). https://ec.europa.eu/eurostat/databrowser/bookmark/2cb2b5ed-6e65-42d1-90611300f2690f90?lang=en. Accessed 18 June 2022
19. The Diplomat, ADR and UTCN organized a public consultation on the first national artificial intelligence strategy. https://www.thediplomat.ro/2022/04/05/adr-and-utcn-organized-a-public-consultation-on-the-first-national-artificial-intelligence-strategy/. Accessed 19 June 2022
20. Macro Polo, America's Got AI Talent: US' Big Lead in AI Research Is Built on Importing Researchers. https://macropolo.org/americas-got-ai-talent-us-big-lead-in-ai-research-is-built-on-importing-researchers. Accessed 24 June 2022
21. CSET, Keeping Top AI Talent in the United States. https://cset.georgetown.edu/publication/keeping-top-ai-talent-in-the-united-states/. Accessed 24 June 2022
22. Reich, Y., Hatchuel, A., Shai, O.: A theoretical analysis of creativity methods in engineering design: casting and improving ASIT within K-C theory. J. Eng. Design **23**(1–3), 137–158 (2012)
23. Dobrusskin, C., Belski, A., Belski, I.: On the effectiveness of systematized Substance-Field analysis for idea generation, TFC 2014, 29-31 October, Lausanne, pp. 123-127 (2014)
24. Petrov, V., Brad, S.: TRIZ for developing innovative businesses and related strategies. Acta Technica Napocensis **6**(23), 451–460 (2021)
25. Petrov, V.: Laws of system evolution: TRIZ. Independently published, p. 57 (2019)
26. Iuga, M.: Towards A Romanian AI Strategy, Understanding The Status Quo. https://www.clujit.ro/wp-content/uploads/2021/12/Towards-a-Romanian-AI-Strategy.pdf. Accessed 20 June 2022

TRIZ Applications

Analysis of Tools Used for Implementation of a Knowledge Base Based on an Ontology for a Service Robot in a Kitchen Environment

Grzegorz Kuduk(✉), Maciej Bekas, Barbara Wąsowska, and Piotr Pałka

Warsaw University of Technology, Warsaw, Poland
{Grzegorz.Kuduk.stud,Maciej.Bekas.stud,Barbara.Wasowska.stud,
Piotr.Palka}@pw.edu.pl

Abstract. The work presents the process of systematic invention in relation to the design of a robot companion component. The component is responsible for knowledge management in the kitchen environment and its automatic use. Based on the TRIZ method, in particular the Contradiction Business Matrix 3.0, the Inventive Principles are listed. Then, an analysis is carried out regarding potential tools for the implementation of inventive principles. The results of the analysis carried out on the KnowRob and Armor tools in terms of: documentation quality, difficulty of installation, usability and performance. The analysis was carried out to determine the superior tool in knowledge processing for robots. KnowRob is a popular tool in this area. Armor is a young and interesting tool with a potential to become widely used. These kinds of tools need to respond quickly and guarantee reliability. For this purpose, installations and configurations of both environments were performed and documented. Then, a set of queries in Prolog and SPARQL were prepared and tested. The ontology used in testing is based on Web Ontology Language (OWL). Our findings indicate that KnowRob is the superior tool in the tested areas.

Keywords: KnowRob · Armor · OWL · Robotics · Ontology

1 Introduction

The subject of this work is the process of systematic invention in relation to the design of a component of the robot companion. The goal of the component is to manage the knowledge that is acquired by a robot and the provision of that knowledge to other components of the robotic system. We assume that the robot is supposed to be used to support the work of an elderly person in the kitchen environment. For this purpose, it must efficiently navigate the concepts and dependencies concerning objects used and activities performed. One of the used solutions [9] is equipping the robotic system with a knowledge database. It

Published by Springer Nature Switzerland AG 2022
R. Nowak et al. (Eds.): TFC 2022, IFIP AICT 655, pp. 315–327, 2022.
https://doi.org/10.1007/978-3-031-17288-5_27

allows the robot to efficiently navigate the concepts and relations described by the ontology [17] used. It should understand rules prevailing in the kitchen environment and descriptions of procedures provided to it. Furthermore, it should act as a "companion" for an elderly person, helping them in their daily duties.

The proposed research method is based on two methods. First, is the TRIZ-based analysis of the problem of robot companion systems working in the kitchen environment. Second, is an experimental analysis of a set of IT tools that support ontology management and allow query automation.

2 Systematic Invention Using TRIZ

This section describes the innovation-oriented TRIZ-based method for solving the problem. According to the aforementioned assumptions, consisting of the fact that the companion robot must efficiently navigate the concepts and dependencies concerning objects used and activities performed, the **specific problem** consists in equipping the robot with a knowledge base from which it will derive domain knowledge.

After having the specific problem formulated, we match it to an **abstract problem**. From the TRIZ toolkit, we select the Contradiction Business Matrix 3.0 [13], which is the revised method developed by Darrell Mann, designed to solve problems in contemporary technological conditions. As the robotic system is in the development phase, only the system features (25–32) are taken into account.

- Improving features:
  - (25) amount of information - the robot should have as much knowledge as possible
  - (28) adaptivity/versatility - the robot should be able to perform different tasks
- Worsening feature:
  - (29) system complexity

After analysing the Contradiction Business Matrix 3.0, four following Inventive (Business) Principles are selected: (15) Dynamics, (25) Self-service, (40) Composite Structures, and (28) Another Sense.

The principle 40 points using composite structures for managing the knowledge in the robotic system. There are specific notations, like OWL (Ontology Web Language) [8] for automatic processing of knowledge. The robotic system should allow self-servicing (principle 25) in the sense that it should add, and modify pieces of knowledge, check dependencies among the concepts (e.g. having the information about the positions of a cup and a table, infer the knowledge that the cup is on the table). According to principle 28 (another sense), the robotic system should be equipped with sensors (cameras, artificial skin) to sense the environment. Principle 15 means that the robotic system should be able to work in a dynamic (changing) environment.

The above analysis points to the need for utilizing the knowledge-based system working together with the robot companion. In the further section, the analysis of known tools for knowledge management is described.

3 Overview of the Tools

The tools used for implementing the knowledge database described in cited articles [9,16,20] are badly documented, which can cause severe problems for potential users. This article aims to take a look at the installation and configuration of the KnowRob and ARMOR (A ROS Multi-Ontology Reference) frameworks, while examining their difficulty of usage and quality of their documentations, as well as test performance using a set of custom queries.

3.1 Service Robot

Creation of autonomous agents was aimed at creating self-sufficient agents cooperating with other agents in a changing environment.

The scope of the ontology in autonomous robots must include the definition of objects, the map of the environment, accessibility and influence factor, actions and tasks, activity and behavior, planning and methodology, abilities, skills, hardware components (Unified Robotics Description Format (URDF) area), programming components (Robot Operating System (ROS) area), software and communications.

3.2 Ontology

The first approaches to create an ontology were made in 1993. The official definition and discussion of the problem took place in 2009 [12]. Ontologies were defined as a logical theory made up of sets of formulas. It is understood as a representation of vocabulary from a given field or a theory of this field that defines objects, properties and relations between objects. The goal of an ontology is to explicitly formalize the domain language so that it can be used and interpreted unambiguously. The resulting formulas are intended to represent the concept of the world as logically as possible. Most often ontologies are based on First-Order Logic (FOL) [14] or Web Ontology Language (OWL). Formulas are constructed using units, classes, relationship functions, and axioms. Units are used to map the objects that the ontology deals with, classes - sets of features for defining units, functions allow to identify and bind units, relations describe connections and dependencies between those units, and axioms are expressions connecting all the above-mentioned elements.

Ontologies can be divided according to the classification of language into strongly informal, semi-informal, semi-formal and strictly formal [21]. The informal ones include those that have no formal semantics associated with them. That includes Resource Description Framework (RDF), Unified Modeling Language (UML) or Business Process Model and Notation (BPMN). However, the following article discusses formal ontologies - those that are related to the formal semantics of the language, because it enables their formal interpretation. Such languages include First Order Logic (FOL) and Web Ontology Language (OWL), which contain clear and comprehensive rules of syntax and semantics.

This group is considered to be one of the most reliable languages in the world of technology.

Ontologies can also be categorized according to the scope they cover. We distinguish high-level, reference, domain and application ontologies [15]. Higher-level ontologies describe a wide range of the world, such as Suggested Upper Merged Ontology (SUMO) [5] which covers descriptions of objects, events, high-level relationships, states, belonging, and quality. Reference ontologies focus on a given domain and the description of its components. It is used in the fields of medicine [10], engineering [18] and entertainment [19]. When an ontology focuses on an even more limited scope (e.g. tourism production), it belongs to the domain group. Application ontology deals only with the description of the theory used in a given application, such as CAD/CAM or ERP.

The scope of inference that the robot must perform is recognizing, categorizing, decision-making, perceiving and assessing situations, anticipating and observing, problem solving and planning, as well as reasoning and holding beliefs, performing actions, interacting and communicating, remembering, reflecting and teaching.

Frameworks used in autonomous robots using the knowledge representation approach include for example, KnowRob, IEEE-ORA, ROSETTA, CERESSES [7] and RehabRobo-Onto [11].

3.3 Ontology Standardization

Robotic systems need to meet both hardware and functional standards. Many of them result from safety or operational standards established by international organizations. Robot ontology standards such as "Robot Standards and Reference architecture" or "Ontology-based unified robot knowledge for service robots in indoor environments" [20] have also been designed, but they are not widely used.

3.4 ROS

ROS (Robot Operating System) is one of the most widely used middleware in robotics. It transforms software components and communication graphs into nodes, each of which listens to or sends messages and offers a service invoked by other nodes. Messages are described with a defined syntax and generated for a given language syntax, e.g. Python, C++ by ROS.

3.5 KnowRob

The basic framework used in robotic knowledge databases is the popular KnowRob system. It is a system designed for service robots. The main programming language used in the KnowRob system is SWI Prolog. This language belongs to the First-Order Logic group of languages and includes a library for managing RDF (Resource Description Framework) tuples. In the form of these tuples, the KnowRob system represents knowledge written as facts in an OWL ontology.

The next generation of KnowRob - KnowRob2 - focuses on simulation, rendering techniques and a hybrid knowledge processing architectures.

The analysis and understanding of knowledge consists in collecting the transmitted facts and data and then integrating them. It uses virtual knowledge databases, thanks to which it processes information on the basis of knowledge structures and relations, the processing of which is defined in the adopted ontology.

One of the flaws of the KnowRob system is the shallow representation of the symbolic principle of the behavioral approach to robots. Another much bigger disadvantage is the lack of the proposed representative standards and, although it is one of the most widely used systems, the user community is very small, so without sufficient documentation and with emerging problems, each user relies on their own knowledge or searching untested sources.

3.6 ARMOR

Armor is a framework for managing one or more ontologies within ROS. It allows users to use an ontology in robotics without knowing Java, which is needed to use AMOR.

4 Documentation

4.1 KnowRob

Framework KnowRob has documentation available in two different places: on its own website [3] and on the github repository site [4]. Documentation which is placed on the github repository site has an advantage in the form of its topicality - it is always compatible with the available framework version. However, this documentation contains very limited information and does not provide descriptions of functions' advanced usage. Framework elements widely described on KnowRob's page are mostly deprecated and do not apply to newer versions. Information might be partially usable, but its proper functionality is not guaranteed. For example, function owl_parse\1 which is included in documentation is not available in the latest version.

4.2 ARMOR

The ARMOR framework's disadvantage is its documentation. ARMOR does not have a documentation aside from the one posted on the github repository. It is very limited and contains little information beyond basic usage of the methods. ArmorPy's documentation is located on an external website [1]. It's also very limited and does not contain any use cases.

Table 1 shows documentation quality evaluation based on authors' experience installing, configuring and experimenting described in this article.

Table 1. Tools' documentation quality. '–' signifies a low-quality documentation, '– –' signifies a very low-quality documentation or its absence.

Tool	Documentation quality
KnowRob	–
ARMOR	– –

5 Used Technologies and Configuration

5.1 KnowRob

KnowRob system installation was carried out according to the instructions in the github repository [4]. As required, ROS, SWI Prolog, Mongo DB server and Rosprolog were installed first, however the Rosprolog package turned out to be problematic. Absence of the SWI-Prolog package was being detected during the catkin_make command execution, which was causing errors. It was an unexpected problem, because the missing package was previously installed. The solution to the problem turned out to be the removal of the workspace folder's contents and reinstalling them the same way.

ROS. ROS's melodic version is required, which forces Ubuntu users to use 18.04 version of the OS. Due to two new released stable system versions, it is very unlikely that 18.04 will be the primary system on a user's computer. For this reason the user is forced to use a shared system or a virtual machine, which causes the loss of performance and ongoing support for the OS.

In the description of the installation process on the official ROS website [6], both required and additional steps are included, although they lack clear distinction. For a person installing this tool for the first time it may result in mistakes during the process, which might result in a situation where a full re-installation is necessary.

5.2 ARMOR

The installation of the ARMOR framework was conducted according to the information published on author's website [1]. As described, ROS (kinetic version), RosJava (kinetic version), AMOR and extension package to AMOR: AMOR services, were installed. Additionally, ARMOR message interface and ARMOR Python API were installed. Unfortunately ARMOR has not been updated since 2019, which resulted in the need of usage of a system on which the framework still works - Ubuntu 16.04 Xenial Xerus.

ROS. ROS is not officially listed as required for ARMOR's installation, however its first requirement is the RosJava package, which cannot be installed without it. Since ARMOR framework has not been updated, kinetic version of ROS is necessary. The user is being led through the installation process efficiently and instructions do not contain glaring inaccuracies. However, the description, same as with the melodic version, does not distinct between required and additional steps clearly enough.

RosJava. RosJava is a ROS implementation in Java. It is a required component in the ARMOR installation process. The instructions are mostly correct, but the user is mislead in the following part while executing the command:

```
rosdep install --from-paths src -i -y
```

RosJava authors suggest to ignore warnings about some packages' absence, but use of the ARMOR framework is not possible without them, making manual installation of listed packages necessary:

```
ros-kinetic-move-base-msgs
ros-kinetic-world-canvas-msgs
ros-kinetic-scheduler-msgs
ros-kinetic-rocon-tutorial-msgs
ros-kinetic-rocon-interaction-msgs
ros-kinetic-rocon-device-msgs
ros-kinetic-rocon-app-manager-msgs
ros-kinetic-gateway-msgs
ros-kinetic-concert-service_msgs
ros-kinetic-concert-msgs
ros-kinetic-yocs-msgs
ros-kinetic-ar-track-alvar-msgs
```

ARMOR. Armor is an AMOR framework extension, therefore installation of both environments is required. Many problems are caused by the general lack of installation instructions. It is not mentioned which github repositories are required to launch the environment and which are not. This resulted in downloading all packages listed in the instruction. Instruction was placed on github repository [2] describing ARMOR functioning examples. After downloading all packages and installing by the catkin_make command, the environment is ready for use.

5.3 Installation Difficulty Evaluation

Table 2 aggregates installation difficulty level of tools used for this article.

5.4 Configuration

Applied Knowledge Base Configuration. Both KnowRob and ARMOR were used along with the KnowRob2 ontologies - unit.owl and knowrob.owl, which contain OWL, RDF-Schema, URDF, Qudt, Quantity, Dimension, SOMA and IOLite ontologies references.

Table 2. Tool installation difficulty level. '+' signifies moderately easy installation, '–' signifies moderately hard installation.

Tool	Installation difficulty
KnowRob: ROS	+
KnowRob: RosProlog	–
ARMOR: ROS	+
ARMOR: RosJava	–
ARMOR: ARMOR	–

KnowRob. In KnowRob's case unit.owl, knowrob.owl and related ontologies are autoloaded by roslaunch server startup or on CLI mode startup using rosrun thus no additional configuration is needed.

ARMOR. To launch ARMOR, the roscore service needs to be ran in the first terminal window using the command 'roscore'. Roscore enables ROS nodes communication. Afterwards, 'rosrun armor execute it.emarolab.armor.ARMORMainService' has to be ran in another terminal window for the ARMOR service to launch. In a third terminal window the ontology should be loaded, which can be done in two ways:

1. By terminal and rosservice call.

```
rosservice call /armor_interface_srv "armor_request:
  client_name: 'terminal'
  reference_name: 'ref1'
  command: 'LOAD'
  primary_command_spec: 'FILE'
  secondary_command_spec: ''
  args: ['/home/user/catkin_ws/src/Mydir/owl/test.owl',
  ↪ 'http://www.IRIs.org/test'
  ]"
```

2. By a Python program using functions provided by ARMOR client interface.

```
client = ArmorClient("client", "reference")
client.utils.load_ref_from_file(path + "test.owl",
↪ "http://www.IRIs.org/test")
client.utils.mount_on_ref()
client.utils.set_log_to_terminal(True)
```

6 Experiments

The experiments focused on the creation of a simulation environment, passing an example state of the environment, and the analysis of the knowledge base through analysis of responses to queries about objects in the environment. For all these cases an example part of a typical kitchen environment was created and knowledge about a pair of kitchen objects, a cup and a cupboard, in a form of triples was added. The knowledge structure is presented on Fig. 1.

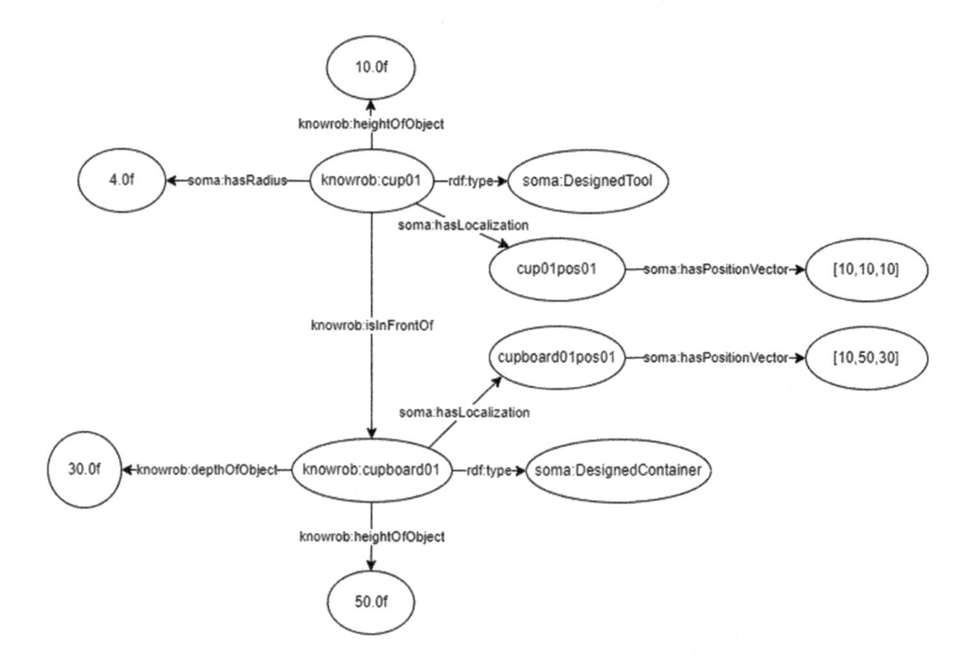

Fig. 1. Applied ontology describing a part of a typical kitchen environment.

To carry out the aforementioned experiments 3 queries were made and repeated 1000 times each in order to check the consistency of the performance. Query 1. checks the class of the cup object 'cup01'. Query 2. checks the relation between 'cup01' and the cupboard 'cupboard01'. Query 3. checks the position of the cupboard 'cupboard01' in the environment.

6.1 KnowRob

The information about the kitchen environment was written down as RosProlog queries and input into the knowledge base by execution of a python script. For this operation KnowRob's function 'rdf_db:rdf_assert()' was used. Afterwards, different RosProlog queries were ran 1000 times, also via the KnowRob framework, in order to check the state of the environment and the response times.

```
1. instance_of('cup01', A)
2. holds(knowrob:'cup01', B, knowrob:'cupboard01')
3. holds(knowrob:'cupboard01', soma:'hasLocalization', _L),
    holds(_L, soma:'hasPositionVector', Vec)
```

Table 3 presents the results of the queries to the KnowRob framework.

Table 3. The results of running 1000 queries to KnowRob (ms).

Query	Average time	Standard deviation	Variance
1	1.089	0.262	0.069
2	6.040	6.599	43.555
3	7.857	9.136	83.460

Query 1. asked only for the class of 'cup01' and took only about 1 ms to finish, while queries 2. and 3., which check the relations between objects and the location of one of them, took several times longer to complete.

6.2 ARMOR

The ARMOR framework allows for running queries in two ways:

1. Via terminal and rosservice call with SPARQL queries.

```
rosservice call /armor_interface_srv "armor_request:
  client_name: 'terminal'
  reference_name: 'ref'
  command: 'QUERY'
  primary_command_spec: 'SPARQL'
  secondary_command_spec: ''
  args: ['
PREFIX knowrob: <http://knowrob.org/kb/knowrob.owl#>
PREFIX soma: <http://www.ease-crc.org/ont/SOMA.owl#>
SELECT ?a ?b WHERE {
  ?a ?b soma:DesignedContainer
}'
]"
```

2. Via a python script using functions of the ARMOR client's interface.

```
client.query.check_ind_exists("knowrob:cup01"))
client.query.dataprop_b2_ind("knowrob:isInFrontOf",
↪   "knowrob:cup01"))
client.query.dataprop_b2_ind("soma:hasLocalization",
↪   "knowrob:cupboard01"))

print(client.query.check_ind_exists("knowrob:cup01"))
```

Queries which were ran through the terminal in the SPARQL language caused suspension of the terminal and the knowledge base. Because of this, the Python module was chosen as the method for running queries. The Python module received the data correctly, but when it came to querying it, it was behaving nondeterministically, which was caused by the suspension of execution of the Python script. Sometimes the whole script would be executed, but most of the time the execution of the script and the knowledge base were getting suspended and only the beginning of the script was being properly executed. This behaviour made collection of the necessary comparison data impossible.

7 Conclusion

The conducted TRIZ analysis indicates what the specific needs in relation to the designed companion robot system are. Analyzed tools: KnowRob and Armor were identified by reducing the problem to a generic form and formulating generic solutions. Table 4 presents the overall evaluation of the tools' components based on what has been experienced through tests in this work.

Table 4. Tools' components' evaluation. '+ +' signifies a high-quality component, '–' signifies a low-quality component, '– –' signifies a very low-quality component.

Tool	Documentation	Installation	Query responsiveness
KnowRob	–	–	+ +
ARMOR	– –	–	– –

Due to implementation problems, lacking documentation and nondeterministic nature of the ARMOR framework, tests could not be carried out on both frameworks and compared. The presented results refer to only the KnowRob framework, which allowed for repeated querying. Queries in the ARMOR framework resulted in both returning valid responses and invalid execution errors. Due to this reason, quantitative analysis was impossible.

KnowRob carries out queries correctly. In response to them, it returns the correct class of an object, relations with another object and relational position, where querying the relation between objects takes considerably longer than a simple class checking query. In the context of carried out tests, queries are very fast, but in a real-life scenario a service robot faces a very large quantity of both simple and more complex queries, making the responsiveness a very important aspect of the framework.

In terms of documentation and available support, the KnowRob framework has a considerably bigger amount of materials available. The amount of users also contributes to the overall shared knowledge and experience, making implementation less time consuming and better tested.

Documentation of the ARMOR framework is lacking. The framework, which had the potential of being more widely used, looks to had been abandoned and is now being used mainly by users strongly determined or those who are forced to for other reasons.

References

1. Armor API documentation. http://emarolab.github.io/armor_py_api/armor_api.html
2. Armor GitHub. http://emarolab.github.io/armor_py_api/armor_api.html
3. KnowRob documentation. http://knowrob.org/doc
4. KnowRob GitHub. https://github.com/knowrob/knowrob
5. Ontology portal - sumo. http://www.ontologyportal.org/
6. ROS ubuntu installation guide. http://wiki.ros.org/melodic/Installation/Ubuntu
7. Caresses project. http://caressesrobot.org/en/
8. Antoniou, G., van Harmelen, F.: Web ontology language: OWL. In: Staab, S., Studer, R. (eds.) Handbook on Ontologies. INFOSYS, pp. 67–92. Springer, Heidelberg (2004). https://doi.org/10.1007/978-3-540-24750-0_4
9. Beetz, M., Beßler, D., Haidu, A., Pomarlan, M., Bozcuoğlu, A.K., Bartels, G.: Know rob 2.0 - a 2nd generation knowledge processing framework for cognition-enabled robotic agents. In: 2018 IEEE International Conference on Robotics and Automation (ICRA), pp. 512–519 (2018). https://doi.org/10.1109/ICRA.2018.8460964
10. Burgun, A.: Desiderata for domain reference ontologies in biomedicine. J. Biomed. Inform. **39**(3), 307–313 (2006). https://doi.org/10.1016/j.jbi.2005.09.002. https://www.sciencedirect.com/science/article/pii/S1532046405000997. Biomedical Ontologies
11. Dogmus, Z., Erdem, E., Patoglu, V.: RehabRobo-Onto: design, development and maintenance of a rehabilitation robotics ontology on the cloud. Robot. Comput.-Integr. Manuf. **33**, 100–109 (2015). https://doi.org/10.1016/j.rcim.2014.08.010. https://www.sciencedirect.com/science/article/pii/S0736584514000714. Special Issue on Knowledge Driven Robotics and Manufacturing
12. Haidegger, T., et al.: Applied ontologies and standards for service robots. Robot. Auton. Syst. **61**(11), 1215–1223 (2013). https://doi.org/10.1016/j.robot.2013.05.008. https://www.sciencedirect.com/science/article/pii/S092188901300105X. Ubiquitous Robotics
13. Mann, D.: Business Matrix 3.0: Solving Management, People & Process Contradictions. IFR Press (2018)
14. Mendelson, E.: Introduction to Mathematical Logic (1987). https://doi.org/10.1007/978-1-4615-7288-6
15. Menzel, C.: Reference ontologies - application ontologies: either/or or both/and? (2003)
16. Olivares-Alarcos, A., et al.: A review and comparison of ontology-based approaches to robot autonomy. Knowl. Eng. Rev. **34**, e29 (2019). https://doi.org/10.1017/S0269888919000237
17. Olszewska, J.I., et al.: Ontology for autonomous robotics. In: 2017 26th IEEE International Symposium on Robot and Human Interactive Communication (RO-MAN), pp. 189–194 (2017). https://doi.org/10.1109/ROMAN.2017.8172300
18. Ruy, F., Guizzardi, G., Falbo, R., Reginato, C., Dos Santos, V.A.: From reference ontologies to ontology patterns and back. Data Knowl. Eng. **109**, 41–69 (2017). https://doi.org/10.1016/j.datak.2017.03.004

19. Sikos, L.: VidOnt: a core reference ontology for reasoning over video scenes. J. Inf. Telecommun. **2**, 1–13 (2018). https://doi.org/10.1080/24751839.2018.1437696
20. Suh, I.H., Lim, G.H., Hwang, W., Suh, H., Choi, J.H., Park, Y.T.: Ontology-based multi-layered robot knowledge framework (OMRKF) for robot intelligence. In: 2007 IEEE/RSJ International Conference on Intelligent Robots and Systems, pp. 429–436 (2007). https://doi.org/10.1109/IROS.2007.4399082
21. Uschold, G.: Ontologies: principles methods and applications. Knowl. Eng. Rev. **11**(2), 93–136 (1996)

Combing TRIZ and LCA for a Better Awareness of the Sustainability of a Technical Solution

Christian Spreafico(✉), Davide Russo, and Daniele Landi

University of Bergamo, Viale Marconi 5, 24044 Dalmine (Bg), Italy
christian.spreafico@unibg.it

Abstract. A reflection about the evaluation of the environmental impacts arising from a technical solution obtained by applying some of the most common TRIZ (Russian acronym for Theory of Inventive Problem Solving) strategies is provided in this study. In fact, some of them provide suggestions to minimize the resources and make a device work better without adding additional substances or energy flows. However, the contained shortcomings for improving the environmental sustainability can only be fully understood only when applying a quantitative assessment such as Life Cycle Assessment (LCA). This was done in this study, by considering a selection of TRIZ strategies and collecting their pros and cons about environmental sustainability by applying LCA. To do this, the discussion of each strategy was supported by exemplary case studies about Comparative LCA, collected from the scientific literature. The intent of the authors is not to bring experimental evidence, but to provide a further and preliminary judging method to select the TRIZ strategies. In this way, problem-solvers can also base their choice on environmental sustainability.

Keywords: TRIZ · Eco-design · Life Cycle Assessment (LCA)

1 Introduction

For several years, many researchers around the world have been trying to propose the use of TRIZ [1] and its tools within eco-design. Some authors (e.g. [2]) improved the design phase through mass reduction and geometry modification, while others (e.g. [3]) reduced the consumption during the use phase. Still others (e.g. [4]) ameliorated the product disassembly and dismantling during the end-of-life. Some efforts have also been spent in customizing TRIZ to favor its integration with eco-design. As consequence, frameworks (e.g. [5]), customizations of its tools (e.g. [6]) and novel usage modalities of the entire methodology (e.g. [7]) were proposed. [8] provided a new interpretive key of the TRIZ methodology from an environmental point of view, by distinguishing which tools and principles are most suitable for Eco-design and which must be properly adapted. This study was a first attempt to customize the TRIZ tools in relation to some typical eco-design problems and to integrate specific knowledge research databases to support eco-design (e.g. materials database). [9] and [10] proposed a new modality to use TRIZ

Published by Springer Nature Switzerland AG 2022
R. Nowak et al. (Eds.): TFC 2022, IFIP AICT 655, pp. 328–339, 2022.
https://doi.org/10.1007/978-3-031-17288-5_28

evolutive trees and Macro-micro trend to organize the knowledge about different pyrolysis techniques. The relation between the level of technological development and the environmental sustainability was investigated in these studies. The obtained framework could be a base for a problem-solving activity based on TRIZ, by providing the main elements to formulate the more strategic contradictions from an environmental perspective. [11] introduced a rigorous ontology to support the application of a specific TRIZ strategy to solve a specific problem, trying to make the user aware of the environmental consequences of her/his design choices. This ontology was organized in steps, following the LCA structure about the lifecycle phases (i.e. pre-manufacturing, manufacturing, use and end-of-life). The items generating the environmental impacts (e.g. logistic, maintenance, packaging) were also included, as well as TRIZ trends and principles (e.g. segmentation, local quality, trimming). The result of this work is a set of 59 guidelines that support designers in applying TRIZ in a targeted manner during eco-design. [12] provides a quantitative evaluation of some TRIZ-strategies based on LCA and involving only case studies, even if without providing a depth discussion about the method application in eco-design.

However, despite the best efforts, some gaps in the literature still need to be filled. The identification of the real benefits arising from TRIZ application on the reduction of the environmental impacts has yet to be specified. According to [2, 13] and [14], the application of TRIZ in eco-design can also lead to substantial environmental benefits even if some environmental criticalities emerge. This is because general-purpose idea generation tools do not usually show any specific preference to sustainable aspects. Therefore, the attention to sustainability is random and demanded to designers' sensibility about environmental and social issues.

However, without an objective assessment of the results of TRIZ application, its advantages in eco-design cannot be confirmed. Only very few authors expressed about this issue, mainly by proposing qualitative or partial judgments. [15] positively assessed the application of TRIZ in small and medium-sized enterprises. Experiments aimed at demonstrating its ability to find notoriously greener solutions were proposed in this study. [3] confirmed TRIZ ability to reduce the consumption during the functioning of the product even if without determining the resulting reduction of the environmental impacts. In addition, methods for measuring eco-innovation also supported by non-TRIZ methods (e.g. [16]) were also proposed, although without quantifying the environmental impacts of the resulting solutions. Consequently, the main limitation of the discussion regarding the applicability of TRIZ in eco-design is the lack of a metric and indicators to effectively evaluate whether and how TRIZ is able to reduce environmental impacts.

The main novelty of this paper is the introduction of the Life Cycle Assessment (LCA) methodology as a tool for evaluating the efficacy of some TRIZ strategies in eco-design. Their main advantages and disadvantages in reducing environmental impacts are discussed in relation with items (e.g. materials, energies) and phases (e.g. pre-manufacturing, manufacturing, use, end-of-life) of LCA. For each TRIZ strategy, an example was chosen from among the many available to highlight the specific aspects related to LCA, which should be considered to carry out its eco-evaluation. All the examples were selected in such a way as to provide a broad analysis overview both on the fields of application and on all the topics of LCA (e.g. physical features, energy flows, logistic, electricity mix).

The main objectives of this study are two: (1) Verifying whether the results of the strategies are actually also the most sustainable in relation to the entire life cycle. (2) Investigating the correlations between the environmental advantages ensured by the strategies, the lifecycle phases and the environmental impact indicators (e.g. global warming, acidification, particulate matter formation).

2 Methodology

In this study, rather than reviewing all the TRIZ tools individually, a selection of strategies was analyzed, mostly transversal to multiple tools, which can be more geared towards eco-design. The considered TRIZ strategies are: (1) Improving the local quality, (2) Exploiting the resources, (3) Changing the state of aggregation, (4) Increasing control (macro-micro), (5) Increasing control (nesting), (6) Dematerialization/ideality.

Each strategy has been analysed by following the same 3-step divided methodology, summarized in Fig. 1.

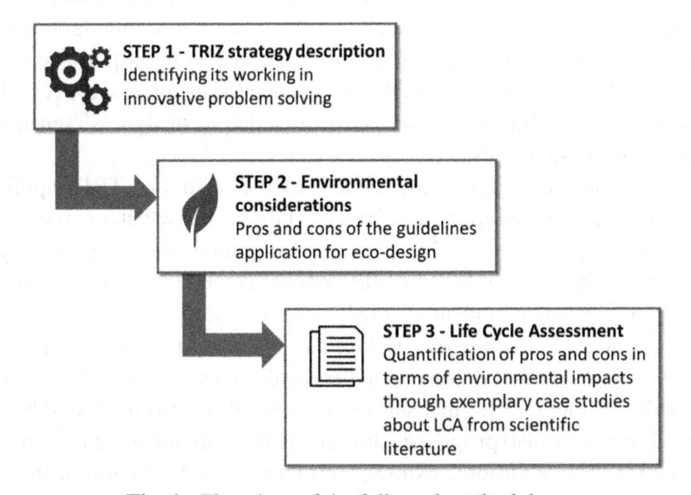

Fig. 1. Flowchart of the followed methodology.

The steps of the followed methodology are:

1. Providing a brief description of its founding philosophy and its modality of application in the context of innovative problem-solving in general.
2. Investigating the specific repercussions of the TRIZ strategies in eco-design with reference to the advantages and disadvantages on the reduction of environmental impacts. These latter were found in the LCA studies from the scientific literature, associated with the considered TRIZ strategies.

3. Presenting a case study of Comparative LCA from the scientific literature, in which two products performing the same task are compared, with reference to the same functional unit and in the same operational scenario. Among them, the most advanced product was seen as the result of the application of one of the TRIZ strategy on the other product. The typical LCA metric with its most common indicators is used to evaluate the effectiveness of the TRIZ strategy in relation to the objective of environmental sustainability: climate change (CO_2 eq.), Terrestrial and marine acidification (SO_2 eq.) and particulate matter formation (PM_{10}).

The case studies from the literature were carefully selected by following these criteria:

- All the considered papers propose a Comparative LCA between at least two different products.
- Exhaustive descriptions of the compared products, including all the data used during the LCA, were provided in the papers.
- The two considered products perform the same function in the same operating scenario, with a common and well-defined functional unit.
- The same calculation procedure was followed to assess the environmental impacts.
- The results were expressed through the same set of impact indicators.

The association of the results from the literature and the different TRIZ strategies were performed by the same authors. In this case, we exploited the TRIZ ontology (i.e. the Minimal Technical System model) as a linkage to research the application of the TRIZ strategies to describe the differences between the two compared products. In other words, the second product is seen as an improvement of the first one resulting from the application of one of the considered TRIZ strategies.

In the following sections the different strategies are described in detail.

3 Evaluated TRIZ Strategies

3.1 Improving the Local Quality

This strategy derives mainly from the TRIZ Principle N. 3 - Local quality, by also involving the concept of the operational area and the Principle of Separation in Space to favor its more targeted application during eco-design. The improvement of the local quality is used for reducing mental inertia when considering space and materials too homogeneous and continuous. It suggests to "Change an object's structure from uniform to non-uniform, make each part of an object function in conditions most suitable for its operation, and make each part of an object fulfil a different and useful function".

Environmental Consideration. The advantages related to the increase of the local quality of a component may depend on the reduction of the mass and the energy or resources used by the same component. The more targeted and optimized use of the component only where and when is needed is also more economically convenient.

The reduction of the environmental impacts arising from the reduction of the expended energy and/or the mass to be produced, moved, and disposed. In particular, the impacts of disposal can be particularly significant especially in the case of auxiliary materials such as lubricating and cooling fluids, oils, and paints. In addition, the improvement of the local conditions of an auxiliary material can also lead to benefits for environmental sustainability. In fact, the useful life the product can be increased and the consumptions can be reduced.

On the other hand, through the application of this strategy, the introduction of additional processes to increase the quality of the component, e.g. a longer milling in the areas where better roughness must be ensured, can increase the impacts. In addition, higher energy consumption during the product use should also be considered to ensure the local increase its performances. For instance, a more powerful pump to increase the flow speed of a lubricating fluid in the area to be cooled. Sometimes, to increase local quality, inserts or additives, difficult to be disposed, are also introduced into the technical system, e.g. the abrasive water jet which contains solid particles.

Example: " cutting fluids". The evolution of cutting fluids in recent years points towards increasing their heat removal performances in the cutting area. This is possible through the increase of the local flow velocity and pressure, to concentrate their action more effectively within the cutting area.

The comparative LCA of two cutting fluid systems (conventional vs high pressure), proposed by [17], shows the validity of the increase of the local quality of the second compared to the first one. A reduction of 25% of global warming potential (CO2 eq.), 30% of acidification potential (SO2 eq.), and 25% of the equivalent energy consumption (MJ) over the entire life cycle were estimated. These advantages were due to the drastic reduction of the impacts of cutting, since through the introduction of the more advanced cutting fluid system, the productivity of the machinery increases. At the same time, the energy consumption is maintained, which abundantly compensates for the increase in the impact of electricity consumption required by the new high-pressure pump of the cutting fluid.

3.2 Exploiting the External Resources

This strategy mainly derives from the TRIZ Law of Technical System of Evolution N. 6 - Switch to supersystem, which points out the use of the resources available in the external environment. From an environmental point of view, the product reaches the minimum consumption of its energy resources, drastically reducing its environmental impact. The use of external resources usually does not contribute to increase the environmental impacts since they are already present in the environment and they are unexploited.

According to this law, when a technical system evolves by exhausting the possibilities of further significant improvements is included in a super-system as one of its parts. As a result, new development of the system become possible. When thinking of resources, in this strategy the energy consumed during either the product use or its manufacturing was highlighted. However, TRIZ considers more resources than those typically considered according to the environmental perspective. A special attention is provided also to to those resources that are commonly considered as negative, e.g. hot exhaust fumes.

Environmental Considerations. The environmental advantages arising from a greater exploitation of the resources and directly ascertainable through an LCA study are different. Among them, one of the most obvious concerns the reduction of energy consumption, by replacing or integrating the most impacting energy sources with natural resources. Examples are the solar cooling cycles with trigeneration and the reuse of the waste energy in order to reduce energy consumption in turbocharged engine and the cogeneration. Furthermore, there is a reduction/elimination of the conventional energy system and its components, e.g. the boiler size in a domestic heating system that also combines solar thermal. Finally, there is also the reduction of the impacts of materials if resources are used to reduce the mass of the product, e.g. plastic components loaded with shavings and scrap metals to replace the virgin metal.

On the contrary, the introduction of components dedicated to the exploitation of the resources, especially the natural ones, in some cases may 'also include materials more difficult to be disposed. For instance, the impacts of photovoltaic modules could be significant and difficult to be mitigated, especially for some specific indicators, e.g. toxicity. On the other side, they can also lead to large reductions of other indicators, such as CO2 eq. Furthermore, the exploitation of renewable resources is still not sufficient for many systems to fully perform their functions, which must therefore resort to less optimized and sometimes more impactful back-up systems than an ad hoc system.

Example: "Wastewater Elimination". In the context of wastewater elimination from manure, a recent evolution of traditional digestion systems is based on gas injection inside the silos where the manure is stored. This is possible through the use of algae activated by sun which have a more effective chemical action compared to the injected gas [18].

The comparative LCA of the two wastewater elimination systems in a pig farming proposed by [19] resulted in a reduction, going from the first to the second technology, of 31% of the global warming potential and of about 5% of the marine and terrestrial eutrophication, over the entire life cycle. These advantages mainly depend by the drastic reduction of the environmental impacts arising from the preparation of the digesters, passing from the gaseous mixture, prepared ad hoc, to algae, available in nature and processed in a much more sustainable way. Secondly, a further advantage is obtained through the elimination of the gas injection circuit and the energy consumption for its handling, by using solar radiation to activate the algae. The disadvantages to switch to the algae concern the partial reduction of the fertilizing power of the manure at the end of the process. This requires an overproduction of the same to fulfil the same function and involves a minimum increase of the impacts, which are below 1% in both impact indicators.

3.3 Changing the State of Aggregation

This strategy embodies part of the founding philosophy of TRIZ by suggesting the designer to exploit the resources obtainable from the change in the state of aggregation of a substance. This concept is included in the TRIZ Principle N. 35 - Parameters change and in TRIZ Principle N. 36 - Phase transition. Its application does not necessarily require a phase change, since Principle N. 36 suggests to exploit also the effects deriving from the phase transition, such as the latent heat released, but is also limited only to the simple substitution. For instance, a hard shell that does not adequately absorb shocks can be replaced by bubble wrap.

Environmental Considerations. From an environmental point of view, the advantages of the change of the state of aggregation depend by different aspects. The reductions of the volume of the product can optimize the logistics and reduce its impact. The energy consumption during manufacturing can be reduced, e.g. obtaining a component from foundry casting is cheaper than machining it from solid to machine tools. The environmental contamination can be reduced, e.g. if the oil escaping from a tanker stranded in the rocks could be solidified, it would undoubtedly be easier to recover. On the other hand, in the analysis of the impacts, the energy spent on heating or cooling during the phase transition must also be considered, if it is assumed to use this option. In addition, it could be an increase in environmental contamination, e.g. a solid fuel is usually more explosive than a liquid one. Finally, the construction of plants, structures and the energies dedicated to maintaining a substance in a state of aggregation other than the natural one in the environment should also be considered. For instance, cryogenic tanks with large thicknesses and coatings and refrigeration cycles.

Example: "CO2 sequestration". In recent years, the evolution of CO2 sequestration systems, inside the special collection tanks, has seen the alternation of solid membrane adsorption systems with gaseous oxidizing agents. The main advantage deriving from the adoption of gaseous oxidizing agents concerns the increase in efficiency in intercepting the CO2 molecules within the entire volume of the tank. While the solid membranes, which, however densely arranged, always leave empty volumes. In addition, with gaseous oxidizing agents, processing times are also reduced since the use of gas avoids the replacement of the membranes once saturated, ensuring continuous cycle operation. The comparative LCA of the two systems proposed by [20], confirms the advantages of the technology with gaseous oxidizing agents also from the environmental point of view. A reduction of the impacts, compared to the membrane system, of 73% of CO2 eq., of 36% of SO2 eq., and of 36% of PM10 eq. Over the entire life cycle can be reached in this way. The reasons for this reduction in impacts are mainly due to the elimination of the production of the membranes, which consist of particularly impacting material (e.g. activated carbon). In this case, the avoided impacts abundantly compensate for those deriving from the energy spent for the production and the supply of the gaseous oxidant inside the reactor, as well as the construction of its recirculation plant.

3.4 Changing the State of Aggregation

This strategy summarizes the philosophy of the static TRIZ laws of technical system of evolution, which suggests that the evolution of a system is often tied to the evolution of its control. According to this law, any working system must have four parts: the supply, the transmission, the tool (working organ) and the control element. The supply generates the needed energy, the transmission guides this energy to the tool, which ensures contact with the outside world (processed object), and the control element makes the system adaptable. According to the same law, among the four elements, the control element is where innovation can make the biggest leap. One way to increase control is by miniaturizing the tool to reduce its interaction with the object on which it acts, from a macro level to a micro level. This is possible by reducing the interface between tool and object and by changing the interaction mode, from mechanical to electromagnetic. In this way, the action of the tool on the object can go from superficial/massive to molecular: for example, the drilling with a laser tip instead of a common drill bit.

Environmental Considerations. Often, increasing control over the system and the tool, or replacing the operating principle of the system with a more controlled one, could require the technological complication of the system. This is due to the introduction of a new structure that has a more impact on the environment. Although, in most cases, the increase in environmental impacts is justified by the greater efficiency of the use phase which is more controlled in execution and therefore less energy intensive. When the miniaturization of the tool suggests the transition from a mechanical to an electromagnetic mode. This presupposes the change of the energy source from fossil to electric, then the sustainability of the latter in relation to the operating scenario must be carefully considered. In fact, unlike fossil sources (e.g. natural gas or oil) which have emission factors due to their extraction and distribution substantially stable in various parts of the world, in the case of electricity the specific impact per kWh produced is much more variable from state to state, regions and provinces. It depends on the electricity mix, i.e. on the characteristic distribution of the means of electricity production (i.e. coal, natural gas, nuclear, renewable). Other factors to be considered are the technological level of the used plants, the environmental characteristics and the political factors. However, recently, much progress has been made to reduce electricity generation emissions and forecasts call for its greater use in the future as a replacement for fossil energy sources.

Examples: "Gas Hob vs Induction Hob". Induction hobs are a more advanced than gas hob and are believed to replace them in the coming decades. However, today their diffusion in many states is comparable or even much higher than the first ones.

The induction hob, thanks to its operating principle, i.e. electromagnetic induction generated by wound electric coils, can guarantee a much more precise control on the energy consumption. This is due to higher heating/cooling ramps, as well as greater efficiency in energy transmission. The comparative LCA proposed by [21] between gas hob and induction hob, shows the environmental limitations of the induction technology (compared with gas) in Italy. The reason is the disadvantageous electricity mix, since the energy consumption of the induction hob is significantly lower. The authors determined an increase in the impacts on the entire life cycle of the induction hob compared to gas hob, by 51% of CO2 eq., in this scenario. However, the situation is different in France, where the electricity mix is clearly more favorable. In this case, the induction hob is more sustainable than the gas hob over the entire cycle of life, with a reduction of CO2 eq. Equal to 58%.

3.5 Increasing Control (Nesting)

Another way to increase the control over the technical system is to introduce additional elements with active function in the realization of the function and/or sensors with monitoring function. Often when there is no other possibility, their introduction must be studied carefully. Among the suggestions that TRIZ provides us in this regard, there is also Principle N. 7 - Nesting.

Environmental Considerations. From the environmental point of view, the introduction into the system of additional components and sensors must deal with the impacts due to their production, use, disposal. In addition, also control or data acquisition structure must be considered. Furthermore, the introduction of additional components within the materials of the product makes its disposal more impactful, due to the complications arising from its disassembly. As consequence, there is the possibility of invalidating the recycling of the constituent materials, especially when particles are embedded within the material and welded by chemical bonds (e.g. doping). Therefore, in order to achieve environmental sustainability through the application of this strategy, the beneficial control action carried out by additional components must reduce the impacts on the product life cycle to compensate for their increase due to their introduction. This can therefore be ensured by the control over the operating mode of the product, to increase efficiency and reduce consumption or to reduce waste, by extending the operational life of the product.

Example: "Cheese Packaging". In the field of cheese packaging and, more generally, the packaging of highly perishable food products, various strategies have been implemented to increase the shelf life of the product, so as to reduce waste. An evolutionary line points towards active and sensorized packaging, including bactericides (e.g. [22]) and/or micro-sensors of light, pressure and temperature (e.g. [23] within the thickness of the package and in contact with the product. The LCA studies of [24, 25] and [26] showed that the increase in the environmental impacts resulting from the modification of the packaging to increase the shelf life of the contained compensate for those resulting from the waste of food. This fact was confirmed by the introduction of both bactericides and sensors, where the extra impacts associated with them are due both to their production and to the greater difficulty in disposing of the packaging. In many cases, bactericides consist of metal oxides and they affect the recyclability, the decomposition and the compostability of the packaging. The missed impacts due to the extension of the shelf life and therefore to the reduction of food waste were instead calculated by quantifying the missed overproduction and distribution that would also be necessary.

3.6 Dematerialization/ideality

This strategy is mainly related to the kinematic laws of evolution. Among them, TRIZ outlines the steps toward ideality and the dematerialization of the product, which can be translated into the concept of zero impact. The evolution of the answering machine is a good example. The product was eventually dematerialized into a service that is now provided without the need for a dedicated system. According to TRIZ, the dematerialization of a product coincides with its Ideality. The TRIZ Ideal Product is that which does not exist while continuing to perform the function for which it was conceived. It is not associated with any energy flow, material, packaging, etc., and cannot produce environmental impacts. It consists in eliminating all useless product components, reducing toward zero auxiliary materials, consumable. When the complete elimination of the structures is not possible, a partial elimination is then suggested, through a greater involvement of the fields in replacement of some structures, e.g. magnetic coupling.

Environmental Considerations. From an environmental point of view, the most obvious advantages of dematerialization concern the elimination of product components and the impacts related to their production. This is possible though the replacement of the components that are much smaller and with a lower impact during production. Furthermore, the behaviour of the dematerialized technology is generally more efficient, especially when electromagnetic fields are involved. This is because the typical inertias of mechanical mechanisms and fluid-dynamic systems is avoided. This can significantly limit the consumption and the impacts arising from the use phase. However, the new technology could include a fluid in place of a mechanical means, which, if not properly managed, could have a lot of impact during disposal. This is the case of medium and high voltage switches that use Sulphur hexafluoride (SF6) gas to extinguish the electric arc instead of compressed air. In this case, SF6 has a very high impact of over 23500 kg of CO2 eq.

Example: "Metal Cutting". For many years, the evolutionary trend of cutting materials has been outlined with the elimination of mechanical systems (i.e. blades) in favor of the use of electromagnetic waves, with electro-erosion and lasers. The study of [27] compares the environmental impacts of the various technologies, based on an analysis of several LCA studies proposed in the scientific literature. It emerges from it that, although there are significant differences in terms of environmental impacts in relation to the type of material processed, over the entire life cycle, both EDM and laser prove to be a more sustainable alternative in many cases compared to the mechanical cutting.

Considering the LCA impact categories, the electromagnetic cutting systems are more energy-intensive during the cutting operation, mainly due to their poor operating efficiency. However, the lack of impacts of the elimination of the mechanical tool, of its replacement and maintenance as well as of the dedicated machinery are such as to compensate for this aspect.

4 Evaluated TRIZ Strategies

In this paper, six macro directions of intervention inspired by TRIZ laws and principles have been presented, which more than others help to improve a product by making the use of resources more efficient. According to this logic, an improvement in its environmental sustainability should also be achieved. This impression was also confirmed through a careful analysis about the modalities to present the results in those papers proposing Eco-design methods based on problem-solving, where the link between sustainability and energy efficiency is considered almost automatic. However, the situation is much more complex. In fact, this article demonstrates with a series of practical cases how the evaluation of a problem-solving activity could be challenging, especially if a specific evaluation metric for sustainability, such as LCA is not applied.

In order to contribute to increase the level of awareness in the problem-solver about the effects that a new solution can produce on the product sustainability, a double level of conviction was proposed. First TRIZ suggestions were contextualized according to an environmental perspective. Then, through LCA a product and a variant obtained through the application of a TRIZ suggestion were compared. The planned future developments of this work consist of an extensive analysis of a large number of contributions in order to quantitatively verify whether the observations identified in this study can be confirmed. In addition, new TRIZ strategies and new observations for each of them will be considered to increase the knowledge basis.

References

1. Altshuller, G. S. Creativity as an exact science: the theory of the solution of inventive problems. Gordon and Breach (1984)
2. Vidal, R., Salmeron, J.L., Mena, A., Chulvi, V.: Fuzzy cognitive map-based selection of TRIZ (theory of inventive problem solving) trends for eco-innovation of ceramic industry products. J. Clean. Prod. **107**, 202–214 (2015)
3. Lim, I.: The effectiveness of TRIZ tools for Eco-efficient product design. In: Chechurin, L. (ed.) Research and Practice on the Theory of Inventive Problem Solving (TRIZ), pp. 35–53. Springer, Cham (2016). https://doi.org/10.1007/978-3-319-31782-3_3
4. Cherifi, A., Dubois, M., Gardoni, M., Tairi, A.: Methodology for innovative eco-design based on TRIZ. Int. J. Interact. Des. Manuf. (IJIDeM) **9**(3), 167–175 (2015). https://doi.org/10.1007/s12008-014-0255-y
5. Livotov, Pavel, et al.: Environmental problems and inventive solution princi-ples in process engineering. In: Proceedings of the 18th TRIZ Future Confer-ence "Towards automated inventions for smart industries (2018)
6. Yang, C.J., Chen, J.L.: Forecasting the design of eco-products by integrating TRIZ evolution patterns with CBR and Simple LCA methods. Expert Syst. Appl. **39**(3), 2884–2892 (2012)
7. Chou, J.R.: An ARIZ-based life cycle engineering model for eco-design. J. Clean. Prod. **66**, 210–223 (2014)
8. Russo, D., Serafini, M., Rizzi, C.: Is TRIZ an ecodesign method? In: Setchi, R., Howlett, R.J., Liu, Y., Theobald, P. (eds.) *Sustainable Design and Manufacturing* 2016. SIST, vol. 52, pp. 525–535. Springer, Cham (2016). https://doi.org/10.1007/978-3-319-32098-4_45
9. Russo, D., Peri, P., Spreafico, C.: TRIZ applied to waste pyrolysis project in Morocco. In: Benmoussa, R., De Guio, R., Dubois, S., Koziołek, S. (eds.) TFC 2019. IAICT, vol. 572, pp. 295–304. Springer, Cham (2019). https://doi.org/10.1007/978-3-030-32497-1_24
10. Spreafico, C., Russo, D., Spreafico, M.: Investigating the evolution of pyrolysis technologies through bibliometric analysis of patents and papers. J. Anal. Appl. Pyrol. **159**, 105021 (2021)
11. Russo, D., Spreafico, C.: TRIZ-based guidelines for Eco-improvement. Sustainability **12**(8), 3412 (2020)
12. Spreafico, C.: Quantifying the advantages of TRIZ in sustainability through life cycle assessment. J. Cleaner Prod. **303**, 126955 (2021)
13. Maccioni, L., Borgianni, Y., Rotini, F.: Sustainability as a value-adding concept in the early design phases? Insights from stimulated ideation sessions. In: Campana, G., Howlett, R.J., Setchi, R., Cimatti, B. (eds.) SDM 2017. SIST, vol. 68, pp. 888–897. Springer, Cham (2017). https://doi.org/10.1007/978-3-319-57078-5_83

14. Bovea, M.D., Pérez-Belis, V.: A taxonomy of ecodesign tools for integrating environmental requirements into the product design process. J. Cleaner Prod. **20**(1), 61–71 (2012)
15. Feniser, C., Burz, G., Mocan, M., Ivascu, L., Gherhes, V., Otel, C.C.: The evaluation and application of the TRIZ method for increasing eco-innovative levels in SMEs. Sustainability **9**(7), 1125 (2017)
16. Baran, J., Janik, A., Ryszko, A., Szafraniec, M.: Making eco-innovation measurable-are we moving towards diversity or uniformity of methods and indicators. In: SGEM2015 Conference Proceedings (2015)
17. Pusavec, F., Krajnik, P., Kopac, J.: Transitioning to sustainable production–Part I: application on machining technologies. J. Cleaner Prod. **18**(2), 174–184 (2010)
18. Tiron, O., Bumbac, C., Manea, E., Stefanescu, M., Lazar, M.N.: Overcoming microalgae harvesting barrier by activated algae granules. Sci. Rep. **7**(1), 1–11 (2017)
19. Wu, W., Cheng, L.C., Chang, J.S.: Environmental life cycle comparisons of pig farming integrated with anaerobic digestion and algae-based wastewater treatment. J. Environ. Manage. **264**, 110512 (2020)
20. Zakuciová, K., Carvalho, A., Štefanica, J., Vitvarová, M., Pilař, L., Kočí, V.: Environmental and comparative assessment of integrated gasification gas cycle with CaO looping and CO_2 adsorption by activated carbon: a case study of the Czech Republic. Energies **13**(16), 4188 (2020)
21. Favi, C., Germani, M., Landi, D., Mengarelli, M., Rossi, M.: Comparative life cycle assessment of cooking appliances in Italian kitchens. J. Clean. Prod. **186**, 430–449 (2018)
22. Dannenberg, G.D.S., Funck, G.D., Cruxen, C.E.D.S., Marques, J.D.L., Silva, W.P.D., Fiorentini, Â.M.: Essential oil from pink pepper as an antimicrobial component in cellulose acetate film: Potential for application as active packaging for sliced cheese. *Lebensmittel-Wissenschaft+[ie und] Technologie* (2017)
23. Emanuel, N., Sandhu, H.K.: Food packaging development: recent perspective. J. Thin Films Coat. Sci. Technol. Appl. **6**(3), 13–29 (2020)
24. Williams, H., Wikström, F.: Environmental impact of packaging and food losses in a life cycle perspective: a comparative analysis of five food items. J. Clean. Prod. **19**(1), 43–48 (2011)
25. Marcuzzo, E., Peressini, D., Sensidoni, A.: Shelf life of short ripened soft cheese stored under various packaging conditions. J. Food Process. Preserv. **37**(6), 1094–1102 (2013)
26. Gutierrez, M.M., Meleddu, M., Piga, A.: Food losses, shelf life extension and environmental impact of a packaged cheesecake: a life cycle assessment. Food Res. Int. **91**, 124–132 (2017)
27. Gamage, J.R., DeSilva, A.K.M.: Assessment of research needs for sustainability of unconventional machining processes. Procedia CIRP **26**, 385–390 (2015)

Eco-Design Pilot Improvement Based on TRIZ and ASIT Tools

R. Benmoussa(✉)

ENSA Marrakesh, Cadi Ayyad University, Marrakesh, Morocco
benmoussa.ensa@gmail.com

Abstract. Eco-design is a relevant research topic because of the need of powerful and confirmed systematic methods that can support companies in their efforts to develop green products within acceptable time and costs. Eco-design Pilot, developed by ADEME and Vienna University of Technology, is one of the tools that seeks this purpose. It proposes measures depending on the nature of the environmental impact of the product. However, the assessment phase of the measures proposed by this tool is both subjective and does not stimulate the development of creative solutions. To overcome this major drawback, this paper seeks to rely on TRIZ and ASIT tools to better assess the proposed measures and develop creative ideas within a new framework for Eco-design Pilot measures assessment. This framework has been tested first on the Eco-design Pilot Improvement Objective and Strategy "Reduction Packaging". As result, 31 realization ideas, that could be used by packaging designers in green context, have emerged from this application. A non-exhaustive research on packaging industry practices has shown that some realization ideas have already been applied to produce innovative solutions. In perspective, we will discuss other Eco-design improvement objectives and strategies in a quest for the framework generalization.

Keywords: Eco-design · Eco-design Pilot · TRIZ · ASIT · Reduction Packaging

1 Introduction

Fast industrial growth and technological advances have recently raised many environmental concerns. As a result, industrial companies are expected to become more environmentally responsible and reduce their negative environmental impact by applying new technologies and especially by delivering eco-friendly products. Additionally, environmental regulations developed recently by the European Union called energy using product, have motivated engineers and designers to more heavily focus on how to create more environmentally friendly products.

Several methods exist for eco-design, to assess and improve environmental impacts. [6–10]. The International Organization for Standardization ISO issued numerous norms, guidelines, and tools. For example, the ISO14040:2006 describes the principles and framework for life cycle assessment (LCA), ISO14044:2006 provides LCA guidelines,

Published by Springer Nature Switzerland AG 2022
R. Nowak et al. (Eds.): TFC 2022, IFIP AICT 655, pp. 340–351, 2022.
https://doi.org/10.1007/978-3-031-17288-5_29

and ISO14006:2011 provides guidelines to implement Eco-Design as part of an environmental management system (EMS) within companies. A number of methods and tools have been developed to support the process of eco-innovation in the last two decades. To the best-known methods belong Eco-Compass, Life Cycle Design Strategy (LiDS Wheel), Sustainability Circle, Eco-design PILOT, Eco-Ideation Tool, Value Mapping Tool, Design for Environment (DfE) and Quality Function Deployment for Environment (QFDE), EcoASIT, Eco-ideation stimulation meso-mechanisms ESMs, 12 Principles of Green Engineering (American chemical society), and other methods. In the field of process engineering should be mentioned in first place Green Process Engineering and Process Intensification (PI), Process Design for Sustainability (PDfS), and other approaches [2].

Despite these countless approaches, green design remains a relevant research topic because of the lack of powerful and confirmed systematic methods that can support companies in their efforts to develop green products within acceptable time and costs. To overcome this lack, several works have addressed this issue. This paper focus on the ones that deal with TRIZ (a Russian acronym of "Theory of Inventive Problem Solving"). TRIZ, as a theory for inventive problem resolution has proven its effectiveness in stimulating designer's creativity in several areas. In green design, several TRIZ utilization attempts exist [1–3, 11–17].

The references [2] and [3] have largely addressed this issue. The paper [2] focuses on the comparison of the ideation mechanisms during the eco-ideation phase, to help users generate relevant ideas with a strong potential of environmental impact reduction. In addition to that, some case studies are performed to compare the adapted creativity tool for eco-innovation, regarding its performance. This research also validated the need to place greater emphasis on the ideation phase in the process of eco-innovation. Indeed, environmental knowledge, ideation mechanisms and the structuring of the session are interdependent factors to consider in order to optimize the eco-ideation session and to be closer to the industrial reality. The paper [3] presents a systemic literature review of TRIZ use in eco-design. Indeed, the paper provides information about what is currently performed in connection to creativity methods, when TRIZ is combined with LCA, LCE Eco-efficiency and other integrated methods for eco-design process.

Even if the existing scientific contributions of TRIZ in eco-innovation cited above tackle different Triz tools and Eco-design methods, they share a common purpose. They effectively address the following main research question: "How can TRIZ tools enhance the ideality of green-design methods and tools?". To contribute to this common purpose, this paper aims to present an exploratory analysis of the contribution of TRIZ tools to green design methods systematization based on the case of Eco-design Pilot [6]. In fact, the main goal of this paper is to analyze Eco-design Pilot in order to improve its ideality through TRIZ tools utilization. Eco-design Pilot proposes measures to improve the environmental impact of products and services depending on the nature of the environmental impact of the product. However, the assessment phase of the measures proposed by this tool is both subjective and does not stimulate the development of creative solutions. To overcome this major drawback, this paper seeks to develop a new framework that rely on TRIZ and ASIT tools to better assess the Eco-design Pilot measures and develop creative ideas of realization. Our methodology is to test this framework first on the "improvement objective and strategy for basic type C (transportation intensive product)", namely "Reduction Packaging" and to generalize dealing with other Eco-design Pilot improvement objectives and strategies.

This paper is organized as follows. Section 2 introduces Eco-design Pilot approach and limitations. Section 3 presents the measures of the improvement objective and strategy "Reduce packaging". It presents also the new framework application to this objective. A conclusion and perspectives are provided in Sect. 4.

2 Eco-Design Pilot

2.1 Presentation

Eco-design Pilot [6] is a web tool developed by ADEME and Vienna University of Technology. It has been proposed as a guide to promote and facilitate the integration of eco-design in businesses in any sector of activity. Starting with the PILOT along product development strategies enables the practitioners to improve an existing product by means of appropriate ECODESIGN measures[1]. In order to improve the environmental performance, each product requires specific measures depending on its environmental impact at different stages of its service life. Thus, different measures have to be taken for products with the main environmental impact at the use stage (use intensive) than for products with the main impact during manufacture (manufacture intensive). What measures are appropriate for which product? Working through the checklists will result in a complete list of ECODESIGN measures particularly suited for the product. Eco-design Pilot divides products in Five types:

- Type A: Raw material intensive.
- Type B: Manufacture Intensive.
- Type C: Transportation Intensive.
- Type D: Use Intensive.
- Type E: Disposal intensive.

Figure 1 shows the list of controls used to estimate the relevance and achievement of the assessment[2] question for the measure "Reduce material input for packaging" of the implementation objective and strategy " Reduction Packaging".

2.2 Measures Assessment Approach

The classical approach of Eco-design Pilot measures assessment is as follow [6]:

1. Relevance: Rate the relevance of the assessment question with a view to your product. (10…very important for my product; 5…less important for my product; 0…not relevant for my product).

[1] A measure for ECODESIGN Pilot is equivalent to a conceptual solution for TRIZ.

[2] "Assessment" is the process that seeks to know if a measure (conceptual solution) is relevant for a given product and a given strategy.

Checklist for ECODESIGN analysis

Product

Does the distribution of the product require only a minimum of packaging?

Can packaging or at least parts of it be avoided? How can packaging of the product be minimized by an optimization of the system product - packaging?

Relevance (R)	Fulfillment (F)	Priority (P)
very important (10)	yes (1)	
less important (5)	rather yes (2)	
not relevant (0)	rather no (3)	
	no (4)	P = R * F

Measure	Reduce material input for packaging LEARN
Idea for Realization	
Costs	more / same / less — because
Feasibility	difficult / easy — because
Action	at once / later / never — Responsibility / Deadline

Fig. 1. Checklist of "Reduce material input for packaging" measure [6]

2. Fulfillment: Estimate the fulfilment of the assessment questions using one of the four possible answers (yes/rather yes/rather no/no); the additional questions support understanding of the assessment question and need not be answered.
3. Priority: Select ECODESIGN tasks with high priority (P) and continue only with these.
4. Idea for Realization: Find ideas to realize these ECODESIGN tasks. The content of the learning part with its examples shall assist you in doing that.
5. Feasibility: Evaluate the feasibility of the suggested ideas (difficult/easy).
6. Costs: Compare the costs of the new ideas with a reference situation (higher/same/lower) and give reason for that.
7. Action: Decide when to carry out the ECODESIGN tasks (at once/later/never) and determine the person or department that shall be in charge of further steps in the realizing the product improvements and fix a deadline.
8. Save: Save the checklist to document the ECODESIGN assessment.

The main limitations of this assessment approach are:

- The Relevance and Achievement of the measure are subjectively estimated. We risk thus missing the right measure.
- The generation of the ideas of realization is not structured. It is in fact based on the intrinsic competence of business experts.
- The generation of ideas of realization does not cover the potential of integrating several measures at the same time.

2.3 Framework Presentation

To overcome the limitations of Eco-design pilot's measures assessment approach, a new Framework has been developed (Fig. 2) based on the use of TRIZ and ASIT tools.

The following section describes the original steps of this Framework, compared to the classic approach, through its application to the Eco-design Pilot improvement objective "Reduction Packaging".

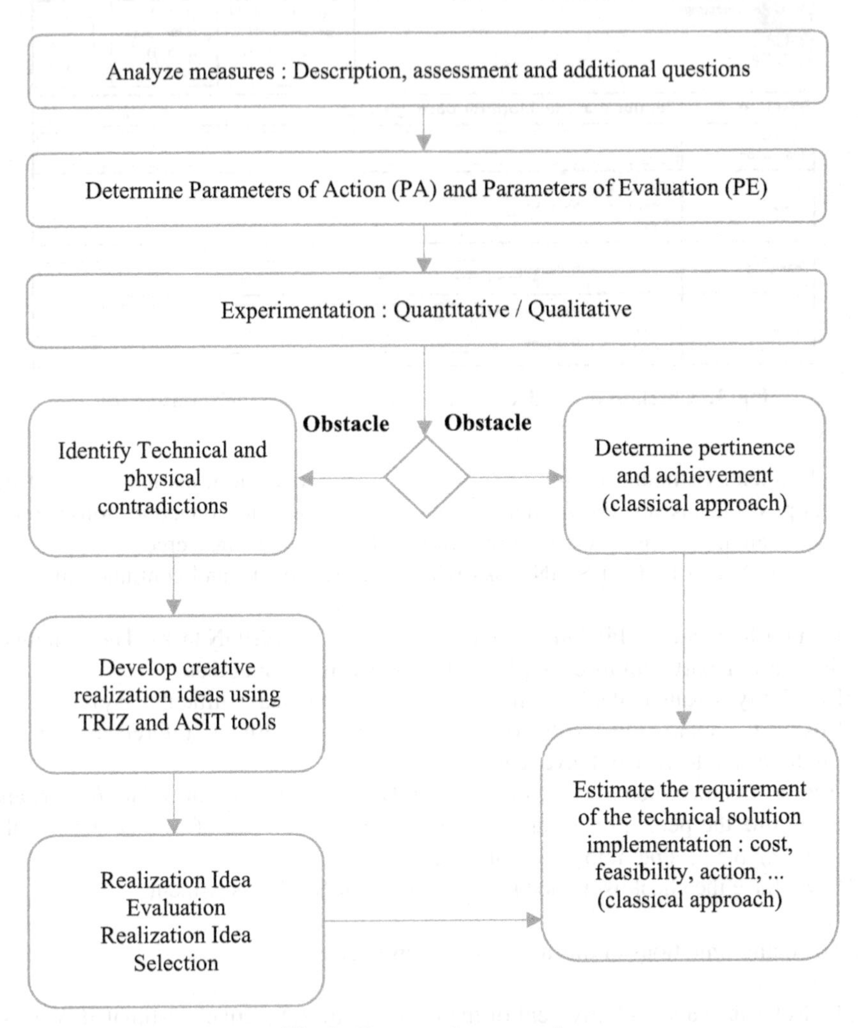

Fig. 2. Framework presentation

3 Framework Application to "Reduction Packaging" Measures

3.1 "Reduction Packaging" Measures Analysis

As packaging material is useful only for a limited period of time (unless it is returnable) the type and quantity of material used for packaging have to be optimized. Especially with products that have to be transported over long distances, the weight of the packaging material, too, has a great influence on the overall consumption of resources. Is it possible to reduce the weight of the packaging material? Is it possible to use returnable packaging or renewable or recycled materials for packaging?

Eco-design Pilot identify five measures to reduce packaging [6]:

1. Reduce material input for packaging: Minimizing packaging with a view to weight can be realized either by optimization of packaging or by appropriate product design, for instance by casings that endure transportation without or with only a minimum of packaging.
2. Preferably use reusable packaging: Robust packaging designed for multiple use reduces the overall input for packaging. Returnable packaging is particularly advantageous in the case of direct delivery. The supplier can take back and subsequently re-use packaging material.
3. Preferably use renewable raw materials for packaging: The use of renewable raw materials (i.e. non fossil materials, usually made from plant material) not only constitutes an adequate solution for the disposal of packaging material but also takes into account the issue of resources (renewables as an important criterion for sustainability).
4. Preferably use recycled materials or packaging materials suitable for established recycling processes: Using recyclable materials reduces the consumption of primary materials as well as the amount of waste generated. Materials for which there are already well established recycling channels facilitate recycling of packaging materials.
5. Use environmentally acceptable packaging materials: The materials used for packaging should be environmentally acceptable not only as far as manufacture is concerned but also at the time of disposal. Therefore, one has to determine at what point in the whole life cycle the issue of disposal of packaging arises and how disposal can be realized there.

3.2 Determine Parameters of Action and Evaluation

The parameters of action (PA) are those on which we can act to improve the environmental impact as well as the functional quality of the packaging. The analysis of "Reduction Packaging" measures makes it possible to identify the following PA:

- The shape of the packaging (morphology) (S).
- The Material Type (M): reusable, recyclable, with little environmental impact,... Etc.
- The thickness of the surface (T).

The parameters of evaluation (PE) are those that measure the environmental performance as well as the functional quality of the packaging. The analysis of "Reduction Packaging" measures makes it possible to identify the following PE:

- The resistance to shocks during transport, storage and use (R).

- The quantity of substance for one use of the packaging (mass (Q)): it is obtained by dividing the total quantity of substance by the number of uses of the packaging during its life cycle. The environmental impact is quantitatively calculated or qualitatively estimated according to the CO_2 and non-renewable energy emissions perceived during the life cycle of the packaging depending on the quantity of substance and the type of material.

3.3 Experimentation

The experiment consists first of all to:

- Formulate the desired objectives for the PE: $(R > r)$ & $(Q < q)$.
- Identify all the solutions available in the business and feasible in the context of the study (involvement of business experts) for PA.
- Then test (quantitatively or qualitatively) all the possible combinations of PA (Table 1).

Table 1. Specification of experimentations

S	M	E	R	Q
1	1	1	R1	Qn
...	...	...	...	...
n	n	n	Rn	Qn

3.4 Experimentation Analysis

In the case that the experimentation table presents scenarios that make it possible to achieve the objectives set for the (PE), the relevance and achievement for each of these scenarios can be determined using the same rating proposed by the classic approach of Eco-design Pilot, by answering evaluation and additional questions of measures. It is nevertheless more relevant because it will be estimated directly in relation to the action parameters of the Idea of Realization (Conceptual Solution). The estimation of the costs and the feasibility of the realization idea (Technical Solution) selected could be done similarly as the classic approach.

In the case that the experimentation table does not present any scenario that makes it possible to achieve the objectives set for the PE, the relevance and the achievement lose their classic meaning of Eco-design Pilot and enforce the search for creative ideas that are not obvious for business experts. In this case, TRIZ and ASIT tools are helpful and can be exploited to generate creative ideas of realization. Thus, Sect. 3.5 presents the realization ideas generated from TRIZ tools. Section 3.6 presents the realization ideas generated from ASIT tools. Without being exhaustive, Table 2 presents some technical solutions already used by the packaging industry that belong to some realization ideas proposed by this new framework.

Table 2. Realization ideas implemented by packaging industry

Solution	*Realization Idea*	*Example*
Spherical packaging	IR1, IR8	
Home delivery without packaging	IR11, IR15, IR24	 
Bulk distribution with personal packaging	IR11, IR15, IR24	
Packaging with parts that have different amounts of material	IR2, IR13, IR8, , IR9, IR10, IR27	
Packaging reusable several times with or without deposit	IR12, IR15	
Packaging made from renewable materials	IR12, IR27	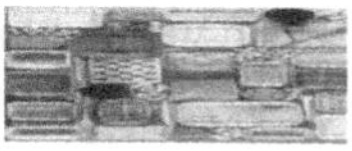
Packaging made from recyclable materials	IR12, IR27	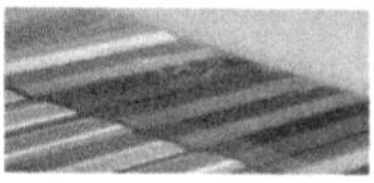
Packaging made from materials with little environmental impact	IR12, IR27	
Garbage can for sorting waste	IR22	

3.5 Generate Realization Ideas with TRIZ Tools

The analysis of the obstacles in relation to the objectives set for the PE (Resistance and Quantity of substance) makes it possible to identify the technical contradiction between these parameters. It also makes it possible to highlight among the PA (S, M, E) those which present a physical contradiction. The 40 inventive principles as well as the principles of separation in time and space [4] could be used to develop innovative realization ideas.

3.5.1 Generate Realization Ideas with the 40 Inventive Principles

According to the matrix of technical contradictions [4], the contradiction between the quantity of substance (Factor to improve PT26) and resistance (Factor that degrades PT14) suggest the following principles: 14, 35, 34, 10.

IR1: principle 14 (spheroidability), suggests for packaging to replace rectilinear parts, surfaces, or forms with curvilinear ones, move from flat surfaces to spherical ones; from parts shaped as a cube (parallelepiped) to ball-shaped structures.

IR2: principle 35 (Parameter changes), suggests to change the physical state of the packaging, to change the concentration or consistency of the packaging material, to modify the degree of flexibility of the packaging, to modify the temperature of the packaging.

IR3: principle 34 (Discarding and recovering), suggests to make portions of the packaging that have fulfilled their functions go away (discard by dissolving, evaporating, etc.) or modify these directly during operation.

IR4: principle 10 (Preliminary action), suggests to perform before it is needed, the required change of the packaging (either fully or partially) or to pre-arrange objects (packaging, products) such that they can come into action from the most convenient place and without losing time for their delivery.

3.5.2 Generate Realization Ideas with the Separation Principles

The principles of separation in time and space make it possible to solve the problem at the level of physical contradictions involving the parameters of action: the shape, the material and the thickness of the surface of the packaging. The packaging life cycle consists of the following phases: product packaging, transport, storage, use and end of life.

The separation in time principles suggest the following realization ideas:

IR5: the packaging has a different shape during its life cycle.

IR6: the packaging is composed from different materials during its life cycle.

IR7: the packaging has a different surface thickness during its life cycle.

The separation in space principles suggest the following realization ideas:

IR8: the parts of the packaging have a different shape.

IR9: the parts of the packaging are made from different materials.

IR10: the parts of the packaging have different surface thickness.

3.5.3 Generate Realization Ideas with the Intensification Tool

Intensification [4] is also a TRIZ tool that allows the resolution of contradictions at the physical level. The application of this tool suggests the following realization ideas:

IR11: the packaging has an infinitely (large,…) or infinitely (small,…) shape.

IR12: the packaging has an infinitely (resistant, flexible, removable, recyclable, etc.) or infinitely (fragile, rigid, non-disposable, non-recyclable, etc.) material.

IR13: the packaging has an infinitely large or infinitely small thickness.

3.5.4 Generate Realization Ideas with ASIT Tools

ASIT [5] is a method of solving inventive problems resulting from TRIZ. It has two rules and 5 tools which can also be used to develop innovative realization ideas.

Before applying ASIT tools to generate realization ideas, we need first to deploy the rule of the "closed world":

- Undesirable phenomenon: the consumer throws away the packaging after use, thus producing waste.
- Objects of the Problem: Consumer, Packaging.
- Objects of Environment: Garbage can.
- Desired Action (DA): Prevent the consumer from throwing away the packaging.

The application of ASIT tools to the Eco-design Pilot's objective "Reduction packaging" suggests the following realization idea.

From Unification tool:

IR14: the consumer will (DA).

IR15: the packaging will (DA).

IR16: The garbage can will (DA).

From Multiplication tool:

IR17: an object of the same type as "the consumer" will (DA).

IR18: an object of the same type as "the packaging" will (DA).

IR19: an object of the same type as "the garbage can") will (DA).

From Division tool:

IR20: the object "consumer" will be divided into [list of parts] and will be reorganized in time or space to (DA).

IR21: the object "packaging" will be divided into [list of parts] and will be reorganized in time or space to (DA).

IR22: the object "the Garbage Can" will be divided into [list of parts] and will be reorganized in time or space to (DA).

From Suppression tool:

IR23: Remove the "consumer" will (DA).

IR24: Remove the "packaging" will (DA).

IR25: Remove the "garbage can" will (DA).

From Break symmetry tool:

The object {"the consumer", "the packaging", "the Garbage can"} has [list of characteristics]. For packaging, the list of characteristics is given in Sect. 3.2: shape, material, thickness.

Symmetry in space:

IR26: in different places of "the consumer", there will be different values for the chosen characteristic.

IR27: in different places of "the packaging", there will be different values for the chosen characteristic (shape, material, thickness).

IR28: in different places of "the garbage can", there will be different values for the chosen characteristic.

Symmetry in time:

IR29: at different times, there will be different values for the chosen characteristic of the "consumer".

IR30: at different times, there will be different values for the chosen characteristic of the "packaging".

IR31: at different times, there will be different values for the chosen characteristic of the "garbage can".

4 Conclusion

The literature has shown the existence of several attempts to improve the effectiveness of Eco-design methods using TRIZ tools. However, this paper presents Eco-design Pilot's first attempt to this focus. This work aimed first to study Eco-design Pilot, then to state a new framework for the ideation phase to fill the limits of this tool. Several results emerge from this study. In the technical level, 31 realization ideas were developed using tools from TRIZ and ASIT concerning reduction packaging. Non-exhaustive research on the web has shown that some of these ideas have already been implemented in the industrial context. The other ideas remain a potential conceptual solutions for packaging reduction to be explored and translated into technical solutions by the designers. Scientifically, some tools from TRIZ and ASIT have generated similar ideas. This is the case of IR11 and IR24 for example. This confirms that ASIT is a method derived from TRIZ.

In light of these results and the exploratory approach conducted through the "Reduction Packaging" improvement objective and strategy, several perspectives emerge:

- Conduct other studies to confirm the effectiveness of Eco-design Pilot new framework for the evaluation phase, dealing with other improvement objectives and strategies.
- Make a comparison between the eco-design methods that attempt to adapt the TRIZ tools to eco-design with our approach.
- Conduct a case study that aims reduction of packaging in a reel industrial context.

References

1. Livotov, P., Sekaran, A.P.C., Law, R., Reay, D., Sarsenova, A., Sayyareh, S.: Eco-innovation in process engineering: contradictions, inventive principles and methods. Therm. Sci. Eng. Prog. **9**, 52–65 (2019)

2. Tyl, B., Legardeur, J., Millet, D., Vallet, F.: A comparative study of ideation mechanisms used in eco-innovation tools. J. Eng. Des. **25**, 10–12 (2015)
3. Buzuku, S., Shnai, I.: A systematic literature review of TRIZ used in Eco-Design. J. Eur. TRIZ Assoc. 02-2017 **4**, 20–31 (2004)
4. TRIZ. Theory of Inventive Problem Solving, Vladimir Petrov. Springer (2019). https://doi.org/10.1007/978-3-030-04254-7
5. ASIT, méthodes pour des solutions innovantes, Dr Roni Horowitz, SolidCreativity (2004)
6. Eco-design pilot homepage. http://pilot.ecodesign.at/pilot/ONLINE/ENGLISH/INDEX.HTM. Accessed 04 Jan 2021
7. ECOFAIRE homepage. https://www.ademe.fr/ecofaire-loutil. Accessed 04 Jan 2021
8. Liu, C.C., Chen, J.L.: Development of product green innovation design method. In: Proceedings Second International Symposium on Environmentally Conscious Design and Inverse Manufacturing, pp. 168–173 (2001)
9. Sakao, T.: A QFD-centred design methodology for environmentally conscious product design. Int. J. Prod. Res. **45**(18–19), 4143–4162 (2007)
10. Vezzetti, E., Moos, S., Kretli, S.: A product lifecycle management methodology for supporting knowledge reuse in the consumer packaged goods domain. Comput. Des. **43**(12), 1902–1911 (2011)
11. Yang, C.J., Chen, J.L.: Accelerating preliminary eco-innovation design for products that integrates case-based reasoning and TRIZ method. J. Clean. Prod. **19**(9–10), 998–1006 (2011)
12. Yang, C.J., Chen, J.L.: Forecasting the design of eco-products by integrating TRIZ evolution patterns with CBR and Simple LCA methods. Expert Syst. Appl. **39**(3), 2884–2892 (2012)
13. Russo, D., Bersano, G., Birolini, V., Uhl, R.: European testing of the efficiency of TRIZ in eco-innovation projects for manufacturing SMEs. Procedia Eng. **9**, 157–171 (2011)
14. Chou, J.-R.: An ARIZ-based life cycle engineering model for eco-design. J. Clean. Prod. **66**, 210–223 (2014)
15. Ben Moussa, F.Z., Essaber, F.E., Benmoussa, R., Dubois, S.: Enhancing Eco-design methods using TRIZ tools: the case of ECOFAIRE. In: Benmoussa, R., De Guio, R., Dubois, S., Koziołek, S. (eds.) TFC 2019. IAICT, vol. 572, pp. 350–367. Springer, Cham (2019). https://doi.org/10.1007/978-3-030-32497-1_29
16. Russo, D., Regazzoni, D., Montecchi, T.: Eco-design with TRIZ laws of evolution. Procedia Eng. **9**, 311–322 (2011)
17. Vidal, R., Salmeron, J.L., Mena, A., Chulvi, V.: Fuzzy Cognitive Map-based selection of TRIZ (Theory of Inventive Problem Solving) trends for eco-innovation of ceramic industry products. J. Clean. Prod. **107**, 202–214 (2015)

Innovation Portfolio Management: How Can TRIZ Help?

Nikhil Shree Phadnis(✉)

Lappeenranta-Lahti University of Technology, 53850 Lappeenranta, Finland
nikhil.phadnis@lut.fi

Abstract. Prioritising the right innovation projects is necessary for effective strategic roadmaps and portfolios in a dynamic market with ever-increasing resource constraints. Innovation portfolio management is a process that informs decision-makers to help them focus and prioritise innovation projects to capitalise on their investments. On the other hand, the applicability of TRIZ in management is an ever-growing area proving its systematic and scientific capabilities. This paper builds on the foundation of expanding TRIZ and identifies several opportunities for TRIZ tools in the innovation portfolio management process. The study extracts data from the scientific literature to study innovation portfolio management as a process to find two distinct approaches to innovation portfolio creation: Emergent and deliberate innovation portfolio processes. The study investigates the pros, cons and gaps in these processes to potentially strengthen them with modern TRIZ tools. The research explores and explains which TRIZ tools can enhance the output of innovation portfolio management. This research finds TRIZ tools such as Main parameters of Value analysis, function analysis, S-curve analysis, Trends of evolution, contradiction analysis and quantum economic analysis critically benefit the portfolio process and can provide better innovation portfolio outputs through new unexplored insights.

Keywords: Innovation portfolio management · TRIZ · Innovation management

1 Introduction

"There are two ways for a business to succeed in new products: doing projects right and doing the right projects", a statement by Cooper et al. paved the way for research in Portfolio management [1]. It is defined as a complex and dynamic process of distributing limited resources to a few prioritised projects to achieve maximal value, balance and a strategic fit with the firm. The innovation portfolio is concerned with creating a unified portfolio strategy and the maturation and selection of project candidates, known as concepts [2]. While the innovation portfolio process has been thoroughly examined in the past, no attempts have been made to combine diverse innovation methodologies to enhance its output. In this case, the theory of inventive problem solving (TRIZ). Even though various ideation techniques are available today, practitioners and academic researchers consider TRIZ a powerful but time-consuming method for systematically

Published by Springer Nature Switzerland AG 2022
R. Nowak et al. (Eds.): TFC 2022, IFIP AICT 655, pp. 352–366, 2022.
https://doi.org/10.1007/978-3-031-17288-5_30

and scientifically improving engineering systems [3]. However, recent TRIZ research is expanding to new domains and the paper sets to explore its applicability in innovation management.

Recently, research into the applicability of TRIZ in Innovation management has been a rapidly developing area of interest [4–6]. Much research in this area concerns TRIZ tools, i.e. micro-level perspectives, for example, contradiction matrix applied to business and non-technical problems or adapting an existing cluster of TRIZ tools repurposed for creating strategic business roadmaps [5, 7–11]. This research adds to the "breadth" of knowledge by describing how TRIZ can enhance the output of an innovation portfolio, making it more robust and reducing failure rates of new product development. Consequently, this paper investigates the possible opportunities for enhancing the innovation portfolio process. As a result, the study question we are attempting to answer is "How can TRIZ tools potentially assist in approaches utilised in Innovation portfolio analysis?".

Firstly, the study identifies a research gap in the literature and provides an understanding of the fuzziness between innovation portfolio management and project portfolio management. Academic literature interchangeably "brands" these processes differently and creates an issue of clarity, portraying that both areas, innovation portfolio management and project portfolio management are the same. We address this issue by a systematic literature review and highlighting studies on innovation portfolio management and project portfolio management that establish a boundary to distinguish between them. Secondly, the literature review also explores a distinct portfolio process that outlines its stages in a well-defined manner. Our study finds that innovation portfolio management has two practical approaches with a small yet vital difference in the innovation strategy development of firms. This paper outlines both identified approaches to highlight their pros and cons. Lastly, an integrative literature review describes the potential of modern TRIZ tools to enhance the output of an innovation portfolio. This research adds novelty in two ways to existing research. Firstly creating a clear distinction between two innovation portfolio approaches and their different sequential phases. Secondly by proposing potential directions of improvement using Modern TRIZ and GEN-TRIZ (proprietary TRIZ tools used by TRIZ consulting companies) tools.

1.1 Innovation Portfolio Management

Most academic research confuses innovation portfolio management and project portfolio management, whilst some papers recently create a distinct separation between the two. Innovation portfolio analysis or management (IPM) is not to be mixed with project portfolio management (PPM). IPM is responsible for creating a unified portfolio strategy. In contrast, PPM is concerned with executing and delivering a group of projects known as a project portfolio. Matthews, 2010 outlines several differences between PPM and IPM, considering dimensions such as process type, objectives, planning horizons, technology readiness level, et cetera. IPM is a process that takes place before PPM, i.e. at an early stage initiation with a weakly defined strategic intent, in high uncertainty conditions [2].

Perhaps the most critical difference between them is in the deliverables of the two processes. IPM delivers a well-articulated strategy aligned with a few promising concepts that enable them to be released in a project portfolio. In comparison, PPM efficiently allocates a firm's resources aligned to its strategies. Strategies that were an output of the innovation portfolio process. PPM ensures the execution of the plans within the budget, scope, and required quality. The innovation portfolio analysis approach produces a cluster of concepts and tailored strategies for each group of concepts, which emerge over time and evolve during the analysis. IPM is more flexible and allows a concept to be substantially morphed by merging with another concept, partitioned into divergent concepts, or shelved altogether. Moreover, unlike the project portfolio's sequential stage-gate procedure, idea detail can be added to the innovation portfolio at any moment and in any sequence [2, 12].

To summarise, innovation portfolio management should perhaps be referred to as "innovation portfolio analysis" as it is an analytical and strategic planning phase in the fuzzy front end as a preprocess to project portfolio management [13]. This phase allows decision-makers to shape their innovation strategy based on the ideation portfolio to provide strategic insights at a holistic level. Some researchers call it ideation portfolio management, referring to procedures that aid in producing and selecting valuable and relevant ideas and concepts for inclusion in the innovation project portfolio [14].

This paper concerns innovation portfolio analysis rather than project portfolio management since research has already established the proficiency of TRIZ in project management through the TRIZ assisted stage-gate process, its problem-solving capabilities with various TRIZ stools [15–17].

2 Methodology

We conduct a systematic critical literature review as an explorative study with a deductive approach to identify several phases in innovation portfolio analysis [18]. We used Web of science as a search engine to collect secondary data and analyse it. The initial search string "innovation portfolio management" without the exact search function focused on the Web of science core collection, i.e. peer-reviewed articles and publications. The initial hit yields 1418 papers from the Web of Science collection containing de-duplicated records. Next, we added a date range and filtered the results from 2010 to 2022 producing 1178 papers. The paper mainly focuses on the process perspective of the innovation portfolio analysis; therefore we decided to exclude mathematical or computer science studies that pertain primarily to algorithms and decision support systems. The search yields 1034 articles, including book chapters, journals, open access publications and proceedings. The papers were further filtered based on top-quality peer-reviewed publishers such as Elsevier, Emerald publications, Springer, Taylor and Francis, MDPI and IEEE to yield 698 articles. Lastly, papers were filtered by titles to produce 43 relevant publications, out of which 35 were available to the authors for analysis and synthesis. Figure 1 illustrates the systematic literature review process, justifying the filter mechanisms, the number of hits(n), excluded and included studies to be analysed.

The final studies selected are analysed with two objectives. First, to investigate an explicit diagrammatic representation of the innovation portfolio process, outlining all of

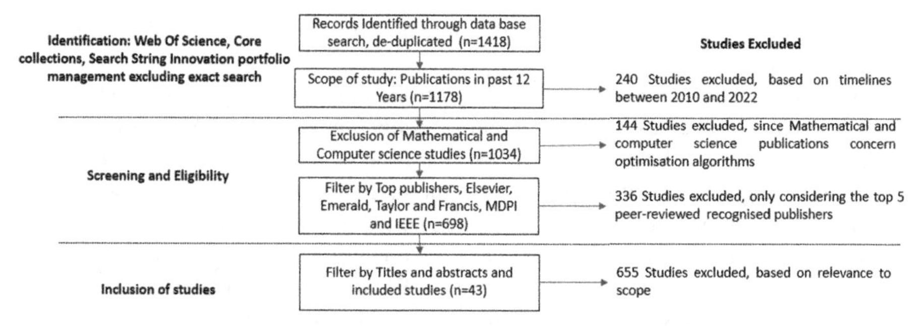

Fig. 1. Literature search and studies included

its stages or descriptions of the innovation portfolio management process activities. The systematic setting of the innovation portfolio management process may reveal research gaps and opportunities for TRIZ tools to fill these gaps. Secondly, identification of studies that attempt to integrate other innovation methods into the innovation portfolio process to enhance its output.

3 Results

The systematic literature review results in novel insights on innovation portfolio management as a topic and highlights the cloudiness between IPM and PPM. The results can be summarised in two complementing directions. First, general trends and statistics on innovation portfolio management, the state-of-the-art and the cloudiness of IPM and PPM. And secondly, two distinct innovation portfolio management processes, i.e., a deliberate innovation portfolio and an emergent innovation portfolio [19].

Our research reveals that 65% of the included studies explicitly mention innovation portfolio management as a process but do not study them from a process perspective. Instead, these studies focus on specific stages of the portfolio process. But only 22% of the studies have attempted to integrate, incorporate or hybridise an innovation method, tool and technique aimed toward enhancing the innovation portfolio output. Additionally, only 25% of the studies highlight innovation portfolio management as a separate process from project portfolio management.

To elaborate, most research considers the integration of an ideation phase before PPM to rebrand it as innovation portfolio management, strategic foresight, strategic innovation planning, ideation portfolio planning and many other names [13, 14, 20–27]. The addition of an ideation phase has also been linked with strategic innovation and strategic foresight as a pre-step to project portfolio management and supports emerging strategies from the project portfolio. Additionally, research indicates scenario planning methods such as backcasting, forecasting, and technology road-mapping in the ideation phase allowing the innovation strategy to be formed at the end of the process. This process is known as an emergent innovation strategy. In traditional literature studies, emergent strategies are defined as processes that disclose emerging patterns in a project portfolio and inform the strategy formulation [19]. Some research studies claim that IPM is an emergent strategy formulation process; however other studies contradict this theory and argue that deliberate strategy implementation is as successful as emergent strategies [19].

Our research shows that the innovation portfolio management process can have two strategic control processes, with emergent or deliberate strategies. More importantly, the literature review IPM process revealed that 54% of the papers mention deliberate strategies to align the resulting portfolio to the organisation's goals, whereas 25% of the reports claim emergent strategies are more suited to the adaptability and growth of a firm. Four studies claim both approaches are equally crucial for strategic decision-making during the innovation portfolio process. We will outline both the emergent and deliberate processes in Sect. 3.1, providing advantages, disadvantages and opportunities to strengthen them using TRIZ tools.

3.1 Innovation Portfolio Management Approaches

Emergent Innovation Portfolios

The term "emerging strategy recognition" refers to discovering unexpected and unplanned events. Such incidents are far more likely to occur in a dynamic business atmosphere. Emergent strategy recognition is a sensing capability that allows companies to recognise and adapt to environmental changes. Compared to incremental innovation projects, radically new product development or disruptive innovations benefit from a more emergent approach. Creativity and adaptability are necessary for effective solutions, especially in a volatile environment, because they are less planned, irreversible, and evolutionary [19, 28].

Figure 2 illustrates the process of an emergent innovation portfolio. Like any other strategic management approach, it initiates with a vision and mission statement of the firm. This is followed by an ideation stage, targeted towards the opportunity recognition and discovery to identify ongoing and new portfolio candidates. The ideation phase is exploratory and consists of sub-phases several sub-phases. While the phase may be exploratory, organisations establish boundary conditions under which a sufficient amount of creative and valuable ideas can be generated. Therefore, there exists a "loosely defined" strategic intent for this stage, within certain specifications of innovation search. The search fields typically concentrate on customer needs in the future, new markets, megatrends, convergence of upcoming technologies or competing technologies that could enable the organization to create new offerings. The deliverables of the ideation phase are conceptual directions that could transform into portfolio candidates for products or product lines that could potentially meet the future market needs [13, 14].

Subsequently, the research phase focuses on synthesis and analysis to substantiate and provide reasons to believe in conceptual directions generated in the ideation phase. Therefore, this phase is intensive and requires substantial support from subject matter experts to estimate its initial feasibility from the technical and business perspective. This step has also been called qualitative coarse screening with a rough order of magnitude estimate of value [2, 12]. The subsequent selection phase evaluates and compares potential candidates to eliminate projects strategically. Most research in this domain concerns evaluation methods, criteria, and risk assessment that inform decision-makers to establish a probability of market winning success. Portfolio balancing, interrelationship analysis, and synergistic selection of projects are critical in these phases. They provide insight into the prioritisation of projects and innovation strategies built around the result of the synergies between the business and the projects. Lastly, innovation strategies evolve or become a byproduct of project landscape and product strategies based on available data and insights. Therefore, this process is also known as the "bottom-up approach". The emergent nature of this process allows firms to re-evaluate their competencies, strategy and vision-based practicality oriented towards the future in a relatively objective way.

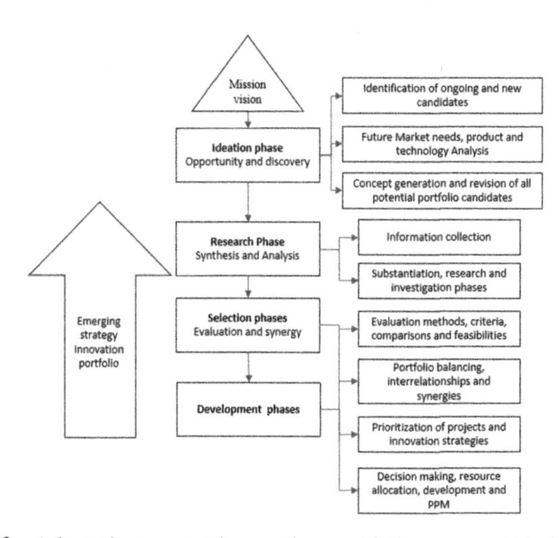

Fig. 2. Adapted emergent innovation portfolio process [19, 26, 29]

Kopmann et al. 2017, highlights the need for an IPM approach for emerging strategies in dynamic and uncertain market conditions; in contrast, the study claims that deliberate strategies are best used when short term predictability is near perfect [19, 28]. However, this process comes with its own set of disadvantages. The research argues that the ideation phase without focus generates too many irrelevant and non-implementable ideas instead of high-quality ideas. Moreover, studies claim that a systematic set of the ideation search field may increase the firm's innovation competencies, which could be debatable since more ideas do not necessarily mean better innovation competencies [19, 28]. Lastly, most tools used in this phase such as scenario planning, forecasting and backcasting, and technology road mapping occur at an abstract phase using brainstorming as one of the primary methods to create such futuristic scenarios [2, 13, 14, 21, 28]. To put it simply, there exists a lack of tools and methods used to strengthen the ideation phase of the emerging innovation portfolio process.

Deliberate Innovation Portfolios

The process of consciously cascading the defined corporate strategy to the project level in a top-down manner is called deliberate strategy implementation [19, 28]. This is also known as the planned strategy approach, wherein the innovation, organisational, and business strategies are predefined and "pushed" down to projects and used to align the resulting innovation portfolio.

Our systematic literature review found that deliberate strategies were the more dominant ones in academic literature from 2010 to 2016. Kopmann et al. 2017, argue while stable long-term strategies facilitate a deliberate strategy implementation, they also require high accuracy of long-term market forecasts and technological advancements [19, 30]. Therefore, the market's level of turbulence plays an important role in the selection of a deliberate innovation portfolio strategy. The more instability in a company's market and applicable future technologies, the less accurate these long-term forecasts will be, and the more frequently the strategies will need to be altered.

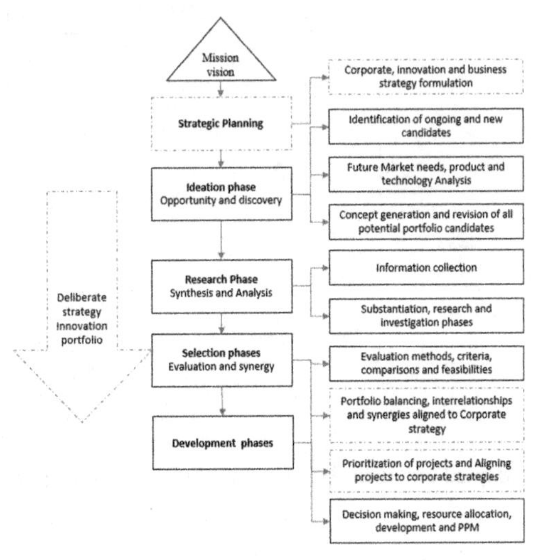

Fig. 3. Adapted deliberate innovation portfolio process [29, 31–33]

Figure 3 illustrates the alternative deliberate innovation portfolio strategy and its stages. It adds a stage of strategic planning before the ideation and discovery phase. In this phase, stakeholders and decision-makers formulate the business strategy and corporate and innovation strategy for their firm, which serves as an input into the ideation phase. These predefined strategies act as boundary conditions that, to a certain extent, restrict creativity and reject ideas misaligned with the fit of the organisation. Moreover, the top-down strategy cascades to the project landscape and eliminates projects that do not conform to the organisational strategy. Thereby leaving little room for long-term creative exploration of radical innovations in new business areas but an enormous scope for short to mid-term exploitation of existing portfolios and competencies. The level of strategic control over a deliberate innovation portfolio is significantly higher and goal-oriented. The deliberate innovation portfolio process plays a big role in the selection phases and prioritisation of projects aligned to predefined strategies wherein the top management controls strategic communication and oversight. The most obvious disadvantage of this process is a lack of flexibility and lost opportunities due to its un-explorative nature. For example, it's possible that a specific technological competency could potentially be used in a distant area of application far from the current market that this company serves, resulting in new markets and business lines.

To summarise, there seems to be a debate in the academic literature on emergent and deliberate innovation portfolio processes and a new area of "planned emergence". Wherein, both approaches are hybridised to complement each other's weaknesses [19]. However, TRIZ can effectively help in both the aforementioned approaches through its various tools and approaches.

4 Discussions

4.1 Enhancing the Innovation Portfolio Analysis Process with TRIZ

In Sect. 3.1, we analyse approaches to innovation portfolio management to reveal their strengths and weaknesses. Section 4.1 consists of an integrative literature study that assesses TRIZ tools to fill the gaps identified in both methods presented. TRIZ in innovation management literature has seen some popularity however most practitioners limit its applicability to engineering systems. TRIZ provides the practical hands-on tools to link business problems with engineering systems such as this one in innovation portfolio processes. We systematically outline each stage of the innovation portfolio process and the applicable TRIZ tool or approach to highlight potential impacts reduce the risk of innovation failure and create strategic innovation agendas.

Ideation Phase

The objective of the ideation phase in the innovation portfolio process is to generate ideas and concepts that are future-oriented i.e. discover future market needs. Scenario planning is a popular tool used by practitioners to forecast scenarios to help them identify events that are likely to happen in line with existing trends. In contrast, backcasting involves scenarios that start from a future vision to explore the resources necessary to create the envisioned future [22]. However, both approaches lack appropriate methods to forecast the future and create abstract scenarios based on obvious megatrends. Typical approaches for forecasting and backcasting techniques is heavily reliant on the voice of the customer (VOC). However, VOC can be insufficient or inaccurate to forecast the future market needs since they do not always reflect the customer's true wants and needs appropriately [34].

Additionally, technological forecasting or road mapping also involves identifying technologies and evaluating their ability to deliver potential product features. This step requires precise knowledge and trends rather than abstract assertive statements. The ideation phase in the Innovation portfolio process relies on brainstorming techniques to create scenario plans and technological forecasts. Therefore, TRIZ compensates for this weakness using the main parameters of value (MPV) analysis approach and trends of engineering systems evolution (TESE) to provide additional insights from the voice of the product and help us establish approximate developmental limits of the product lines [35, 36]. These processes help decision-makers reduce the risk of innovation failure significantly and act as an additional screening mechanism in the portfolio selection phases.

MPV Analysis and TESE
Main parameters of value are defined as a key feature of an unsatisfied offering that influences the customer's purchase decision, making it a crucial tool for developing new products [34]. One may argue that MPV analysis is a repurposed process consisting of function analysis, S curve analysis and trends of engineering systems evolution. But, MPV analysis provides the voice of the product, which reflects a product's "needs and wants" to supplement the voice of the customer as additional insight. Put another way, it is as if we question the existing product or product line and allow it to show us its direction of evolution.

Typical tools used within MPV analysis initially begin with high-level function analysis with two different perspectives of the function model. Customer-oriented function model (macro or high-level model), wherein we consider different scenarios and use cases of the product highlighting its functional disadvantages super systems to reveal latent or hidden needs of the customer [34, 35]. And secondly, a product-oriented function model or regular function analysis to discover functional disadvantages within the product isolated from the super system. Based on the results of both function models, we perform either qualitative regular S-curve analysis (based on action principle) or pragmatic S curve analysis for each potential MPV to determine their current performance level and their estimated developmental limits [36]. This process allows us to establish the feasibility and the effort of improvement needed to improve the product.

Parallelly, TESE analysis is performed to analyse evolutionary directions for the identified MPVs. This step is different from other classical scenario planning methods that are based on creating futuristic scenarios for the entire product. Instead, we use each MPV of the product to forecast its evolution resulting in more precision. Since the technology forecasting is based on the MPV of the product, it automatically accounts for competing technologies that meet the minimally acceptable level of performance for the MPV. For example, if the MPV is long-range for electric vehicles, our analysis would include all competing technologies that meet the MPV of delivering long-range in a benchmarking study. In other words, state-of-the-art and upcoming new technologies are inbuilt into this approach. Lastly, S curve analysis coupled with TESE provides essential insights into innovation activity to improve the MPV. The organisation could get insights into types of innovation projects to be undertaken, such as adding new functionalities, cost reduction, switching to a new action principle, et cetera, to improve the engineering system effectively. Abramov, 2016 showcases the process of MPV discovery and outlines the process of discovering the latent needs of a product [35]. The MPV and TESE analysis result in new conceptual directions, ideas or evolutionary directions for product lines that become new potential candidates included in the innovation portfolio to be investigated in the subsequent phases.

Selection Phases
The selection phase of the innovation portfolio process primarily has two goals. First, act as a screening mechanism and eliminate projects or concepts that do not create sufficient value for the organisation. Secondly, to help identify synergies between projects and concepts because at this stage, portfolio candidates can be merged, broken down into many divergent concepts or shelved altogether. Research suggests countless methods, evaluation criteria, and techniques to strategically select projects quantitatively

and qualitatively to address the issue of effective screening [37–40]. However, none of the methods considers the S-curve positions of the potential new product, market and organisation simultaneously to estimate their level of fit. This may be achieved by a new Gentriz tool, Quantum Economic Analysis. Whereas addressing the synergy and assessing the interdependency of projects on each other can be resolved through contradiction analysis.

Quantum Economic Analysis Screening
Quantum economic analysis (QEA-screening) is a new approach, which considers the combined stages of development of the company, its product and its target market to indicate a successful business set [41]. By considering S-curve positions of the product, market, and company simultaneously, this research provides 13 successful business combinations that claim a better chance of business success than others, irrespective of the company's financial prowess. Consequently, research suggests that if a business combination does not fall within the allowed set, its chances of failure are significantly higher regardless of the company or the product [7, 8, 42]. Recently, QEA based screening has been used in the TRIZ assisted stage-gate process and in the new product development to provide a promising 80% success rate [8]. In some cases, QEA screening has also been used as a selection mechanism for open innovation strategies and innovation management [42].

While the approach is relatively new and is not a part of modern body of knowledge in TRIZ, it may still be utilised for research purposes to anticipate and quantify its impact on innovation portfolio success. The addition of QEA-based screening in innovation portfolio selection phases could allow decision-makers to converge on a set of projects and select suitable open innovation strategies at the early stage for better resource management and parallelisation of business development activities in the stage-gate process.

Contradiction Analysis
"How many birds can you get with one stone?" is an analogy for synergy. It explains how solutions developed for one product may benefit another. Achieving portfolio balance and interrelationships between projects are necessary for project prioritisation phases and product strategies. While aligning potential projects to corporate strategy is necessary, it is equally essential to consider the synergies between several projects [17]. That is, to explore interdependencies and interrelationships across the portfolio. The technical synergy between projects explains how solutions developed for one product may benefit another, causing projects to be merged, divided, prioritised or eliminated entirely from the portfolio.

Identifying key contradictions at the macro level can be used to identify technical synergies and challenges across various product families and business units [43]. Similarly, identifying key problems or contradictions that could potentially impact the largest number of products in a portfolio could be prioritised. To elaborate, there could be four types of technical synergies possible.

Verbitsky and Casey, 2008 outline a framework for four types of synergistic situations in an innovation portfolio, as seen in Fig. 4. Considering an innovation portfolio, a firm must assess and analyse all four types of synergies to exploit their existing competencies while exploring new potential opportunities that arise from these synergies. It is

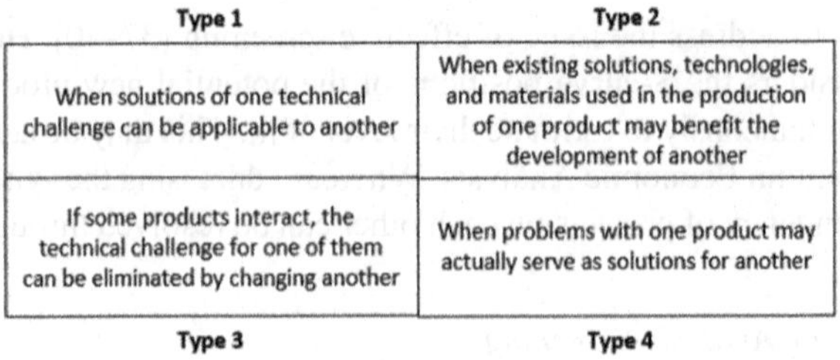

Type 1	Type 2
When solutions of one technical challenge can be applicable to another	When existing solutions, technologies, and materials used in the production of one product may benefit the development of another
If some products interact, the technical challenge for one of them can be eliminated by changing another	When problems with one product may actually serve as solutions for another
Type 3	Type 4

Fig. 4. Framework for synergies in innovation portfolios

challenging for multinational companies with hundred plus product lines to identify key contradictions; however, it is possible to estimate a key contradiction at the macro level. Decision-makers and TRIZ practitioners can cluster a group of products based on similar bottlenecks, thereby having an additional set of screening criteria in the selection phases and maximising value. The key contradiction could enable a breakthrough technology or a solution that could be replicated across multiple product lines, business units or the entire portfolio. Therefore, potential portfolio candidates must be carefully studied and researched to look out for key contradictions that may have a potentially large impact across a range of products. Moreover, resolving major bottleneck contradictions may also create super effects that can impact other product lines. Lastly, resolving key contradictions allows for replicability of solutions and adaptation of existing solutions instead of inventing new ones, similar to the clone problems approach. Table 1 summarises the applicability of TRIZ tools in innovation portfolio processes and its potential impacts.

Table 1. Summary of TRIZ tools and their impact on innovation portfolio process.

TRIZ tools	Phase of portfolio process	Potential impact on innovation portfolio
MPV analysis (combination of function analysis, TESE and S-curve analysis)	*Ideation*	• *Identifies of products voice and uncover latent needs* • *Better assessment of product and technology evolution* • *Identifies all competing technologies meeting a future MPV* • *Discovers better opportunities for disruptive innovation* • *Links future market requirements with the potential of the product to deliver on MPVs*

(*continued*)

Table 1. (*continued*)

TRIZ tools	Phase of portfolio process	Potential impact on innovation portfolio
QEA-screening (combination of S-curve positions of product, company and market simultaneously)	*Portfolio selection*	• *Insights into fit conditions of Market, product and organization* • *Selection and resource allocation wrt open innovation strategy* • *Recommends changes to either product, market or company to meet synergize the strategic fit* • *Acceleration of early-stage business development activities and faster time to market*
Contradiction analysis	*Portfolio balancing and prioritization*	• *Identifying key breakthrough contradictions impacting multiple products* • *Acts as an additional screening mechanism aiding in project prioritization* • *Resolving secondary problems that may produce super effects over other projects* • *Enables replicability of solutions through clone problems over many product lines*

5 Conclusions and Future Scope of Work

This paper analyses innovation portfolio management, qualitatively to discover two distinct approaches, emergent and deliberate innovation portfolios and describes their pros and cons. The paper systematically outlines the process to highlight their weaknesses through existing literature and suggests improvements to enhance the portfolio output using TRIZ methods. While researchers suggest several perspectives on innovation portfolios, it was surprising to investigate a relatively small amount of papers indexed in Web of science. Similarly, few approaches integrate innovation methodologies such as Ecodesign, design thinking, creativity, technology road mapping and strategic foresight with innovation portfolio processes. However, very few approaches consist of systematic methods that follow scientific approaches such as TRIZ. This study contributes to the theoretical body of knowledge and encourages practitioners to use TRIZ in management domains.

On the one hand, TRIZ is expanding its applicability to innovation management, yet studies are still scarce and consider classical TRIZ tools to be applied to business problems [4, 44–46]. In this paper, we provide an overview of potential TRIZ tools and new approaches such as MPV analysis, TESE, S-curve analysis, quantum economic analysis and contradiction analysis to potentially enhance the output of the innovation portfolio. We highlight the stages in which specific TRIZ tools can help provide additional insights to aid in the selection of the right innovation targets. Currently, this research hypothesizes TRIZ tools to be used in IPM, however, it does not provide tangible concrete examples to demonstrate the claims in practice. Therefore, paving the way for further research to quantify the impact of utilising the proposed tools in IPM and empirically test the TRIZ tools.

Acknowledgement. We thank Dr Simon Litvin, Dr Oleg Abramov and Dr Alex Lyubomirsky (GENTRIZ) for providing valuable insights on GENTRIZ tools and knowledge that aided this research.

References

1. Cooper, R.G., Edgett, S.J., Kleinschmidt, E.J.: New problems, new solutions: making portfolio management more effective. Res. Technol. Manag. **43**(2), 18–33 (2000). https://doi.org/10.1080/08956308.2000.11671338
2. Mathews, S.: Innovation portfolio architecture. Res. Technol. Manag. **53**(6), 30–40 (2010). https://doi.org/10.1080/08956308.2010.11657660
3. Shealy, T., Hu, M., Gero, J.: Neuro-cognitive differences between brainstorming, morphological analysis, and TRIZ. In: ASME International Design Engineering Technical Conferences (2018). http://mason.gmu.edu/~jgero/publications/Progress/18ShealyHuGero-ASME.pdf
4. Teplov, R.: What is known about TRIZ in innovation management? In: ISPIM Conference, June 2014. http://search.proquest.com/openview/7dcfbd982d2ea05341092a69ae6fe2ae/1?pq-origsite=gscholar
5. Livotov, P., Ruchti, B.: TRIZ-based innovation principles and a process for problem solving in business and management. Educating the Edisons of the 21st Century View project Advanced Innovation Design Approach-AIDA View Project (2001). www.diwings.ch. Accessed 04 Apr 2021
6. Kaliteevskii, V., Bryksin, M., Chechurin, L.: TRIZ application for digital product design and management. In: Borgianni, Y., Brad, S., Cavallucci, D., Livotov, P. (eds.) TFC 2021. IAICT, vol. 635, pp. 245–255. Springer, Cham (2021). https://doi.org/10.1007/978-3-030-86614-3_20
7. Abramov, O.Y.: Generating new product ideas with TRIZ-derived 'Voice of the Product' and quantum-economic analysis (QEA). In: 17th International TRIZ Future Conference, pp. 80–87 (2017)
8. Abramov, O., Markosov, S., Medvedev, A.: Experimental validation of quantum-economic analysis (QEA) as a screening tool for new product development. In: Koziołek, S., Chechurin, L., Collan, M. (eds.) Advances and Impacts of the Theory of Inventive Problem Solving, pp. 17–25. Springer, Cham (2018). https://doi.org/10.1007/978-3-319-96532-1_2
9. Abramov, O.Y., Medvedev, A.v., Tomashevskaya, N.: TRIZ fest 2019 entering adjacent markets is a widely accepted marketing strategy for developing business without, pp. 1–9, September 2019

10. Chai, K.H., Zhang, J., Tan, K.C.: A TRIZ-based method for new service design. J. Serv. Res. **8**(1), 48–66 (2005). https://doi.org/10.1177/1094670505276683
11. Otavă, E., Brad, S.: TRIZ-based approach for improving the adoption of open innovation 2.0. In: Cavallucci, D., Brad, S., Livotov, P. (eds.) TFC 2020. IAICT, vol. 597, pp. 452–464. Springer, Cham (2020). https://doi.org/10.1007/978-3-030-61295-5_34
12. Mathews, S.: Innovation Portfolio Architecture-Part 2: Attribute Selection and Valuation. A "sufficiently simple" valuation philosophy quickly identifies the most valuable concepts in an innovation portfolio while minimizing analytical time and cost (2011)
13. Heising, W.: The integration of ideation and project portfolio management - a key factor for sustainable success. Int. J. Project Manage. **30**(5), 582–595 (2012). https://doi.org/10.1016/j.ijproman.2012.01.014
14. Kock, A., Heising, W., Gemünden, H.G.: How ideation portfolio management influences front-end success. J. Prod. Innov. Manag. **32**(4), 539–555 (2015). https://doi.org/10.1111/jpim.12217
15. Abramov, O.Y.: Innovation funnel of modern triz: experimental study to show the efficacy of the TRIZ-assisted stage-gate process (2018)
16. Abramov, O.Y.: Innovation funnel of modern TRIZ: experimental study to show the efficacy of the TRIZ-assisted stage-gate process. In: TRIZfest 2018, pp. 105–110, September 2018
17. Abramov, O.Y.: TRIZ-assisted stage-gate process for developing new products. J. Financ. Econ. (2014). https://doi.org/10.12691/jfe-2-5-8
18. Saunders, M., Lewis, P., Thornhill, A.: Research Methods for Business Students. Pearson, London (2016)
19. Kopmann, J., Kock, A., Killen, C.P., Gemünden, H.G.: The role of project portfolio management in fostering both deliberate and emergent strategy (2017). https://doi.org/10.1016/j.ijproman.2017.02.011
20. Brook, J.W., Pagnanelli, F.: Integrating sustainability into innovation project portfolio management - a strategic perspective. J. Eng. Technol. Manage **34**, 46–62 (2014). https://doi.org/10.1016/j.jengtecman.2013.11.004
21. Hadjinicolaou, N., Kader, M., Abdallah, I.: Strategic innovation, foresight and the deployment of project portfolio management under mid-range planning conditions in medium-sized firms. Sustain. (Switz.) **14**(1), 80 (2022). https://doi.org/10.3390/su14010080
22. Villamil, C., Schulte, J., Hallstedt, S.: Sustainability risk and portfolio management—a strategic scenario method for sustainable product development. Bus. Strateg. Environ. **31**(3), 1042–1057 (2022). https://doi.org/10.1002/bse.2934
23. Randhawa, K., Nikolova, N., Ahuja, S., Schweitzer, J.: Design thinking implementation for innovation: an organization's journey to ambidexterity. J. Prod. Innov. Manag. **38**(6), 668–700 (2021). https://doi.org/10.1111/JPIM.12599
24. Paula Pinheiro, M.A., Jugend, D., Dematte Filho, L.C., Armellini, F.: Framework proposal for ecodesign integration on product portfolio management. J. Clean. Prod. **185**, 176–186 (2018). https://doi.org/10.1016/j.jclepro.2018.03.005
25. Ram Irez, R., Roodhart, L., Manders, W.: How Shell's domains link innovation and strategy. Long Range Plan. **44**(4), 250–270. https://doi.org/10.1016/j.lrp.2011.04.003
26. Oliveira, M.G., Rozenfeld, H.: Integrating technology roadmapping and portfolio management at the front-end of new product development. Technol. Forecast. Soc. Change **77**(8), 1339–1354 (2010). https://doi.org/10.1016/j.techfore.2010.07.015
27. Phaal, R., Simonse, L., den Ouden, E.: Next generation roadmapping for innovation planning. Int. J. Technol. Intell. Planning **4**(2), 135–152 (2008). https://doi.org/10.1504/IJTIP.2008.018313
28. Clegg, S., Killen, C.P., Biesenthal, C., Sankaran, S.: Practices, projects and portfolios: current research trends and new directions. Int. J. Project Manage. **36**(5), 762–772. https://doi.org/10.1016/j.ijproman.2018.03.008

29. Pashley, D., Tryfonas, T., Crossley, A., Setchell, C., Karatzas, S.: Innovation portfolio management for small-medium enterprises. J. Syst. Sci. Syst. Eng. **29**(5), 507–524 (2020). https://doi.org/10.1007/s11518-020-5467-z
30. Cooper, R.G., Kleinschmidt, E.J., Edgett, S.J.: New product portfolio management practices and performance. J. Prod. Innov. Manage. **16**(4), 333–351 (2004)
31. Xu, Q., Zhao, X., Shaohua, W., Chin, J.: Competence-based innovation portfolio. In: Proceedings of the 2000 IEEE International Conference on Management of Innovation and Technology, vol. 1, pp. 134–139 (2000). https://doi.org/10.1109/ICMIT.2000.917303
32. Klingebiel, R., Rammer, C.: Resource allocation strategy for innovation portfolio management. Strateg. Manag. J. **35**(2), 246–268 (2014). https://doi.org/10.1002/smj.2107
33. Kock, A., Georg Gemünden, H.: Antecedents to decision-making quality and agility in innovation portfolio management. J. Prod. Innov. Manage. **33**(6), 670–686 (2016). https://doi.org/10.1111/JPIM.12336
34. Abramov, O.Y.: Product-oriented MPV analysis to identify voice of the product. In: TRIZfest 2016, Beijing, People's Republic of China, 26–28 July 2016. Introduction: Existing MPV Analysis is Insufficient to Identify the VOP, August 2016
35. Abramov, O., Medvedev, A., Tomashevskaya, N.: Main parameters of value (MPV) analysis: where MPV candidates come from. In: Borgianni, Y., Brad, S., Cavallucci, D., Livotov, P. (eds.) TFC 2021. IAICT, vol. 635, pp. 391–400. Springer, Cham (2021). https://doi.org/10.1007/978-3-030-86614-3_31
36. Lyubomirskiy, A., Itvin, S., Ikovenko, S., Aduka, R., Thurnes, C.M.: Trends of Engineering Systems Evolution (TESE) TRIZ paths to innovation (2014). http://library1.nida.ac.th/termpaper6/sd/2554/19755.pdf
37. Flechas Chaparro, X.A., de Vasconcelos Gomes, L.A., Tromboni de Souza Nascimento, P.: The evolution of project portfolio selection methods: from incremental to radical innovation. Revista de Gestao **26**(3), 212–236 (2019). https://doi.org/10.1108/REGE-10-2018-0096/FULL/PDF
38. Montajabiha, M., Khamseh, A.A., Afshar-Nadjafi, B.: A robust algorithm for project portfolio selection problem using real options valuation. Int. J. Manag. Proj. Bus. **10**(2), 386–403 (2017). https://doi.org/10.1108/IJMPB-12-2015-0114
39. Klingebiel, R., Rammer, C.: Resource allocation strategy for innovation portfolio management. Strateg. Manage. J. **35**, 246–268 (2014). https://doi.org/10.1002/smj.2107
40. Browning, T.R., Yassine, A.A.: Managing a portfolio of product development projects under resource constraints. Decis. Sci. **47**(2), 333–372 (2016). https://doi.org/10.1111/deci.12172
41. Topchishvili, G., Katsman, Y., Schneider, A.: Science to win in investment. Management and Marketing (2003). http://1drv.ms/1D5ikaA%5Cnleader/nauka_pobezhdat_k_1.djvu
42. Phadnis, N.: Integrating open innovation strategies with quantum economic analysis (2021). https://lutpub.lut.fi/bitstream/handle/10024/162408/MastersThesis_Finalversion1.pdf?sequence=1&isAllowed=y
43. Altshuller, G.S.: The innovation algorithm: TRIZ, systematic innovation and technical creativity, p. 312. Technical Innovation Center, Inc. (1999). http://www.amazon.com/dp/0964074044
44. Abramov, O., Sobolev, S.: Current stage of TRIZ evolution and its popularity. In: Chechurin, L., Collan, M. (eds.) Advances in Systematic Creativity, pp. 3–15. Springer, Cham (2019). https://doi.org/10.1007/978-3-319-78075-7_1
45. Souchkov, V.: Breakthrough thinking with TRIZ for business and management: an overview (2007). www.xtriz.com. Accessed 04 Apr 2021
46. Chechurin, L., Borgianni, Y.: Understanding TRIZ through the review of top cited publications. Comput. Ind. **82**, 119–134 (2016). https://doi.org/10.1016/j.compind.2016.06.002

Development of an Ontology of Sustainable Eco-friendly Technologies and Products Based on the Inventive Principles of the TRIZ Theory (OntoSustIP) – Research Agenda

Claudia Hentschel[1], Kai Hiltmann[2], Norbert Huber[3], Pavel Livotov[4], Horst T. Nähler[5], Christian M. Thurnes[6](✉), and Agata M. Wichowska[4]

[1] HTW Berlin, University of Applied Sciences, 10318 Berlin, Germany
[2] Coburg University of Applied Sciences and Arts, 96450 Coburg, Germany
[3] Weihenstephan-Triesdorf University of Applied Sciences, 91746 Weidenbach, Germany
[4] Offenburg University of Applied Sciences, 77652 Offenburg, Germany
[5] c4pi - Center for Product Innovation c4pi, 36088 Hünfeld, Germany
[6] Hochschule Kaiserslautern (UAS), 66482 Zweibrücken, Germany
christian.thurnes@hs-kl.de

Abstract. Rising societies' demands require more sustainable products and technologies. Although numerous methods and tools have been developed in the last decades to support environmental-friendly product and process development, an interdisciplinary knowledge base of eco-innovative examples linked to the eco-innovative problems and solution principles is lacking. The paper proposes an ontology of examples for eco-friendly products and technologies assigned to the Inventive Principles (IPs) of the TRIZ methodology in accordance with the German TRIZ Standard VDI 4521. The examples of sustainable technologies and products build a database for sharing and reusing eco-innovation knowledge. The ontology acts as a tool for systematic solving of specific environmental problems in typical life cycle phases, for different environmental impact categories and engineering domains. Finally, the paper defines a future research agenda in the field of the TRIZ-based systematic eco-innovation.

Keywords: Sustainability · Eco-innovation · Eco-design · Knowledge-based innovation · Ontology · TRIZ

1 Introduction

1.1 Sustainability in a Nutshell

It has become common knowledge that the term sustainability originates from forestry, where it was meant to describe that no more should be cut down in a forest than will grow back. In agriculture, there is a direct connection between sowing and harvesting, but in contrast to classical agriculture, this connection was particularly pronounced in

Published by Springer Nature Switzerland AG 2022
R. Nowak et al. (Eds.): TFC 2022, IFIP AICT 655, pp. 367–381, 2022.
https://doi.org/10.1007/978-3-031-17288-5_31

forestry: both in terms of time and space, it takes a longer time to regrow a forest than e.g. a vegetable, and moreover, regrowing a forest had to be undertaken at another place. Since forestry is no longer the main economic sector of industrial countries, this understanding of sustainability has become too narrow. Current economic and business activity is considered as affecting a variety of (not only) natural resources, be they near and/or far away. The entire tendency for long-term economic and business developments to be unsustainable is now being questioned, e.g., when short-term success is strongly rewarded, whereas e.g., landscape consumption or air pollution in a distant place is ignored.

1.2 Sustainability and Artificial Intelligence - A Twofold Approach

The term of sustainability has therefore gained new attention, even the European Community has been discussing new definitions for the term, as can be seen by the "European Green Deal" and the EU Taxonomy Regulation [1]. New general approaches - be they temporally or spatially staggered - follow the direction that no negative externalities should be connected to an economic measure: Sustainability should not be a marketing buzzword, but should be about long-term oriented economic activity.

This raises the question what kinds of sustainable products, technologies, and services exist and may be conceived. The concept of sustainability alone not necessarily leads to sustainable products and technologies: There are solutions, be it products or processes, that are to be considered sustainable in the long term, but harmful and cost-intensive in the short term – and vice versa (education policy can be taken as an example for cost-intensity on the short term, but sustainable on the long term).

This pacing problem (e.g., also highlighted by Matthews, Stamford and Shapira [2]) triggers the conflict thinking approach of TRIZ, which is why we are searching for solution examples for sustainable product and technology development soon and envisage a long-term Artificial Intelligence (AI) approach later.

The automation of business processes has come a long way, because today it is possible to automate any process whose sequence is precisely defined in the form of algorithms. It is more challenging with sustainability processes that require a complex assessment, and are characterized by uncertainty, missing information, or changing evaluation criteria, for example. Such processes are rather difficult to map with conventional means, for example, using IF…THEN…ELSE applications. The upcoming challenges of sustainability are not that easy to be evaluated today and thus require unconventional approaches. For example, how can a product developer or designer assess if a technology is likely to fail on low eco-friendliness in the future? Assuming that the actual evaluation is fully sufficient to service actual evaluation criteria, it is apparent that many characteristics are ambiguous and may change over time. Under this premise, the AI-application for anticipatory sustainable product and process development can be envisaged in a meaningful way.

Our approach to cope with this pacing problem is therefore twofold. In a first step, solution examples of sustainable products and technologies shall be collected and categorized not only according to the contradictory approach regarding the various categories of natural resources they are affecting. These solutions will also be assigned to their inherent Inventive Principle(s) of the TRIZ methodology. As a second step, this allows to define future research topics in the field of eco-friendly product and process design, paving the way for future systematic and AI-supported eco-innovation.

2 The TRIZ 40 Inventive Principles and Their Relation to Sustainability

The development of technical systems can have various goals. Beside the very basic goal of simple functionality, additional development goals may be efficiency, cost, simplicity, robustness, or other attributes of machines and devices. With a growing world population, the impact of technical systems on the environment as a foundation for life becomes more and more important.

According to [3], the TRIZ concepts of Ideality, resource-oriented and compromise-free problem-solving fit in perfectly with the strategy of sustainable eco-innovation. The contradictions oriented TRIZ thinking helps to identify secondary problems in the sustainable solutions and to creatively solve such eco-contradictions in advance. TRIZ also helps to mobilize resources of the existing products and processes and to reduce the negative environmental impact of technologies without efficiency losses.

A recent systematic review by one of the co-authors identifies more than 60 papers on eco-innovation methods using TRIZ elements or tools [3]. Many of them apply the 40 Inventive Principles (IPs) as one of the most important and frequently used TRIZ components and propose tools for IPs-selection for eco-innovation based on the concept of the 39 $\times$ 39 contradictions matrix. As well known, the classical TRIZ 39 $\times$ 39 contradiction matrix recommends best fitting IPs to couples of contradicting common generalized engineering parameters, an approach that seems appropriate for sustainability goals as well. For example, Chen and Liu propose a matrix containing 39 TRIZ engineering parameters and 7 eco-efficiency elements [4]. Kobayashi supports eco-innovative product design with 39 $\times$ 39 TRIZ contradiction matrix and 40 inventive principles [5]. Fitzgerald et al. conceptualize a 21 $\times$ 6 innovation matrix with 6 eco-goals and 21 functional eco-parameters [6]. Pokhrel et al. adapt 39 $\times$ 39 contradiction matrix to a 14 $\times$ 14 matrix with 3 eco-engineering parameters and 8 solution principles [7]. Ko and Chen propose eco-contradiction matrix between customer needs and eco-efficiency elements, recommending TRIZ inventive principles [8]. To this end, environmental impacts – positive or negative – also need to be considered and analyzed in order to protect the environment and ensure sustainable life.

The authors therefore suggest developing an ontology of sustainable eco-friendly technologies and products and on this foundation to statistically find correlations of IPs and environmental parameters that are affected by inventions. For this purpose, a large number of inventions will need to be analyzed. It is planned to do this analysis in an open innovation fashion by involving the entire community of inventors and scientists interested in IPs and environment.

Ontologies are broadly used in engineering for representation and classification of knowledge and for definition of relations between categories, concepts, and other entities of a specific domain. In accordance with [9] three different types of ontologies can be distinguished based on their levels of generality: a) generic or foundational ontologies, which capture general domain independent knowledge; b) domain ontologies, which capture the knowledge in a specific engineering domain; c) application ontologies, which capture the knowledge necessary for a specific application. One of the first attempt to conceptualize an ontology for the whole specific domain of TRIZ dates back to 2009 [10]. Another conceptual work [11] proposes an ontology as an advantageous approach to bridge the problems in engineering, formulated with the help of TRIZ in terms of contradictions, with similar trade-offs in solutions in biology. The proposed ontology concept is seen as first step towards a complete AI-based problem solving system using biomimetics. Finally, the WUMM TRIZ Ontology Project [12] cites the previous research on TRIZ ontologies and aims to present the TRIZ body of knowledge as a machine-readable Resource Description Framework for the Semantic Web.

The authors advocate a hypothesis that the application of TRIZ in the field of eco-innovation requires a hierarchical structure of general engineering and ecological parameters, which comprehensively and precisely help to define various engineering contradictions in products or technologies in all phases of their life cycle. Thus, the proposed research project will try to find out how the list of classical 39 TRIZ engineering parameters must be replaced, extended or amended in the field of eco-innovation. This leads to the question of what are the relevant environmental parameters? On the one hand, these shall cover all needs of environmental protection in order to allow for all 17 sustainable development goals (SDG17) listed by the UN [13]. On the other hand, the parameters shall be detailed enough to generate a meaningful output. The SDG17 themselves do not seem to be very suitable as categories since these goals are strongly cross linked and contain many social aspects that depend on technical aspects again.

A more comprehensive way suitable for the world of engineering seems to follow the concept of life cycle assessment, LCA, [14] that is part of the environmental management described in ISO 14001 [15]. The role of life cycle assessment in a so-called Life Cycle Sustainability Assessment is further described by Klöpffer and Grahl [16]. Summed up in a few words, this concept starts with the prerequisites of manufacturing a product, where materials, land and energy are needed. It continues with the manufacture of the product again with land use, energy use, and many more aspects of environmental impairment. The same holds during the use of the product and at the end of its life when recycling and reuse should replace the waste of the product. As a suggested state-of-the-art method in LCA [14], an extended set of environmental impact categories consists of 5 main categories and in total 22 subcategories listed in Table 1. Another list of 14 environmental impact parameters for process engineering is presented in [3]. The empirically identified relationship between the eco-parameters helps engineers to see how one improved eco-parameter can affect the other eco-parameters either positively or negatively.

Derived from existing lists of eco-parameters, a reduced set of the following 10 categories will be used for initial assessment of sustainable inventions in a first step: Material consumption, Energy consumption, Land use, Water protection, Air protection, Climate protection, Waste avoidance, Safety, Protection of animals and plants, Cost reduction of sustainable product or technology. Additional categories or sub-categories can be easily added if necessary.

Table 1. Environmental impact categories according to Klöpffer & Grahl [16].

Consumption of resources	Abiotic resources (water and others)
	Biotic resources
	Land use
Chemical Emissions	Climate change
	Stratospheric ozone depletion
	Formation of photo-oxidants
	Acidification
	Eutrophication
	Human toxicity
	Eco-toxicity
	Odour
Physical Emissions	Noise
	Radioactivity
	Waste heat
Biological Emissions	Impacts on Ecosystems
	Impacts on Humans
Further categories	Casualties
	Health at the working place
	Drainage, erosion and salting of soils
	Destruction of landscapes
	Ecological systems, biodiversity
	Solid waste

3 The OntoSustIP Project at a Glance

3.1 Objectives

The primary aim of the OntoSustIP project is to identify examples of sustainable eco-friendly technologies and products, to assign them to the underlying Inventive Principles of TRIZ (according to VDI guideline VDI 4521 [17]) and to other categories of the systematic eco-innovation, such as environmental impact, eco-engineering contradictions, engineering domains, physical or chemical operating principles, life cycle phases, and others, as illustrated in Fig. 1. The empirical analysis examples of sustainable technologies and products will help to build an application ontology for sharing and reuse of eco-innovation knowledge.

The further objective of the project is the identification of strongest TRIZ IPs and inventive sub-principles to counteract specific negative impact on environment according to pre-defined eco-parameters. Such statistically strongest IPs can be in a targeted way applied to resolve typical eco-engineering contradictions. The eco-engineering contradiction is defined here as a situation in which the improvement of one eco-parameter (e.g. reduction of energy consumption) implies a deterioration of other engineering or ecological parameters (e.g. complexity of manufacturing or amount of waste) within a system or eco-system.

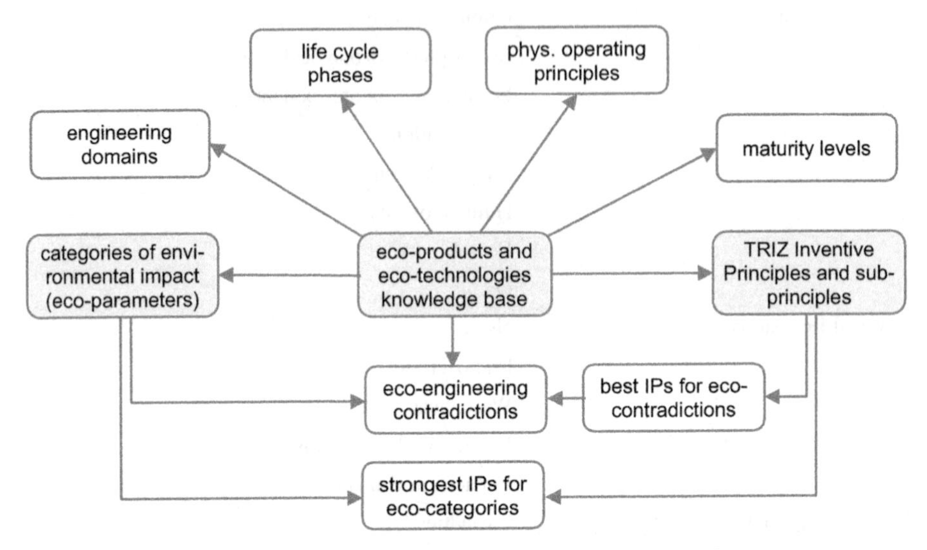

Fig. 1. Basic structure of the ontology in OntoSustIP project.

Finally, the project will create a base to define a future research agenda in the field of the TRIZ-based systematic eco-innovation. For example, it can help to evaluate the potential of TRIZ extension with the specific eco-inventive operators.

3.2 Web-Based Questionnaire for Data Collection

The first thought for creating this application ontology was related to the typical way in TRIZ of using a broad and extensive patent analysis as a starting point. Appropriate approaches have proven to be very insightful in the past (see e.g. [3, 6]). On the other hand, however, an enormous number of resources is required to perform these analyses for a large number of example solutions.

This is not only due to the process of a patent analysis per se, but also due to the peculiarity that the sought-after feature "eco-friendly" is not directly recognizable in patents. Furthermore, the basic problem is that "eco-friendly" is not an absolute characteristic of a product or technology:

- Products or technologies can be viewed from different perspectives and can therefore be assessed differently with regard to the parameter "eco-friendly". This becomes evident in the current European discussion on the assessment of the sustainability of various sources of energy. For example, there are very different opinions on the eco-friendliness of nuclear power.

- Moreover, the question of whether an innovation in a product or technology is eco-friendly can often only be answered in comparison with other solutions. These, in turn, can be relevant in different ways depending on the framework conditions and the situation – especially when comparable problem solutions influence different sustainability dimensions to different degrees.

Therefore, an open approach is used in the OntoSustIP project. A web-based questionnaire searches for examples of eco-friendly products and technologies. The assessment of eco-friendliness is deliberately left to the respondents in order to collect as much data as possible. Even though some categories and selection options prescribe the eventual structure of the results, the basic evaluation of the sustainability impact of the example is left to the participants who do the entries, as illustrated in Fig. 2. This approach leads to a higher number of expected entries, but also requires intensive post-processing of the entries by the project team.

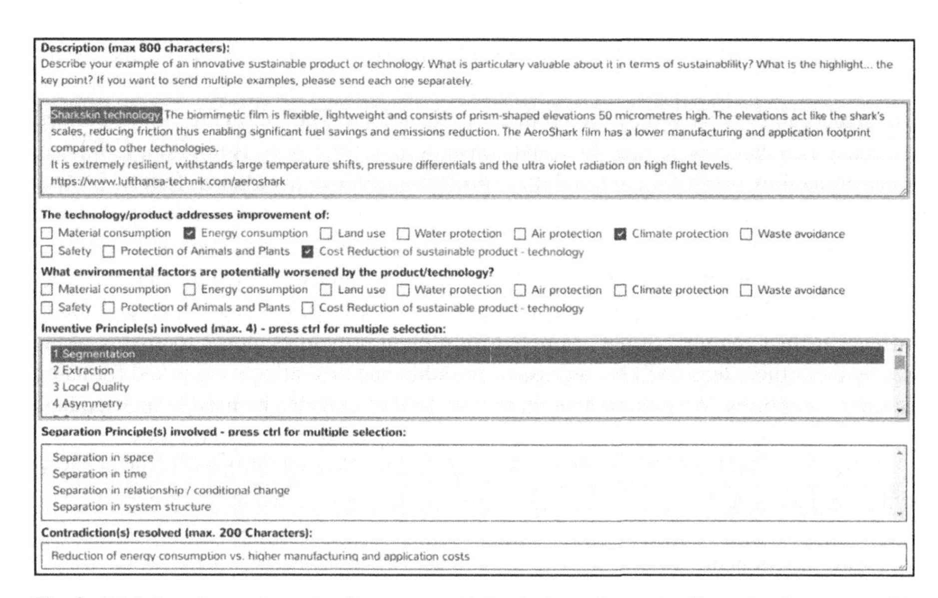
Description (max 800 characters):
Describe your example of an innovative sustainable product or technology. What is particulary valuable about it in terms of sustainablility? What is the highlight... the key point? If you want to send multiple examples, please send each one separately.

Sharkskin technology The biomimetic film is flexible, lightweight and consists of prism-shaped elevations 50 micrometres high. The elevations act like the shark's scales, reducing friction thus enabling significant fuel savings and emissions reduction. The AeroShark film has a lower manufacturing and application footprint compared to other technologies.
It is extremely resilient, withstands large temperature shifts, pressure differentials and the ultra violet radiation on high flight levels.
https://www.lufthansa-technik.com/aeroshark

The technology/product addresses improvement of:
☐ Material consumption ☑ Energy consumption ☐ Land use ☐ Water protection ☐ Air protection ☑ Climate protection ☐ Waste avoidance
☐ Safety ☐ Protection of Animals and Plants ☑ Cost Reduction of sustainable product - technology

What environmental factors are potentially worsened by the product/technology?
☐ Material consumption ☐ Energy consumption ☐ Land use ☐ Water protection ☐ Air protection ☐ Climate protection ☐ Waste avoidance
☐ Safety ☐ Protection of Animals and Plants ☐ Cost Reduction of sustainable product - technology

Inventive Principle(s) involved (max. 4) - press ctrl for multiple selection:
1 Segmentation
2 Extraction
3 Local Quality
4 Asymmetry

Separation Principle(s) involved - press ctrl for multiple selection:
Separation in space
Separation in time
Separation in relationship / conditional change
Separation in system structure

Contradiction(s) resolved (max. 200 Characters):
Reduction of energy consumption vs. higher manufacturing and application costs

Fig. 2. Web-based questionnaire (fragment with basic data of sustainable technology example).

3.3 Gathering Information for the Knowledge Base

A web-based questionnaire will be used to create the knowledge basis, on which the ontology will be developed. Since one of the goals of the OntoSustIP project is to collect a large number of examples of eco-friendly products and technologies, not only TRIZ experts but the eco-innovators and specialists are asked to contribute. The survey must also be addressed to people who have little or no knowledge of TRIZ – so there is a contradiction that the questionnaire has to fit for respondents with high TRIZ-experience and also for respondents with no TRIZ-experience. To solve this contradictory requirement, the questionnaire was segmented into a mandatory part to be completed by all respondents and another part that is open to all respondents but requires the expertise of TRIZ experts. In addition, the project team has decided to post-process all entries. This post-processing may (but does not have to) include a more detailed consultation with a respondent or other experts at a later point in time.

Descriptions of Products and Technologies. The core of the survey is the search for eco-friendly, sustainable products or technologies. As mentioned above, the assessment of whether an artifact is eco-friendly or not is left to the respondents themselves in order to receive an open and broad selection of entries. A continuous text should describe the product or technology. The knowledge and skills of the respondents will be very different – therefore, the expected entries may range from merely naming artifacts to entries that contain distinct technical information. In order to be able to include the subjective evaluation aspect, the questionnaire also asks respondents to state where they can identify particular aspects of sustainability in the respective example.

Categories: The 40 IPs and Environmental Impacts. When asked about the example, respondents are given the opportunity to tick the sustainability field in which positive effects can be achieved by the product or technology. The following 10 pre-defined eco-categories can be selected: Material consumption, Energy consumption, Land use, Water protection, Air protection, Climate protection, Waste avoidance, Safety, protection of Animals and Plants, and Cost reduction. An input field for free text permits to propose and assign additionally or alternatively any other eco-parameter or category.

The additional input options provide up to four Inventive Principles per example. Moreover, also four TRIZ separation principles can be assigned to a sustainable solution. Since an artifact can serve as an example for different principles, post-processing of the data by the project team will be necessary. Possible negative effects are asked for using the same categories. We assume that these possibilities can also be used by participants who are familiar with the corresponding sustainable products or technologies.

A free text field allows specification of contradictions which have been resolved by the product or technology. This field is also primarily directed at TRIZ experts. The formulation of the contradiction can then be checked for consistency with the innovation principles mentioned.

Action Plan and Time Frame. The project in its first phase has initially been limited to the duration of one year (01/22 to 01/23). During this time, the following activities are planned:

- until 02/22: conception of the approach and the website for the collection of contributions.
- 01/22–04/22: conception of the structural design of the ontology
- 02/22–07/22: pretests of the web-based questionnaire and first collection of contributions (example products and technologies) for the ontology
- 07/22 onwards: collection of contributions, editorial work and additions, incorporation into the structure to be developed and preparation of the regular publication
- 12/22: publication of the intermediate data

- 01/23 onwards: editing of the contributions received, insertion into the ontology and publication (every 3 months)
- 12/23: Decision or conception of continuation of the activities after the end of the first phase of project.

Support the Project with your Experience. The authors invite interested persons and organizations to bring their broad knowledge, experience with innovative sustainable products or technologies, and TRIZ skills to the OntoSustIP project. Please participate in collecting example cases and classifying or evaluating them according to given criteria on the following project website: https://www.hs-kl.de/en/betriebswirtschaft/aktivitaeten/kompetenzzentrum-opinnometh/ontosustip-project.

4 Future Use-Cases for Sustainable Eco-Friendly Design Based on TRIZ

4.1 Possible Outcomes of OntoSustIP

The results of this project are expected to offer a number of possible use cases. With a growing number of classified examples, several outcomes are conceivable. However, a solid and unambiguous assignment of the examples to the respective principles is crucial, highlighting the importance of the editing phase of this project. For this reason, a solid base for the description of 40 Inventive Principles with extended list of 160 operators or sub-principles presented in [3] will be used.

As depicted in the Fig. 3, the database entries give examples that improve certain environmental impacts (eco-parameters) and Inventive Principles that can be identified within an example. On a most basic level, the database will provide a substantial collection of sustainability examples for the TRIZ Inventive Principles. The evaluation of probable relations between the eco-parameters and the IPs has a substantial value for further application of the ontology, such as identification of the strongest IPs for eco-categories and eco-engineering contradictions.

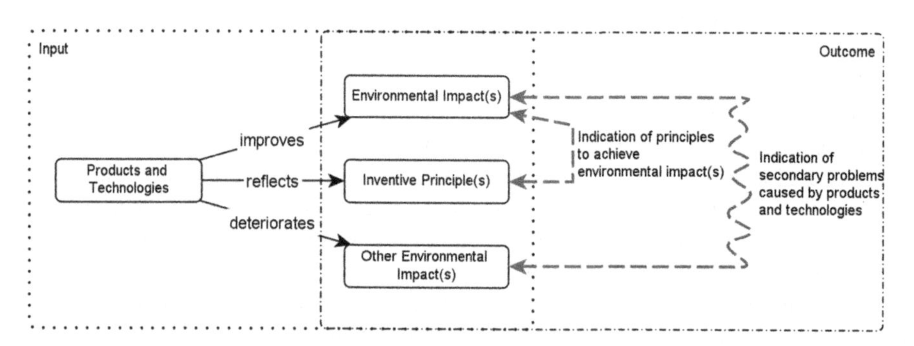

Fig. 3. Database entries given by the contributors and outcome.

For this purpose, the statistical frequency of Inventive Principles used to improve a certain sustainability factor (eco-parameter) have to be assessed. Given a sufficient number of examples are entered into the database to justify a statistical evaluation and the principles have been assigned with high confidence, several possible outcomes are depicted in Fig. 4.

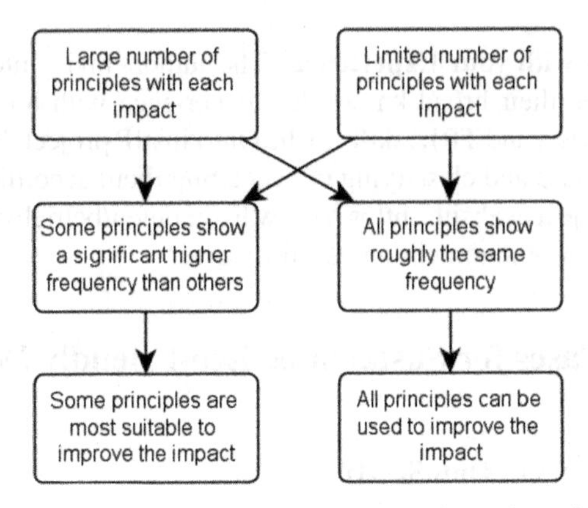

Fig. 4. Evaluation of relation of impact and inventive principles

In terms of usefulness, it will be most helpful if a distinct set of principles can be identified to address specific environmental impacts that need to be improved in new problem-solving situations. In this way the users (engineers, specialists, or ideation teams) can focus on the principles most probably able to solve their problems, enriched by examples that have been entered initially into the database. If all principles seem to be equally useful, the users will at least dispose of numerous examples to spark idea generation and support the creative process by building analogies to the examples at hand.

With the additional collection of environmental impacts that deteriorate due to the solution given by the product or technology entered, several secondary problems can be identified to be addressed in the next step. Thus, another outcome of the OntoSustIP project could be a repository of eco-engineering contradictions and challenges for further eco-innovation.

Table 2 illustrates the top 10 inventive operators (sub-principles) with corresponding parent IPs for reduction of energy consumption and losses in process engineering. The inventive operators are sorted by their occurrence frequency in the analyzed 58 thermal process engineering operations and has been derived from data presented in [3]. In general, the identification of strongest inventive operators for different environmental parameters and eco-problems specific to various engineering domains appears to be a more efficient and precise ideation technique.

The OntoSustIP project might even result in an updated contradiction matrix for eco-innovation. Contributors can enter an engineering and/or physical contradiction that is solved by the example. However, the quality and informative value of the formulated contradictions highly depend on the experience of the contributor and might be subject to a downstream editorial process.

Table 2. Top 10 inventive operators (sub-principles) for reduction of energy consumption in process engineering equipment [3].

Pos	Inventive operator	Ranking	Parent TRIZ Inventive Principle
1	2a. Take out disturbing parts	0,053	2. Leaving out / Trimming
2	29e. Heat transfer and exchange	0,052	29. Pneumatic or hydraulic constructions
3	25b. Utilize waste resources	0,041	25. Self-service
4	2e. Extract useful element	0,036	2. Leaving out / Trimming
5	5b. Combine functions	0,034	5. Combining
6	10a. Prior useful function	0,034	10. Prior action
7	22a. Utilize harm	0,034	22. Converting harm into benefit
8	35d. Change temperature	0,034	35. Transform physical and chemical properties
9	35a. Change aggregate state	0,033	35. Transform physical and chemical properties
10	28a. Use electromagnetic fields	0,032	28. Replace mechanical working principle

4.2 First Impressions of Exemplary Data Collection and Evaluation

The following section illustrates a preliminary analysis of 50 eco-friendly products and sustainable technologies inspired by nature [18], systematically accessed form the AskNature database of the Biomimicry Institute [19] and some other sources. There is a reason to believe that the eco-friendly biomimetic technologies show a lesser secondary negative environmental impact in comparison with purely engineering inventions. The selected eco-products and technologies comprise, among others, building insulation, water filtration, underwater adhesive, air-purifying billboard, life-saving surgical superglue, self-cooling buildings, shark skin technology, biodegradable packaging. For example, the shark skin-inspired coatings and films (IP30. Flexible shells or thin films) improve aerodynamic performance and protect surfaces against biofouling thus enabling significant reduction in fuel or energy consumption and emissions [20].

The initial results of the study are briefly presented in the Table 3 and 4 in accordance with the study performed by a co-author [18]. Table 3 indicates the percentage of environmental parameters improved by a product or technology. Among 50 analyzed examples, the solutions related to climate protection (36%), energy consumption (34%), material consumption (16%), and waste avoidance (14%) constitute the majority of all eco-innovation goals, whereas 8% of eco-examples could not be assigned to any pre-defined eco-parameter. The evaluation of the eco-parameters, as defined in more detail in Table 3, allowed a multiple assignment of parameter to the eco-solutions with on average 1.6 eco-parameters improved by product or technology.

Table 4 presents the Inventive Principles most frequently "used" in 50 analysed eco-innovations. These principles show a significant higher frequency of application than

Table 3. Frequency of mentioning of environmental eco-parameters improved in 50 sustainable products or technologies, inspired by natural solutions (multiple choice).

Eco-parameter (eco-category)	Description of eco-parameter	Freq.
1. Material consumption	Raw material usage, material consumption, material losses, consumables	16%
2. Energy consumption	Amount of energy consumed, energy losses, waste energy, e.g., waste heat	34%
3. Land use	Sustainable use of land, amount of used land, land protection, drainage, erosion and salting of soils, destruction and loss of landscapes	2%
4. Water protection	Groundwater pollution chemicals, eutrophication, pollution and excessive usage of water	10%
5. Air protection	Air pollution, fly ash, dust, smog, micro- and nano particles, acidic gases emissions	4%
6. Climate protection	Global warming, greenhouse gas emissions, ozone layer depletion, deforestation	36%
7. Waste avoidance	Solid waste, packaging waste, non-recyclable waste, plastic, toxic waste	14%
8. Safety	Safety and technogenic risks, radioactivity, hazardous materials, safety in case of natural disasters	10%
9. Protection of animals and plants	Preservation and protection of animals, plants, and their habitats	8%
10. Cost reduction of sustainable product or technology	Making an existing or new sustainable products or technologies more affordable and attractive for customers, industry and society	12%

others. Moreover, 20 out of 40 IPs have a frequency of mentioning between 0 and 2%. A multiple assignment of IPs to the eco-solutions results in on average 2.4 IPs per product or technology. Further analysis should identify the statistically strongest sub-principles and the relationships between the Inventive Principles and improved environmental parameters.

Table 4. Frequency of mentioning of top 10 Inventive Principles in 50 sustainable products or technologies, inspired by natural solutions (multiple choice).

Frequency	Identified IPs		Frequency	Identified IPs
28%	30. Flexible shells or thin films		18%	31. Porous materials
24%	25. Self-service / Use of resources		14%	3. Local quality
22%	40. Composite materials		14%	28. Replace mechanical working principle
22%	35. Transform physical and chemical properties		10%	15. Dynamism and adaptability
20%	17. Shift to another dimension		8%	14. Sphericity and Rotation

5 Discussion and Summary

When new technologies are used, their unintended side effects may not be recognized until they have become so widespread that they are difficult to reverse. This pacing problem that in literature is well-known by the term "the Collingridge Dilemma" [21] is also valid for sustainability and eco-friendliness: awareness awakens when negative side-effects appear to a wider public, but an evaluation only becomes accessible when a comparative value can be used. Moreover, this evaluation should be possible at the early stage of product and process development. The ontology of examples for eco-friendly products and technologies assigned to the TRIZ Inventive Principles should enhance systematic creativity and accelerate eco-innovation.

The proposed OntoSustIP project firstly tries to evaluate existing eco-solutions but is directed at facilitation of the early assessment of eco-friendly products and technologies in the near future. Nevertheless, even an evaluation will not hide the fact that every new technology has its downsides. It shall be taken into consideration that new ideas shall be given enough space to develop and improve. The OntoSustIP project provides a guideline on how TRIZ methodology can accompany here.

The initial evaluation of already collected eco-friendly products and technologies demonstrates significant time expenditures of data post-processing by the project team and requires a standardized protocol to avoid any misinterpretation of results. Therefore, we advocate the idea that the application of machine learning algorithms can considerably speed-up the evaluation procedure and to assure its high quality. A development and optimization of AI-based methods must consequently be subject of future research activities.

Future work shall also aim at developing of recommendations for systematic solving of specific environmental problems in various life cycle phases of technical systems, for different environmental impact categories and engineering domains.

References

1. European Union: Regulation (EU) 2020/852 of the European Parliament and of the Council of 18 June 2020 on the establishment of a framework to facilitate sustainable investment and amending Regulation (EU) 2019/2088. https://eur-lex.europa.eu/legal-content/EN/TXT/PDF/?uri=CELEX:32020R0852&from=EN. Accessed 28 Apr 2022
2. Matthews, N.E., Stamford, L., Shapira, P.: Aligning sustainability assessment with responsible research and innovation: towards a framework for constructive sustainability assessment. Sustain. Prod. Consumption **20**, 58–73 (2019). https://doi.org/10.1016/j.spc.2019.05.002
3. Livotov, P., et al.: Eco-innovation in process engineering: contradictions, inventive principles and methods. Therm. Sci. Eng. Prog. **9**, 52–65 (2019). https://doi.org/10.1016/j.tsep.2018.10.012
4. Chen, J.L., Liu, C.C.: An eco-innovative design approach incorporating the TRIZ method without contradiction analysis. J. Sustain. Prod. Design **1**(4), 263–272 (2001)
5. Kobayashi, H.: A systematic approach to eco-innovative product design based on life cycle planning. Adv. Eng. Inform. **20**, 113–125 (2006). https://doi.org/10.1016/j.aei.2005.11.002
6. Fitzgerald, D.P., Herrmann, J.W., Schmidt, L.C.: A conceptual design tool for resolving conflicts between product functionality and environmental impact. J. Mech. Design **132**(9), 091006-1 - 091006-11 (2010) https://doi.org/10.1115/1.4002144
7. Pokhrel, C., Cruz, C., Ramirez, Y., Kraslawski, A.: Adaptation of TRIZ contradiction matrix for solving problems in process engineering. Chem. Eng. Res. Des. **103**, 3–10 (2015)
8. Ko, Y.T., Chen, M.S., Lu, C.C.: A Systematic-Innovation design approach for green product. Int. J. Constr. Eng. Manag. **5**(4), 102–107 (2016)
9. Guarino, N.: Formal ontology and information systems. In: Guarino, N. (ed.). Proceedings of the Formal Ontology in Information Systems Conference, FOIS'98, pp. 3–15. IOS Press, Amsterdam (1998)
10. Cavallucci, D., Rousselot, F., Zanni, C.: An ontology for TRIZ. Procedia Eng. **9**, 251–260 (2011)
11. Vincent, J., Cavallucci, D.: Development of an ontology of biomimetics based on altshuller's matrix. In: Cavallucci, D., De Guio, R., Koziołek, S. (eds.) TFC 2018. IAICT, vol. 541, pp. 14–25. Springer, Cham (2018). https://doi.org/10.1007/978-3-030-02456-7_2
12. The WUMM Project on a TRIZ Ontology. Basic Concepts (2021). https://wumm-project.github.io/Texts/WOP-Basics.pdf. Accessed 02 June 2022
13. UNric Homepage. https://unric.org/en/united-nations-sustainable-development-goals/. Accessed 28 Apr 2022
14. DIN e.V. (ed.): DIN EN ISO 14040:2006 Environmental management - Life cycle assessment - Principles and framework. Umweltmanagement – Ökobilanz – Grundsätze und Rahmenbedingungen. Beuth-Verlag, Berlin (2006)
15. CEN (ed.): ISO 14001:2015 Environmental management systems – Requirements with guidance for use. CEN-CENELEC, Brussels (2015)
16. Klöpfer, W., Grahl, B.: Life Cycle Assessment (LCA): A Guide to Best Practice. Wiley, London (2014)
17. VDI e.V. (ed.): VDI Standard 4521 Inventive problem Solving with TRIZ. The Association of German Engineers (VDI), Beuth, Duesseldorf, (2018–2021)
18. Wichowska, A.M.: Ontology of Sustainable Eco-friendly Technologies for Process Intensification based on the Biomimetics and Inventive Principles of the TRIZ Methodology. Master Thesis, Offenburg University (2022), in work
19. AskNature database of the Biomimicry Institute. https://asknature.org/. Accessed 28 Apr 2022

20. Domel, A.G., Saadat, M., Weaver, J.C., Haj-Hariri, H., Bertoldi, K., Lauder, G.V.: Shark skin-inspired designs that improve aerodynamic performance. J. R. Soc. Interface (2018). https://doi.org/10.1098/rsif.2017.0828
21. Collingridge, D.: The Social Control of Technology. Palgrave Macmillan, Basingstoke (1981)

Eco-Feasibility Study and Application of Natural Inventive Principles in Chemical Engineering Design

Mas'udah[1(✉)], Pavel Livotov[2], Sandra Santosa[1], Arun Prasad Chandra Sekaran[3], Anang Takwanto[1], and Agata M. Pachulska[2]

[1] Politeknik Negeri Malang, Jl. Soekarno Hatta No. 9, 65141 Malang, Indonesia
masudah@polinema.ac.id

[2] Offenburg University of Applied Sciences, Badstr. 24, 77652 Offenburg, Germany

[3] iFluids Engineering, 20, Seventh Cross Street, Shenoy Nagar West, Chennai 600030, India

Abstract. The early stages of the front-end process development are critical for the future success of projects involving new technologies. The application of eco-inventive principles identified in natural systems to the design of chemical processes and equipment allows one to find ways to mitigate or avoid secondary ecological problems such as, for example, higher consumption of raw materials or energy, generation of hazardous waste and pollution of the environment by toxic chemicals. However, before implementing a new technology in a real operational environment, it is necessary to completely investigate its undesirable ecological impact and to evaluate the future viability of this technology. Therefore, the research paper presents a study of ecological feasibility of an innovative process design utilising natural eco-inventive principles and analyses the correlations between applied inventive principles. Such eco-feasibility study can be considered as an important decision gate to determine whether the technology implementation should be moved forward. Furthermore, the study evaluates the practicability of natural inventive principles to the eco-friendly process design and is illustrated with an example of a sustainable technology for nickel extraction from pyrophyllite.

Keywords: Eco-inventive principles · Nature-inspired innovation · Feasibility study · Chemical engineering · Process design · TRIZ

1 Introduction

The field of chemical engineering is essential for human life, but it also contributes to the deterioration of ecosystem products and services that are important for maintaining the activity of all human beings. For sustainability, chemical processes must be designed to use raw materials, energy and water as efficiently and economically as possible in order to avoid the generation of hazardous waste and conserve raw material reserves [1]. In addition to raw material, energy and water consumption, all aspects of chemical processing must include appropriate health and safety standards. Aqueous and atmospheric

Published by Springer Nature Switzerland AG 2022
R. Nowak et al. (Eds.): TFC 2022, IFIP AICT 655, pp. 382–394, 2022.
https://doi.org/10.1007/978-3-031-17288-5_32

emissions must not be harmful to the environment and solid waste in landfills must be avoided [2].

Today, learning from nature is a promising approach for sustainable innovations, where chemical processes and operations can be designed by taking more inspiration from the existing biological systems. Nature has always been an inspiration for innovation and has led to a number of scientific design approaches. A number of new green chemical technologies have been developed using eco-innovation approaches and databases such as Ask Nature from the Biomimicry Institute [3] available for design inspiration. Unfortunately, many examples of what is popularly referred to as biomimicry or biomimetics resort to superficial analogies, duplicating nature, creating similar geometries, or reproducing isolated features that appear attractive, but all too often ignore the actual underlying mechanisms or context. As a result, the copying nature leads to suboptimal results. Not surprisingly, substantial biomimetic methods are supported by numerous analysis and creativity tools, for example tools derived from TRIZ, inventive problem-solving theory [4, 5]. The TRIZ methodology [6, 7] provides abstract solution principles and methods for identifying and eliminating engineering contradictions and helps to dramatically improve the inventive abilities of engineers.

The Nature-Inspired Chemical Engineering (NICE) methodology, developed in the recent decade by Coppens and co-workers [1, 8], is based on nature-inspired fundamental mechanisms to solve chemical engineering problems. The NICE approach provides a systematic platform to innovate and share the transforming technologies needed to tackle major challenges, including sustainability. Assisted by theory and experimentation, NICE aims to innovate, guided by nature, but does not mimic nature. The principal four themes for NICE consist of (T1) hierarchical transport networks; (T2) force balancing; (T3) dynamic self-organisation; and (T4) ecosystems, networks and modularity [8]. These themes include concepts related to fundamental mechanisms that are ubiquitous in nature, grouped by properties that are also critical to applications. The recent paper on this nature-inspired solution (NIS) methodology provides more examples relevant to chemical Process Intensification [8]. It facilitates the scaling of bubbling gas-solid fluidized beds, making their operation smoother and more controllable regardless of scale. Fluidized beds are widely used in chemical processes such as catalyst manufacturing, incineration, gasification, drying, coating and other industrial processes.

Similar to the NICE methodology, our recent studies [2, 9] have shown that working with nature can pave the way to a greener, more competitive, energy- and resource-efficient economy. Observing natural systems helps us discover the enablers that underpin the behaviour of complex systems. There, the principles can be applied to the design of industrial processes to achieve resilience and efficiency. The research work highlighted that the natural principles identified in natural ecosystems and innovative bio-inspired designs mainly utilise natural processes that increase the level of biodiversity, use microorganisms and biodegradable substances for various tasks instead of using hazardous chemicals. The study [2] was illustrated with an example of development of a sustainable process design for nickel extraction from pyrophyllite. It showed that the application of inventive principles identified in natural systems for chemical processes helped to improve the ecological problems faced by current technology. However, the study had not yet fully explored actual positive environmental impacts as well as had not investigated the ecological feasibility (eco-feasibility, for short) of the proposed experimental design in practice.

The eco-feasibility study should help to anticipate and to avoid secondary environmental problems of a novel technology. It can be performed in one of the early phases of the front-end process development and is crucial for the sustainability and future success of projects. Furthermore, besides evaluating the environmental impact and viability of the proposed technology, the eco-feasibility study can also be used to investigate whether the natural eco-inventive principles implemented in the process design can solve all the environmental issues faced by the prior technology without inflicting secondary ecological problems. In addition, the results of the eco-feasibility study can be used to analyse the relationships between the natural principles applied in the design. From there, it can be determined which combination of natural principles addresses most environmental issues. Also, the practicability of applying natural eco-inventive principles to process design can be assessed.

Therefore, the main objective of this paper is to present an ecological feasibility study of the process design using natural eco-inventive principles and to analyse the correlations between the applied inventive principles throughout the production process. This research is a follow-up to our previous work [2] and is illustrated with an example of further development of nickel extraction from pyrophyllite using a froth flotation process that implements the natural eco-inventive principles. Even though the case study in this research uses an example of nickel recovery from pyrophyllite ores, it also can be applied to any chemical process with similar eco-problems. The ecological impacts of any innovative process design must be fully investigated to assess the feasibility of the proposed technology before its realisation in a specific operating environment. Thus, the eco-feasibility study can be considered as an important decision gate to determine whether the technology implementation should be moved forward.

2 Eco-Feasibility Study of Application of Natural Eco-Inventive Principles in Chemical Engineering Design

2.1 A Case Study – Froth Flotation of Nickel from Pyrophyllite Ores

Froth flotation is a process used to recover base metals such as nickel, copper, zinc, and molybdenum from sulphide ores. A chemical agent such as solvent extraction is usually added to the minerals prior to the froth flotation process. The solvent is used to separate metals from minerals. As our recent study [2] has shown, the use of chemical agents in froth flotation is a major threat to sustainable development. Fewer than 20 metal mines worldwide such as nickel mines dispose of their waste in the ocean (known as Deep Sea Tailings Disposal or DSTD), which is home to the greatest diversity of coral and reef fish. In fact, underwater waste disposal is a harmful and outdated practice that is destroying marine life and destroying the livelihoods of communities that depend on fishing.

In this case study, we have analysed the environmental issues in the existing process technology and applied natural inventive principles for designing the new complete production of sustainable nickel extraction from pyrophyllite. The problems analysis and selection of appropriate natural principles presented in Table 1 and Table 2 was carried out based on a problem-driven bio-inspired design process [2]. The problem analysis in the current technology as presented in Table 1 began with a comprehensive analysis of environmental issues, including an understanding of the basic functions of the equipment, its operation, the environment and operating conditions using Root conflict analysis RCA +, Function Analysis and Process Mapping [10]. The identified eco-problems in the current process technology then were translated into environmental impact categories in accordance with [11]: Acidification, Air pollution, Chemical waste disposal, Depletion of abiotic resources, Energy consumption, Eutrophication, Ozone layer depletion, Photochemical oxidation, Radioactivity, Raw material intensity, Safety risks, Solid Waste, Toxicity, Water pollution and others. With the help of correlation matrix of ecological requirements, an empirical tool proposed by the authors in [11], the classification of the identified eco-problems can help to select relevant natural eco-inventive principles and also to forecast possible secondary ecological concerns.

Table 1. Eco-problems in the existing technology of nickel recovery from pyrophyllite.

System level	Eco-problems	Category
Super system	Toxicity of chemical collectors	Toxicity
System	High chemical waste generation in the tailing solutions	Chemical waste disposal
Sub systems	High energy consumption for aeration due to long process	Energy consumption
	High dust generation during the process	Air pollution
	Ocean acidification due to DSTD system	Acidification
	High possibility of dust inhalation during the process	Safety risks
	Mining for minerals/contamination of resources	Depletion of abiotic resources
	Waste disposal containing phosphate contaminates the water which potentially causes eutrophication	Eutrophication, Water pollution
	Chemical agents such as hydrogen cyanide can volatilize during the process and potentially react with the sunlight and cause photochemical oxidation	Photochemical oxidation, Ozone layer depletion
	Harmful underwater waste disposal	Water pollution
	High reagent consumption	Raw material intensity
	High solid waste in tailing product	Solid waste

The selection and application of relevant natural eco-inventive principles, as presented in Table 2, was based on the identification of biological solutions and the extraction of natural eco-inventive principles using the modified solution-driven process [9].

Table 2. Experimental results of applying natural eco-inventive principles for nickel recovery.

Eco-problem categories	Applied natural eco-inventive principles	Experimental realisation	Experimental results
Toxicity	A. Use natural materials B. Use microorganisms	Use organic solvent or microorganisms instead of hazardous chemical agent	Very low ecological impact, no mitigation of eco-impacts required
Chemical waste disposal	A. Use natural materials C. Utilise waste resources	Use water for slurry instead of chemicals Reuse the water for feed in	Very low ecological impact, no mitigation of eco-impacts required
Energy consumption	D. Use in parallel different technologies	Use parallel different technologies to speed up the process	Moderate ecological impact, minimal and effective mitigation of eco-impact possible
Air pollution	E. Simultaneous absorption of substances from gas/fluid	Absorbing dust from the air	Low ecological impact, minimal and effective mitigation required/possible
Acidification	C. Utilise waste resources	Recycle tailings back to the feed and tunnels the ores came from (material cycle)	Very low ecological impact, no mitigation of eco-impacts required
Safety risks	E. Simultaneous absorption of substances from gas/fluid	Absorb dust from the air	Low ecological impact, minimal and effective mitigation required/possible
Depletion of abiotic resources	F. Use different parts of an object for (competing) operations	Extract part of minerals containing desired metal	Moderate ecological impact, minimal and effective mitigation of eco-impact possible
Eutrophication, Water pollution	A. Use natural materials C. Utilise waste resources	- Use organic material instead of hazardous chemicals - Recycle tailings back into the feed and tunnel where the minerals come from	Low ecological impact, minimal and effective mitigation required/possible

(*continued*)

Table 2. (*continued*)

Eco-problem categories	Applied natural eco-inventive principles	Experimental realisation	Experimental results
Photochemical oxidation, Ozone layer depletion	A. Use natural materials B. Use microorganisms	Use organic solvent or microorganisms instead of hazardous chemical agent	Low ecological impact, minimal and effective mitigation required/possible
Water pollution	C. Utilise waste resources	Recycle tailings back to the feed and tunnels the ores came from (material cycle)	Low ecological impact, minimal and effective mitigation required/possible
Raw material intensity	A. Use natural materials B. Use microorganisms	Use organic solvent or microorganisms instead of hazardous chemical agent	Low ecological impact, minimal and effective mitigation required/possible
Solid waste	C. Utilise waste resources	Recycle tailings back to the feed and tunnels the ores came from (material cycle)	Low ecological impact, minimal and effective mitigation required/possible

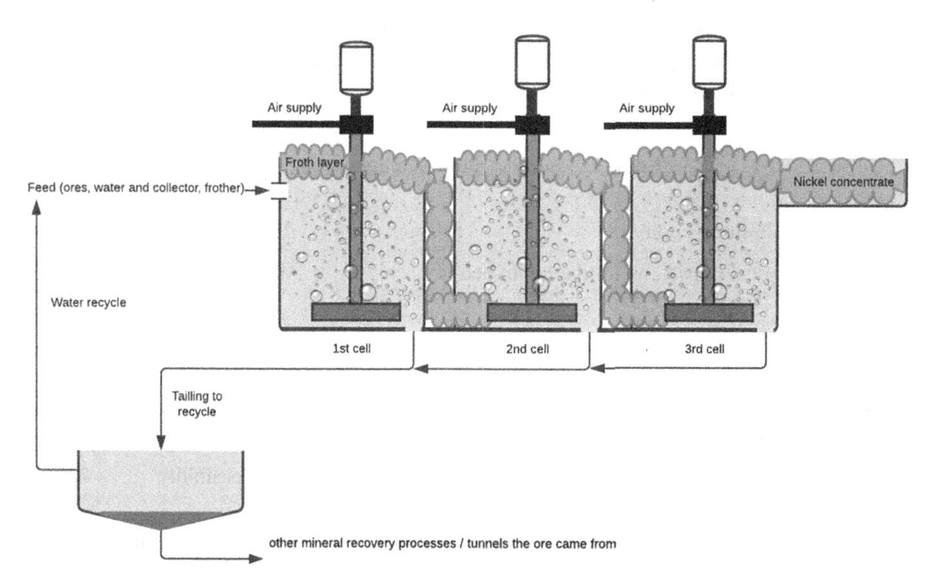

Fig. 1. Process flow diagram of nickel extraction applying natural inventive principles.

The natural principles, assigned to the pre-defined eco-categories, were incorporated into a nature-inspired solution concept and then into a formulated process design for an intended use. The last step includes the creation and optimization of the sustainable solution concepts, followed by prototyping with experimental realisation according to the Advanced Innovation Design Approach AIDA [10].

Figure 1 illustrates the process flow diagram of the experimental embodiment for the froth flotation of nickel from pyrophyllite ores after applying natural eco-inventive principles. The sludge from the mills was mixed with organic solvents instead of dangerous chemicals. Then the flotation was carried out in three different parallel chambers to speed up recovery and increase the quality and quantity of nickel products. In addition, the tailings were recycled to be used as feed and any remaining minerals were returned to the source ores.

2.2 Ecological Feasibility Evaluation of the Proposed Process Design

The ecological feasibility study of the proposed process design (Fig. 1) is based on experimental results. The experiment was carried out on a pilot scale under real operating conditions. The eco-feasibility assessment was focused primarily on the eco-problems presented in Table 1 and took into account all aspects that enable environmental permits, such as strategies to mitigate the specific impacts, environmental regulations, and standards. The results were then compared to current technology. Evaluation criteria and ranking for ecological feasibility assessment are given in Table 3. The final ecological feasibility refers to the average number of assessments of overall environmental impacts.

It is also essential to consider the viability and feasibility of the process design project in the context of sustainable technology development. If this proves to be viable and estimations suggest that the project indicates signs of being ecologically feasible in the future, a broader perspective related to technical and economic aspects must be considered to reach an informed decision.

Table 3. Eco-feasibility ranking criteria.

Ecological feasibility criteria	Ranking	Score
Very low ecological impact, no mitigation of eco-impacts required	Very high feasibility	5
Low ecological impact, minimal and effective mitigation required/possible	High feasibility	4
Moderate ecological impact, minimal and effective mitigation of eco-impact possible	Moderate feasibility	3
Significant ecological impact, limited mitigation of eco-impact possible	Low feasibility	2
High ecological impact, little/no mitigation of eco-impact possible	Very low feasibility	1

A comparative analysis of the ecological assessment and a scored ranking comparison between prior and new process of nickel extraction are presented in Table 4. It can be seen that the application of natural eco-inventive principles in the new technology of froth flotation is rated as ecologically high feasible (low to very low ecological impact) with average score of 4.1 out of 5.

Table 4. Eco-feasibility assessment of a new nickel extraction process based on natural eco-inventive principles compared to the prior technology.

No	Eco-parameters	Prior process	New process (natural principles applied)
1	Energy consumption	2	3
2	Air pollution	1	4
3	Acidification	1	5
4	Safety risks	1	4
5	Chemical waste disposal	1	5
6	Depletion of abiotic resources	1	3
7	Toxicity	1	5
8	Eutrophication	2	4
9	Photochemical oxidation	2	4
10	Ozone layer depletion	2	4
11	Water pollution	1	4
12	Raw material intensity	1	4
13	Solid waste	2	4
Average ecological feasibility score (scale: 1 - very low … 5 - very high)		1.4 (Very Low)	4.1 (High)

The ecological assessment of the impacts identified for the proposed process design of froth flotation indicates that most of the environmental impacts could be effectively mitigated. All eco-parameters, except for *Energy consumption* and *Depletion of abiotic resources*, in the nickel extraction operations have a 'low' to 'very low' ecological impact. The feasibility ranking for eco-parameters of *Energy consumption* and *Depletion of abiotic resources* in the new process technology is slightly lower (moderately feasible) than in the prior process because the ores grinding process here was not directly integrated into the proposed design. However, the overall ecological feasibility scores are deemed 'feasible' due to the generally low ecological impacts.

3 Correlations Between Natural Eco-Inventive Principles

The analysis of the correlations between the six natural eco-inventive principles A–F implemented in the process design (Table 2) is based on the experimental test of the proposed solution illustrated in Fig. 1 and on the eco-feasibility assessment presented in Table 4. From this it can be determined which combination of inventive principles addresses the most important environmental issues. The relationships between the eco-parameters and natural eco-inventive principles applied here are shown in a correlation matrix in Table 5.

The correlation matrix presents interactions between 13 eco-parameters and contains the most relevant combinations of natural eco-inventive principles to solve primary and secondary eco-problems. This 13x13 correlation matrix is a further development of our

Table 5. Correlations matrix of eco-parameters and applied natural inventive principles (A-F).

Primary eco-parameter to be improved		Secondary eco-parameter to be improved												
		1	2	3	4	5	6	7	8	9	10	11	12	13
1	Energy consumption		D E	D C	D E	D A C	D F	D A B	D A C	D A B	D A B	D C	D A	D C
2	Air pollution	E A		E C	E	E A C	E F	E A B	E A C	E A F	E A F	E C	E A	E C
3	Acidification	C D	C E		C E	C A	C F	C A B	C A	C A B	C A B	C	C A	C
4	Safety risks	E D	E	E C		E A C	E F	E A B	E A C	E A B	E A B	E C	E A	E C
5	Chemical waste disposal	A C D	A C E	A C	A C E		A C F	A C B	A C	A C F	A C B	A C	A C	A C
6	Depletion of abiotic resources	F D	F E	F C	F E	F A C		F A B	F A C	F A B	F A B	F C	F A	F C
7	Toxicity	A B D	A B E	A B C	A B E	A B C	A B E		A B C	A B	A B	A B C	A B	A B C
8	Eutrophication	A C D	A C E	A C	A C E	A C	A C F	A C B		A C B	A C B	A C	A C	A C
9	Photochemical oxidation	A B D	A B E	A B C	A B B	A B C	A B F	A B	A B C		A B	A B C	A B	A B C
10	Ozone layer depletion	A B D	A B E	A B C	A B E	A B C	A B E	A B	A B C	A B		A B C	A B	A B C
11	Water pollution	C D	C E	C	C E	C A	C F	C D B	C A	C A B	C A B		C A	C
12	Raw material intensity	A D	A E	A C	A E	A C	A F	A B	A C	A B	A B	A C		A C
13	Solid waste	C D	C E	C	C E	C A	C F	C A B	C A	C A B	C A B	C	C A	

previous studies [9, 11] and can be used to select natural eco-inventive principles. For instance, for reduction of *(1) Energy consumption* as a primary eco-problem and avoiding *(2) Air pollution* as a secondary eco-problem the matrix recommends the inventive principles of *(D) Use in parallel different technologies* and *(E) Simultaneous absorption of substances.* A look at other applications of principles D and E for more secondary problems in the first matrix line shows that they would additionally improve the eco-parameter *(4) Safety risks.* In other words, the application of a combination of two natural eco-inventive principles can improve one primary and one or more secondary eco-parameters. In its concept the correlation matrix is similar to the Contradiction Matrix known in TRIZ [6, 7] and can be used for resolving eco-engineering contradictions. The eco-engineering contradiction is defined here as a situation in which the improvement of a primary eco-parameter causes the worsening of a secondary eco-parameter.

The identification of the primary eco-parameters and eco-problems can usually be carried out at the phase of problem definition. Thus, the natural inventive principles can be selected according to the estimated priority of main ecological parameters to be improved. In more sophisticated problem situations, the Importance-Satisfaction Analysis for chemical and process engineering [12] can help to determine environmental parameters to be addressed first. The Importance-Satisfaction Analysis can be also applied for requirement ranking or for the objective identification of the high priority environmental and process intensification problems.

4 Concluding Remarks and Outlook

Nature provides us with numerous examples of extraordinary properties that far surpass those of current technologies and are relevant to addressing ongoing challenges that require transformative change, such as those related to sustainable development. Nature's handling of multi-objective optimization under boundary conditions brings a perfect example of chemical process intensification and sustainable process development. The research paper presents an eco-feasibility study of process design using natural eco-inventive principles and analyses the correlations between the inventive principles applied to mitigate the ecological impact. This is an important early-decision point in determining whether the selected technology or process should be implemented.

The paper gives a feasibility assessment example of application of natural eco-inventive principles to the sustainable process design. In accordance with the Nature-Inspired Chemical Engineering approach [1, 8] these natural eco-inventive principles should help to innovate guided by nature but without mimicking nature. The authors advocate a hypothesis that the eco-inventive principles inspired by nature should lead to the sustainable solutions with a lower secondary environmental impact. Therefore, the future research should be focused on systematic identification of new natural inventive principles or sub-principles for eco-innovation, their classification and assignment to the eco-parameters, eco-engineering contradictions and eco-innovation domains under consideration of environmental, technological, and economical requirements.

Table 6 illustrates the current state of research based on literature reviews, our previous [2, 9] and ongoing [13] studies. A comparison to the classical TRIZ inventive principles outlines that some natural eco-inventive principles give a more precise recommendation on how to create sustainable products or processes.

Table 6. Natural eco-innovations principles in progress (fragment).

No	Natural eco-inventive principle	Corresponding TRIZ inventive principle
1	Use in parallel different technologies, for example to block harmful effect [9]	5. Combining
2	Simultaneous absorption of substances from gas/fluid [9]	5. Combining; 29. Pneumatics and hydraulics
3	Use different sides or parts of an object for (competing) operations [9]	Separation in space
4	Use natural materials [9]	25. Self-service/Use of resources
5	Utilise waste resources [9]	25. Self-service/Use of resources
6	Use microorganisms [9]	25. Self-service/Use of resources
7	Attract and use bio-resources [9]	25. Self-service/Use of resources
8	Apply biodegradable waste to collect and remove harmful substances [9]	25. Self-service/Use of resources; 22. Converting harm into benefit:
9	Isolate sensitive processes from hostile environment in time and space [9]	3. Local quality; Separation in time and space
10	Use non-regular 3D reinforcement structures [9]	4. Asymmetry; 17. Shift to another dimension
11	Dynamic equilibrium [14] and force balancing [1]	15. Dynamism and adaptability
12	Dynamic self-organisation and self-adaptability [1]	15. Dynamism and adaptability; 25. Self-service
13	Performing critical process phase in advance (under more favourable conditions)	10. Prior useful action
14	Accumulation of energy or substances in advance	10. Prior useful action
15	Energy harvesting and micro-harvesting	25. Self-service/Use of resources
16	Substance or material harvesting	25. Self-service/Use of resources
17	Use of chemically diverse multi-structure	40. Composite materials
18	Use a variety of resources: rapid, flexible, and reversible resources management	25. Self-service/Use of resources
19	Use the same system to collect and store substance or energy	6. Universality
20	Use large surface with short distance to desirable substance or object	3. Local quality

Furthermore, the proposed eco-feasibility valuation should be compared with other existing approaches to sustainability assessment of technologies in the early phases of product and process development [15]. Moreover, it should be investigated how the natural eco-inventive principles correlate with the modern TRIZ approaches to eco-innovation, for example, presented in [16, 17].

The identification and classification of natural inventive principles in biological systems is a challenging time-consuming task that requires expert knowledge and experience. Therefore, for this purpose, the authors recommend the application of machine learning algorithms and AI-based methods as the subject of future research activities.

References

1. Trogadas, P., Coppens, M.-O.: Nature-inspired chemical engineering: a new design methodology for sustainability. In: Szekely, G., Livingston, A. (eds.) Sustainable Nanoscale Engineering, pp. 19–31. Elsevier, Amsterdam (2020). https://doi.org/10.1016/B978-0-12-814681-1.00002-3
2. Mas'udah, Santosa, S., Livotov, P., Chandra Sekaran, A.P., Rubianto, L.: Nature-inspired principles for sustainable process design in chemical engineering. In: Borgianni, Y., Brad, S., Cavallucci, D., Livotov, P. (eds) Creative Solutions for a Sustainable Development. TFC 2021. IFIP Advances in Information and Communication Technology, vol 635, pp. 30–41. Springer, Cham (2021). https://doi.org/10.1007/978-3-030-86614-3_3
3. Ask Nature database of the Biomimicry Institute. https://asknature.org/. Accessed 02 Apr 2022
4. Cohen, Y.H., Reich, Y.: Biomimetic Design Method for Innovation and Sustainability. Springer, Cham (2016). https://doi.org/10.1007/978-3-319-33997-9
5. Vincent, J.: Biomimetics - a review. Proc. Inst. Mech. Eng. Part H: J. Eng. Med. **223**(8), 919–939 (2009)
6. VDI Standard 4521: Inventive problem Solving with TRIZ. Fundamentals, Terms and Definitions. The Association of German Engineers (VDI), Beuth Publishers, Duesseldorf, Germany (2016)
7. Altshuller, G.S.: Creativity as an Exact Science. The Theory of the Solution of Inventive Problems. Gordon & Breach Science Publishers, New York (1984)
8. Coppens, M.-O.: Nature-inspired chemical engineering for process intensification. In: Annual Review of Chemical and Biomolecular Engineering, vol. 12, pp. 187–215 (2021). https://doi.org/10.1146/annurev-chembioeng-060718-030249
9. Livotov, P., Mas'udah, Chandra Sekaran, A.P.: Learning eco-innovation from nature: towards identification of solution principles without secondary eco-problems. In: Cavallucci, D., Brad, S., Livotov, P. (eds.) TFC 2020. IFIP AICT, vol. 597, pp. 172–182. Springer, Cham (2020). https://doi.org/10.1007/978-3-030-61295-5_14
10. Casner, D., Livotov, P.: Advanced innovation design approach for process engineering. In: Proceedings of the 21st International Conference on Engineering Design (ICED 17), vol. 4, pp. 653–662. Design Methods and Tools, Vancouver (2017)
11. Livotov, P., Chandra Sekaran, A.P., Mas'udah, Law, R., Reay, D. et al: Eco-innovation in process engineering: contradictions, inventive principles and methods. In: Thermal Science and Engineering Progress 9, pp. 52–65, Elsevier (2019) https://doi.org/10.1016/j.tsep.2018.10.012
12. Mas'udah, Livotov, P., Chandra Sekaran, A.P.: Sustainable innovation in process engineering using quality function deployment approach and importance-satisfaction analysis of requirements. In: Benmoussa, R., De Guio, R., Dubois, S., Koziołek, S. (eds.) New Opportunities for Innovation Breakthroughs for Developing Countries and Emerging Economies. TFC 2019. IFIP Advances in Information and Communication Technology, vol 572, pp. 269–281. Springer, Cham (2019). https://doi.org/10.1007/978-3-030-32497-1_22

13. Pachulska, A.M.: Systematic nature inspired eco-innovation in process engineering. Master Thesis, Offenburg University (2022). (in work)
14. Vincent, J., Cavallucci, D.: Development of an ontology of biomimetics based on Altshuller's matrix. In: Cavallucci, D., De Guio, R., Koziołek, S. (eds.) TFC 2018. IAICT, vol. 541, pp. 14–25. Springer, Cham (2018). https://doi.org/10.1007/978-3-030-02456-7_2
15. Matthews, N.E., Stamford, L., Shapira, P.: Aligning sustainability assessment with responsible research and innovation: towards a framework for constructive sustainability assessment. In: Sustainable Production and Consumption 20, pp. 58–73, Elsevier, (2019). https://doi.org/10.1016/j.spc.2019.05.002
16. Russo, D., Spreafico, C.: TRIZ-based guidelines for eco-improvement. Sustainability **12**(8), 3412 (2020). https://doi.org/10.3390/su12083412
17. Spreafico, C.: Quantifying the advantages of TRIZ in sustainability through life cycle assessment. J. Cleaner Prod. **303**, 126955, Elsevier (2021). https://doi.org/10.1016/j.jclepro.2021.126955

Focusing the First Phase – An Interdisciplinary Approach to Modeling an Interactive System on the Use-Case Indoor-Blind-Navigation

Ken Pierre Kleemann(✉), Nadine Schumann, Veronika Heuten, and Hans-Gert Gräbe

Leipzig University, Leipzig, Germany
kleemann@infai.org

Abstract. Concepts such as "resource" or "system" pose a challenge when it comes to adapting to the context. So the main problem is to adapt the concepts and the development process to the requirements of the users. The user is not only the methodological foundation of the concepts, but also a resource and a component of the system. Therefore, the initial phase of the development process becomes crucial. In order to explore the basic requirements of the product and the requirements of the potential use, it is necessary to combine several methods for interactive modeling. By linking qualitative research methods and formal iteration processes, new insights are generated. For handling the complexity of the problem, the experiences of AppPlant, Gräbert GmbH and a team at Leipzig University are applied. The project "Outdoor and Indoor Navigation for Blind and Visually Impaired People" (IBN) combines systems research with the application of social methods to improve successful human-machine interaction. The developed solution is based on the term "interactive system" of ISO 9241–11 as a methodological starting point. The key concept for an interactive system is usability, which describes the extension to "which a system, product or service can be used by specified users to achieve specified goals with effectiveness, efficiency and satisfaction in a specified context of use". In order to focus on user requirements, the iterative approach of Design Thinking, TRIZ Modeling and Mixed Methods of social sciences were combined.

Keywords: Human-machine interaction · Design thinking · Qualitative research · Outdoor-Indoor-Blind-Navigation · TRIZ and requirements analysis

1 Introduction

In general, developers must face several problems in the first phase of development processes. Especially when it comes to reconciling the products requirements with demands on functions of the potential use. The aim of this text is to present an interdisciplinary approach to modeling an interactive system on the use case Indoor-Blind-Navigation. On the one hand we show the enhancement of the initial phase and how the management of the whole process can be adapted to the user-machine-interaction on the other.

Published by Springer Nature Switzerland AG 2022
R. Nowak et al. (Eds.): TFC 2022, IFIP AICT 655, pp. 395–404, 2022.
https://doi.org/10.1007/978-3-031-17288-5_33

Design Thinking works with associative power of brainstorming, where TRIZ, instead, favors systematic modeling. In fact, both strategies are helpful for the determination of requirements and contexts. Considering our special user group, defining the specific requirements is problematic at first. Therefore, this first phase (Define phase in [1]) depends on the identification of the requirements of blind and visually impaired people as well as on the determination of the categories for conceptual definition. The specific demands and conditions of the respective environment as well as the interrelations of the product to be developed are predominantly influential.

We combine a mixture of technological possibilities and adaptation to the user as well as innovation methods. For the determination of the categories for the conceptual definition we used methods of qualitative social sciences, which lead us to a better understanding of the user both as system and component for a user-centered modeling at different systemic levels. The Define phase is therefore enhanced and also not complete with a first model, but complemented with newly discovered demands of the user over the whole time of the project. The linking of different approaches of innovation methodology and qualitative social sciences research is considered promising for successful human-machine-interaction in an indoor-outdoor-navigation-system.

2 Problem

Concepts such as "resource" or "system" pose a challenge when it comes to adapting to the context. Accordingly, the main problem is to adapt the concepts and the development process to the requirements of the users. The user is not only the methodological foundation of the concepts, but also a resource and a component of the system. Therefore, the initial phase of the development process becomes crucial. In project management this initial phase is called strategic orientation, in software engineering requirements analysis, in TRIZ it is Part I of ARIZ-85C. In our case it is all the same, but our first phase is longer, more intensive, focused on iteration expansion and on continuous productive intervention even beyond the normally seen first phase. Our first phase is therefore a different concept. In order to explore the basic requirements of the product and the requirements of its potential use, it is necessary to combine several methods for interactive modeling. By linking qualitative research methods and formal iteration processes, new insights are generated, which are a remarkable progress for users with special interests.

To cope with the complexity of the problem, the experiences of AppPlant, Gräbert GmbH and a team at Leipzig University are applied. The project "Outdoor and Indoor Navigation for Blind and Visually Impaired People" (IBN) [2] combines systems research with the application of social methods to improve successful human-machine interaction. The special interest of this group can be channeled and brought to the development of functions which are suitable for the special needs.

While such instruments for visually impaired people in the outdoor area based on GPS already exist and a smartphone-based solution is offered by the partner AppPlant, navigation in the indoor area is more difficult. Up to now there are no established solutions comparable to GPS. Instead, island solutions based on different technologies are used, which require special instrumentation of the respective buildings (beacons, induction loops) or the technical signatures of fixed digital devices (such as WLAN routers) for orientation.

Accordingly, dealing with today's possibilities of machine guidance in buildings is a concern that is important both for the systemic development of user-centered applications and for the lifeworld possibilities of participatory and inclusive design of society.

To highlight the problems complexity the experience of the process management, available at AppPlant and the team at Leipzig University, is used. The project outdoor and indoor navigation for blind and visually impaired people combines system research and the observation of user-machine interaction.

The digital evolution of today's progress requires not only a non-linearity of its decentralized infrastructure in the physical sense, but a decentralized and non-linear way of working. This more comprehensive infrastructure has been dominated for a long time by concepts of an agile and flexible nature, in which accessibility, interoperability and usability are in the central focus. This means the close alignment of consumer, producer, and device makes an iterative procedure a straightforward obligation. Consequently, innovation methods against this background cannot be limited to promoting creativity or technical efficiency. Rather, they are set in a complex infrastructure, which in turn affects the application and design of innovation methods.

However, there are experiences of relevant innovation methods which, on the one hand, can and have to meet the complex infrastructure requirements and, on the other hand, have followed the spectrum of free association and systematic invention. On the one hand, the established innovation methodology of design thinking combines both iterative work organization with creative unsystematic brainstorming. On the other side, the large complex of TRIZ approaches cultivates iterative procedures more relying on systematic modeling for creativity control.

Taken together, both methods meet a research interest that wants to understand the connection between technical progress and innovation as well as the comprehensive infrastructural context of modern work organization. A sufficient system of outdoor and indoor navigation for visually impaired and blind people needs the straight interaction of researchers, developers and users.

3 Methods

The solution to be developed is based on the term *interactive system* of ISO 9241–11 as a methodological starting point. [3] Human-machine-interaction is here defined as interactive system (ISO 9241–11:2018: 3.1.5), namely, as "combination of hardware and/or software and/or services, and/or people that users interact with in order to achieve specific goals" (ibid.: 3.1.5). By contrast, the term *system* is outlined as a "combination of interacting elements organized to achieve one or more stated purposes" (ibid.: 3.1.4). The key concept for an interactive system is *usability*, which describes the extension to "which a system, product or service can be used by specified users to achieve specified goals with effectiveness, efficiency and satisfaction in a specified context of use". To focus on the users demands the iterative approach of Design Thinking, TRIZ Modeling and Mixed Methods of social sciences were combined.

3.1 Interactive System

In the first step the interactive system is seen to be built of two linked systems, based on:

1) User-Requirements, speech interaction concept (SIC)
2) Infrastructure, processed data /BIM (BIM – Building Information Modeling), Conceptualization process

The application is conceptualized as a complex system of combined resources and as well a concept of management. In our view the user is not only a part of the supersystem, organizing the interaction between user and device as bi-system, but also a component in that system. Therefore our user centered-modeling cannot follow a pure formal method of defining the needs in the first phase. We need to develop a methodology which includes the requirements, models and developed infrastructure.

Our approach combines in the second step Design Thinking, TRIZ Modeling and Mixed Methods of social sciences in the IBN use case. This form of interdisciplinarity has clear advantages for the development of innovative applications because user expectations and technical possibilities are thought together from the very beginning. We emphasize the importance of a tight feedback loop between requirements elicitation and rapid prototyping within an iterative TRIZ based development process. The initial phase of TRIZ and Design Thinking (DT) is extended by mixed methods of social sciences.

3.2 Design Thinking

Design Thinking is a powerful process of problem solving. This solution based approach provides a well-known analysis methodology that plays a selective role in teaching at Leipzig University, and which is used in the project for a classical requirements analysis with special consideration of the specific target group [4]. Design Thinking stands out in its function as an innovation method especially through the combination of two characteristics. These are interdisciplinarity and a continuous iterative evaluation of the project stages. In groups from different disciplines, experts are to jointly identify and develop innovations in several stages, constantly evaluate them, and constructively incorporate these results into the work process. The first step involves defining the so-called "Design Challenge". This is the most important starting point at the beginning of the process and is constantly changing due to the various findings of the respective phases and the correlation of these. In the following step, the previously defined design challenge must be understood. Here, users are depicted in typical situations, analyzed, and as much information as possible is collected on the topic. These findings are fed back into the first phase and help to realign the design challenge and better frame the topic. The third phase is to help to define the viewpoint. This essentially means that the results of the previous phases are processed, interpreted and weighted. On this basis, typical user profiles are created. These created profiles are tested again and readjusted if they were not defined precisely enough. In the step of idea generation, creative methods are used to realize the needs already identified. These ideas are developed into prototypes, which are then tested by the already defined target group. The final step involves integrating the prototype into the product or service. The phases are always compared and fed back to use the latest results for the specification of the previous phase. The combination of rapid prototyping and user centered design is the part of this method which was adapted by the team of Leipzig University. It was transformed to an iteration process which enhances the first stage of DT with tools and methods from TRIZ for creating a permanent feedback loop for the whole project work. The aim of our approach is to systematize brainstorming, rather than just using it for inspiration.

3.3 TRIZ

DT usually starts with brainstorming and empathy. This is replaced by context modeling and TRIZ functional analysis of users demands. Here, TRIZ and methods from social sciences are applied and lead to useful knowledge generation [5]. We do not use TRIZ as a formal method, but as a use-case related methodology.

TRIZ already played an important role in the GDR inventor schools in the 1980s. The reappraisal of these experiences was the starting point for LIFIS, since 2016, to develop a corresponding focus on innovation management and systematic learning. In the WUMM project, we have been focusing our activities on the Central Germany region since mid-2018, for contradiction-oriented systematic innovation methodologies in the field of management consulting and for academic teaching in the region as well [6]. The academic core of the WUMM project is the development of a multilingual dataset of established TRIZ concepts and methods as well as of metadata of corresponding research activities in an Open Data Space. Using established concepts of semantic technologies is the seed of a more comprehensive research data infrastructure on the topic of Systematic Innovation Methodologies.

Just like Design Thinking, the iterative and self-referential process is the crucial point. In addition, however, a lot of time and methodical effort is spent on the initial phase. There, from a systematic point of view and through effective alignment to the existing infrastructure, modeling is used instead of brainstorming. There are also different approaches for this modeling of a so-called ideal final result. Common to all of them, however, is to integrate a product- and consumer-specific orientation into this modeling of the ideal outcome. Thus, this ideal final result becomes the variable basis and the motor of further project development. By the analysis of problems and final contradictions differences and levels of penetration of the systematic possibilities as well as limitations of the project and of the product can be received.

This trait of the innovation method serves on the one hand to obtain an associative creativity, but on the other hand the systematic processing of this creativity should get a context and framework of the feasible. Directly, technical progress and innovation are achieved through agile work organization and systematic contextualization. Contradictions can be recognized and uncovered here. This enables not only problem solving, but also an awareness of contradictions, which leads to further insights and approaches in an iterative process. In this sense, a very close connection to the needs and requirements of the users can be generated; a prosumer [7] approach can be systematically incorporated into the development. Thus, indoor navigation for the visually impaired and blind not only provides insights into a complex methodological work, but also a possibility of crucial starting points for the development of this technical field itself.

First, the purpose, principle and problems of this system are determined as a black box. During its use, the application evolves from a formal-functional to a living technical system with its own systemic development logic, which appears in the context of the project as intended functionalities in the requirements analysis and in the system test with test persons of the target group as proof of concept. The modeling of the application system assumes a bi-system of user and device. Users use the application as the top-level systemic structure (application system), which in turn is embedded in one or several social supersystems. At this system level, both users as a whole group of subjects and AppPlant as app-maintainer can be addressed to understand the relationship and synergetic effects of the interaction of these two stakeholder classes (such as the improvement of the data basis and data pool of all users). In addition to the classes, user instances and application instances are to be distinguished. Next, the main useful function is formulated for the app and therefore for the project as system.

The project context comprises three components (the project partners as black boxes, since they organize their internal processes autonomously in each case). The main useful function as emergent outcome results (ideally) from the design of the interactions between these components.

As a black box, the project context is characterized by:

- Useful product: The app as a technical artifact (its operation is not part of it).
- Useful principle: Development of the app
- Problem: Organization of the interaction based on the division of labor

A further distinction must be made between systemic developments within the context of the project, of which the app is a product, and systemic developments that arise while using the app.

The components user and application as subsystems of the bi-system are now modeled in more detail at the level of prototypical instances (i.e., in the language of OO programming their attributes and functions are modeled).

In addition to project-centered systemic modeling, a product-centered systemic modeling is also required. Specific attribute characteristics are important for the performance of a more detailed requirements analysis in contact with potential users, using preserved sociological methods.

We use therefore an iterative approach of system transformation (TRIZ) and rapid prototyping (DT) to further detail the functional analysis. This repeated evaluation of experienced results leads to a significant improvement of the system design.

3.4 Mixed Methods

In order to determine the requirements of the technical assistance system from the perspective of the end user, a participant observation was conducted at the very beginning to evaluate the special needs and hurdles of blind and visually impaired people. The methodological approach of participant observation is on the one hand pluralistic, as different methods can be combined (triangulation and mixed methods) [8]. And on the other hand it is directively, as second-person participant observation is carried out in a team. In the phase of data collection, different sources are considered: participant observation, open interview, video recording, field notes. Data collection in participant observation consists of a one-to-two-hour interaction with the subjects and two participant observers. The inspection of the building, subsequent interview and video documentation are done in the areas of selected partners. The data are then processed, that means transcriptions of the interviews were made, field notes were taken, and videos were edited (of the most striking situation, e.g., in the entrance area). The data are analyzed according to the data form afterwards and the contents are typified. Data interpretation and typing are done in a team with the aim of filtering out the main hurdles and needs.

Thus, different interpretations of the data set from different perspectives can be considered. During this, a joint evaluation is carried out in the sense of clustering the specific main points. In the following, the individual clusters are sorted by frequency and then prioritized. The prioritization is purely quantitative and is then grouped thematically. This results in general categories that run from user to device and from device to user. The most common categories are the need for security, demand and communication. The device is seen by the user as an aid to greater autonomy and independence. This leads to a prioritization of the general requirements. A decisive prerequisite for the realization of such an application is the transmission and warning of immobile obstacles, additional information about the location (e.g., information about reception and info point) and changing environments (ground texture), which could occur as stress factors regarding the device-user interaction. Thus, the main task of the application is to provide security and thus reduce stress that can arise from being overwhelmed with an unknown situation. On the one hand, this resulted in the desire for a preparation mode in which the respective route can be displayed and practiced independently of the current location of the user. On the other hand, the desire for a graduated level of detail depending on the user configuration emerged. This points to the importance of the profile configuration of the application before its use.

Our analyses shows that at least three modes are to be distinguished: Preparation, Orientation, and Navigation. When implementing the language concepts, it is imperative to consider the life world of blind people. This leads to an approach that should consider the verbalization of information by blind people for blind people. It must be noted that the transition from indoor to outdoor, and the accompanying change in acoustics, presents a particular challenge. Another requirement for a digital orientation aid is compatibility with other applications. The identified requirements for the technical assistance system will be passed on to the developers. In our use case, a preparation mode for navigation was derived and incorporated into the technical architecture at an early stage. The architecture is currently in a process of adaptation and the results are monitored through a combination of development modeling and social science analysis. Here also Social Research, Design Thinking and TRIZ-modeling go interdisciplinary hand in hand.

3.5 Modeling in the Triangle

An ideal modeling can be derived from the results of the preceding analyses. The TRIZ-modeling serves to clarify the interaction of infrastructure and user function, which are brought together in the execution of the assistance system. As it turns out, the different modes already discussed are of particular importance. The differences of the preparation, navigation and orientation modes result from the target group specific requirements and the processuality of the configuration space.

This is used in a further step for the extraction of detailed scenarios of the individual modes. On the one hand, the specific requirements and the conditions of the infrastructure serve as a starting point. On the other hand, the user functions are used according to their respective importance in relation to the infrastructure. These user functions are derived from the functionalization. In the user function the mutual dependence of user-device and device-user perspectives takes effect. Due to the process-like sequence of the individual steps, which refer to themselves as well as to each other, a reciprocal relationship results in the application. When considering the user function description, this distinction and their interaction is taken up again in specific terms. The specific stakeholder requirements reflect user requirements and the specific infrastructural conditions. Here, infrastructure is understood not only as a material prerequisite, but also as a processual system that changes through use and as an environment (user-technology world) [9]. Here the two linked systems as interactive system are realized.

The sociological determination of a case oriented to the application partner resulted in a first basic scenario, which is functionalized as an example for the scenarios for the different modes. With these scenarios, the flow of the assistance system is to be tested, which will be verified in the user-device usage by the iteration with the prototype. Further steps will be realized in the frontend design. This also includes the Voice Over function, which is essential for blind and visually impaired people. In further socio-scientific investigations an ideal description form of the Voice Over is to be examined. This includes the scientific foundation as well as further elaboration of the speech interaction concept and an imaginable description form, which considers the specifics of a linguistic implementation in the concept blind for blind. In a future version, a dynamic user-specific extension of the list of recurring situations should be provided.

After the latest developments, it is becoming apparent that the further work of the team at Leipzig University up to the end of the project will focus on testing a constantly evolving prototype with subjects from different cohorts. These tests will set up a permanent feedback loop, which will be used to compare the further technical development of the app with the requirements and the status of the processing of issues. In the sense of a consolidation of systems in cooperative action, the requirements and the glossary in a digital RDF based semantic web structure will also be updated. The transcripts of the test runs are also used as a source for the further development of the speech interaction concept (SIC). For this purpose, speech interaction units (SIU) are extracted as intents from the transcripts, to have a sufficiently large pool of such units, from which generalized patterns (templates) can be derived. Similar to the transcripts, a table structure for text will be developed, which can be transferred to a CSV file that is digitally processable. Parallel to this, the interaction scenarios already available from the first test runs are prepared according to this scheme and will be matched to the BIM data. In contrast to previous approaches, a co-evolution of use and existing statically considered building data is used here. Thus, information from a long-term perspective of building development can be included. A reprocessing of already existing structured data is possible. The alignment between use and user is improved by a continuous iterative process. The only thing that remains unclear is exactly which data transformations must be performed for the new purposes and how these inventory data (orientation-driven) are to be interleaved with specific user data (navigation-driven).

Accordingly, indoor navigation must take questions of location, orientation, and control seriously. The scientific evaluation and feedback to a user-oriented development is on the one side an aid to clarify the basics and on the other side an integral part for an efficient orientation of the use of the application to be created. This refers to the needs and requirements of the user group. For the clarification of this connection and for the integration of systematic innovation methods an interdisciplinary and problem-oriented view is to be won from the patterns of the development and the already existing market positions.

This will be applied to the particular subject of navigation for the blind. On the one hand, a usage-user comparison is to be made comprehensible and usable, which demands very special requirements for sensory back bindings of localization and orientation. On the other hand, navigation and especially control requires much more than a simple manual determination of points of interest can guarantee. Especially the application to the navigation of blind people in indoor areas offers insights into the sensory requirements of localization and orientation as well as into the possibilities of adaptive navigation and control. Today's developments in the field of indoor navigation must therefore focus on a mixture of technological possibilities and user adaptations as well as innovation methods. These must be embedded in a coordinated, agile, and flexible way of working to meet the iterative demands. A combination of different approaches to innovation methodology is considered promising for this purpose.

4 Solution

Our strategy in combining Design Thinking, TRIZ and social science leads to the following three outcomes:

a. The initial phase of TRIZ and Design Thinking were combined and extended. Design Thinking usually starts with brainstorming and empathy, which is now replaced by context modeling and functionalizing the users demands. Here the TRIZ method and mixed methods coming from social sciences lead to additional useful knowledge generation.
b. This iterative process is not only sufficient concerning starting points but also for the whole agile development process. At this point the user's requirements and the developers' technical know how come together. The entanglement results in a first prototype evaluation and iterative adaption. Regarding the use case IBN, a preparation mode for navigation was invented and introduced in the technical architecture.
c. Interpreting, evaluating and adapting the research outcomes underline the need of a constant feedback loop. Therefore, the management of the process doesn't only need content modeling but also the interconnection of a wide scale term use of resource and system.

Interdisciplinary approaches to plural process management methods are a new way of combining innovative product development and methodological use of system research. Both Design Thinking and TRIZ meet the requirements of today's work organization through their iterative approach and the interdisciplinary nature of the development work. Additionally, what both methods have in common is that they carry out technical development and innovation by closely aligning them with a broad concept of infrastructure. Although there may be arguments about incompatibility issues of these methods, a deeper analysis reveals the relatedness of the approaches rather than a competitive relationship. Only in the different design of the initial phase do the two ways of putting creativity to work for the development of comprehensive innovations differ. Design Thinking relies on the associative power of brainstorming and repeated rapid prototyping, whereas TRIZ, on the other hand, favors systematic modeling. In fact, both perspectives can have a positive effect on the identification of requirements and contexts. The specific requirements and conditions of the respective environment as well as the context of the product to be developed remain decisive. Both methods benefit from the usage of qualitative research in social science. Since a present-day requirement of the digital transformation is the close contact of prosumer, developer and device, a contextualization of the possible development trends becomes an unavoidable necessity [10]. Neither one nor the other method can claim absolute advantages for itself. User requirements are needed for a systematic modeling and for a contextualized brainstorming. The initial phase of TRIZ and Design Thinking is extended by social science research.

Initial phase is crucial and more than an intuitive brainstorming. The combination of methods is not only a possibility for a wide range perspective but also for a way to challenge the interaction of producer and consumer to a new level.

References

1. Mann, D.L.: Hands-On Systematic Innovation in Business & Management. IFC Press (2007)
2. Project Website. http://graebe.informatik.uni-leipzig.de/IBN-Web/IBN.html
3. VDI (2018): ISO 9241–11
4. Afflerbach, T., Ducki, A., Glasener, K.: Design Thinking, Digitalisierung und Diversity Management: Ein Praxisleitfaden für die Lehre. Berlin, Toronto (2019)
5. Altschuller. G.S.: Creativity as exact science. Moscow (1979)
6. Project Website. https://wumm-project.github.io
7. Toffler, A.: The third Wave. New York (1980)
8. Nadine, S.: Zur Methodologie der Zweiten-Person-Perspektive. Kritik der experimentellen Psychologie und Neurophysiologie unter besonderer Beachtung phänomenologischer Zugangspositionen. Würzburg (2020)
9. Hans-Gert, G.: Technical systems and purposes. In: Oliver, M., (ed.) Proceedings TRIZ Anwendertag 2020, pp 1–13 (2021). Springerhttps://doi.org/10.1007/978-3-662-63073-0_1
10. Schumann, N., Du, Y.: Machines in the triangle: a pragmatic interactive approach to information. Philos. Technol. **35**(2), 1–17 (2022)

Ideal Final Result for Agriculture: Striving for Sustainability

Valery Korotchenya(✉)

IEEE, 3 Park Avenue, 17th Floor, New York, NY 10016-5997, USA
valor99@gmail.com

Abstract. In this paper, we used TRIZ to resolve the existing contradictions between agriculture and the environment, which would make agriculture sustainable. Through the identification of strategic types of agriculture and their goals, we formulated ideal final result for agriculture. This formulation includes all the contradictions that exist between the strategic types. We proposed a strategy for the balanced development of agricultural technologies, which aims to realize the ideal final result for agriculture by resolving all contradictions between all the strategic types of agriculture. The balanced development means that in the resulting system of agriculture, the goals of all the strategic types of agriculture are achieved simultaneously. Regarding the issue of making agriculture sustainable, in accordance with the TRIZ methodology, administrative, technical and physical contradictions arising between agriculture and the environment were stated and examined. Some solutions to the physical contradictions were proposed. Based on these solutions, two main factors of making agriculture sustainable were determined: 1) making green technologies highly productive; and 2) the humane control of the growth of the world population.

Keywords: Strategic types of agriculture · Theory of inventive problem solving (TRIZ) · Ideal final result · Kinds of contradictions in TRIZ · Physical contradictions · Sustainable agriculture · Balanced development of agricultural technologies

1 Introduction

There can be different strategies for developing agricultural technology. Usually the economic motive behind them consists in boosting economic efficiency (profitability) or winning the competitive struggle, thereby increasing sales and the market share. At the same time, as a matter of fact agriculture harms the environment by changing natural ecosystems, exhausting biological resources, leading to the loss of biological diversity, polluting the soil with pesticides and mineral fertilizers, contributing to global greenhouse gas emissions and so on [1–5] (some data on these issues will be presented below). Therefore, economic reasons cannot be a sole driver for bettering agricultural technology. A more balanced way of technological development can be based on the approaches offered by the theory of inventive problem solving (TRIZ).

Published by Springer Nature Switzerland AG 2022
R. Nowak et al. (Eds.): TFC 2022, IFIP AICT 655, pp. 405–416, 2022.
https://doi.org/10.1007/978-3-031-17288-5_34

According to TRIZ, with evolution, technical systems become closer to ideal [6, p. 228]. Being the embodiment of ideality, ideal systems while performing their useful functions do not expend any energy/resources. Ideality also means that the harmful effect of a technical system in question must be zero [7, p. 59]. This is what we need as regards ideal agriculture.

In general, we should analyze agricultural technology in order then to formulate ideal final result for agriculture, which is supposed to involve different contradictions. Though the main contradiction implied here is between humanity and Mother Nature, we will present as many contradictions in agricultural technology as possible. The ultimate purpose is to put forward some ways of resolving the contradictions between agriculture and the environment that will make it possible to create sustainability and realize the ideal final result for agriculture in what we term "the balanced development of agricultural technologies".

This balanced development is presumed to be a holistic strategy for developing agricultural technology because the proposed approach suggests that the contradictory goals of strategic types of agriculture (presented below) are achieved simultaneously.

Also, our approach expands the application of TRIZ. With regard to environmental issues, the TRIZ methodology is most often used for the eco-design of technical systems [8]. However, in this paper, TRIZ is applied to the entire industry.

One of the closest to our work is the paper [9], where the concept of sustainability is studied through the prism of TRIZ (for example, the notion of ideal final result is used). However, our study is entirely devoted to the subject of agriculture, while in [9] agriculture is not touched upon. Besides, in addition to the sustainability of agriculture, we consider other aspects of agricultural technology.

We should also highlight the paper [10], where TRIZ is used to design eco-innovations in biological systems such as agriculture, and the authors present their BioTRIZ axioms for eco-innovation. The difference between our work and [10] is that our approach reflects a macro approach: we look at the problem from the perspective of the entire agricultural sector, while the authors of the paper [10] study the eco-design of biological systems involving some technological features, which constitutes a micro approach.

The starting point for achieving the purpose of our research is specifying strategic types of agriculture, which is necessary for analysis of agricultural technology.

2 Strategic Types of Agriculture and Ideal Final Result for Agricultural Production

Based on the classification of aspects of agricultural technology, similar to the innovations in agriculture according to form [11, p. 209], we identify eight *strategic types of agriculture* (Table 1).

Note that our understanding of organic agriculture is that this strategic type is completely about the healthcare-related aspect, while the conservation of natural resources in agricultural production is the core of green agriculture.

The combination of all the strategic goals presented in Table 1 can be regarded as *ideal final result* [7, pp. 59–60] *for agriculture*, i.e. the latter is the agriculture that

Table 1. Strategic types of agriculture.

Strategic type of agriculture	Extremely significant aspects in agricultural technology	Strategic goals (elements of the ideal final result for agriculture)
Classical agriculture (conventional agriculture): represented by mechanized agriculture based on scientific agronomy, with the use of fertilizers, pesticides, and selective breeding (but without genetic engineering)	agronomic agrochemical biological mechanical	- food security of large masses of people
Digital agriculture (smart agriculture)	informational (cyber-physical)	- maximum efficiency of agricultural production (achieved by the use of data) - full automation and intellectualization of agricultural production
Organic agriculture	healthcare-related	- food safety for human health
Green agriculture (sustainable agriculture)	ecological	- environmental sustainability of agricultural production
Climate-smart agriculture	climatic	- high adaptability to climate change and climate neutrality
Biotechnology-based agriculture (includes genetic engineering)	biotechnological	- creation of varieties of agricultural plants and farm animal breeds with the best properties, subject to the limitations of complete safety for human health and the environment
Controlled-environment agriculture (in greenhouses and buildings)	agronomic	- independence from weather and climatic conditions
Agricultural biofuel production as an alternative energy source	energy-related	- the maximum possible contribution to the development of alternative energy

ensures the food security of large masses of people, maximum efficiency of agricultural production, full automation and intellectualization of agricultural production, and so on.

The contradictoriness of the ideal final result is shown in Table 2 by revealing contradictions between the strategic types of agriculture.

Analyzing the strategic types of agriculture and the corresponding ideal final result, we acknowledge the presence of serious contradictions in them—especially between

Table 2. Contradictions between the strategic types of agriculture and the corresponding contradictory nature of the ideal final result for agriculture.

Contradictory pair of strategic types of agriculture		Contradictions between the strategic goals of the pair in question
Classical agriculture	Organic agriculture	Providing food security of large masses of people by using mineral fertilizers, pesticides and other chemicals in classical agriculture deteriorates food safety for human health
Classical agriculture	Green agriculture	The necessity of feeding a large number of people (classical agriculture) often leads to harmful effects of agriculture on the environment and makes impossible environmental sustainability of agricultural production
Classical agriculture	Climate-smart agriculture	Ensuring food security of large masses of people is not climate-neutral and leads to climate change, which contradicts the strategic goal of climate-smart agriculture
Organic agriculture	Classical agriculture	Achieving the strategic goals of organic agriculture, green agriculture, controlled-environment agriculture, and digital agriculture (i.e. full automation and intellectualization of agriculture) can cause a significant increase in agricultural prices and worsen the food security of large masses of people
Green agriculture	Classical agriculture	
Controlled-environment agriculture	Classical agriculture	

(continued)

Table 2. (*continued*)

Contradictory pair of strategic types of agriculture		Contradictions between the strategic goals of the pair in question
Digital agriculture	Classical agriculture	
Biotechnology-based agriculture	Organic/green agriculture	By improving food security and creating better plant varieties and animal breeds, biotechnology can deteriorate the safety of agricultural products for human health and the environment (achieving the pair's goals at the same may become impossible)
Agricultural biofuel production as an alternative energy source	Classical agriculture	The production of agricultural biofuels can result in a deterioration in food security (the strategic goal of classical agriculture may not be achieved)

agriculture and the environment. Because of both the great importance and extreme difficulty of achieving this ideal final result in practice, we call this task a *grand challenge.* The striving to obtain a practical solution should be fulfilled as a *mission-oriented policy* of the world community [12].

As an attempt at finding this solution, we propose a *strategy for the balanced development of agricultural technologies.* It includes two stages: 1) developing each strategic type of agriculture separately to an advanced degree, thereby achieving separately the eight parts of the ideal final result for agriculture, which are represented by the eight strategic types of agriculture (Table 1; the third column); and 2) combining strategic types of agriculture into viable agricultural systems by resolving the corresponding contradictions (for example, combining classical agriculture, digital agriculture, organic agriculture, green agriculture, climate-smart agriculture, and biotechnology-based agriculture), which ideally would mean the complete realization of the ideal final result for agriculture, or, in other words, achieving all the goals of all the strategic types of agriculture at the same time.

The second stage is the most difficult, because no one has yet managed to find an optimal way of farming, which does not cause damage to the environment or humans. Therefore, this study is focused on identifying a large scientific and social problem and trying to offer its own solutions with the help of TRIZ.

Figure 1 depicts the general idea of the proposed strategy.

In the next section of our paper, using the TRIZ methodology, we will present some ways how to resolve contradictions between ensuring the food security for large masses of people and the need to preserve the environment. How to make agriculture sustainable is our goal and challenge. Hence, we will make an effort to combine classical agriculture

Stage 1: Separate development of each strategic type of agriculture and the corresponding part of the ideal final result

Contradictions between the strategic types are purposefully not taken into account.

Stage 2: Combination of the strategic types of agriculture into viable agricultural systems (the corresponding parts of the ideal final result are also combined)

A purposeful search for ways to resolve contradictions between the strategic types is carried out.

The result of stage 2 in case of resolution of all the contradictions: the complete realization of the ideal final result

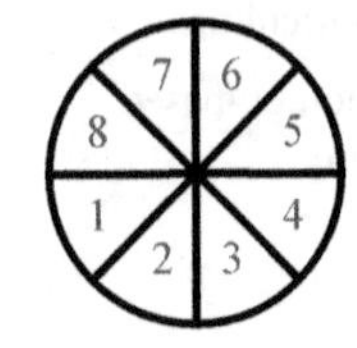

All the contradictions are resolved. The ideal final result is completely realized. All the goals of the strategic types of agriculture are achieved simultaneously.

Each segment represents one of the eight strategic types of agriculture.

Fig. 1. Strategy for the balanced development of agricultural technologies: a general idea.

and green agriculture as part of stage 2 of our approach to the balanced development of agricultural technologies.

3 Striving for Sustainability of Agriculture

TRIZ [7, pp. 76–77] distinguishes three kinds of interrelated contradictions: administrative, technical, and physical. Let us reveal their essence in relation to providing the sustainability of agriculture and offer ways of resolving them.

With regard to making agriculture sustainable, *the administrative contradiction*, a contradiction between a certain need of a person or a group of people and the ability to satisfy it, can be formulated as follows: agriculture, while providing food for large masses of people, unintentionally harms the environment, and it is not known how to solve this ecological problem. Here, the indicated contradiction is expressed as opposition between the need to prevent an undesirable effect (harm to nature) and not knowing how to do it.

The technical contradictions (contradictions between the parameters of the technical system in question, when the improvement of some parameter(s) leads to the deterioration of (an)other parameter(s); they underlie the administrative contradiction) will be the following:

the need to feed the growing population of the planet causes an increase in agricultural land and a simultaneous reduction in the area of wildlife habitat with a tremendous loss of biodiversity[1];

ensuring food security for a large number of people gives rise to climate change (due to greenhouse gas emissions and the conversion of non-agricultural land such as forests to farmland[2]);

obtaining high yields through the use of mineral fertilizers and pesticides causes environmental pollution[3];

intensive agriculture, in pursuit of high yields, degrades soil, reducing its fertility [4, 5][4].

We should note that as regards different ways of resolving technical contradictions in accordance with the contradiction matrix by Altshuller [13], we could not find any appropriate solutions. In our problem of making agriculture sustainable, we have the following two pairs of contradictory characteristics (the first characteristic improves, but the other deteriorates): area of a stationary object—harmful factors developed by an object; and capacity/productivity—harmful factors developed by an object.

The first contradictory pair represents extensive agricultural practices and covers the first two technical contradictions mentioned above. The characteristic "area of a stationary object" is the area of agricultural land, and its improvement means the increase in the farmland. Our purpose is to feed large masses of people. In order to achieve that we increase our agricultural land, which in turn creates more harmful impact on the environment. Hence, our agricultural practices become unsustainable.

On the other hand, the second contradictory pair represents intensive agricultural practices and concerns the last two technical contradictions mentioned above. Here, to feed people, we increase the productivity of agriculture, but intensive methods exacerbate environmental harm. Again, the problem of unsustainability arises.

In both of these situations, the ways of resolving technical contradictions that can be found in the contradiction matrix [13] are not appropriate. For example, it seems inconceivable to successfully apply the "Convert Harm into Benefit" principle in relation to the environmental damage of agriculture. Apparently, these contradictions are impossible to solve on the technical level. We should go deeper—down to the level of physical contradictions.

Finally, the following assertions constitute *the physical contradictions* (imposing diametrically opposing requirements to any property of some part of the technical system in question; they are the cause of technical contradictions):

---

[1] According to FAO data, the share of land used in agriculture in the total land area of all countries increased from 34.3% in 1961 to 36.7% in 2019 (the maximum of 37.6% was reached in 2001, and the minimum for this period was in 1961). Continued growth of the world's population will contribute to the expansion of farmland and will require a constant increase in agricultural productivity. Agriculture is considered one of the main causes of biodiversity loss [1, 2] and changes in terrestrial and freshwater natural ecosystems [1].

[2] About a quarter of the global greenhouse gas emissions come from land clearing and agriculture, with 75% of them coming from animal husbandry [1, p. XXXII].

[3] In [3], it was found that 64% of the agricultural land in the world (about 24.5 million km^2) is at risk of pesticide pollution, and 31% is at high risk.

[4] According to experts [5, p. 52], globally, the loss of soil is 24 billion tons per year.

1. in order to feed a large number of people, it is necessary to have a large area of agricultural land, but in order to preserve the wildlife habitat, the agricultural land area must be small; the same holds for the contradiction between food security and climate neutrality;
2. to increase crop yields, it is necessary to use mineral fertilizers and pesticides, but in order not to pollute the environment, mineral fertilizers and pesticides cannot be applied; the same holds for the contradiction between intensive agriculture and the need to preserve soil fertility (to increase crop yields, it is necessary to apply intensive technologies, but to preserve soil fertility, intensive technologies cannot be used).

These physical contradictions can be formulated even more succinctly: the current extensive and intensive agricultural practices, which are necessary to ensure food security, harm the environment.

As specified by TRIZ [7, p. 79], resolving a physical contradiction is done by separating the opposite properties in space, time, structure, or according to condition.

Now let us present some solutions to the physical contradictions in relation to agriculture and the environment, organizing them in accordance with the above two-element list of the physical contradictions, and separating the opposite properties:

1. **in space:** the creation of protected nature reserves helps to preserve natural ecosystems [14];
   in space and structure: relocating cropland to a territory with the most favorable conditions for growing agricultural plants can significantly reduce the size of necessary cropland and environmental impact of agriculture, including carbon emissions [15];
   according to condition: subject to compliance with the requirements of agricultural technology and post-harvest management, there is a decrease in crop losses [16, 17];
   in structure and according to condition: promoting healthy lifestyles and sport activities can help tackle the problem of obesity, which in turn will lead to consuming less food [18];
   according to condition: upon reaching the threshold values in the world population growth rate, humane mechanisms that mitigate the problem of overpopulation should be engaged (at the global level, this solution requires close international cooperation) [19];
   in structure and space: the transition of humanity to a vegetarian lifestyle would reduce the amount of greenhouse gases emitted by agriculture [20];
2. **in space and structure:** variable rate fertilizer and pesticide application, a technique used in precision agriculture, reduces environmental pollution [21, 22];
   in structure and space: zero and minimum tillage contribute to soil conservation and carbon sequestration, thus preventing respectively soil erosion and global warming [23–25];
   in time and space: fallowing and crop rotation as farming techniques can be very helpful in restoring soil fertility and reducing the harm of intensive agriculture [26];

according to condition: in the event of overexploitation of agricultural land or damage to the environment, measures of government support for the restoration of ecosystems, as well as penalties, should be provided [27, 28];
according to condition: a significant contribution to resolving the contradiction between humans and nature can be made by purposeful improvement of agricultural technology, when there is high agricultural productivity without environmental degradation (for example, through digitalization/intellectualization [29]; substituting physical, biological and other eco-friendly weed control systems for herbicides [30]).

The list of solutions is far from exhaustive. Nevertheless, despite the fact that the majority of these solutions are well-known and represent only a mitigation of the anthropogenic impact of humanity on Mother Nature (with the exception of zero and minimum tillage, fallowing and crop rotation, and purposeful improvement of agricultural technology), in our opinion, the presented methods still provide understanding and point the way to resolving the contradiction between agriculture and the environment. Let us demonstrate this in the context of our strategy for the balanced development of agricultural technologies, which was discussed in Sect. 2. Our purpose is to combine classical agriculture and green agriculture by making agriculture sustainable.

First, agricultural technologies must be improved as much as possible so that they become truly sustainable, or green (*this is included in the first stage of our proposed strategy*). If it is difficult, impossible or too expensive to ensure the sustainability of certain aspects of agricultural technology, mitigation measures should be used, examples of which are given above. We do not consider perfecting classical agriculture here because this type of agriculture has been developing and improving for centuries.

Second, in order to ensure environmental sustainability and be able to feed large masses of people, our green technologies must be made highly productive, and we as a society should establish the humane control of the growth of the world's population (*these measures are part of the second stage of our strategy*). High productivity is achieved through the direct, purposeful improvement of green technologies.

As regards both the control of the growth of the global population and productivity of green agriculture, it is necessary to set up a long-term experiment on the application of sustainable agriculture in a relatively large area in order to test green technologies and experimentally determine the *agricultural carrying capacity of the environment* (similar to the human carrying capacity of the environment, or ecological footprint [31, 32]). By the agricultural carrying capacity of the environment, we mean the maximum number of people who can be fed from a given territory, and at the same time not permanently damage the environment. In the scientific literature, there are different estimates of the human carrying capacity of the entire planet (up to one trillion people or even more [32, p. 1; 33]). However, we believe that it is better to establish the agricultural carrying capacity *experimentally*, taking into account the productivity of the green technologies and the soil and climate conditions in the area in question.

Also, we should take into account an opportunity of relocating agricultural production to land more suited for agriculture, which will give us an advantage of reducing our farmland and environmental footprint [15].

It is necessary to achieve both the sustainability of agricultural technologies and a high level of agricultural carrying capacity, although the latter is more dependent on nature itself. The idea behind all this is very simple: the use of highly productive green technologies in agriculture and the humane control of the growth of the world's population is what we need to ensure long-term sustainability of our food systems, thereby resolving contradictions between agriculture and the environment. This is how, on a fundamental level, we combine classical agriculture and green agriculture, which is done in the second stage of our strategy.

Before we create efficient green technologies, it is advisable to use the mitigation measures listed above or similar to those.

At the end of this section, we want to emphasize that the approach proposed in this paper expands the application of TRIZ: the concept of ideal final result is applied not to a technical system, but to an entire industry. This macro approach and the micro approach to the environmental problems of agriculture available in the literature, embodied, for example, in [10], can be used together.

4 Conclusion

The major conclusion of our paper is that TRIZ is a convenient, suitable tool for finding ways to resolve contradictions between strategic types of agriculture, when the solution found opens the way to a balanced development of agricultural technologies. The key category here is ideal final result, which is applied to agriculture as a whole.

Thus, after defining eight strategic types of agriculture, the application of TRIZ allowed us to state ideal final result for agriculture, identify the contradictions—administrative, technical, and physical—contained in it, and point the way to solving the fundamental contradiction between agriculture and the environment. It was found that making green technologies highly productive and humane control of the growth of the world's population are those factors that ensure sustainability of agriculture.

Such a process of resolving contradictions between all the strategic types of agriculture was called in the paper a strategy for the balanced development of agricultural technologies.

The limitation of our research is that we considered the resolution of contradictions only between two strategic types of agriculture—classical and green (though we indicated contradictions between all the types). The combination of all the eight strategic types into a viable agricultural system was not our task, but can be a purpose for future research. Due to the complexity of the issue, it is appropriate to perform this task sequentially, adding one strategic type at a time. In the case of successful resolution of all contradictions between all the strategic types, which is expressed in obtaining a viable, organically coherent system of agriculture, there is a complete realization of the ideal final result for agriculture. This is the most difficult part of a strategy for the balanced development of agricultural technologies presented in this paper.

But even the combination of conventional and green technologies presents a challenge. The current state of development of agricultural technologies is such that there is now an integration of classical agriculture and digital agriculture, which is in line with the economic-driven goal of maximizing the efficiency of agricultural production, while green technologies reflect an alternative paradigm that has not yet become mainstream. However, sooner or later humanity will come to the inevitability of sustainable agriculture on a large scale.

We hope that this work will make its contribution to the creation of a kind, harmonious world, where both humanity and Mother Earth would live a happier and more sustainable life.

References

1. IPBES: Global assessment report of the Intergovernmental Science-Policy Platform on Biodiversity and Ecosystem Services. IPBES secretariat, Bonn (2019)
2. Kehoe, L., Romero-Muñoz, A., Polaina, E., Estes, L., Kreft, H., Kuemmerle, T.: Biodiversity at risk under future cropland expansion and intensification. Nat. Ecol. Evol. **1**, 1129–1135 (2017). https://doi.org/10.1038/s41559-017-0234-3
3. Tang, F.H.M., Lenzen, M., McBratney, A., Maggi, F.: Risk of pesticide pollution at the global scale. Nat. Geosci. **14**, 206–210 (2021). https://doi.org/10.1038/s41561-021-00712-5
4. Kopittke, P.M., Menzies, N.W., Wang, P., McKenna, B.A., Lombi, E.: Soil and the intensification of agriculture for global food security. Environ Int **132**, 105078 (2019). https://doi.org/10.1016/j.envint.2019.105078
5. United Nations Convention to Combat Desertification: Global land outlook. 1st edn. Secretariat of the United Nations Convention to Combat Desertification, Bonn (2017)
6. Altshuller, G.S.: Creativity as an Exact Science: The Theory of the Solution of Inventive Problems. Gordon and Breach Publishers, Amsterdam (1984)
7. Petrov, V.: TRIZ Theory of Inventive Problem Solving: Level 1. Springer, Cham (2019). https://doi.org/10.1007/978-3-030-04254-7
8. Buzuku, S., Shnai, I.: A systematic literature review of TRIZ used in eco-design. J. Eur. TRIZ Assoc. **4**, 20–31 (2017)
9. Mishra, U.: Understanding secrets of sustainability through TRIZ philosophy (2014). https://doi.org/10.2139/ssrn.2421392. Accessed 19 June 2022
10. Bogatyrev, N., Bogatyreva, O.: BioTRIZ: a win-win methodology for eco-innovation. In: Azevedo, S.G., Brandenburg, M., Carvalho, H., Cruz-Machado, V. (eds.) Eco-Innovation and the Development of Business Models. GINS, vol. 2, pp. 297–314. Springer, Cham (2014). https://doi.org/10.1007/978-3-319-05077-5_15
11. Sunding, D., Zilberman, D.: The agricultural innovation process: research and technology adoption in a changing agricultural sector. In: Gardner, B., Rausser, G. (eds.) Handbook of Agricultural Economics, vol. 1, pp. 207–261. Elsevier, Amsterdam, New York (2001)
12. Mazzucato, M.: Mission Economy: A Moonshot Guide to Changing Capitalism. Harper Business, New York (2021)
13. Altshuller, G.S.: The Innovation Algorithm: TRIZ, Systematic Innovation and Technical Creativity, 2nd edn. Technical Innovation Center Inc, Worcester (2007)
14. Geldmann, J., Manica, A., Burgess, N.D., Coad, L., Balmford, A.: A global-level assessment of the effectiveness of protected areas at resisting anthropogenic pressures. PNAS **116**(46), 23209–23215 (2019). https://doi.org/10.1073/pnas.1908221116
15. Beyer, R.M., Hua, F., Martin, P.A., Manica, A., Rademacher, T.: Relocating croplands could drastically reduce the environmental impacts of global food production. Commun Earth Environ **3**, 49 (2022). https://doi.org/10.1038/s43247-022-00360-6
16. Chakraborty, A.C.: Pre- and post-harvest losses in vegetables IVI. In: Singh, B., Singh, S., Koley, T.K. (eds.) Advances in Postharvest Technologies of Vegetable Crops: Postharvest Biology and Technology, pp. 25–87. Apple Academic Press, Waretown, NJ : Apple Academic

Press, 2018. | Series: Postharvest biology and technology (2018). https://doi.org/10.1201/978 1315161020-2
17. Sawicka, B.: Post-harvest losses of agricultural produce. In: Filho, W.L., Azul, A.M., Brandli, L., Özuyar, P.G., Wall, T. (eds.) Zero Hunger, pp. 654–669. Springer International Publishing, Cham (2020). https://doi.org/10.1007/978-3-319-95675-6_40
18. Frey, S., Barrett, J.: Our health, our environment: the ecological footprint of what we eat. In: International Ecological Footprint Conference. BRASS, Cardiff University, Cardiff (2007)
19. Lidicker, W.Z., Jr.: A scientist's warning to humanity on human population growth. Glob. Ecol. Conserv. **24**, e01232 (2020). https://doi.org/10.1016/j.gecco.2020.e01232
20. Xu, X., et al.: Global greenhouse gas emissions from animal-based foods are twice those of plant-based foods. Nat. Food **2**, 724–732 (2021). https://doi.org/10.1038/s43016-021-003 58-x
21. Bongiovanni, R., Lowenberg-Deboer, J.: Precision agriculture and sustainability. Precis. Agric. **5**, 359–387 (2004). https://doi.org/10.1023/B:PRAG.0000040806.39604.aa
22. Majumder, D., et al.: Precision input management for minimizing and recycling of agricultural waste. In: Bhatt, R., Meena, R.S., Hossain, A. (eds.) Input Use Efficiency for Food and Environmental Security, pp. 567–603. Springer, Singapore (2021). https://doi.org/10.1007/ 978-981-16-5199-1_19
23. Komissarov, M.A., Klik, A.: The impact of no-till, conservation, and conventional tillage systems on erosion and soil properties in Lower Austria. Eurasian Soil Sc. **53**, 503–511 (2020). https://doi.org/10.1134/S1064229320040079
24. Ogle, S.M., et al.: Climate and soil characteristics determine where no-till management can store carbon in soils and mitigate greenhouse gas emissions. Sci. Rep. **9**, 11665 (2019). https:// doi.org/10.1038/s41598-019-47861-7
25. Krauss, M., et al.: Reduced tillage in organic farming affects soil organic carbon stocks in temperate Europe. Soil Tillage Res. **216**, 105262 (2022). https://doi.org/10.1016/j.still.2021. 105262
26. Nadeem, F., Nawaz, A., Farooq, M.: Crop rotations, fallowing, and associated environmental benefits. In: Oxford Research Encyclopedia of Environmental Science. https://doi.org/10. 1093/acrefore/9780199389414.013.197. Accessed 19 June 2022
27. Coria, J., Sterner, T.: Natural resource management: challenges and policy options. Annu. Rev. Resour. Econ. **3**, 203–230 (2011). https://doi.org/10.1146/annurev-resource-083110-120131
28. Peake, L., Robb, C.: Saving the ground beneath our feet: establishing priorities and criteria for governing soil use and protection. R. Soc. Open Sci. **8**, 201994 (2021). https://doi.org/10. 1098/rsos.201994
29. Hrustek, L.: Sustainability driven by agriculture through digital transformation. Sustainability **12**, 8596 (2020). https://doi.org/10.3390/su12208596
30. Hasan, M., Ahmad-Hamdani, M.S., Rosli, A.M., Hamdan, H.: Bioherbicides: an eco-friendly tool for sustainable weed management. Plants **10**, 1212 (2021). https://doi.org/10.3390/pla nts10061212
31. Chakraborty, T., Thakur, B.K.: Ecological footprint and sustainable development: a two-way approach. In: Filho, W.L., Azul, A.M., Brandli, L., Salvia, A.L., Wall, T. (eds.) Affordable and Clean Energy, pp. 303–311. Springer International Publishing, Cham (2021). https://doi. org/10.1007/978-3-319-95864-4_41
32. Siegel, F.R.: The Earth's Human Carrying Capacity: Limitations Assessed, Solutions Proposed. Springer International Publishing, Cham (2021). https://doi.org/10.1007/978-3-030-73476-3
33. Franck, S., von Bloh, W., Müller, C., Bondeau, A., Sakschewski, B.: Harvesting the sun: new estimations of the maximum population of planet Earth. Ecol. Modell. **222**(12), 2019–2026 (2011). https://doi.org/10.1016/j.ecolmodel.2011.03.030

Analyzing the Role of Human Capital in Strengthening National Innovation System Through University-Industry Research Collaboration: A TRIZ-Based Approach

Abeda Muhammad Iqbal[1](✉), Narayanan Kulathuramaiyer[1], Adnan Shahid Khan[1,2], and Johari Abdullah[2]

[1] Institute of Social Informatics and Technological Innovation, Universiti Malaysia Sarawak, Malaysia, 94300 Kota Samarahan, Sarawak, Malaysia
miabeda@unimas.my

[2] Faculty of Computer Science and Information Technology, Universiti Malaysia Sarawak, Malaysia, 94300 Kota Samarahan, Sarawak, Malaysia

Abstract. University-industry research collaboration (UIRC) is seen as a key measure in formulating national-level indicators of research and innovations and economic growth. Despite the extensive availability of shreds of evidence indicating the importance of such areas of collaboration in developed and developing countries, existing literature on the strengthening of the National innovation system (NIS) via UIRC is still scarce. Literature has highlighted the impact of human capacity as having a strong influence on researchers' innovative activities as well as on the NIS. Moreover, in strengthening the NIS, it is considered mandatory to model human capacity, specifically, in the aspects of universities- industries personnel. In this paper, we explore the usage of the TRIZ thinking models together with the adaptation of its toolkit to shed insights on the influence of human capacity on NIS. Using the TRIZ heuristic modeling paradigm, a systems model of human capacity and its significant influence on NIS has been demonstrated. Thus, the findings of this research suggest the tremendous potential of employing TRIZ as a systems modeling tool for explorative analysis of intangible outcomes. The potential of using TRIZ as an explanation module for systems thinking will be a game-changer for knowledge-based modeling of national innovation systems.

Keywords: University-industry research collaboration · National innovation system · Human capital · Theory of Inventive Problem Solving (TRIZ) approach

1 Introduction

It is usually accepted that technological change and innovation are key requirements to ensure sustained economic growth. Specifically, to overcome the economic crisis and for winning the future, all countries need to boost their capacity to strengthen the national innovation system. Moreover, many studies provide pieces of evidence for the

Published by Springer Nature Switzerland AG 2022
R. Nowak et al. (Eds.): TFC 2022, IFIP AICT 655, pp. 417–428, 2022.
https://doi.org/10.1007/978-3-031-17288-5_35

strategic importance of the University-Industry research collaboration that is producing a very huge impact on the national innovation system [1–7]. Regardless of extensive indicators of the significance of UIRC, the rate of technological innovation from UIRC is seen to be not satisfactory in several developing countries [8–10]. Several studies were conducted to explore the factors that can enhance such rates of technological innovation through efforts in minimizing the barriers of UIRC. Nevertheless, most works focused on university-industry orientation factors which include: conducting workshops & seminars and hiring educated, trained and skilled personnel, which usually act as a symptomatic way out of the problem [11–15].

Furthermore, universities and industries are seen as the primary components of the national innovation system (NIS) which directly perform technological innovation [16–18]. In this regard, the authors [2] have emphasized that if the aim is to strengthen NIS, it is advisable to investigate the influence of UIRC on NIS. However, comprehensive literature about the factors relating to the UIRC and their influences on NIS is still scarce.

Secondly, the main limitation as highlighted in current literature is the lack of a systemic approach, which has led to the limited predictability of outcomes [19, 20]. Moreover, NIS maintain their existence through the mutual interaction of their primary parts ("Universities and industries") that demand a systemic approach for its evaluation [21, 22]. Thus, a systemic approach will be required to overcome fundamental weaknesses in consequential mapping owing to the reliance on its linear model [23].

This study, therefore, aims to investigate the influence of UIRC on the NIS.

In addition, this study proposes the usage of the TRIZ model of systems thinking and key instruments from its toolkit. The theory of Inventive Problem Solving or TRIZ provides a collection of visual thinking instruments that takes a two-stage approach and thereby helps in systematically resolving complex problems [24, 25]. The realization that the presence of a contradiction that is worth to be identified, represented, and solved algorithmically, serves as a powerful problem-solving toolkit [26]. In this research, we have explored preliminary TRIZ instruments, the Cause and Effect Chain Analysis (CECA), Function Modelling, and Contradictory Analysis, to provide insights into the complexities of modeling abstract relationships. The use of the TRIZ instruments that have been formulated by its founder Genric Altshuller as elements of productive thinking in innovative engineering is evaluated as a bridging method in addressing the concerns addressed in these papers [27–29].

For example, the use of CECA not only identifies the root causes of the problem but also provides the pathway to the optimum solution. Thus, by utilizing TRIZ instruments as a series of problem formulation and guided solutions outlining pathways, an enriched perspective has been presented.

Rest of the paper is structured as follows, Sect. 2 briefly discussed literature review and hypothesis development. Section 3 provides detailed methodology followed by conclusion and future direction in section 4.

2 Literature Review and Hypothesis Development

It is generally accepted that human capital (HC) is an essential part of any national innovation system and a central element of economic growth theory. An innovation system and economy with a larger total stock of human capital experience faster growth [30]. However, it is observed that when UIRC is formed, imbalanced personnel in universities and industries is one of the major constraints between them. Imbalanced personnel here refers to the inequality in education and skills, exchange of knowledge, expertise, and the bits of advice among the research organizations to resolve the issues and problems that arise during such research and innovation processes [31]. Similarly, authors of [32], identified that highly educated and trained personnel are able to highly contribute to the development of research and innovations. It is observed that when university-industry research collaboration is formed, HC is one of the major constraints between them [33].

On the other side, the majority of firms that collaborate with universities to recruit personnel who can fulfil the needs of the industries. Universities in developing countries focus on their curriculum to produce graduates who are able to meet the needs of the industry [34]. Furthermore, the authors of [35], defined that education and training institutions must have a research-based curriculum rather than a theory-based one. Theoretical knowledge is insufficient for producing capable human capital and the development of research and innovations which is also the major bottleneck hindering the innovative interactions between universities and industries. Human capital in universities and industries is embodied in skilled and experienced personnel who provide professional training and skills via exchanging of information and expertise with each other and increase the levels of technological abilities, which lead to the R&D partners' satisfaction and performance and eventually NIS performance [17, 36–40].

The literature describes knowledge transfer of best practices from developed countries as workable solutions for mapping to apparently similar conditions in local contexts. However, such over-generalised outside-in prescribed solutions tends to be a misfit at times. For example, the overall standard of education and training system in developing countries needs be carefully treated in an attempt to produce the required human capital to strengthen the NIS. In addition, not taking a systems approach is the key disadvantage, whereby a black-box method is taken, with non-connected constructs tends to be proposed in isolation. In addition, methods that can fully map the current state of the knowledge base are required to enable a visual depiction of components of a National Innovation Systems modeling.

3 Methodology

In this study, TRIZ models and instruments have been utilized to depict the high-level macro systems. A functional analysis was firstly used to reverse engineer the knowledge flows based on the review of the literature and the past research as carried out by the team of researchers [5, 6, 14, 17]. Subsequently, a Functional model was formulated, followed by the Cause and Effect Chain Analysis and Engineering contradiction modelling. In this preliminary works that we have undertaken, an elaborate TRIZ-based Engineering systems model has not been formulated. Future works will address such an extension where flows models of TRIZ will be taken into consideration. Section 4 then illustrates the more comprehensive view of the national innovation model with a discussion on what can the use of TRIZ provide.

3.1 Reverse Engineering System Model

In Fig. 1, the function model is showing a smooth way of strengthening the NIS. The function model provides a systematic approach to generating an inventive solution to the problem. The constructs are based on our past quantitative models. Figure 1 represents the annotated functional flows that have been explicitly stated as a function name.

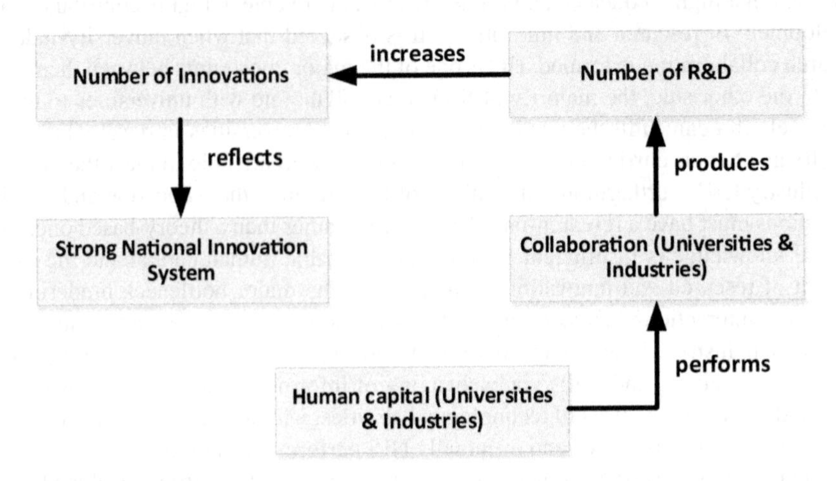

Fig. 1. Functional model depicting the current state of UIC as an enabler for a strong NIS

This step has been instrumental to highlight the use of visual modelling as a way of augmenting tacit knowledge as extracted from past system thinking models. The functional modelling tools has presented a way of adopting a knowledge-based approach to characterize and depict the knowledge flows and interactions.

3.2 Functional Analysis

Based on the reverse engineering stage in providing a deeper understanding of the system thinking models, we present a function model describing the generic structure of systems thinking instruments. We have identified the main useful function as a system approach in informing policymakers and in the simulation of the knowledge-based economy. The detailed functional model with their influences is illustrated in Fig. 2.

We have further defined a meta-level function model that describes a need for the TRIZ-based visual knowledge system as shown in Fig. 3.

3.3 Cause and Effect Chain Analysis

In adopting the Cause and Effect method for the problem modelling of strengthening National Innovation systems, a collaborative brainstorming approach was used. The structuring of the Cause and Effect chain analysis is maintained, while a multi-disciplinary team is engaged. The method used is outlined as follows:

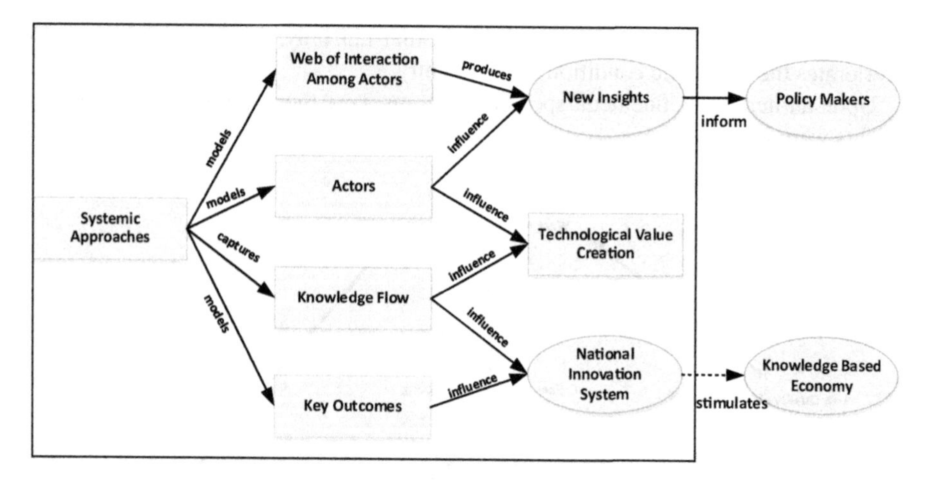

Fig. 2. Understanding of systems relationship

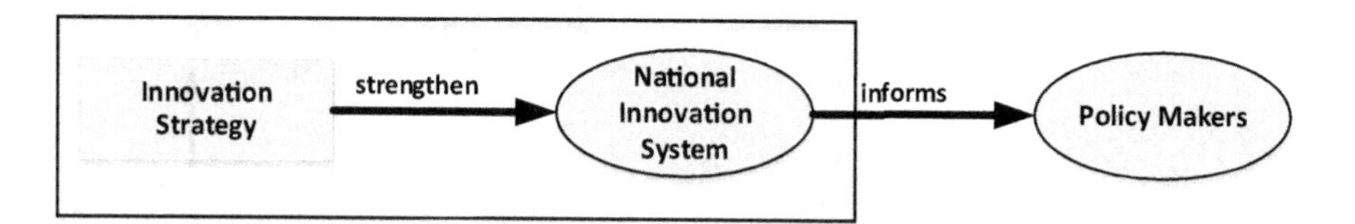

Fig. 3. TRIZ-based explainable systems model

1. We start the session with the formulation of the goal of the system. The goal of this CECA activity has been to strengthen the NIS.
2. The second step has been the definition of target disadvantage to serve as the starting node in the cause and effect chain. In our situation, this was defined as "Weak NIS" as the target to be addressed.
3. We then proceeded with an iterative questioning as to why this happens. There are hypotheses established, which are then verified by either a literature review or a consultation with relevant experts.
4. The brainstorming element in this process is key as the various modelling possibilities are evaluated and a coherent flow is accepted.
5. The process goes on until a key disadvantage is identified. The involvement of decision-makers and policymakers adds value to this stage as the process itself becomes co-created with stakeholders.

The CECA model in Fig. 4 is showing some challenges to the function model that universities and industries specifically in developing countries always face the challenge of low-quality of human capital, and the reason behind this is the quality of education that is not up to the standard of developed countries. However, changes in the index of education can provide a well-educated workforce to the universities and industries. Nevertheless, after nurturing an educated workforce in the country, developing countries normally face the challenges of a brain drain that not only affects the NIS but also deteriorates the economic condition of the country.

Upon an iterative refinement, specific programs were then outlined, and additional insights were gained.

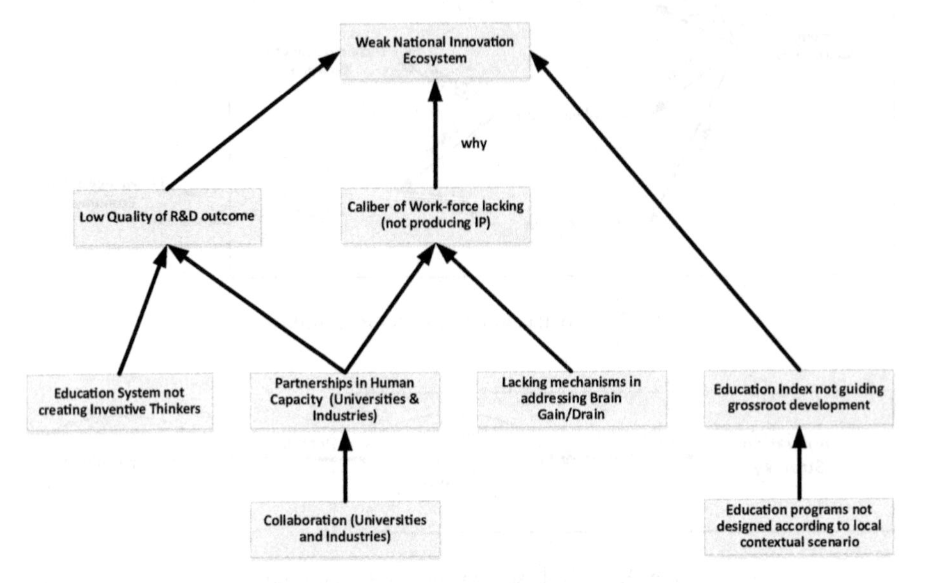

Fig. 4. Cause and Effect Chain Analysis

3.4 Engineering Contradiction Model

Subsequently, an Engineering Contradiction model was formulated for the macro-level problem at hand for strengthening the NIS. The preliminary outcomes and the potential of using the tool will be evaluated. Based on the models that have been produced it is seen that both micro and microlevel heuristic models can be built. In engaging experts in systems modelling, policymakers, and TRIZ thinkers, it was possible to construct alternative models as opposed to the state-of-the-art reliance on black-box models.

IF:	Standard Initiatives are targeting the increase of the Quality of primary education
THEN:	Universities and industries will have access to qualified (human capacity) expertise
BUT:	Brain drain becomes a major concern

Program–level model micro encoding that defines the current approach to dealing with NIH modeling.

IF:	Standard Initiatives are targeting in increasing the Quality of primary education
THEN:	A competent Workforce will be nurtured
BUT:	Participatory and Access gaps for grooming local leaders/champions still not addressed

Contradiction Analysis. Based on the improving and worsening parameters, we obtained the suggested inventive principles from the contradiction matrix. For the domain of strengthening the NIS, it was necessary to map the substances of a typical technical system to human dimensions of increase in human capacity quality and brain drain indicators respectively.

The inventive principles were formulated through an extensive study of the innovative solution from past patents. These inventive principles give an insight into the potential approach that can be utilized to solve a challenge. Table 1 shows the inventive principles for improving and worsening parameters.

Table 1. Remodeling solutions based on inventive principles

Recommended inventive principles	Inventive trigger as a guideline	Potential solutions from the suggested inventive principles
# 6 Universality	Make an object perform multiple functions	Creating avenues for professionals/thought leaders by diversifying their roles
# 3 Local quality	Make an object suitable for every condition	Create a conducive environment with sufficient support for systematic brain gain programs (goal-driven) and averting brain drain
# 10 Prior action	Perform the required change in advance	Create support systems in advance to make it favorable for brain gain/brain drain
# 24 Mediator	Perform as an intermediary	Establish an intermediary agency as a UIC to reverse the outflow of talent

3.5 TRIZ Based Framework for Explorative System Thinking

Based on the worsening (loss of substance) and improving parameters (amount of substance) from the TRIZ contradiction matrix, Table 2 shows 4 inventive principles {(# 3 local quality, # 6 universality, # 10 prior action, and # 24 Mediator)}.

Table 2. Summary of the solution

TRIZ Tool	Proposed solution
1 Function Model	Input: Human capital in universities and industries Output: Strong NIS
2 Cause and Effect Chain Analysis	Low quality of human capital in universities and industries
2 Engineering Contradictions	
Challenge 1	
Root causes:	Standard of Education
Solutions: # 3 Local Quality #10 Prior Action	Changes in the index of Education
Challenge 2	Loss of human capital
Root cause	Brain Drain
Solutions: #6 Universality #24 Mediator	Number of local employment

Figure 5 shows two parameters "standard of education and brain drain" as the constraints, concurrently, outlining parameters "index of education and number of employees" as the suggested solution to diminish the constraints and to enhance the innovative capabilities of HC in universities and industries and consequently strengthen the NIS. Based on the developed framework, this model further incorporates relationships from previous literature that human capital not only improves the technological competencies of universities and industries but also strengthens the NIS by improving the quality of R&D and the number of innovations. Thus, from the findings of this research the TRIZ-based framework has concluded that human capacity is the key factor to enhance the outcome of NIS.

TRIZ's approach consequently provides the solution to diminish the constraints by indicating the factors of educated and skilled personals (ESP) as the critical factors of UIRC. An extensive and exhaustive discussion is elaborated in Sect. 2.

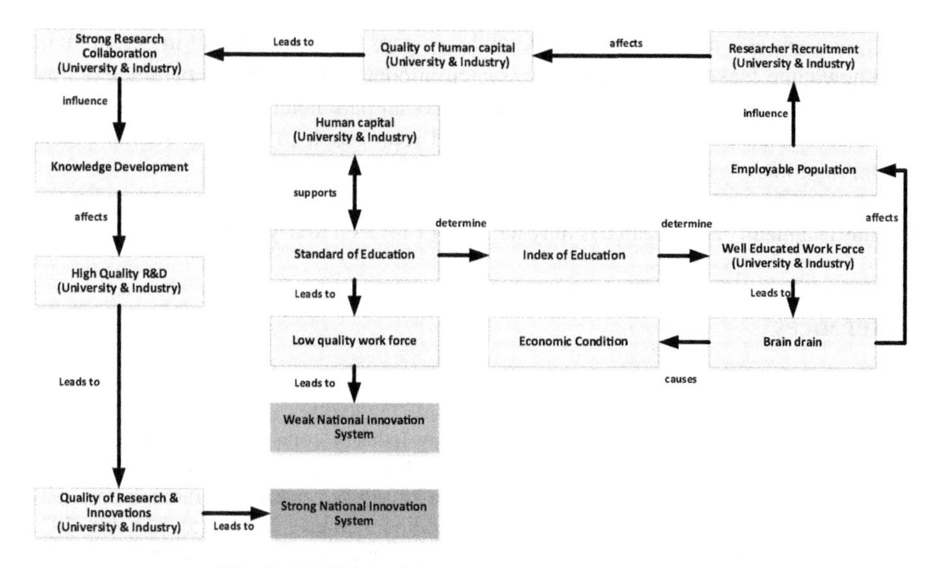

Fig. 5. TRIZ-based framework to strengthen the NIS

4 Conclusion and Future Direction

In transferring best practices by adopting an outside-in approach and employing indicators from de-contextualized scenarios, intrinsic aspects of local system conditions tend to be overlooked. As opposed to using a specialised set of parameters for the domain of Human Resource [41], this research has adopted the original parameters of Alshuller which we used as triggers for suggesting human capacity related indicators. Our goal was to explore the TRIZ-based methods in themselves as a means of augmenting and enriching the systems modelling ability. The role of TRIZ as a means of knowledge representation whereby a heuristic modelling of knowledge flows was then visualised. Our paper has demonstrated TRIZ models as a way to bridge the gap between research and innovation organizations and policymakers by illustrating the influence of HC on NIS. The TRIZ approach has helped policymakers to see the root causes of the problem and the ways to overcome it [42, 43].

As opposed to taking an entirely black-box model, TRIZ practitioners, consider an analytical approach where human capacity aspects are structured as components or even as building blocks in the formulation of a national-scale innovation system. We find that when an empirical study is adopted there is a tendency to specialize the study to a geographical region, with the main difficulty in repurposing the models constructed. While these studies highlight complexities of context, we have explored the capacity of TRIZ in augmenting these abstract systems thinking models with a representation of intrinsic concepts within an analytical supportive framework.

Reverse Engineering and visualization as function models has been a crucial component of the proposed methodology to systematically augment black-box models. CECA is then adapted to suit the design of the framework for an explainable systems model that drills down into the knowledge-flows and interaction pathways. The use of engineering contradictions as a means to identify knowledge gaps provided support in externalizing the knowledge locked up in black-box system models. The technical parameters however needed to be customized and mapped as social parameters. This paper has shown that TRIZ-based methods can be adapted and repurposed to serve as social systems modelling tools.

Acknowledgment. This research is fully funded by Universiti Malaysia Sarawak.

References

1. Lin, J.Y., Yang, C.H.: Heterogeneity in industry-university R&D collaboration and firm innovative performance. Scientometrics **124**(1), 1–25 (2020)
2. Messeni Petruzzelli, A., Murgia, G.: University-Industry collaborations and international knowledge spillovers: a joint-patent investigation. J. Technol. Transfer **45**(4), 958–983 (2020)
3. Jones, S.E., Coates, N.: A micro-level view on knowledge co-creation through university-industry collaboration in a multi-national corporation. J. Manag. Dev. **39**(5), 723–738 (2020)
4. Ting, S.H., Yahya, S., Tan, C.L.: Importance-performance matrix analysis of the researcher's competence in the formation of university-industry collaboration using smart PLS. Public Organ. Rev. **20**(2), 249–275 (2019). https://doi.org/10.1007/s11115-018-00435-z
5. Iqbal, A.M., Khan, A.S., Senin, A.A.: Reinforcing the national innovation system of Malaysia based on university-industry research collaboration: a system thinking approach. Int. J. Manag. Sci. Bus. Res. **4**(1), 6–15 (2015)
6. Iqbal, A.M.: Influence of national innovation system on university-industry research collaboration. Doctoral thesis, Universiti Teknologi Malaysia (2018)
7. Iqbal, A.M., Khan, A.S., Iqbal, Abdullah, J, Kulathuramaiye, N., Senin, A. A. , Blended system thinking approach to strengthen the education and training in university-industry research collaboration. Technol. Anal. Strategic Manag. **34**(4), 447–460 (2021)
8. Parmentola, A., Ferretti, M., Panetti, E.: Exploring the university-industry cooperation in a low innovative region. What differences between low-tech and high-tech industries? Int. Entrep. Manag. J. **17**, 1469–1496 (2021)
9. Tseng, F.C., Huang, M.H., Chen, D.Z.: Factors of university-industry collaboration affecting university innovation performance. J. Technol. Transf. **45**(2), 560–577 (2020)
10. Bohin, P.: Effectiveness of innovative policies to enhance university-industry collaboration in developing countries. Towards technical university-industry links in Ghana. Br. J. Educ. **6**(2), 54–70 (2018)
11. Chen, K., Lu, W., Wang, J.: University-industry collaboration for BIM education: lessons learned from a case study. Ind. High. Educ. **34**(6), 401–409 (2020)
12. Dooley, L., Gubbins, C.: Inter-organisational knowledge networks: synthesizing dialectic tensions of university-industry knowledge discovery. J. Knowl. Manag. **23**(10), 2113–2134 (2019)
13. Brazile, T., Hostetter, S.G., Donough, C.M., Citters, D.W.: Promoting innovation: enhancing transdisciplinary opportunities for medical and engineering students. Med. Teach. **40**(12), 1264–1274 (2018)
14. Iqbal, A.M., Khan, A.S., Iqbal, S., Senin, A.A.: Designing of success criteria-based evaluation model for assessing the research collaboration between university and industry. Int. J. Bus. Res. Manag. **2**(2), 59–73 (2011)
15. Iqbal, A.M., Khan, A.S., Parveen, S., Senin, A.A.: An efficient evaluation model for the assessment of university-industry research collaboration in Malaysia. Res. J. Appl. Sci. Eng. Technol. **10**(3), 298–306 (2015)

16. Iqbal, A.M., Khan, A.S., Bashir, F., Senin, A.A.: Evaluating national innovation system of Malaysia based on university-industry research collaboration: a system thinking approach. Asian Soc. Sci. **11**(13), 45 (2015)
17. Iqbal, A.M., Khan, A.S., Senin, A.A.: Determination of high impact evaluation metrics for evaluating the University-industry technological linkage. Int. J. Phys. Soc. Sci. **2**(4), 111–122 (2012)
18. Chen, Y., Han, J., Xuan, Z., Gao, W.: Higher education's role in Chinese national innovation system: a perspective of university-industry linkages. In: 17th International Conference on Scientometrics and Informatics, ISSI 2019 - Proceedings, vol. 1, pp. 573–583 (2019)
19. Kafouros, M., Wang, C., Piperopoulos, P., Mingshen, Z.M.: Academic collaborations and firm innovation performance in China: the role of region-specific institutions. Res. Policy **44**(3), 803–817 (2015)
20. Iqbal, A. M., Khan, A.S., Iqbal, S., Senin, A.A.: A novel cost-efficient evaluation model for assessing research-based technology transfer between university and industry. Sains Human. **64**(2), 87–91 (2013)
21. Razorenov, Y.I., Vodenko, K.V.: Innovative development of the national university system in Russia: trends and key elements. Int. J. Sociol. Soc. Policy **41**(1/2), 245–262 (2021)
22. Wilson, L.E., Gahan, M.E., Lennard, C., Robertson, J.: Why do we need systems thinking approach to military forensic science in the contemporary world? Aust. J. Forensic Sci. **52**(3), 323–336 (2020)
23. Allender, S., Brown, A.D., Bolton, K.A., Fraser, P., Lowe, J., Hovmand, P.: Translating systems thinking into practice for community action on childhood obesity. Obes. Rev. **20**(S2), 179–184 (2019)
24. Blackburn, T.D., Mazzuchi, T.A., Sarkani, S.: Using a TRIZ framework for systems engineering trade studies systems engineering. Syst. Eng. **15**, 355–367 (2012)
25. Kasravi, K.: Applications of TRIZ to IT: cases and lessons learned. In: TRIZCON 2010; The Altshuller Institute for TRIZ Studies, Dayton, OH, USA (2010)
26. Czinki, A., Hentschel, C.: Solving complex problems and TRIZ. In: TRIZ Future 2015 (TFC 2015), pp. 27–32. Elsevier, Berlin (2015)
27. Chechurin, L.: TRIZ in science Reviewing indexed publications. Proc. CIRP **39**, 156–165 (2016)
28. Altshuller, G.: 40 Principles Extended Edition: TRIZ Keys to Technical Innovation. Technical Innovation Center Inc., Worcester (2005)
29. Orloff, M.A.: Inventive Thinking Through TRIZ: A Practical Guide, 2nd edn., p. 351. Springer, Heidelberg (2006). https://doi.org/10.1007/978-3-540-33223-7
30. Santos-rodrigues, H., Dorrego, P.F. Jardon, C.F.: The influence of human capital on the innovativeness of firms. Int. Bus. Econ. Res. J. **9**, 53–63 (2010)
31. Cohen, W.M., Nelson, R.R., Walsh, J.P.: Links and impacts: the influence of public research on industrial R&D. Manage. Sci. **48**(1), 1–23 (2002)
32. Rebne, D.: Faculty consulting and scientific knowledge—a traditional university-industry linkage. Educ. Adm. Q. **25**(4), 338–357 (1989)
33. Brimble, P., Doner, R.F.: University-industry linkages and economy development: the case of Thailand. World Dev. **35**(6), 1021–1036 (2007)
34. Sarriot, E., Morrow, M., Langston, A., Weiss, J., Landegger, J., Tsuma, L.: A causal loop analysis of the sustainability of integrated community case management in Rwanda. Soc. Sci. Med. **131**, 147–155 (2015)
35. Kaymaz, K., Eryiğit, K.Y.: Determining factors hindering university-industry collaboration: an analysis from the perspective of academicians in the context of entrepreneurial science paradigm. Int. J. Soc. Inquiry **4**(1), 185–213 (2011)

36. Iqbal, A.M., Aslan, A.S., Khan, A.S., Iqbal, S.: Research collaboration agreements: a major risk factor between university-industry collaboration. An analysis approach. In: Third International Graduate Conference on Engineering, Science and Humanities, UTM, Johar Bahru, Malaysia (2010)
37. Iqbal, A.M., Aslan, A.S., Khan, A.S.: Innovation oriented constraints between university-industry technological collaboration. In: Proceeding ICPE-4 2010, vol. 4, p. 24 (2010)
38. Mallana, M.F.B.A., Iqbal, A.M., Iqbal, S., Khan, A.S., Senin, A.A.: The critical factors for the successful transformation of technology from developed to developing countries. J. Teknol. **64**(3), 105–108 (2013)
39. Iqbal, S., Iqbal, A.M., Khan, A.S., Senin, A.A.: A modern strategy for the development of academic staff based on university-industry knowledge transfer effectiveness & collaborative research. Sains Human. **64**(3), 35–38 (2013)
40. Iqbal, A.M., Kulathuramaiyer, N., Khan, A.S., Abdullah, J., Khan, M.A.: Intellectual capital: a system thinking analysis in revamping the exchanging information in university-industry research collaboration. Sustainability **14**, 6404 (2022)
41. Ersin, F.: Implementation of TRIZ methodology in human capital. Master thesis, İstanbul, Bahcesehir University (2009)
42. Terninko, J., Zusman, A., Zlotin, B.: Systematic Innovation: An Introduction to TRIZ (Theory of Inventive Problem Solving). CRC Press, USA (1998)
43. Belski, I., Baglin, J., Harlim, J.: Teaching TRIZ at university: a longitudinal study. Int. J. Eng. Educ. **29**, 346–354 (2013)

TRIZ-Based Remodeling of Body Enclosure for Corpse

Ashley Edward Roy Soosay[1,3](✉), Muhammad Hamdi Mahmood[1,3], Mohd Saiful Bahari[2], and Narayanan Kulathuramaiyer[3]

[1] Faculty of Medicine and Health Sciences, Universiti Malaysia Sarawak, 94300 Kota Samarahan, Sarawak, Malaysia
sashley@unimas.my

[2] Faculty of Applied and Creative Arts, Universiti Malaysia Sarawak, 94300 Kota Samarahan, Sarawak, Malaysia

[3] Institute of Social Informatics and Technological Innovations, Universiti Malaysia Sarawak, 94300 Kota Samarahan, Sarawak, Malaysia

Abstract. This project started as an effort to gauge the epidemiological study for Road-Traffic-Accident (RTA). The data collected enabled us to make a Personalised product for management of the deceased. The deceased-on-site need to be handled with respect, compassion and safety in mind. For this purpose, a body bag is required. The nature of the death requires this body bag to work efficiently in various situations for example during accidents, arson, infectious disease death and other criminal acts. Currently the available body bag is a one-size-fits-all and at the same time the body bag poses a health hazard of exposure to harmful microorganisms for the forensic workers. This obvious contradiction suggests an inventive problem. The technical and physical contradiction in this particular problem allowed us to use TRIZ to better understand and discover a solution for the existing problem. Using TRIZ we were able to come up with Body Enclosure for Corpse (BEC). This product stands out as a solution for forensic workers and psychologically supports the family members of the deceased.

Keywords: Body bag · Body Enclosure for Corpse · TRIZ

1 Introduction

Our research team started with observing increasing Road-Traffic-Accident (RTA) statistics. In early 2020, this sharp rise was observed, indeed it was alarming that the number of deaths was high. Accidents lead to trauma, disconnected support, challenging the capacity for adequate care and support for the close family members. This observation then led to an exploration for providing a better level of care for the deceased. The handling of deceased at points of sudden life-changing scenarios is subjected to social, cultural and emotional balance. As we follow through the process of handling the deceased at the local mortuary; we noticed that a decent covering material is the initial interaction point between an user and the body [1]. This led us to enhance the existing body bag. A

Published by Springer Nature Switzerland AG 2022
R. Nowak et al. (Eds.): TFC 2022, IFIP AICT 655, pp. 429–439, 2022.
https://doi.org/10.1007/978-3-031-17288-5_36

body bag is often referred to as cadaver pouch or human remains pouch, is an enclosure designed to contain a human dead body. The body bag mainly functions as a mean to transport or store the corpse. Historically the bags are made of heavy plastic material and generally black in colour. However, in modern day, multicoloured bags are often used. Lightweight white colour bags are popular due to its background colour enabling easy identification of exogenous material within the enclosure. Regardless of its colour, body bags are usually made from thick plastic material with full-length zipper [2]. The zipper design may vary from straight down the middle zipper to a "J" or "D" path zipper design. Handles are included for facilitating lifting while transporting the deceased. Furthermore, some body bags are made with surface that are friendly to marker inscriptions detailing the vital information regarding the corpse. Additionally, transparent label pocket facilitates name-card insertion and the corpse toe-tag maybe secured in one of the handles. As body bags are generally a single use items, the bags are required to be discarded or incinerated upon being used.

Death may occur due to variety of causes, and in some cases can be due to pathogenic microorganism such as *Mycobacterium tuberculosis*, Hepatitis virus, Human Immunodeficiency Virus (HIV) and Severe Acute Respiratory Syndrome Corona Virus-2 (SARS-CoV-2). Beyond the regular road death scenarios the present novel corona virus-2019 (COVID-19) pandemic has made the world experience a dramatic death toll [3]. Due to COVID-19 pandemic a sudden surge in suicide cases has been observed [4].

The COVID-19 pandemic adversity has challenged the Frontliners who are handling the deceased. The mortuary processing the COVID-19 death are on heighten alert due to the severity of this disease. How long SARS-CoV-2 can be detected in body fluid is yet to be determined. In these circumstances a proper and well-designed body bag would minimize the risk of contracting COVID-19 for Frontliners handling the deceased body or body fluid [2, 5, 6]. SARS-CoV-2 can transmit in the environment and can be detected outside the host. van Doremalen indicated that SARS-CoV-2 are still viable in aerosol within 3 h and stable on surfaces till 72 h [7]. Furthermore, SARS-CoV-2 RNA can be detected in infected people's cabin up to 17 days [8]. In the case of known cause of death, it is easier for Frontliners to be prepared but sometimes death may be caused by unknown yet virulent factors. Therefore, a well-designed body bag is essential.

Apart from this, managing of the deceased requires respect, compassion and it must not add more gloominess particularly to the grieving family members. Irrespective of situations, a proper body bag with personalized touch is required. Although cost is required for production of a better body bag, care and detailed multi-faceted framework for dealing with sensitivities is greatly needed. At present, it is difficult to find a single tailor-made body bag for the deceased.

The current situation restricts the Frontliners the use of a body bag with one-size-fits-all type and a single stereotypical colour. Frontliners when using this body bag is at risk of exposure to harmful microorganism [2]. Minimizing the contact with the deceased who may or may not be infected with COVID-19 is pertinent to cut the transmission chain [3].

As numerous contradictions were observed such as matching the size of the bag must be aligned with the corpse, the presentation/visibility need to support the needs of the health-care-workers/the-family-members and the possible health hazard posed by the corpse to the immediate users especially during this pandemic. Therefore, the existing body bag does not protect users, inform users and safeguard corpse adequately. The apparent use of TRIZ become a logical alternative. TRIZ allowed us to model our problem as an inventive problem. TRIZ, founded by Genrich Altshuler in the former Soviet Union is apt for solving wicked problems [9]. Using TRIZ the body bag has been developed into Body Enclosure for Corpse (BEC). This product stands out as a solution for psychological support to the family members of the deceased and enhancing safety measures for the Frontliners handling the corpse. Collectively, BEC as a high potential tangible product, will be the leader for personalized forensic service. This paper delivers the application of TRIZ in conceptualizing the problem and using the Engineering Contradiction together with Physical Contradiction to elicit a solution to the existing body bag.

2 Research Methodology

The application of a systems approach with functional analysis has been explored as a means to characterise the problem of our Body Enclosure for Corpse (BEC). The lack of proper body bag during this pandemic has been adopted as the target problem. This problem required an immediate practical solution due to the gravity of the current situation.

The existing body bag scenario was studied and this was followed by the formulation of the Function Analysis. Figure 1 shows a typical body bag. Then system analysis and contradictory analysis directed us to conceptualize the problem [10]. The tools render us insights into the nature of the problem. The overall methodology of the BEC project is illustrated in Table 1.

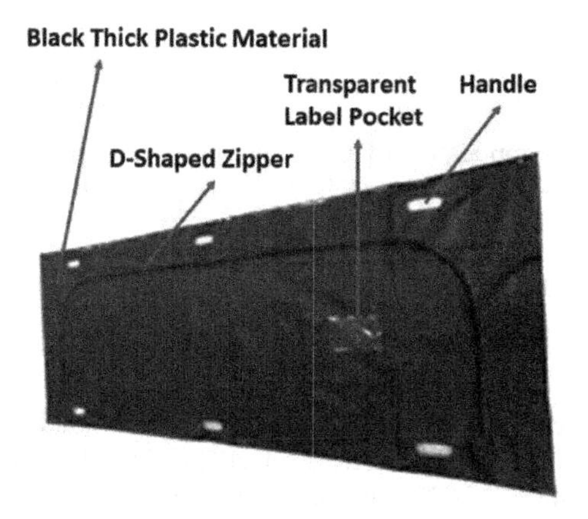

Fig. 1. A typical body bag

Table 1. Overall concept methodology of the project.

Research methodology	Tools
Review of related work	Review covered articles with "body bag", "corpse bag", "human cadaver pouch" and "cadaver bag" in reliable databases
Problem modelling based on function analysis	Function model was formulated to model domain knowledge
System analysis as inventive problem solving model	Engineering contradiction as a knowledge based approach was utilized to elicit directions
Contradictory analysis as a deep knowledge model of inventive solution	Physical contradiction served as an instrument for deepening our understanding
Designing BEC	Utilising designing tools to work out a practical model of BEC. Fruitful discussion with the end user of BEC (Frontliners) for feedback and further enhancements

2.1 Function Analysis

As a problem step, the main useful function as an engineering system of BEC is captured in Fig. 2. BEC's main functions are to transport or store the corpse. The preliminary component analysis (see Table 2) indicates that the system is composed of the user and the deceased. This system interacts intimately with the super system which is subject to spatial and temporal changes, which in return will influence the body within the product. The product in our component analysis is the BEC. Approximately four minutes after a person has been pronounced dead, the decomposition process starts and this continues until the body breaks down into simple organic matter [11]. Here the body is seen as a system which interacts with the environment. On the other hand, the deceased is managed by healthcare workers. Here, interactions between the deceased and another human takes place. The application of a systems approach to prove clarifies the problem model through the use of function model.

2.2 Engineering Contradiction

Based on the initial analysis the following Engineering systems were formulated.

a. **IF** BEC is one-size fits all **THEN** it supports a basic solution **BUT** it may pose health hazard. The improving parameter was selected as 39: "Productivity". The BEC is able to be very functional as it becomes productive with multipurpose usage for various size corpse. One the other hand the worsening parameter was selected as 31: "Object-generated harmful effect", which suggest that the BEC due to its versatility may render a health hazard to the end user in the hospital or at the site of the mishap
b. **IF** BEC is one-size fits all **THEN** it supports a basic solution **BUT** it may affect the corpse. The improving parameter was selected as 39: "Productivity". The BEC

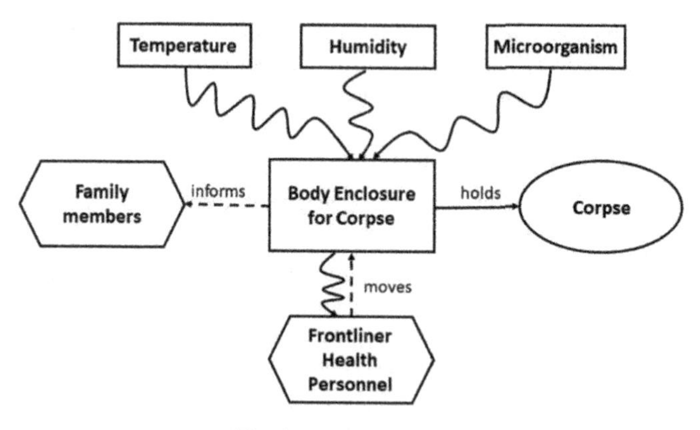

Fig. 2. Function model

Table 2. Preliminary component analysis.

Components	Tools	System	Product
Items	Environment	Frontliners deceased	Body enclosure for corpse

is able to be very functional as it becomes productive with multipurpose usage for various size corpse. One the other hand the worsening parameter was selected as 30: "Object-affected harmful effect", which suggest that the BEC due to its versatility may damage the corpse/content and as a result loses its value.

c. **IF** BEC is one-size fits all **THEN** it supports basic solution **BUT** it may affect enclosure-size compatibility. The improving parameter was selected as 39: "Productivity". On the other hand, the worsening parameter was selected as 4: "Length of stationery object", which suggest that the BEC due to its length, affects harmfully the content of BEC.

The inventive principles provided an insight in the modelling of BEC by suggesting intuitive strategy in designing and modelling a purposeful yet safer BEC for handlers. Based on this model, the possible approaches are illustrated as shown in Table 3a, 3b and 3c.

3 Physical Contradictions (PC)

PC1

The body bag must be long and short to accommodate the entire body or just the components.

PC2

The body bag must have a protective layer to protect the people nearby but must have protective layer to enable body visibility.

Table 3a. Engineering contradiction based on BEC with considering the improving parameter as "Productive" and the worsening parameter as "31: Object-generated Harmful Effect" were contemplated and its possible solutions postulated.

Inventive principles	Possible solutions
18 Mechanical vibration	The BEC fabric is tough to resist mechanical stress. Strong and durable for body recovery at sea, land or air accident
22 Blessings in disguise	There is no local industry that manufactures body bags in Malaysia. BEC was designed in collaboration with forensic/mortuary staff. This opens network interactions with local manufacturer thus improves the local economy
35 Parameter changes	Addition of six handle providing holding support for either 2, 4 or 6 persons to carry the deceased within the BEC. Furthermore, a transparent window which is different from the surrounding opaque BEC coverings. Furthermore, although there has been body bag made from fully transparent material, it is not a perfect choice. Here the transparent window has been designed wide enough for body identification and at the same time limiting/ hiding family members as well as those non-related people from seeing defects such as body markings or autopsy artefacts
39 Inert Atmosphere	High-Density Polyethylene (HDPE) is inert and low hazard plastic material

Table 3b. Engineering contradiction based on BEC with considering the improving parameter as "Productive" and the worsening parameter as "30: Object-affected Harmful Effect" were contemplated and its possible solutions postulated.

Inventive principles	Possible solutions
13 The other way around	In a disaster situation, which is when the local mortuary 4°C individual storage facility is unable to cater the current cases; the deceased in the BEC can be kept in larger size temporary facility such as refrigerated cold room/tent
22 Blessings in disguise	Introducing innovation in the management of the deceased, which removes the stereotypical thinking about mortuary
24 Mediator	BEC can be used as an intermediate transport device from site/filed/clinic to the mortuary and body release
35 Parameter changes	Gloomy colour can be changed to accommodating pastel colour. Providing uplifting mood for family member of the deceased. Furthermore, by lifting the flap minimizes exposure to harmful bodily discharge

Table 3c. Engineering contradiction based on BEC considers the improving parameter as "Productive" and the worsening parameter as "Length of Stationary Object", were contemplated and its possible solutions postulated.

Inventive principles	Possible solutions
7 Nested doll	Foldable to accommodate adult; children; infant/ body parts
14 Curvilinearity	Minimal design curve D-shape zipper has been selected
26 Copying	BEC shape copy & fit body sizes giving it an ergonomic design. Once the deceased is inside the BEC, frontliner can flip a Velcro flap copying the D-shape to see inside without opening the zipper
30 Flexible material	Made from HDPE tough yet flexible

4 Results

In order to validate the concept, our group had an engagement to test the mock-up of the BEC. This is captured in Fig. 3. This was followed by several other discussions with the frontliners, which is the part of design thinking approach. This led us to the function and component analysis.

Fig. 3. Engagement with the Frontliners for testing the pilot BEC model.

The application of a systems approach with functional analysis has been used to understand the problem of body bag. Figure 4 depicts the Body Enclosure for Corpse. At the super system level, having a body bag creates a separation between the corpse and the Frontliners. Here the corpse is seen as a system which interacts with the environment. Figure 5 depicts the interaction analysis of the BEC. Taking into account the deceased is managed by healthcare workers, the safety and wellbeing of the Frontliners is paramount. Body Enclosure provides a separation barrier preventing negative interactions from those environmental factors.

On the contrary, complete covering prevent body identification. Although a complete barrier can protect health workers, and hide physical deformities it can hinder body identification. This is an important process in order to correctly release the body to the family member. The solution is to provide a transparent window adequate for identification by authorities, whilst the rest of the enclosure can be coloured in pastel instead of black. On top of the transparent window, a Velcro-bound covering is placed for securing privacy purposes. This enhancing local quality of the enclosure results in a more personalized product. In addition, by having foldable enclosure, the product can be used for adult, children and body parts, known as the nested doll. Inventive principle of Inert atmosphere is used in such a way to create a humidity control by desiccants and oxygen absorbers. Within the BEC a humidity indicator is placed to keep track of the humidity.

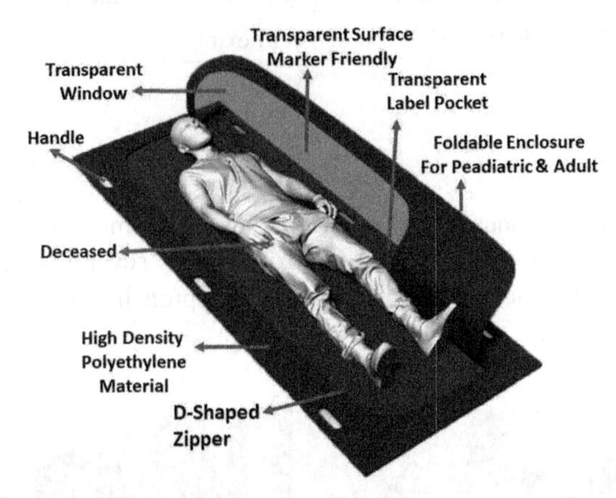

Fig. 4. Body Enclosure for Corpse

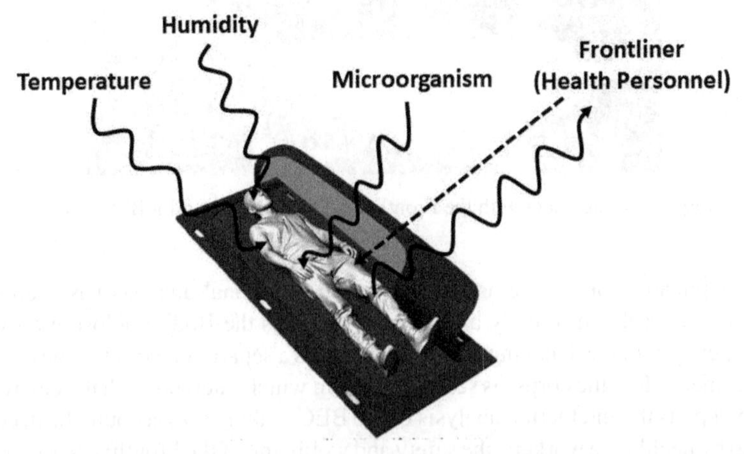

Fig. 5. Body Enclosure for Corpse interaction analysis

5 Discussion

Ideality is affected by variables such as Useful Function (UF), Harmful Effects (HF) and Resources [12]. Therefore, Ideality can be adjusted by increasing or decreasing these variables. These changes will reflect on the quality and performances of the product. In our project, BEC is the product therefore the Ideality equation for BEC can be written as shown in Eq. 1.

$$\text{Ideality} = \frac{[\text{Transport BEC}][\text{Storing BEC}]}{[\text{Temp}][\text{Hum}][\text{mbes}][RES]} \quad (1)$$

where transport and storing BEC are Useful Functions whereas the effect of temperature (Temp), humidity (Hum), microorganisms (μbes) and resources (RES) are the Harmful Effects

Decreasing the effect of temperature so that the body does not decay rapidly will certainly preserve the corpse for post death uses. Reducing the humidity within the BEC by creating inert atmosphere via the use of desiccants will enable the corpse to be better preserved. Removing oxygen will hinder aerobic microorganisms from thriving within the corpse. Optimal amount of High-Density Polyethylene (HDPE) material will reduce the cost which also will contribute to a smaller denominator in Eq. 1, in return a bigger Ideality.

Medical equipment is closely related to health and safety. With competition amongst medical manufacturers, a practical approach is needed to tackle the problem on site [3]. Therefore, this present study demonstrates the use of TRIZ in providing a functional design for forensic medical equipment. The use of cadaver bags or body bags in general is a conventional method to store dead body or human remains, for transportation purposes at locations wherever a dead body is found. Existing body bags are either solid colors or translucent, both may have issues related to body identification & protecting the deceased. In a conventional body bag design, viewing the content requires unzipping. This may lead to the spread of contagious diseases to those who come in contact with the body bag including Frontliners and family members. To solve this issue, we have used TRIZ to design BEC with a transparent window underneath a Velcro-bound covering. This enable the corpse to be viewed without unzipping the bag. The principle of nested doll, color change and feedback is combined here for a view only when needed solution, with no need to unzip. So additional principles used is "partial action", local quality and segmentation.

Additional features include, foldability to fit in different sizes, protruding handle allowing variable number of persons to carry the BEC as required. The BEC is made from strong material, and it's colour can be customized according to needs giving it a more personalized touch.

At the end of 2018 we developed MORTuary Information System (MORTIS). MORTIS is a database for managing death-register at medical facility. Extended-MORTIS (e-MORTIS), was developed as an extension to MORTIS. e-MORTIS consists of BEC and unique identifier for the deceased. The unique identifier will carry QR code linking the BEC with MORTIS. Although quick response (QR) technology was initially used in 1994 in Japan by Denso Wave company, in late 2019 due to the COVID-19 pandemic,

QR Code has exploded into human-everyday-use and especially in education [13]. In our future designs, the process of managing the corpse will include scanning a QR code printed on the BEC itself or on a robust card inserted into the transparent pocket of the BEC. The QR code will contain unique identifier that links BEC (with corpse in it) to the MORTIS. The software system will enable MORTIS to track the BEC with the corpse. This will enable us to facilitate the information relevant to the work done on the corpse with the user.

6 Conclusion

The body bags used by hospitals in Malaysia are imported from overseas, hence a locally manufactured product with our own design will ultimately change the conventional body bags to serve more purposes. For example, Inventive principle No. 13; another way round, a perfect design for camouflaged sleeping bag in military use, as well as lowering the cost of import through involvement of the local manufacturing industrial sector. TRIZ tools as an aid for modelling complex problem is demonstrated in this project. The COVID-19 pandemic has provided us with an opportunity to revisit a long-standing problem to enhance the existing body bag. The TRIZ tool has certainly enabled us to improve the body bag to Body Enclosure for Corpse (BEC). TRIZ has allowed us to focus on the key factors involved in resolving the Ideality. The BEC prototype created will certainly allow us to enhance it further based on the solution gearing towards an ideal pathway through TRIZ-based modelling.

References

1. Aiman, M.R., Helmi, A.M., Suffian, I.M.: Epidemiologic profile of fatal road traffic accident (RTA) encountered in forensic medicine service. In: Paraclinical Sciences. Universiti Malaysia Sarawak, Sarawak (2013)
2. Requirements and technical specifications of personal protective equipment (PPE) for the novel coronavirus (2019-ncov) in healthcare settings (2020). https://iris.paho.org/handle/10665.2/51906. Accessed 1 Mar 2022
3. Dijkhuizen, L.G.M., Gelderman, H.T., Duijst, W.L.J.M.: Review: The safe handling of a corpse (suspected) with COVID-19. J. Forensic Leg. Med. **73**, 101999 (2020)
4. Wang, C., et al.: The impact of COVID-19 pandemic on physical and mental health of Asians: a study of seven middle-income countries in Asia. PLoS ONE **16**(2), e0246824 (2021)
5. E.C.D.C, Considerations related to the safe handling of bodies of deceased persons with suspected or confirmed COVID-19. European Centre for Disease Prevention and Control (2020)
6. Patel, M., et al.: Conceptual design of a body bag for preventing infections and safe disposal of deceased from COVID-19 virus. Trans. Indian Natl. Acad. Eng. **5**(2), 429–435 (2020). https://doi.org/10.1007/s41403-020-00135-5
7. van Doremalen, N., et al.: Aerosol and surface stability of SARS-CoV-2 as compared with SARS-CoV-1. N. Engl. J. Med. **382**(16), 1564–1567 (2020)
8. Moriarty, L.F., et al.: Public health responses to COVID-19 Outbreaks on cruise ships - worldwide, February-March 2020. MMWR Morb. Mortal. Wkly Rep. **69**, 347–352 (2020)
9. Altshuller, G.S.: The innovation algorithm: TRIZ, systematic innovation and technical creativity. Technical innovation Center, Inc. (1999)

10. Soosay, A.E.R., Kulathuramaiyer, N.: TRIZ-based remodeling of multiple true-false questions. In: Borgianni, Y., Brad, S., Cavallucci, D., Livotov, P. (eds.) TFC 2021. IAICT, vol. 635, pp. 355–366. Springer, Cham (2021). https://doi.org/10.1007/978-3-030-86614-3_28
11. Vass, A.A.: Beyond the grave—understanding human decomposition. Microbiol. Today. **28**, 190–192 (2001)
12. Maccioni, L., Borgianni, Y.: An ideality-based map to describe sustainable design initiatives. Creative Solutions for a Sustainable Development. In: Borgianni, Y., Brad, S., Cavallucci, D., Livotov, P. (eds.) TFC 2021. IFIP Advances in Information and Communication Technology, vol. 635, pp. 3–13. Springer, Cham (2021). https://doi.org/10.1007/978-3-030-86614-3_1
13. Karia, C.T., Hughes, A., Carr, S.: Uses of quick response codes in healthcare education: a scoping review. BMC Med Educ. **19**(1), 456 (2019)

TRIZ Education and Ecosystem

WTSP Report (5) Catalogs of TRIZ and Around-TRIZ Sites in the World: We Can Learn Full Scope of TRIZ-Related Works in the World

Toru Nakagawa[1](✉), Darrell Mann[2], Michael Orloff[3], Simon Dewulf[4], Simon Litvin[5], and Valeri Souchkov[6]

[1] Osaka Gakuin University, 3-1-13 Eirakudai, Kashiwa, Chiba 277-0086, Japan
nakagawa@ogu.ac.jp

[2] Systematic Innovation Network, Bideford, UK

[3] Academy of Instrumental Modern TRIZ, Berlin, Germany

[4] AULIVE and Innovation Logic, Glen Elgin, Australia

[5] GEN TRIZ, LLC, 31 Gilbert St., Newton, MA, USA

[6] ICG Training and Consulting, Enschede, The Netherlands

Abstract. World TRIZ-related Site Project (WTSP) is a volunteer-based international project working to build Catalogs of TRIZ and Around-TRIZ Websites in the World. It aims at creating a reliable information resource in the field of Creative Problem Solving Methodologies in general. We collect good websites widely and list the selected ones with brief or close annotations in categories. Starting in Dec. 2017, we built the Beta Edition of World WTSP Catalogs in 2020, which have good maintainable and extendible structure. We select websites using multi-aspect criteria, and grade them in 5 levels as: ◎ top about 30 sites, ○ next about 100, □ worthy in World Catalogs, △ worthy in Country Catalogs, and - not listed. We have recently re-visited the Around-TRIZ websites obtained by Internet surveys in 2019, and revised the Basic Catalog (with 30 ◎ sites and 143 ○ sites) and newly built the Extended Catalog (with 244 □ sites). On the other hand, WTSP World Catalogs of TRIZ Sites have been updated very little from the Beta Edition (with 23 ◎ sites and 39 ○ sites), while waiting for contributions of manuscripts from individual countries. We carried out Internet surveys of TRIZ sites in 52 individual countries, and detected 1200 + sites. For 40 (less active) countries of them, we selected 103 □ sites and 43 △ sites. For 12 other countries (including 6 TRIZ-active major countries) we detected 480 sites but not investigated them yet while knowing there must be many good TRIZ websites. Anyway, WTSP World Catalogs already have abundant reliable information of selected websites, and are certainly helpful for many professionals and users to study and apply TRIZ and Creative Problem Solving Methodologies in general.

Keywords: WTSP project · World catalogs of TRIZ sites · World catalogs of around-TRIZ sites · Guide to good websites

R. Nowak et al. (Eds.): TFC 2022, IFIP AICT 655, pp. 443–457, 2022.
https://doi.org/10.1007/978-3-031-17288-5_37

1 Introduction

1.1 Vision and Goals of the WTSP Project

This is the fifth report of the World TRIZ-related Sites Project (WTSP), which started in Dec. 2017 as a volunteer-based international project. (We operate, inside the website "TRIZ Home Page in Japan" [1], our Project base site [2] and our product platform [3].) Its goal is to build Catalogs of TRIZ-related Websites in the World [3] as a reliable and useful information resource. The WTSP Catalogs are expected to be used openly and widely for learning, spreading, and promoting various methodologies which help people in the world to solve numerous difficult problems creatively or innovatively.

Scope of our WTSP Catalogs is 'Creative Problem Solving Methodologies' in general. We know that in the world there are various types and numerous numbers of documents concerning to such methods and their actual cases. However, the flood of information with large diversity and quantity ironically hides really useful information resources. So we want to collect good information/documents widely and show the selected ones in "Catalogs" with annotations.

We started the project in the community of TRIZ (Theory of Inventive Problem Solving) [4]. Besides TRIZ there exist many method such as Brainstorming, Lateral thinking, Value engineering, Six sigma, Lean and agile engineering, Project management, Design thinking, etc. etc. Since all of them have a common purpose to provide methods to solve difficult problems creatively, WTSP handles them together. WTSP Catalogs have two pillars: one is Catalogs of TRIZ Sites, and the other is Catalogs of Around-TRIZ Sites (Note this naming is tentative; we wish to change it when the wide field of our scope get some proper naming).

As the unit of information source, we have chosen Websites instead of webpages. Websites are units larger than webpages, papers, articles, patents, etc. while smaller than books, conference proceedings, journal volumes, paper repositories, etc. Websites usually represent a group of people with a common academic/business approach, and post a series of articles more or less in a systematic manner for many years, actively keeping them up-to-date. They are mostly free of charge and open access. In addition to typical documents of works, we want to include information about organizations, activities, information resources, etc. Websites are flexible to convey these various types of information and play important roles for the site owners to promote their business/job purposes.

Catalogs (of any topic) should fulfill the following three basic requirements:

Req. 1: Collect good items (i.e. websites) widely, select them with evaluation, and show them in categories.
Req. 2: Individual items should be introduced in a proper/fair manner briefly or closely and guide the readers/users to good items.
Req. 3: It should be easy for users to find one or many items which match their current interest/needs inside the Catalog.

In each country people want to learn about websites operated and written (especially in their own national language) in their country before visiting many other websites in the world. Thus it is natural to build Country WTSP Catalogs first and to integrate them into World WTSP Catalogs. The World Catalogs are useful to learn excellent works and websites of higher quality and wider scope.

1.2 Brief History of Building the World WTSP Catalogs

- (Dec. 2017) Nakagawa proposed WTSP as a volunteer-based international project, and got support by many TRIZ colleagues [4].
- (2018) Five TRIZ leaders were invited as Global Co-editors and Nakagawa served as the Project Leader [5]. WTSP Catalogs in Japan was built as a pilot model. 80 members from 30 countries joined (or signed to join) WTSP at TFC2018 [6].
- (2019) Preliminary Edition of World WTSP Catalogs was quickly built [3]: TRIZ Sites (of ◎○ levels) having contributions of 4 countries (Japan, Malaysia, China, and Russian language region) and 2 internet surveys (World in English, USA). Around-TRIZ Sites (of ◎○ levels) having the results of 5 cases of internet surveys. Difficulty has become clear in forming country WTSP Teams to build Country Part of WTSP Catalogs [7].
- (2020) Beta Edition was publicized [3]: World WTSP Catalogs for Print were also made in PDF to be downloaded and used in user's PC. WTSP was presented online at 4 International Conferences (MATRIZ, ETRIA, AI, and ICSI) and 1 domestic conference (Japan) [8].
- (2021) Nakagawa made Internet surveys of TRIZ-related sites in 52 countries individually [9]. Focus of the WTSP activities has been shifted from building Catalogs to promoting usage of Catalogs [10, 11].
- (2022) Results of internet surveys of Around-TRIZ sites were refined to build the Catalogs of sites of ◎○□ levels [12]. Survey results of TRIZ-related sites in individual countries are checked and reported [9].

2 Structure, Function, and Usage of the World WTSP Catalogs

2.1 Evaluation and Description of Individual Sites in the WTSP Catalogs [13]

Evaluation of websites is a delicate-but-necessary step to decide whether to include them in the WTSP Catalogs (and at what level). We use the following multiple criteria, with the stance of finding strong positive aspects of individual sites.

- Quality (reliable, correct, novel, original, comprehensive, up-to-date,…)
- Usefulness (resourceful, reference, handy, practical,…)
- Attractiveness (interesting, easy-to-understand, illustrative,…)
- Accessibility (open access, free/low charge, used by many,…)
- Scope (scope of the theme/field, scope of activities,…)

We grade the sites into the following 5 Levels, representing with the symbol marks. We must evaluate at least tentatively and should refine them later in some public reviewing.

◎ Most important (about top 30) in World Catalogs
○ Important (about next top 100) in World Catalogs
□ Worthy in World Catalogs,
△ Worthy in Country Catalogs,
– Insignificant / irrelevant, not included in the WTSP Catalogs

Descriptions of individual sites are written in 4 levels of formats as follows:

(0) Basic information for Indexing: Site Code, Site name, URL of site domain, Location (Country) of site (main office), Language(s) of site, Single-line description, and Roles of site. – Need to be written first by a surveyor/recommender and revised later by the site owner. To be used in categorizing and arranging the sites, shown in the Index table, and shown as the heading part of site descriptions.
(a) (0) + Brief introduction in free format (about 3–15 lines). –To be written by a surveyor/recommender, using excerpts of top pages, 'About us' pages, and some more main pages. To be used in the Site description part of the WTSP Catalogs.
(b) Standard Site Description written in the WTSP Standard Form (in 1 page in A4). (0) + Application phases, Application fields, Methods, Description of introductions (in free format about 5–10 lines), (optional) Link to (c). – Should be written by the site owners. Mandatory for the ◎○ sites, desirable for the □ sites. Most of the items are to be answered with (ordered) multiple choices using the codes set in our Guidelines [13]. To be placed just below (a) and forms the main body of site introduction, to be used for users to understand the site. Users often scan some of the columns of their interest quickly to find good websites which match with their current interest.
(c) (0) + Optional close site introduction in free format (2–10 pages or more in A4). – To be stored as a separate file inside the WTSP Catalog folder and accessed with a hyperlink from (b). Site owners are encouraged to write this introduction Useful and Attractive for users by using figures, photos, tables, etc.

Samples of these site descriptions exist a lot in the WTSP Catalogs. It is clear that main body of site introductions, i.e., mandatory (b) and optional (c), can be written not by surveyors but by the site owners themselves. Site owners would think them worthy of writing for promoting their thoughts and businesses [14].

2.2 Structure of WTSP Catalogs at the Base Levels [13]

The structure of WTSP Catalogs, at the lowest level, is shown in Fig. 1.

The structure of WTSP Catalogs to be built by a Country Team or an Internet Survey is show in Fig. 2. By virtue of the flexibility of hyperlinks, the websites in the Index page can be arranged (and re-arranged) in various ways of categorization and ordering without any readjustment in the Site Description page. This structure is useful for making the WTSP Catalogs updatable, maintainable, and extendible.

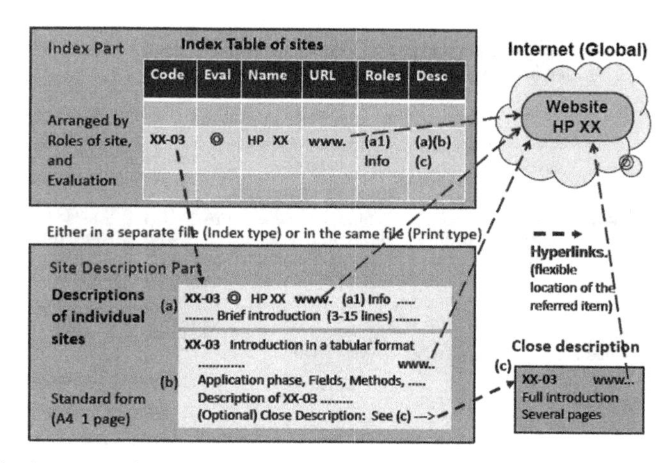

Fig. 1. Structure of a WTSP catalog sample page from the viewpoint of a website.

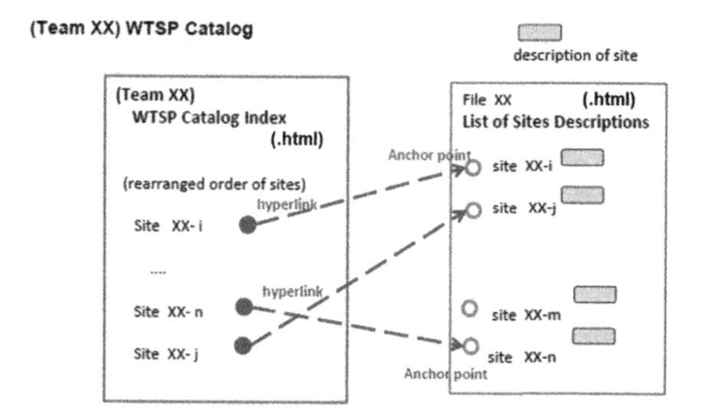

Fig. 2. Structure of a country part of the WTSP world catalogs

2.3 Overall Structure of the WTSP World Catalogs [15]

Overall structure of the whole WTSP World Catalogs are shown in Fig. 3 from the view points of users. In the top page of the WTSP World Catalogs [3], various sets of Catalog Indexes can be accessed. Source Data Catalogs, shown in the left part, are composed of Index pages contributed by Country Teams and Surveyors and have hyperlinks for accessing to their corresponding Site description pages. In the middle and right parts of the figure, we have World Indexes and Catalogs in integrated forms.

We have built a set of World Indexes and World Catalogs for Print, as shown in Table 1. The WTSP Catalogs [3] have two pillars: (A) Catalogs of TRIZ Sites and (C) Catalogs of Around-TRIZ Sites which cover a wider field of Creative Problem Solving Methodologies in general. For the sake of handiness, each pillar has 3 Catalogs containing specified level(s) of sites: Top Catalog (◎), Basic Catalog (◎○), and Extended Catalog (□). Each Catalog serves in two types: Index Type, where the site descriptions are stored separately and accessed interactively, and Catalog for Print type, which has the same Index part but the site descriptions are included in the Catalog file (in .html and .pdf) so that the Catalog may be downloaded, printed, and used interactively on user's PC.

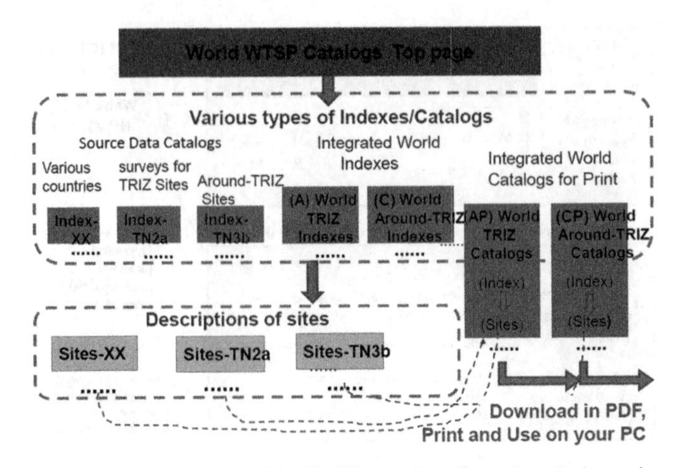

Fig. 3. Overall structure of the WTSP catalogs from users' viewpoints.

Table 1. Full Set of WTSP world indexes/catalogs for print.

Contents	Catalog name	Levels of contents	Code name	
			Index type	Catalog for print
(A)TRIZ sites	Top catalog	◎	(A1)	(A1P)
	Basic catalog	◎ ○	(A2)	(A2P) [16]
	Extended catalog	□	(A3)	(A3P)
(C) Around-TRIZ sites	Top catalog	◎	(C1)	(C1P)
	Basic catalog	◎ ○	(C2)	(C2P) [17]
	Extended catalog	□	(C3)	(C3P)

For building the Index part of World WTSP Catalogs, the Index parts of all the manuscript Catalogs are fist merged into a large Index table (in Excel) with minor adjustment of the hyperlinks so as to reflect the change in folders. After checking and suppressing the duplications of sites, all the sites are sorted with respect to the Evaluation and the (main) Roles of sites (as shown later in Fig. 5).

The World Indexes thus obtained can be used smoothly in the interactive way of referring the site descriptions by clicking hyperlinks. Catalogs for Print [16, 17], on the other hand, are convenient to browse/read the whole Catalog quickly to find sites of one's current interest. Catalogs for Print may be downloaded in PDF (while keeping the hyperlinks active), printed, stored, and used interactively on user's own PC, just as in the way of operating in the WTSP website. All these usage may be done freely without registration nor charge, while its copyrights are reserved by the WTSP Project.

3 Recent Revisions/Enhancement of World WTSP Catalogs

3.1 WTSP World Catalogs of TRIZ Sites: TRIZ Sites in Individual Countries

Since publicizing the Beta Edition (Nov. 2020), we received and posted the site descriptions of only several ◎○ sites. Activities of WTSP Teams have not been reported from any country so far, to our pity.

In this situation, Nakagawa carried out Internet surveys of TRIZ sites in individual countries during May to Nov. 2021. Yahoo!Japan search engine was used with the option of limiting the target sites located in the specified country and the option of outputting only one webpage from any site. Key words of 'TRIZ OR "Systematic Innovation"' were typically used, while adjusting them to match the language of the country. Language in the target site was specified to be 'any'.

The Results of the surveys [9] are summarized in a tabular form in Fig. 4. Countries are shown in groups of regions and arranged alphabetically. 1200 + sites in total were detected in 52 individual countries in the world. For 40 countries among them, their 700 + sites were visited one by one to understand the sites and write brief introductions in the form (a). This work is often hard especially because of the language barriers even with the help of machine translation by Google Translator or later by DeepL. For each country, the number of detected sites is shown first. Then we list the number of TRIZ sites at the levels of △□ (and higher). Next is the number of Around-TRIZ sites, which are detected with the keyword of TRIZ but are found to have very little description about TRIZ while much more about Creative Problem Solving in general. Note that in these two numbers, websites located in the country but operated globally are excluded (e.g., Wikipedia, Google Scholars, LinkedIn, YouTube, etc.). We found 103 TRIZ sites at the levels □ (and higher), but they are not included in the World Catalogs yet due to the missing of Roles of site information.

The survey results of 12 other countries are posted as raw data without further studies and selection yet. The countries include most active TRIZ players (e.g., France, Germany, Italy, UK, Korea, and Taiwan), where many good websites are known in operation. We believe the TRIZ community in such a country can build its Country WTSP Catalogs of TRIZ Sites (as an extension of this survey) much better than any foreign surveyors.

We should also note that 5 other countries (i.e., Japan, Malaysia, China, Russian language countries, and USA) posted their data in Sept. 2019 and it is the time we should update and enhance them.

3.2 WTSP World Catalogs of Around-TRIZ Sites: Extended Catalog (□)

Internet surveys of Around-TRIZ sites were carried out by Nakagawa in 2019 [7]. In the 5 cases of survey, following sets of keywords were used.

- Creative Think Method
- (Creative/innovative/Systematic) Problem Solve (Method/Process/Technique)
- Innovation (Process/Strategy/Method/Management/Technology)
- (Quality/Value/Cost/Productivity)
  (Deploy/Engineering/Management/Control/Analysis)
  (Method/Technique/Theory/Process/"Case Study")

Reg	Country	Det.	TRIZ	Around TRIZ
A-Europe				
	Austria	26	4	0
	Belgium	22	Not Selected yet	
	Czech Republ	24	5	0
	Denmark	14	0	0
	Finland	21	5	0
	France	55	Not Selected; Many good.	
	Germany	68	Not Selected; Many good.	
	Greece	8	1	1
	Hungary	13	2	2
	Iceland	11	1	6
	Italy	50	Not Selected; Many good.	
	Netherlands	33	5	0
	Norway	4	2	0
	Poland	50	Not Selected yet	
	Portugal	15	4	1
	Romania	9	4	2
	Spain	27	7	8
	Sweden	19	Not Selected yet	
	Switzerland	21	6	4
	UK	39	Not Selected; Many good.	
B-Russian				
	Russia		Already shown.	
C-MidEast				
	Iran	64	6	0
	Israel	24	4	3
	Turkey	27	9	1
D-Asia				
	Bangladesh	3	0	0
	China		Already shown.	
	Hong Kong	16	5	8
	India	45	5	10
	Indonesia	37	9	1
	Japan		Already shown.	
	Korea	72	Not Selected; Many good.	
	Malaysia		Already shown.	
	Pakistan	11	3	4
	Philippines	11	0	2
	Singapore	15	7	3
	Sri Lanka	6	Not Selected yet	
	Taiwan	45	Not Selected; Many good.	
	Thailand	31	Not Selected yet	
	Vietnam	24	11	1
E-Oceaania				
	Australia	23	Not Selected yet	
	New Zealand	7	3	1
F - N America				
	Canada	30	1	3
	USA		Already shown.	
G- CS America				
	Argentina	10	1	4
	Bolivia	6	0	3
	Brazil	41	6	1
	Chile	12	1	5
	Colombia	19	2	7
	Costa Rica	12	3	3
	Mexico	23	4	0
	Nicaragua	6	0	1
	Peru	13	2	2
	Uruguay	7	1	1
H- Africa				
	Egypt	8	2	0
	Kenya	6	2	0
	Morocco	13	7	1
	South Africa	9	6	0
Total		725	146	89
Total (Not Sel.)		480		

Fig. 4. Numbers of TRIZ and around-TRIZ sites found by the internet surveys in individual countries in the world.

- (Patent/IP/"Intellectual Property") (Analysis/Protect/Circumvent/Mapping /Strategy)

He detected nearly 1000 sites, visited them one by one, wrote brief introductions in the form (a) with categorization and evaluation, integrated them into the World Catalogs, and finally publicized the WTSP Catalogs of Around-TRIZ Sites in the World (◎○) in 2019.

During Mar.-Jun. 2022, Nakagawa re-visited all these Around-TRIZ websites again to revise the brief introductions, especially in Roles of site and in Evaluation, and newly built and publicized the Extended Catalog (☐) of World Around-TRIZ Sites [12]. The results are summarized in a tabular form in Fig. 5.

The numbers of websites included in the current WTSP World Catalogs [3] are listed in Fig. 5, classified with their Roles of sites and Evaluation. The right side part shows Around TRIZ sites, newly publicized according to the present paper. The left side part, on the other hand, shows TRIZ sites which are posted currently, reflecting very minor updates of the Beta Edition (Nov. 2020). While the sites of Around-TRIZ are listed thoroughly, those of TRIZ are apparently fewer at the ◎○ levels and missing at the ☐ level. This reflects the lack of manuscript contributions from many TRIZ active countries as mentioned above.

4 What We Can Learn with the WTSP Project/Catalogs

4.1 What We Can Learn with the WTSP World Catalogs

In our WTSP website [2] inside "TRIZ Home Page in Japan" [1], you now can see/use the following Catalogs of the World WTSP Catalogs [3]:

(Main) Role of site	WTSP World Catalogs (A) TRIZ Sites (2020)			WTSP World Catalogs (C) Around-TRIZ Sites (2022)		
	◎ Most Important	○ Important	□ World Worthy	◎ Most Important	○ Important	□ World Worthy
(a) Information sending sites	9	4		8	19	12
(b) Promotor organizations	5	4		3	8	6
(c) Public organizations	1	1		3	10	24
(d) Academia	0	0		2	20	33
(e) Developer organizations	5	8		5	36	84
-- (e5) Training organizations	2	7		0	9	19
(f) Publisshing organizations	0	6		4	17	13
(g) Information sharing	1	7		5	12	21
(h) User organizations	0	2		0	12	27
(i) Personal	0	0		0	0	5
(Total number of sites)	23	39		30	143	244

Fig. 5. Numbers of TRIZ and around-TRIZ Websites in the current WTSP world catalogs

- TRIZ sites: (A1) Top Catalog (◎), (A2) Basic Catalog (◎○), and (A1P)(A2P)
- Around-TRIZ Sites: (C1) Top Catalog (◎), (C2) Basic Catalog (◎○), and (C3) Extended Catalog (□), and (C1P)(C2P)

The sites listed at the ◎○ levels are essentially the same as those in the Beta Edition; we omit showing them here again. Some of the points we can learn with these TRIZ and Around-TRIZ Catalogs are:

- There are a large variety of websites in their Roles of sites and in their methods/approaches. Inside TRIZ, for example, we have Classical TRIZ approaches (mostly technological, some pedagogical, etc.), extended TRIZ approaches (for applications to business/management, software/IT, bio-engineering, future predictions, etc.), modernized TRIZ approaches (reforming the framework and proposing new ones, adopting and integrating with some other methodologies, etc.). These different approaches are developed by consulting firms and academia, presented in conferences/journals/websites, posted in the websites for dedicated for information sending (mostly by consulting firms and academia groups), and taught/trained in open/closed lectures/seminars/webinars, etc. We all (professionals/practitioners/users in TRIZ) can learn essence of these different approaches and adopt and use some suitable ones while keeping our knowledge up-to-date as much as possible.
- The fields of Around-TRIZ are much wider than TRIZ. There are a larger variety of methodologies, having complex history of development, intended to use in different application areas/application phases, sometimes overlapping and some other times competing, etc. Under such diversity, however, there seem some representative website(s) exist in each methodology/approach; and our WTSP Catalogs of Around-TRIZ Sites hopefully have listed such representative websites with evaluation ◎○.

(We should, of course, improve much the WTSP Catalogs by getting collaboration of many people in the communities of such wide fields.)

- Large number and diversity of websites in the Catalogs have one serious demerit, i.e., being hard for ordinary users/beginners to learn good and easy websites. We should be careful in evaluating websites from multiple criteria including this point, and in introducing various websites with such criteria. Short articles for introducing/recommending good and attractive websites in the WTSP Catalogs may be useful in various specific viewpoints.
- There are a lot of posting about methodologies and their application procedures in the websites, in the forms of papers, articles, lectures, movies, (free or paid) online courses, etc. Thus people in any country/company/school (even in countries with little TRIZ activities yet) have good chances to learn TRIZ and related methodologies through various websites recommended in the WTSP Catalogs.

4.2 What We Learn with the WTSP Project

Our WTSP Project has been working step by step on a volunteer basis for building the WTSP Catalogs to fulfill the basic 3 requirements for Catalogs (see 1.1). Here we note what we learned and what we wish to do with the WTSP Project.

- First we should review our current WTSP Catalogs with respect to the Basic 3 Requirements:

  Req. 1: Collecting/evaluating/showing the sites: The scheme, process, and actual practice for these work have been developed at a reasonably high level, but they lack the real activities of such a practice in many (including most active) countries.

  Req. 2: Introduction of individual sites: Four forms of introduction are specified: (0) (a) (b) (c). All the sites in the Catalogs have introductions (0) (a). Form (b) is mandatory, but is written only for a few tens of sites. Form (c) is optional, and is written only for several sites. Forms (b) (c) need to be written by the site owners, especially (b) for usefulness and (c) for attractiveness.

  Req. 3: Searchability: Evaluation and Roles of site are used for categorizing and arranging the sites, and helps for better searchability. For improving further, inputs of single-line introduction in Form (0) and full description in Form (b) are desirable. Besides, some search mechanism for finding sites by using Site name and many other keys in Form (0) and (b) is necessary to be installed.
- The review results show the necessity of collaborative work for collecting good websites in various countries (by forming Country WTSP Teams) and individual yet collaborative work for site owners to write the introductions (b)(c) of their own sites [14].
- We have been reporting and posting our WTSP activities [3] and sending communication letters to our about 200 supporters to ask for their collaborative work. However, the problem "many people support and join the WTSP Project, but few of them actually work for it" still remains even now from the start of the Project. Our measures have been (1) showing our vision clearly and making good prototypes, examples, and guidelines, and (2) dividing the jobs into small easy tasks which are capable for individual persons to achieve in parallel. We have been concretizing these measures

steadily, and we now wish a number of people in the TRIZ (and related) community to voluntarily work together with us [11].
- The history of WTSP pushes us to recognize the insufficiency of collaboration in the TRIZ community not only in the global scale but also in the domestic scale in a number of active countries.

4.3 Discussion on Referee's Comments

Aren't There Others' Works to Refer?. –There are various types of websites/webpages/books which list a (large) number of items (e.g., products, papers, organizations, websites, etc.) with/without annotations. But they are not good prototypes for us. For example:

- List of Links (or URLs) to relevant websites (of organizations) – often a limited scale, without annotation
- Catalogs of books/papers in a library/repository – a large number of items, usually without evaluation
- Websites of e-Commerce of various items – different items coming from many sellers, with possible evaluation through the customers' feedbacks
- Catalogs of selected restaurants – often in a limited type and location of restaurants,
- List of Best 100 websites in the world – ranked with the number of visits/visitors
- Handbook of (100) methods/tools for creative problem solving – articles written by each professional, without further evaluation

What we wanted to find is:

- Catalogs of websites in some (academic/industrial) area in the world – not limited to a narrow area, not limited to a small number of countries, selected with clear evaluation criteria, containing various types of websites (in respect to theories/applications, papers/patents/books, (academic/industrial/public/user) organizations, activities, etc.).

But we have never met such an example in our internet surveys (in the scope of WTSP Catalogs and much wider) nor in the response from many followers/readers of the WTSP project. Hence we made self-citation only, with no reference to other prototypical works.

What is the Basis of your Websites Evaluation Criteria (Sect. 2.1)? – We started with the more-or-less standardized criteria for reviewing academic papers, and adjusted them for evaluating websites in our project. We recently recognize Google's Webmaster Guidelines for rating websites [18]. Its rating criteria are, in short:

- Needs met -- taking account of different needs of different users and also different purposes of websites
- Page quality, especially E (Expertise) – A (Authoritativeness) - T (Trustworthiness) of the page authors and website organization – these are mostly evaluated with meta-data of the page,

- Usability – including usability in mobile devices and for handicapped persons

Since Google deals with a huge variety of websites used by numerous people in the world, their criteria are rather vague. Here are some points of our Criteria:

- Needs met: Our purpose is reflected in the Scope (scope of the theme/field, theories/applications/activities, etc.)
- Page quality: Our criteria in this regard (Quality) are clearer and more academic, because we deal with methodologies and their applications. Reflecting practical sides of websites, we put more weights on Usefulness and Attractiveness in comparison with the academic review criteria.
- Usability: Our criteria include Accessibility (open access, free/low charge) and Attractiveness. In this context, we prefer open access websites/webpages to (costly) books, and prefer open free webpages to consultants' hidden documents/stories of methods/tools/case studies. At moment we do not include the usability in small-screen mobile devices nor usability for handicapped persons, while assuming the ordinary use in PCs and tablets.

How do you Actually Grade Websites with the Criteria?. – We use the multiple criteria so as to find strong points of the website depending on their nature/purposes. For example, a webpage describing a new approach in TRIZ method is evaluated with the criteria suitable for academic papers, a web page reporting an application of TRIZ to a manufacturing problem with the criteria suitable for technical/practical documents, an introduction of TRIZ way of thinking for children with the criteria adjusted for educational texts, and so forth.

Grading into 5 levels is done on the basis of experts' experiences. ◎ for top (about) 30 sites, and ○ for next (about) 100 sites. Once we have some examples of ◎○ sites, we may set other websites at the grade similar to them. Websites working in multiple countries or worthy of being known in the world (but not at the level of ○) are graded as □. Websites working in a country and not so unique in the world level are graded as △. All these grading is tentative and under possible revisions. We should have some mechanism of public review or users' feedback in future.

What is the Basis of your Description of Basic Requirements of Catalogs (Sect. 1.1)? – The 3 Requirements were obtained through observation and discussion in our WTSP Project. You may think of any Catalog, for instance Michelin Catalog of Restaurants. You may easily guess what would happen if an excellent restaurant is missed, if a poor restaurant is included at a high rating, if introductions of restaurants are described in an unfair manner due to bribe or not, if you want to find a good Japanese sushi restaurant among 200 restaurant in the Catalog without proper categorization, etc. We think the 3 Basic Requirements of Catalogs are now well defined.

Categorization of Catalogs Seems too Detail (Especially in the Abstract). – As you see in Fig. 5, we now have 400 + websites listed in the WTSP World Catalogs of Around-TRIZ sites, and we will probably have 300 + websites in the Catalogs of TRIZ sites.

Listing so many websites together in each Catalog is not easy for users to read/handle and not easy for the developers to build and maintain. Thus we made Top Catalog listing about 30 ◎ Most important sites only. Basic Catalog having about 30 ◎ sites and about 100 ○ Important sites together is supposed to be most useful for users. Then, 200 + websites □ Worthy in the World Catalogs are listed separately in Extended Catalog in order to keep handy for users and easy to update for developers.

History of developing the WTSP Catalogs is described in the Abstract and in the text by using the numbers of websites in these categories/grades. Such numbers can vividly tell how far our WTSP project has come. It is clear that Catalogs of TRIZ sites are far behind those of Around-TRIZ sites.

Access Statistics Available?. – Access counters to the WTSP Catalogs are not installed yet. We guess the accesses are less than we expected at the present stage. Please come and read/use the WTSP Catalogs! They already have abundant useful information!!

5 Concluding Remarks

The WTSP Project has built and publicized the WTSP Catalogs of TRIZ Sites and Around-TRIZ Sites in the World as a reliable information source in the field of TRIZ and Creative Problem Solving methodologies in general. The current status of the WTSP Catalogs is summarized in Fig. 5. The Catalogs of Around-TRIZ Sites are well developed by using the results of Internet surveys, but the Catalogs of TRIZ Sites are now only partially done, while missing many good websites from TRIZ-active countries and not yet containing websites in the □ level. We should overcome these weak points by some voluntary team work in the TRIZ community in individual countries and by the contributions of individual site owners in writing their own sites. Overcoming these difficulties, we hope to publicize the Third Edition of World WTSP Catalogs by the end of 2022.

The WTSP World Catalogs [3] already have abundant and selected information, and are publicly posted for ready to be read/used. Please use them fully for your jobs/works/studies.

One specific point Nakagawa noticed: In the TRIZ community there are a number of international associations/conferences and many good websites operated by individual consulting firms and academic groups, but there exists no 'representative' place where good TRIZ works are coming from various groups in the world to be published regularly. The TRIZ Journal served as the leading journal for promoting TRIZ in the early stage, but not in recent years. A new website, so to speak "Systematic Innovation Review", is anticipated to be established and operated by some solid international collaboration in the global TRIZ(-related) community.

References

1. Nakagawa, T., (ed.) Website TRIZ Home Page in Japan (in English). https://www.osaka-gu.ac.jp/php/nakagawa/TRIZ/eTRIZ/. Established 15 Nov 1998, Accessed Jul 2022

2. Nakagawa, T., (ed.) Website of World TRIZ-related Sites Project (WTSP). https://www.osaka-gu.ac.jp/php/nakagawa/TRIZ/eTRIZ/eWTSP/. Established Dec 2017, Accessed Jul 2022
3. Nakagawa, T.: World WTSP Catalogs: Top Page. https://www.osaka-gu.ac.jp/php/nakagawa/TRIZ/eTRIZ/eWTSP/eWTSP-WorldCatalog/eWTSP-World%20Catalog-World/World-Catalog-TopPage.html
4. Nakagawa, T.: World TRIZ Sites Project: Its Plan and Invitation: Volunteer Project for Connecting TRIZ Sites in the World, 9 Dec 2017. https://www.osaka-gu.ac.jp/php/nakagawa/TRIZ/eTRIZ/eforum/e2017Forum/eNaka-WorldTRIZSitesProject2017/eNaka-WTSP-Invitation-171207.html
5. Nakagawa, T., Mann, D., Orloff, M., Dewulf, S., Litvin, S., Souchkov, V.: WTSP Appeal: An Appeal for Building Catalogs of TRIZ-related Sites in the World, 25 Jun 2018. https://www.osaka-gu.ac.jp/php/nakagawa/TRIZ/eTRIZ/eWTSP/eWTSP-B2-News2018/eWTSP-News2018-Appeal-180625.html
6. Nakagawa T., Mann D., Orloff M., Dewulf S., Litvin S., Souchkov V., World TRIZ Sites Project (WTSP) for Building and Maintaining Global Catalogs of TRIZ-related Web Sites, Summary slides of ETRIA TFC2018 presentation, 11 Nov 2018, https://www.osaka-gu.ac.jp/php/nakagawa/TRIZ/eTRIZ/epapers/e2018Papers/eNaka-WTSP-TFC2018/eNaka-WTSP-TFC2018-Summary-4P-181023.pdf
7. Nakagawa, T., Mann, D., Orloff, M., Dewulf, S., Litvin, S., Souchkov, V.: World TRIZ Sites Project (WTSP) (2): To Build Catalogs of TRIZ-related Web Sites in the World, Presented at ETRIA TFC 2019, held on 9-11 Oct 2019, at Marrakech, Morocco; https://doi.org/10.1007/978-3-030-32497-1_40, https://www.osaka-gu.ac.jp/php/nakagawa/TRIZ/eTRIZ/epapers/e2019Papers/eNaka-WTSP2-ETRIATFC2019/eNaka-WTSP2-ETRIATFC2019-191016.html
8. Nakagawa T., Mann D., Orloff M., Dewulf S., Litvin S., Souchkov V., World TRIZ-related Sites Project (WTSP) (3): World WTSP Catalogs of TRIZ and Around-TRIZ Sites: First Edition (2019) and Its Further Enhancement, ETRIA TRIZ Future Conference 2020, held online at Cluj-Napoca, Romania, on Oct. 14–16 2020. https://www.osaka-gu.ac.jp/php/nakagawa/TRIZ/eTRIZ/eWTSP/eWTSP-B4-News2020/eWTSP-Presentations2020-WTSP3/eWTSP-Presen-ETRIA-TFC2020-WTSP3-201106.html
9. Nakagawa, T.: Current Status of the Survey Results of WTSP World TRIZ Sites, and Overview of Deployment of TRIZ in the World, 20 Nov 2021. https://www.osaka-gu.ac.jp/php/nakagawa/TRIZ/eTRIZ/eWTSP/eWTSP-B5-News2021/eWTSP-News2021-Results-Surveys4th-211118.html
10. Nakagawa T., Mann D., Orloff M., Dewulf S., Litvin S., Souchkov V.: World TRIZ-related Sites Project (WTSP) (4): World WTSP Catalogs of TRIZ Sites and Around-TRIZ Sites, Let's Make Them Attractive and Useful!, Presented at ETRIA TFC 2021, held on 22-24 Sept 2021, at Free-Universtiy of Bozen-Bolzano, Italy. https://www.osaka-gu.ac.jp/php/nakagawa/TRIZ/eTRIZ/eWTSP/eWTSP-B5-News2021/eWTSP-Appeal-ETRIATFC2021-210927.html
11. Nakagawa T., Mann D., Orloff M., Dewulf S., Litvin S., Souchkov V.: WTSP Appeal 2021: World WTSP Catalogs of TRIZ Sites and Around-TRIZ Sites, Let's Make Them Attractive and Useful! 24 Sept 2021. https://www.osaka-gu.ac.jp/php/nakagawa/TRIZ/eTRIZ/eWTSP/eWTSP-B5-News2021/eWTSP-Appeal2021-ETRIATFC-210924-Slides.pdf
12. Nakagawa, T.: WTSP Catalogs of (C) Around-TRIZ Sites in the World are Revised and Extended, 3 Jun 2022. https://www.osaka-gu.ac.jp/php/nakagawa/TRIZ/eTRIZ/eWTSP/eWTSP-WorldCatalog/eWTSP-World%20Catalog-World/World-C-AroundTRIZSites-Catalogs.html
13. Nakagawa, T.: WTSP (A4) Guidelines for Building World WTSP Catalogs, 20 Nov 2020. https://www.osaka-gu.ac.jp/php/nakagawa/TRIZ/eTRIZ/eWTSP/eWTSP-A4-Guidelines.html

14. Nakagawa, T.: Mission and Philosophy of WTSP, 16 Aug 2021. https://www.osaka-gu.ac.jp/php/nakagawa/TRIZ/eTRIZ/eWTSP/eWTSP-A1-Mission&Philosophy-2108.pdf
15. Nakagawa, T.: Introduction to World WTSP Catalogs of TRIZ Sites and Around-TRIZ Sites, 24 Jan 2022. https://www.osaka-gu.ac.jp/php/nakagawa/TRIZ/eTRIZ/eWTSP/eWTSP-B6-News2022/eWTSP-WorldCatalogsIntroduction-220123.html
16. Nakagawa, T.: (A2P) World TRIZ Sites Basic Catalog (◎○) for Print, 30 Jun 2020. https://www.osaka-gu.ac.jp/php/nakagawa/TRIZ/eTRIZ/eWTSP/eWTSP-WorldCatalog/eWTSP-World%20Catalog-World/World-A2P-TRIZBasic-Catalog-PDF.pdf
17. Nakagawa, T.: (C2P) World Around-TRIZ Sites Basic Catalog (◎○) for Print, 17 Aug 2022. https://www.osaka-gu.ac.jp/php/nakagawa/TRIZ/eTRIZ/eWTSP/eWTSP-WorldCatalog/eWTSP-World%20Catalog-World/World-C2P-AroundTRIZBasic-Catalog.html
18. Google, Webmaster Guidelines. https://developers.google.com/search/docs/advanced/guidelines/webmaster-guidelines. Accessed 05 July 2022

TRIZ Training Within a Continuous Improvement (Kaizen) Event – Exploration and Evaluation

Tony Tanoyo[1](✉) and Jennifer Harlim[2]

[1] Leica Biosystems, Melbourne, Australia
tony.tanoyo@gmail.com
[2] Monash University, Melbourne, Australia

Abstract. This paper discusses the implementation and evaluation of an online-based TRIZ training that was simultaneously conducted within a Kaizen (continuous improvement) event. The purpose of the training was to introduce and show the application of TRIZ tools using a real problem that was undertaken as part of the Kaizen exercise. The problem was a software related problem. The TRIZ tools that were presented in the training were Situational Analysis (SA), Functional Modelling (FM), Cause and Effect Analysis (CEA), Substance-Field (SF) Analysis, 40 Inventive Principles, Patterns of Evolution and Function Oriented Search (FOS). The research questions investigated were: *1) Can combining a TRIZ training within a Kaizen event which was conducted online, result in an effective training session?* and 2) *Which TRIZ tools that were presented during the event were perceived by the participants to be the most useful and which ones were the least useful when delivered in this manner?* It was found that embedding the TRIZ training in the online Kaizen session was well-received and considered effective. It was identified that participants in the study prefer TRIZ tools that assist in problem identification. Areas of improvements were also sought from the participants' responses. The study has implications for organisations wishing to develop in-house TRIZ training.

Keywords: TRIZ training · Engineering professional development · Kaizen · Problem solving · Self-efficacy · Evaluation · Action Research · Online training

1 Introduction

The value of staff training within an organisation is well documented. Training can bring about several benefits including boosting the morale of employees, staff-retention, increasing and improving performance and productivity as well as bringing about innovation and long-term profit for the organisation [1–4]. While the value is undeniable, training and staff development is an investment that can be costly. Thus, it is important that training activities are implemented properly.

This paper describes the implementation of an online-based TRIZ training within a global large-sized company based in Melbourne, Australia. The company has a strong

Published by Springer Nature Switzerland AG 2022
R. Nowak et al. (Eds.): TFC 2022, IFIP AICT 655, pp. 458–469, 2022.
https://doi.org/10.1007/978-3-031-17288-5_38

focus on problem solving and formally includes TRIZ within its recommended problem-solving tools repertoire. This has allowed for opportunities to formalise in-house TRIZ training within the organisation. This study is a continuation of the exploration on how TRIZ training can be effectively implemented within this organisation.

The challenge of deploying successful TRIZ training was discussed by Ilevbare et.al. Who identified that barriers to TRIZ uptake are the complexity of TRIZ tools and the time consuming nature to understand the application [5]. In their literature review on the usage of TRIZ tools, Sojka and Lepšík also reinforced the same challenges in effectively teaching TRIZ [6]. In addition, Gadd found that the uptake of TRIZ can be limited too if the organisation does not provide enough opportunities to utilise the tools [2]. It can be summarised that the challenges for an organisation wishing to implement TRIZ training are: 1) Time; 2) Complexity and; 3) Opportunity to use the tools in real context. The use of a real problem in the Kaizen session within the organisation in this study provided a good opportunity to address these issues.

The aim of this research is to evaluate an implementation of a TRIZ training which was embedded within an online-based Kaizen event. The research questions explored in this paper are:

1. Can combining a TRIZ training within a Kaizen event which was conducted online, result in an effective training session?
2. Which TRIZ tools that were presented during the event were perceived by the participants to be the most useful and which ones were the least useful when delivered in this manner?

2 Embedding TRIZ Training Within a Continuous Improvement – Kaizen Event

Several previous attempts to introduce TRIZ training within the organisation have been undertaken. One such implementation was a survey style online-based training which participants complete at their own time [7]. It was observed that participation was high initially. However, participation dropped off dramatically when the engineers were asked to go through materials and exercises covering the 40 Inventive Principles. In addition, the exercise was not formalised or mandated. One of the main feedbacks received was that more guidance during the training was preferred [7].

Learning from this experience, it was clear to improve the training, formalisation was needed. However, similarly, it was ideal if the training did not take extra time from the engineers but could be incorporated within their current work tasks. Hence, a training within a Kaizen event was trialed. The Kaizen event took place via Zoom due to Covid-19 restrictions. It is expected that the online mode of learning will continue as Covid-19 has changed the way the organisation operates, allowing for remote working arrangements. In addition, given that the organisation involved is a global one, it is expected that the training will involve engineers from multiple sites.

Kaizen is a Japanese word that indicates a process of continuous improvement of the standard way of work [8]. It is a compound word involving two concepts: Kai (change) and Zen (for the better) [9] The terms come from Gemba Kaizen, which means "Continuous Improvement" (CI).

Within the organisation in this research, a Kaizen event is a team-based continuous improvement process to be used at any time to focus the effort of a team when needed to drive improvement or creation of a process. The organisation suggests Kaizen as the mindset of continuous improvement and the belief that one can always do better. It utilises tools to quickly eliminate *Muda* (waste) and standardise the resulting process at *Gemba* (where the work is done).

The concept of Kaizen has been reviewed and mentioned in several literatures. Similarities between the philosophy behind Kaizen and TRIZ have also been identified. Womack and Jones referred to Kaizen as a lean thinking and lay out a systematic approach to help organisations systematically to reduce waste [10]. They described waste as any human activity that absorbs resources but creates or adds no value to the process [10]. Most employees could identify several different types of *Muda* in their workplace, but unfortunately the waste that they identify is only the tip of the iceberg. The authors stated that until these employees have been taught the essentials of lean thinking, they are unable to perceive the waste present in their environment [10]. TRIZ philosophy of ideality is very much aligned with this principle. Similarities between TRIZ, TOC (Theory of Constraints) and Lean are well described by Martin [11] as well as Anosike and Lim [12]. Sojka and Lepšík proposed that lean is described as good for waste elimination and TRIZ enables creativity and foresight [6].

These literatures suggest that TRIZ can work well with Kaizen. However, the investigation described in this paper focuses on the merging of application and training at the same time. The exercise was an effort to embed a formalised training within a Kaizen event to demonstrate how TRIZ can be applied.

3 The Use of a Real Problem

A real problem was used in the previous attempt to deploy TRIZ training in the organisation. The use of a real problem worked effectively as it helped with the initial buy-in from the engineers to participate [7]. Hence, the utilisation of a real problem was again replicated in this study.

The problem used for the Kaizen and TRIZ training was related to the intermittent time jump issue seen by the operator of the instrument in the field. This issue happens once or twice a year and seems to be happening after Daylight Saving Time (DST) and in the region of countries that are observing Daylight Saving Time (DST). Addressing this issue is important because the performance of the instrument relies heavily on the accuracy of time keeping. The problem was a software-based issue.

4 TRIZ Tools Covered in the Training

As the organisation has formally included TRIZ in their problem-solving tools repertoire, the management mandated that the formal training should include the following:

1. Functional Modelling (FM) and analysis
2. Cause-and-Effect (CE) analysis
3. Substance-Field (SF) analysis
4. Technical contradiction
5. Physical contradiction
6. Patterns of Evolution
7. Function-Oriented Search (FOS)

In addition to the prescribed tool sets by the training module of the organisation, the heuristic of Situational Analysis (SA) [13] was also introduced to help during the problem definition stages.

It can be proposed that there are two main categories of TRIZ tools in the training – those that assist with problem identification and those that assist with problem ideation.

5 Methodology

Action Research [14–17] was used in the investigation. This methodology was chosen as it allows the planning, implementation, and data collection processes to be conducted at the same time. It allowed the first author, who conducted the training to adjust the training procedures as needed. To minimise bias, the data analysis was carried out by the second author who is independent from the training and the organisation.

To evaluate the effectiveness of the training, two sets of surveys were conducted. Pre- and post-surveys were conducted prior to and after the Kaizen event. In the pre-event questionnaire, baseline information including the number of years of work experience was collected. In both pre- and post-questionnaires, participants were asked problem solving self-efficacy questions. While the data was self-reported, in her study, Gadd found that self-perception of creative self-efficacy is a useful measure for training evaluation [2]. In his theory of self-efficacy, Bandura [18] proposed that it is a set of beliefs that allows a person to perform in a particular situation. Harlim suggested that self-efficacy is one of the factors that impact problem solving performance [19]. Therefore, the use of self-efficacy measurement is appropriate for the evaluation of the training. A set of problem solving self-efficacy questions developed by Belski [20, 21] were utilised. Participants were asked to rate the degree of their agreement to the following 5 questions using a 5-point Likert-scale (*1 – Strongly disagree, 2 – Disagree, 3 – Neither disagree or agree, 4 – Agree and 5 – Strongly agree*):

1. I am very good at problem solving
2. I am never intimidated by unknown problems
3. I am unable to tackle unfamiliar tasks
4. So far, I have been able to resolve every problem I faced

5. I am certain that I am able to resolve any problem I will face.

In addition to the quantitative questions, qualitative questions were also included in the survey. This was because the number of participants in the study was limited by the number of engineers who were involved in the Kaizen event. A focus on qualitative data can yield in a richer in-depth analysis which are suitable for smaller sample sizes [16–18]. This qualitative question was included in the pre- and post-surveys:

- Do you expect that the way you approach problems may change as a result of this exercise? How do you think it might change or not change?

Additional qualitative questions were included in the post-survey:

- What you think was done well in the training/Kaizen?
- What do you think can be improved in the training/Kaizen?
- Do you think the exercise has helped you to understand how TRIZ tools can be applied to a real problem? If yes, how so. If not, why not.
- Do you think you will be more interested to learn more about TRIZ tools? If yes, why? If not, why?

In the post-survey, participants were also asked to rate the 8 tools that were used during the training in terms of their most-liked (1) and least liked (8).

The Kaizen and training were conducted over 6 half days (5 h with 1 h break) with the last day dedicated for report out. The training modules ran for half hour to one hour followed by team activities and discussions. On the first day, the training covered introduction and Functional Model (FM). On the second day, the training covered Cause and Effect Analysis and Trimming technique. On the third day, it was initially planned for Function Oriented Search (FOS), but the trainer decided to introduce Patterns of Evolution instead. On fourth day, Substance-Field (SF) analysis was covered followed by Technical Contradictions. On the fifth day, Physical Contradictions and FOS were covered. On the last day, the team analysed and ranked the ideas based on the number of efforts, project duration, complexity of the changes and technical risk. The ideas were then sorted with the lowest number of low scores being the top idea for easy implementation and simple changes and low technical risk. Prior to Kaizen and training, the participants completed pre-work on the initial CEA, FM and investigations summary. All the participants were briefed about the problem in the same way. SA was used several times within the different training days, especially to bring about consensus on the problem identification process.

15 engineers took part in the training. 10 were male and 5 were female. The participants were from diverse domain technical knowledge and department (electronics, mechanical, software, system, science, field support, verification). Using the classification of expertise used by Harlim and Belski [22], out of those who participated in the training included 3 novice engineers (under 5 years of work experience), 3 mid-level engineers (5–10 years) and 8 engineers are expert levels (more than 10 years). 1 engineer opted not to answer the question related to work experience. Two of the engineers did

not complete the post-survey. Therefore, for the purpose of this study only 13 responses were included for analysis.

6 Results and Discussion

6.1 Changes in Perception of Self-efficacy in Problem Solving

Table 1 presents the change in the perception of self-efficacy in problem solving by the engineers in the study. For questions 1, 4 and 5, the mean ratings of their problem-solving self-efficacy were reduced compared to their evaluation in the pre-survey. However, for questions 2 and 3, the mean ratings indicate that their confidence towards unknown and unfamiliar problem increased after the training. No statistical significance was observed when the data was tested with the Wilcoxon signed rank test ($p > 0.05$).

Table 1. Changes in perception of self-efficacy in problem solving

No.	Questions	PRE	POST
1	I am very good at problem solving	3.92	3.69
2	I am never intimidated by unknown problems	3.69	3.92
3	I am unable to tackle unfamiliar tasks	2.38	2.00
4	So far, I have been able to resolve every problem I faced	3.31	3.08
5	I am certain that I am able to resolve any problem I will face	3.46	3.31

$N = 13$, Wilcoxon signed rank test $p > 0.05$

The reduced mean rating in questions 1, 4 and 5 could indicate that after the training, participants may have a better understanding of their ability to solve problems hence, correcting their perception of problem-solving efficacy. Bandura suggested four ways of developing self-efficacy: 1) Mastery experiences; 2) Social modelling; 3) Social persuasion and; 4) Psychological responses [18]. Out of the 4, Bandura suggested that mastery experience or previous performance experiences is the most accurate [18]. The self-efficacy questions proposed by Belski [20, 21] covers two of the four sources of self-efficacy. Questions 1, 4 and 5 are questions based on mastery experiences, while Questions, 2 and 3 covers psychological responses. It is suggested through the experience of working with a real problem during the Kaizen event, participants were given a chance to evaluate their own problem-solving performances. This allowed them to develop a more accurate problem-solving self-efficacy. It can be suggested that the increase in confidence mean ratings in Questions 2 and 3 support this.

This can also be further supported by the responses of the participants when asked *Do you expect that the way you approach problems may change as a result of this exercise? If yes, how did it change? If no, why do you think not?* All the participants believed that the exercise had a positive impact on the way they approach problems. Some of these responses are exemplified in Table 2 below.

Table 2. Examples of responses to "Do you expect that the way you approach problems may change as a result of this exercise? If yes, how did it change? If no, why do you think not?"

Responses
"Yes, I will approach the issues and the problems with a different mindset." **(Participant 2, Male, Mid-level)**
"Yes. It's opened my mind to more possibilities for a start. And also given me a toolset to draw from in a methodical way." **(Participant 3, Male, Expert)**
"TRIZ is a structured way of approaching a problem, rather than going by the gut feeling. It also helps me keep an open mind to think in the uncharted territory." **(Participant 5, Male, Unknown)**
"Yes, I think going forward when presented with a problem I will be more inclined to go back to the basic principle of TRIZ about trying to overcome psychological inertia. In a more practical sense, now having used all these TRIZ tools, I am more equipped to approach problems in new and different ways… I am also more confident that if presented with a problem I do not have a strong background in that I can still think through the problem in a way that I can suggest solutions." **(Participant 10, Female, Novice)**

6.2 Things that Worked Well in the Training, Areas for Improvements and Further Interest in TRIZ

Table 3 shows samples of participants' responses in the qualitative post-event evaluation. The general feedbacks indicate that the use of real problem, embedding the training within the Kaizen event, group work and the use of online platform worked well. In identifying areas for improvements, the participants suggested group participation can be improved, more Gemba and topics related to the 40 inventive principles can be explained better. Technological challenges such as internet connection was also raised.

The responses show that the training within the Kaizen event was well-received. The findings support the literatures reviewed by Sojka and Lepšík that TRIZ and Kaizen are compatible [6]. In addition, these data suggest there may be benefits in embedding the training in a Kaizen event. The use of a real problem offered opportunity for the engineers involved to apply TRIZ as they learned it. This may assist in overcoming the barrier of uptake highlighted by Gadd [2] and other barriers of TRIZ uptake due to time and complexity as suggested by Ilevbarre et. al. [5] as well as Sojka and Lepšík [6]. The responses received from the participants indicate further interest in learning more about TRIZ.

Participants mentioned that they would like more Gemba. The Japanese term Gemba refers to "where the action takes place". For this problem, more Gemba would relate to more opportunities for the use of the software on the actual instrument. Unfortunately, this could not be done easily due to the constraints with online interaction and social distancing rules when this training took place. This highlights one of the limitations of an online-based training. Participants also suggested that the information related to 40 Inventive Principles can be better conveyed. This was also reflected in the data presented in the next section.

Table 3. Examples of responses to the additional post-survey questions

Questions	Responses
What you think was done well in the training/Kaizen?	"Splitting into teams and carrying through the exercises as a team, hearing different opinions. Not shooting down ideas was great and created a safe environment." **(Participant 3, Male, Expert)** "Defining question and making sure everyone on same page. Also training on concept and group exercises to apply the concept/tool on current problem was really good approach." **(Participant 6, Male, Mid-level)**
What do you think can be improved in the training/Kaizen?	"Pre-session deep diving into the data, looking at actual data, not hearsay from others. Looking at it together. Going to the Gemba." **(Participant 3, Male, Expert)** "This was the first time this training was being done, and it was also online so there were issues because of that. Especially around team participation." **(Participant 5, Male, Unknown)** "I think better definitions and examples of the 40 inventive principles would be really helpful to decrease time to proficiency of the tool." **(Participant 8, Male, Novice)**
Do you think the exercise has helped you to understand how TRIZ tools can be applied to a real problem? If yes, how so. If not, why not	"Yes, during this training/Kaizen, learning the new tools and practicing it in parallel with real problems showed the tools are helpful." **(Participant 7, Male, Expert)** "I definitely see the utility for TRIZ tools. While they are flexible to applied to many situations the time involved to consider and implement them for a problem means they are likely best suited to events like Kaizens where an issue warrants great discussion and potentially has several unknown solutions which will deliver a great impact to key stakeholders." **(Participant 2, Male, Mid-level)**

(*continued*)

Table 3. (*continued*)

Questions	Responses
Do you think you will be more interested to learn more about TRIZ tools? If yes, why? If not, why?	"Yes, I would like to learn about practicing TRIZ in non-engineering problems and also in ideation. Standard work to run an slimmed down model of TRIZ would also be great and easier to sell to the business." **(Participant 8, Male, Novice)** "Yes but also would like to practice using what we already learnt more too. Why - it is obvious to me that these tools would be quite useful for other problems/research/ideation. Given my background in research and systems and NPI engineering, I think back to all the design and development work where I can see that these tools would have been incredibly useful." **(Participant 10, Female, Novice)**

6.3 Rating of Usefulness of the Tools Presented

Instead of looking at the rating of all 8 tools, only the tools rated 1 – most-liked and 8 – least-liked were considered for the purpose of this paper. Table 4 presents the comparison of the responses of the 13 participants and Table 5 are examples of some of the reasoning for the choices. Out of 13 participants, 12 participants rated tools that fall

Table 4. Rating of the tools presented most-liked vs least-liked

Participant No.	Gender	Expertise level	Most-liked (rated 1)	Least-liked (rated 8)
1	Male	Expert	FM	Technical contradiction
2	Male	Mid-level	FM	Su-Field
3	Male	Expert	Technical contradiction	FOS
4	Male	Expert	FM	FOS
5	Male	Unknown	SA	Patterns of evolution
6	Male	Novice	FM	Technical contradiction
7	Male	Expert	CEA	FOS
8	Male	Novice	FM	FOS
9	Female	Expert	FM	FOS
10	Female	Novice	FM	Physical contradiction
11	Female	Expert	FM	Physical contradiction
12	Male	Expert	FM	Physical contradiction
13	Male	Expert	FM	FOS

Table 5. Examples of reasonings given by the participants for the ratings of the tools.

Questions	Responses
Why did you select the top module as the most-liked compared to the rest	"In my opinion situational analysis is the most practical of all the tools. It asks important questions on what is wrong with the situation, helps us to scope the problem by defining what system we are looking at, identifying what is the actual MUF so that we open our mindset to more possibilities for changes, helps us to think about how we measure success and to consider user and system needs. It reminds me of the traditional system V model in design and development." **(Participant 10, Female, Novice)** "It helps to decompose a system (framework) into cubes and make it possible to explore the interaction among cubes." **(Participant 9, Female, Expert)**
Why did you select the bottom module as the least-liked compared to the rest?	"Requires time to get used to the tools, especially the relating the 40 principles to the problems at hand." **(Participant 6, Male, Novice)** "Still do not fully understand the application of these concepts." **(Participant 12, Male, Expert)**

under the category of tools for problem identification as most-liked. Participant 3 who rated Technical Contradiction, a tool that falls under ideation (40 Inventive Principles) explained his reasoning as:

> "*I felt the contradictions together were most relatable to thinking patterns I already had used before. This made them simpler to understand, build upon and enjoy the process of using them more as a result.*" **(Participant 3, Male Expert)**

Out of 13 participants, 12 participants rated tools related to ideation as least-liked. Participant 2 who rated Su-Field as least liked indicated that his reasoning was because he was not present during the session. Interestingly, looking at the tool that he had rated 7, it was Physical Contradiction which also fell under the category of ideation tools.

The data indicates that there was a preference for tools that help in problem identification. During the training, it was also observed by the trainer that some of the TRIZ tools were less effective than the others for the specific problem presented in the training. The preference for FM as shown in the data may be because the problem was software specific. On the other hand, in general, participants perceived tools that are focused on ideation to be least liked as they were difficult to comprehend. It is acknowledged that TRIZ can be complicated and challenging to teach or train people in [5, 6]. This study identified specific tools of TRIZ that are perceived to be more challenging for learners new to TRIZ to uptake. In the data, the one participant who liked the Technical Contradiction module indicated that his past experiences made it easier to understand. This supports that prior experience has an impact on understanding more complex principles in TRIZ and should be taken into consideration when designing TRIZ trainings.

7 Implications, Research Limitation and Future Research Direction

The results in this study have implications for organisations wanting to develop or implement in-house TRIZ training. It also has implications for global organisations considering TRIZ training remotely via online conferencing tools to cater to their workforce who may be located in multiple sites.

Overall, this research shows that the embedding of TRIZ training within an online Kaizen session worked well and can be effective. The use of a real problem and the merging of training with the Kaizen session, allowed participants to learn and apply TRIZ at the same time. The training allowed participants to develop a more accurate problem-solving efficacy, boosting their confidence for unfamiliar and unknown problems. The investigation also identifies which TRIZ tools posed challenges for deploying effective training such as tools that are commonly used for ideation: FOS, Patterns of Evolution and tools related to the 40 Inventive Principles. The findings suggest that when designing training around these tools, proper consideration of the participants' previous experiences may also need to be considered, on top of the overall training design.

One of the key limitations of this study was the number of participants which was constrained by the engineers in the organisation who were involved in the Kaizen session. However, the research design had taken this into account and included qualitative data collection on top of quantitative ones. The qualitative responses allowed for an in-depth data with the smaller sample. The use of a real problem was beneficial, but it also meant that the practice of application of TRIZ was limited to that one problem during the training. Future research could focus on the expansion the training within other Kaizen sessions in the organisation, increasing the different types of problems to apply on during the training as well as improving data collection with more participants. An investigation on how to design more effective training around some of the ideation tools which were identified as complex can also be carried out.

References

1. Sung, S.Y., Choi, J.N.: Do organizations spend wisely on employees? Effects of training and development investments on learning and innovation in organizations. J. Organ. Behav. **35**(3), 393–412 (2014)
2. Haines-Gadd, L.: Does TRIZ change people? Evaluating the impact of TRIZ training within an organisation: Implications for theory and practice. Proc. Eng. **131**, 259–269 (2015)
3. Kennett, G.: The impact of training practices on individual, organisation, and industry skill development. Aust. Bull. Labour **39**(1), 112–135 (2013)
4. Kwon, K.: The long-term effect of training and development investment on financial performance in Korean companies. Int. J. Manpower (2019)

5. Ilevbare, I.M., Probert, D., Phaal, R.: A review of TRIZ, and its benefits and challenges in practice. Technovation **33**(2–3), 30–37 (2013)
6. Sojka, V., Lepšík, P.: Use of TRIZ, and TRIZ with other tools for process improvement: a literature review. Emerg. Sci. J. **4**(5), 319–335 (2020)
7. Tanoyo, T., Harlim, J., Belski, I.: Facilitation of a creative culture through the implementation and initial evaluation of a triz course within an organisation. In: Borgianni, Y., Brad, S., Cavallucci, D., Livotov, P. (eds.) TFC 2021. IAICT, vol. 635, pp. 379–390. Springer, Cham (2021). https://doi.org/10.1007/978-3-030-86614-3_30
8. Chen, J.C., Dugger, J., Hammer, B.: A kaizen based approach for cellular manufacturing system design: a case study (2001)
9. Palmer, V.S.: Inventory management KAIZEN. In: Proceedings 2nd International Workshop on Engineering Management for Applied Technology, EMAT 2001. IEEE (2001)
10. Womack, J., Jones, D.: Lean Thinking. Simon and Schuster, New York (1996)
11. Martin, A.: TRIZ theory of constraints and lean. TRIZ J. (2010)
12. Anosike, A.I., Lim, M.K.: Integrating lean, theory of constraints and TRIZ for process innovation. In: Short Research Papers on Knowledge, Innovation and Enterprise. KIE Conference Book Series, Berlin (2013)
13. Belski, I., et al.: TRIZ in enhancing of design creativity: a case study from Singapore. In: Chechurin, L. (ed.) Research and Practice on the Theory of Inventive Problem Solving (TRIZ), pp. 151–168. Springer, Cham (2016). https://doi.org/10.1007/978-3-319-31782-3_9
14. Coghlan, D.: Doing Action Research in Your Own Organization. Sage (2019)
15. Business Research Methodology. Action Research. https://research-methodology.net/research-methods/action-research/. Accessed 10 May 2021
16. Koshy, E., Koshy, V., Waterman, H.: Action research in healthcare (2010)
17. Tripp, D.: Action research: a methodological introduction. Educ. Pesquisa **31**(3), 443–466 (2005)
18. Bandura, A.: Self-efficacy: The Excercise of Control, 11th edn. W. H. Freeman and Company, New York (1997)
19. Harlim, J.: Identifying the factors that impact on the problem solving performance of engineers. In: School of Electrical and Computer Engineering - College of Science, Engineering and Health, p. 146. RMIT University, Melbourne (2012)
20. Belski, I.: Teaching thinking and problem solving at university: a course on TRIZ. Creat. Innov. Manag. **18**(2), 101–108 (2009)
21. Belski, I.: TRIZ course enhances thinking and problem solving skills of engineering students. Proc. Eng. **9**, 450–460 (2011)
22. Harlim, J., Belski, I.: The allocation of time spent in different stages of problem solving: problem finding and the development of engineering expertise. In: Proceedings of the 28th Annual Conference of the Australiasian Association for Engineering Education. School of Engineering, Macquarie University, Sydney (2017)

TRIZ Training in Pandemic Time

Malgorzata Przymusiala(✉)

Novismo Sp. Z O.O, Warsaw, Poland
malgorzata.przymusiala@novimo.com

Abstract. Pandemic time forced us to change our life. Suddenly we faced many new problem to solve. One of them was the following contradiction: we have to stay isolated and we have to continue our activities. In case of lectures and training it seemed to be quite easy: go online. The Internet is available almost everywhere and it offers many opportunities to teach/to learn while remaining at home. Unfortunately, many things went wrong and many students and teachers soon started to dream to come back to face-to-face classes. The article is a case study based on own experiences of remote training, which have been highly rated by participants. It is a practical example of using TRIZ at TRIZ training, a suggestion which TRIZ tools are useful at teaching online and explanation how we use them in our everyday practice.

Keywords: TRIZ · Training · Online · Communication · Experience

1 Basic Contradiction: Training Must Be Continued and Training Cannot Be Continued

The SARS Covid-19 pandemic surprised the whole world. Schools and universities suddenly had to be closed, training had to be cancelled. People had to continue their educational activities, they could not pretend they had holidays. How could we describe this situation? From the TRIZ point of view it was a typical physical contradiction: classes must be continued as they cannot be suspended or postponed indefinitely and training cannot be continued because people have to stay at home. The solution seemed quite obvious, even for those not familiar with formulating and solving contradictions: to replace face-to-face meetings with video conferences, with meetings online. Internet is available almost everywhere, most of people has computers or at least smartphones [1], so the idea of lessons/training online is a great solution.

2 Disappointing Remote Classes

After two years of pandemic restrictions, many people claim that working online, especially online courses, is not as great solution as they expected. What is the cause of the disappointment? The list of causes is long: no or too few computers at home (imagine a family of two adults and two or three children of school age), insufficient Internet

Published by Springer Nature Switzerland AG 2022
R. Nowak et al. (Eds.): TFC 2022, IFIP AICT 655, pp. 470–478, 2022.
https://doi.org/10.1007/978-3-031-17288-5_39

bandwidth, no separated spaces for each family member working at the same time, etc. On the other hand, the companies had similar issues: too few computers, no procedures of remote work, no experience with online tools and no knowledge about them. People complained about no direct contact among people, boredom, lack of discipline, etc.

As we have no influence on many of the items mentioned above, we will focus only on the things that we were able to change.

3 What had to Change

The key to a success is to understand that online training is totally different from the standard one and there is a lot of preparation work to do before launching remote classes. It is not just talking to the audience using computer: the content must be structured in a new way as well as the new strategy of classes must be developed. Online tools must be chosen and tested earlier. Talking heads on a computer screen are unacceptable as boring and ineffective [2].

4 Training Before Pandemic

Novismo sp. z o.o. launched TRIZ L1 training in accordance with the International TRIZ Association (MA TRIZ) requirements [3] in 2017. During three-day classes we met the participants at lecture halls, conference rooms, etc. The agenda included 24–30 class h, divided into 1,5 h blocks. Each participant got a set of printed materials covering the whole presentation used during classes. A lot of content to comprehend in a very short time. As we met face-to-face, it was possible to use some props (toothbrushes, watches, school rulers, etc.) as a support to trainers' explanations and a way to make training more attractive for students. In post-class surveys participants rated the training as interesting (mean score 4,5 on a scale from 1 to 5; 328 trainee, 26 groups) and useful (mean score 4,1 on a scale from 1 to 5, 328 trainee, 26 groups), but challenging, and they complained about being exhausted afterwards (81% of 328 trainee).

How could it all be changed to cope with working online? How to convert our standard classes to successful online training? We decided to apply TRIZ tools: to study 40 Inventive Principles and check which recommendations could be useful to improve the online classes.

5 TRIZ Tools/40 Inventive Principles

The classical TRIZ tools are based on patents analysis. Genrich S. Altshuller, the creator of TRIZ, focused on patents relating to solving contradictions and he noticed some regularities: the recurring schemes of conflicts as well as the recurring patterns of solutions. That was the starting point to formulate 40 Inventive Principles–models of solutions [4]. Here are examples how we used some of these models at improving online training. We found the following Inventive Principles especially useful in remodelling our classes: 1, 2, 3, 5, 6, 10, 15, 20, 21, 23 and 25. The below explanations reveal details and give some practical hints.

6 Inventive Principle No. 1: Segmentation

- Divide the object into independent parts.
- Make the object easy to disassemble.
- Increase the degree of fragmentation or segmentation of the object.

6.1 Time–Changes in the Agenda

The 90 min blocks are too long, people are unable to maintain the level of attention for such a long time [5]. The risk of mind wandering is higher when taking into consideration the online conditions: there are many distracting factors, more than in classrooms.

Our first decision was to change the agenda: instead of 90 min block, our online class hour lasts 45 min. It is highly appreciated by the participants: in the post-class survey people claim they are not exhausted, in spite of 8 class h a day. (67% of 328 trainee, 29 groups taught remotely).

As we observed that each activity online takes at least 20% more time than offline, we decided that the L1 training online, including the certification exam, would last four days, instead of three days, as we used to do it before pandemic. It made no sense to plan 10 h of classes a day–we wanted to teach people, not to torment them.

Hint: To manage time well we ask participant to be punctual and we use timers. The first timer, 45 min countdown, is visible for the trainer only. The second one, 10 min countdown, showing the remaining break time is visible for the participants on the screen shared during breaks (we ask participants to stay logged in during breaks). You can use a timer application embedded in Windows system, or choose another one from the variety of timers available online and free of charge [6].

6.2 The Content Divided into Modules

The whole content of the training is divided into modules. When necessary it is easier to adjust the agenda to meet the customers' expectations (e.g. to convert the 4 full days to 8 days, 4 class h each). It is also convenient for participants as each module is accompanied with a separated set of printed materials.

6.3 Changes in Slides

The presentation used during standard offline training had to be modified. In a classroom the participants focus their attention on the trainer and the slides are not crucial. While online, the slides on the screen shared by the trainer must be attractive enough to focus attention and to keep it focused for 45 min. To learn how to remodel our slides, how to do it right we completed additional training for trainers and studied some helpful books [7] and we did our best to implement the gained knowledge. First of all, we try to follow the rule *one piece of information on one page*–it makes the content easier to understand and to memorize. When a slide is full of tables, figures and long texts the audience will try to read it instead of listening to the trainer. It is more distracting than helpful.

In our opinion it is better idea to cut the content into small pieces and not to worry about a big number of slides: our *Function Analysis of Devices* module includes over 200 slides. We remember that we cannot eat an elephant with a single bite, but we can do it piece by piece.

6.4 Breakout Rooms

Each training includes some practical exercises, when the presented theory can be immediately tested. It is the time when we divide the whole group of participants into smaller teams (2–4 persons). It is easy to do in a classroom, but now it is also available online, using breakout rooms offered by Zoom (the option also available in other applications). The trainer, who is a host of the meeting, decides about the number of rooms and assigns people to rooms. The group can be divided into rooms automatically. Even though the team members are assigned automatically, the trainer is able to transfer people among the breakout rooms.

Team members can use their cameras and microphones to communicate within a room. They are able to share their screens and send text messages as well. Using the breakout rooms is also a way to check the presence: to enter the room a person must accept the invitation, so the host can immediately see who is logged in but away. It is possible to set the time how long the rooms are active, they remain open until the host accepts to close them, though. The rooms can be renamed, i.e. *Anna's team.*

7 Inventive Principle No. 2: Separation

- Separate any interfering parts or properties from the object, or single out the only necessary part (or property) of the object.

7.1 No Cameras

Some participants are surprised when they learn that cameras will not be used. In our opinion cameras are one of the most distracting factors: instead of following the lecture people observe themselves, observe each other, observe the interiors behind the other people, etc. We do not use cameras, we ask participant to focus on the screen shared by the trainer. Cameras can be used only at the beginning, just to say 'hello', but they should be turned off immediately afterwards.

7.2 Changes in Slides

We make our best to present the content in the way it is easy to grasp and to memorize. We avoid filling slides with blocks of text, we try to use only the minimum of words, we use symbols, figures and colours. We reduce text as it takes more time to read and to comprehend. On the other hand, the participant can take notes in paper materials as the slides are not overloaded so there is enough space for handwritten remarks.

7.3 Changes in Communication

Good communication is a base of a good remote training. We encourage people to use status icons to communicate: to confirm their presence, to indicate a question or a comment. This way the trainer can immediately check the presence and control activity of each participant.

Hint: We recommend using status icons rather than chatting as reading text takes more time than a reaction to the icon.

7.4 Paper Materials

Each training participant receives a set of paper materials, but they are not just a copy of slides presented by the trainer. The slides on printouts are nearly the same, but we have left some places blank–to be fulfilled by the participant during the training. There are several advantages of such solution: people have to take notes. It means they have to listen carefully to the trainer and react. Handwriting supports memorizing [8], so the effectiveness of classes increases. As during the exam it is allowed to use all the notes, books and other available sources, it is recommended to have the completed set of materials on hand.

8 Inventive Principle No. 5: Merging

- Bring closer together (or merge) identical or similar objects, assemble identical or similar parts to perform parallel operations.
- Make operations contiguous or parallel; bring them together in time.

We merged the online tools into our standard classes. To avoid a boring lecture, we reach for different online tools: quizzes and spreadsheets shared by linking in the Zoom chat window. As they are well rated by the participants, we included them in standard classes–on condition participants have computers connected to Internet. The links and/or files are being sent by e-mail or copied from a pendrive. Previously we had the same exercises in offline versions only (printed in paper materials), now a student can decide if he/she writes the answers on paper or uses the computer.

9 Inventive Principle No. 6: Universality

- Make a part of the object, or the entire object perform multiple functions; eliminate the need for other parts.

9.1 One Set of Printed Materials

To keep the things simple we use the same presentations and the same printed papers for online classes as well as for those offline. The customers can change their minds whether they want an online training or an offline version–there is no need to prepare a different set of materials.

9.2 Using Popular Internet Tools

Some of the companies prepare their own tools to be used and shared online and store them on their own servers. To prevent unauthorised access they build sophisticated structures on their own servers: users have to log in, to enter passwords etc. To avoid any problems with access restrictions, we use popular websites open to the public. We use Google Forms and Google Sheets–the main advantage is their file formats are accessible and readable for Microsoft Office users as well as for Apple systems users. The needed files are just linked in the chat window. The files themselves are blank and useless for non-participants, so there is no risk even in case of unauthorised access or copying.

10 Inventive Principle No. 10: Preliminary Action

- Make any changes in the object (either fully or partially) before such changes are required.
- Pre-arrange objects so that they can be quickly activated without losing time delivering them.

Generally, remote training must be well prepared in advance: it cannot be just a copy of a standard classes, not just a lecture transmitted online. Each detail must be set to achieve the goal of keeping participants focused and active.

10.1 Online Presentation

We strongly recommend to analyze presentations used so far. We observe that in many cases trainers forget that slides must be helpful for participants, not to be just a content reminder for the speaker! If necessary, an additional set of materials must be prepared for the students, but the presentation slides cannot include all the details of content as it is a highly distracting factor: instead of listening, people try to read texts, diagrams and tables on a slide.

10.2 Online Tools

All the tools and files helpful during the training have to be prepared in advance. It takes some time, but the same files can be used many times. The only thing to do before another training is to remove the previous answers from the form or just to make a new copy of previously used form. If anything is being sent by e-mail (links or exercises files), it is good idea to compose the messages (or at least drafts) before and make sure that none of the attachments is left out.

10.3 Introduction to Online Communication

To be sure that every person can fully take part in classes, we have to start the meeting introducing the rules of communication. That is the reason why we have prepared an additional set of several slides just to explain all the technical details. We are aware that there are many kinds of software used for online meetings. They are similar to each other, but there are some differences that must be explained in the beginning to avoid interruptions during training.

11 Inventive Principle No. 15: Dynamization

- Allow changes (or design such changes) in the characteristics of the object, external environment, or process that optimize the object, or that optimize the operating conditions.
- Divide the object into parts capable of moving relative to each other.
- If the object (or process) is rigid or inflexible, make it movable or adaptable.

11.1 Changes in Slides

We dynamized most of slides in our presentations to make them more attractive. We use animations: the objects appear on a slide in a sequence, to illustrate the trainer's words.

11.2 Short Films to Illustrate Examples

As the online mode excludes any props, any object that could be given to participants' hands, we use short films instead–it is a further step on the way of the picture dynamization. Even though a film is not embedded in a slide, it could be easily presented by immediate sending a relevant link to participants–online meeting tools offer a chat window to communicate with written texts and links or to send files.

Hint: When sharing a film, make sure you also share the sound – check the settings of your online meeting tool.

11.3 The Final Test

The final test is an inherent part of a training. The Google tools offer many settings to adjust forms to the individual requirements. The one worth recommendation is the possibility to shuffle the questions as well as to shuffle the answers to each question. This way it is enough to prepare only one set of questions of the final test as each student will get his/her unique version. People do not realize they answer the same set of questions and the risk of cheating is reduced.

12 Inventive Principle No. 20: Continuity of Useful Action

- Carry on work continuously; make all parts of the object work at full load, all the time.
- Eliminate all idle or intermittent actions or work.

12.1 Recordings

We have prepared the recorded lectures for the participants. The whole training content is divided into modules and each person can come back to the chosen module and to watch it online any time. The links are active for 60 days after the training.

12.2 Constant Improving

We thoroughly read the participants' opinions expressed in post-class surveys and try to follow the suggestions what and which way could be improved. Even if the participants are satisfied and the scores are high, we do our best to present the most up-to-date pieces of information, we change the examples. The new examples also facilitate the trainer's work: repeating the same things many times is boring.

13 Inventive Principle No. 21: Skipping

- Conduct a process, or certain stages of it (e.g. destructible, harmful or hazardous operations) at high speed.
  The exam online has two important advantages: firstly, as we use Google Forms the part of trainers' work has been skipped and the test answers are checked automatically. The second advantage is a result of the first one: the results of the test are known immediately, the trainer can share them with the participants.

14 Inventive Principle No. 23: Feedback

- Introduce feedback (referring back, cross-checking) to improve a process or an action.
- If feedback is already used, change it.

14.1 Questions and Answers Any Time

At the very beginning of classes we inform the participants it is not a standard lecture. We reserve the right to ask questions any time, but on the other hand, we encourage the participant to ask their questions any time. This is also another way to verify the presence of each person.

Hint: In larger groups we recommend to use a paper list of participant's names to tick to be sure everybody is asked.

Hint: After each answered question we make sure we are fully understood. It is also good to be prepared for several ways to explain one thing, to have more different examples.

14.2 The Post-class Survey

We ask the participant to share their opinion in a post-class survey. It is also a Google form, but the answers are fully anonymous. It is a good source of information about the audience's expectations and about items to be improved.

15 Inventive Principle No. 25: Self-service

- Make the object serve itself by performing auxiliary helpful functions.
- Use waste resources, energy, or substances.

Many things mentioned above can be understood as an implementation of the Inventive Principle no.25. The participant cannot be just logged in and remain inactive, they are forced to be focused: they have to take notes as there are many blank places in printed materials, they have to answer questions than can be asked any time, they have to enter the breakout rooms and work in small teams. This way we effectively control the presence even though the cameras are off.

16 Summary

For those who know and use TRIZ the remote training is not a surprise. We can say it is a way to dynamize the training–it complies with the Trend of Dynamization. We have just to accept it, whether we like it or not. We have shared our own experiences and prompted some hints as we believe it is evitable to work remotely.

We constantly work on our training to make them as attractive and effective as possible. In our opinion it is a must as there is at least one but a very serious reason to continue training online: cost savings. In case of online classes we can skip the travels, hotels, conference rooms. We save time and money. We can be more flexible to meet the customers' expectations and match their calendars. We expect that many customers will prefer to work remotely, even when the pandemic is over.

References

1. BankMyCell Homepage. https://www.bankmycell.com/blog/how-many-phones-are-in-the-world. Accessed 08 Mar 2022
2. https://www.science.org/content/article/lectures-arent-just-boring-theyre-ineffective-too-study-finds. Accessed 11 Mar 2022
3. International TRIZ Association Homepage. https://matriz.org/wp-content/uploads/2021/01/Appendix-1_Knowlege-Standard-ENGL-2017-03-1.pdf. Accessed 10 Mar 2022
4. Novismo sp. z o. o Homepage. https://novismo.com/altshullers-matrix/?lang=en. Accessed 10 Mar 2022
5. https://doi.org/10.1152/advan.00109.2016. Accessed 11 Mar 2022
6. https://www.online-stopwatch.com/. Accessed 11 Mar 2022
7. Bucki, P.: Porozmawiajmy o komunikacji, 1st edn. Wydawnictwo Słowa i Myśli, Lublin (2017)
8. https://www.sciencenewsforstudents.org/article/handwriting-better-for-notes-memory-typing. Accessed 10 Mar 2022

Author Index

Printed in the United States
by Baker & Taylor Publisher Services